Sensory Ecology
Review and Perspectives

NATO ADVANCED STUDY INSTITUTES SERIES

A series of edited volumes comprising multifaceted studies of contemporary scientific issues by some of the best scientific minds in the world, assembled in cooperation with NATO Scientific Affairs Division.

Series A: Life Sciences

Recent Volumes in this Series

Volume 11 – Surface Membrane Receptors:
 Interface Between Cells and Their Environment
 edited by Ralph A. Bradshaw, William A. Frazier, Ronald C. Merrell, David I. Gottlieb, and Ruth A. Hogue-Angeletti

Volume 12 – Nucleic Acids and Protein Synthesis in Plants
 edited by L. Bogorad and J. H. Weil

Volume 13 – Prostaglandins and Thromboxanes
 edited by F. Berti, B. Samuelsson, and G. P. Velo

Volume 14 – Major Patterns in Vertebrate Evolution
 edited by Max K. Hecht, Peter C. Goody, and Bessie M. Hecht

Volume 15 – The Lipoprotein Molecule
 edited by Hubert Peeters

Volume 16 – Amino Acids as Chemical Transmitters
 edited by Frode Fonnum

Volume 17 – DNA Synthesis: Present and Future
 edited by Ian Molineux and Masamichi Kohiyama

Volume 18 – Sensory Ecology: Review and Perspectives
 edited by M. A. Ali

Volume 19 – Animal Learning: Survey and Analysis
 M. E. Bitterman, V. M. LoLordo, J. B. Overmier, and M. E. Rashotte

Volume 20 – Antiviral Mechanisms in the Control of Neoplasia
 edited by P. Chandra

The series is published by an international board of publishers in conjunction with NATO Scientific Affairs Division

A	Life Sciences	Plenum Publishing Corporation
B	Physics	New York and London
C	Mathematical and Physical Sciences	D. Reidel Publishing Company Dordrecht and Boston
D	Behavioral and Social Sciences	Sijthoff International Publishing Company Leiden
E	Applied Sciences	Noordhoff International Publishing Leiden

Sensory Ecology
Review and Perspectives

Edited by
M. A. Ali
Département de Biologie
Université de Montréal
Montréal, Canada

PLENUM PRESS • NEW YORK AND LONDON
Published in cooperation with NATO Scientific Affairs Division

Library of Congress Cataloging in Publication Data

NATO Advanced Study Institute on Perspectives in Sensory Ecology, Bishop's University, 1977.
Sensory ecology.

NATO advanced study institutes series: Series A, Life sciences; v. 18)

"Lectures presented at the 1977 NATO Advanced Study Institute on Perspectives in Sensory Ecology, held at Bishop's University, Lennoxville, Québec, Canada, July 10–22, 1977, co-sponsored by the Université de Montréal and the National Research Council of Canada."

"Published in cooperation with NATO Scientific Affairs Division."

Includes index.

1. Animal ecology—Congresses. 2. Senses and sensation—Congresses. I. Ali, Mohamed Ather, 1932-   II. Université de Montréal. III. National Research Council of Canada. IV. North Atlantic Treaty Organization. Division of Scientific Affairs. V. Title. VI. Series.

QH540.N17 1977   591.5   78-17597
ISBN 0-306-40024-3

Lectures presented at the 1977 NATO Advanced Study Institute on Perspectives in Sensory Ecology held at Bishop's University, Lennoxville, Québec, Canada, July 10–22, 1977

Co-sponsored by the Université de Montréal and the National Research Council of Canada

Director: M. A. Ali, Montréal

Advisory Committee: H.-J. Autrum, Munich
                     P. Couillard, Montréal
                     F. E. J. Fry, Toronto
                     W. R. A. Muntz, Stirling

©1978 Plenum Press, New York
A Division of Plenum Publishing Corporation
227 West 17th Street, New York, N.Y. 10011

All rights reserved

No part of this book may be reproduced, stored in a retrieval system, or transmitted, in any form or by any means, electronic, mechanical, photocopying, microfilming, recording, or otherwise, without written permission from the Publisher

Printed in the United States of America

Preface

> Never so pleased, sir. 'Twas an excellent dance,
> And for a preface, I never heard a better.
>
> Two Noble Kinsmen, Act III, Sc.5

This volume is based mostly on the lectures delivered at an Advanced Study Institute (ASI) of the same title held in July 1977. One lecture given is not in the volume and three chapters, although not based on lectures delivered, have been added to better balance the book. A chapter on the ecosensory functions in crustaceans could not be put in due to time contingency. This absence is deeply regretted.

The idea to hold an ASI on Sensory Ecology evolved slowly, mainly due to my own research interest in the past and partly to the discussions I had with a number of colleagues, particularly Dr. John Lythgoe of the University of Sussex. The purpose was to interface Sensory Physiology with Ecology so that workers in those fields will develop a greater awareness for each other. Sense organs have of course evolved to keep their possessors aware of the environment and changes in it. Thus, normally one could expect that a study of their functions will be undertaken in relation to environmental parameters. But this has not been the case. Many if not most studies of sense organs have been undertaken either to study their structure or, some aspect of their function in a restricted way. Ecology, of course, deals with the interrelationship between the organism and its environment. This interrelationship obviously cannot exist without some kind of sensory input. However, one sees very little awareness of the importance of sensory functions in Ecology. Even ethologists often tend to ignore the importance of the interrelationship between sensory modalities and environmental factors in moulding the organism's total reaction. Thus, since sensation of some sort is essential to the detection of environmental factors, particularly changes in them, it appears that an imaginative approach to the study of sensory functions with regard to ecological factors on the one hand, and the investigation of environmental parameters in relation of the sensory functions in various taxonomic groups on the other, will provide a great deal of interesting information. The interplay between the physico-chemical organism and the physico-chemical environment is a delicate one and leads to complex behavioural responses such as migrations, rhythms etc. These will have to be examined in relation to sensory functions and their interactions with environmental cues in the light of information obtained by modern methods in both sensory physiology and ecology.

The ASI was structured with the volume in mind. It was divided into two parts, one dealing with ecosensory functions in the various animal groups and the other concerning the adaptive radiation of sensory functions across the animal kingdom. The animal groups, particularly the invertebrate ones, were arbitrarily drawn up on the basis of convenience, availability of lecturers/authors and time. In most cases, the lecturers/authors were chosen and invited on the basis of suggestions made by the members of the advisory committee and some others. Being a NATO-ASI, we also had to make sure that the lecturers were drawn from as many different countries as possible. I asked the authors to prepare lectures of a general nature and encouraged them to be as provocative and speculative as they wanted. Many accepted to do so and indeed during the ASI many hours were spent in lively discussions. On the basis of these discussions, the general reaction of the participants and the content of the others' lectures, the lecturers prepared their final drafts after the ASI. Obviously, in a volume of this sort a certain amount of overlap or repetition are unavoidable, even desirable. I have tried to keep them to a minimum but may not have succeeded entirely since I am certainly not expert in all aspects that this volume deals with. In my capacity as the director of the ASI, I attended every lecture and seminar given and, as the editor I have had to read every chapter in this volume carefully. This has considerably broadened my knowledge of Sensory Ecology and I appreciate having had this opportunity and hope that the readers of this volume will find it worthwhile.

I am grateful to Dr. T. Kester of the NATO-Scientific Affairs Division for the advice and encouragement he gave, without which the ASI could never have been held. I thank Dr. Maurice L'Abbé, Vice-Rector for Research in my university and Professor Jean-Guy Pilon, Chairman of my department for their moral and material support. Financial help was also provided by the National Research Council of Canada.

The authorities of Bishop's University cooperated in every possible way, making our stay enjoyable. I am especially thankful to Monsieur J.L. Grégoire, Vice-Principal and Mr. Ivan Saunders,

PREFACE

Director of Buildings and Grounds for all their help. I thank also the members of my advisory committee for the help they gave and the suggestions they made concerning the choice of lecturers. I particularly thank my colleague Pierre Couillard for the more than the expected role he played in the organisation of the ASI and especially arranging the excursion to Québec city. I wish to note how greatly the assistance of my librarian Miss Margaret Pertwee in putting the volume together is appreciated.

I thank Mademoiselle Marielle Chevrefils for typing most of the camera-ready manuscript. On behalf of the authors and myself I wish to acknowledge the permission given by various authors and publishers for the reproduction of figures. Last but not the least, my thanks to Steve Dyer, Production Editor at Plenum for his patience and for his efficiency in getting the volume out quickly.

Montréal
February, 1978

Contents

1. *GENERAL INTRODUCTION*
   M.A. Ali 3

2. *SURVEY OF ECOSENSORY FUNCTIONS*

   Phylogenetic Survey of Sensory Functions 11
   M.A. Ali, R.P. Croll and R. Jaeger

   Taxes in Unicells, Especially Protozoa 31
   P. Couillard

   Ecosensory Functions in Lower Invertebrates 55
   E.A. Ferrero

   Coelomate Invertebrates (Except Crustacea, Arachnida
   and Insecta) 91
   M.A. Ali

   Ecosensory Functions in Insects (With Remarks on
   Arachida) 123
   M. Gogala

   Fishes: Vision in Dim Light and Surrogate Senses 155
   J.N. Lythgoe

   Ecological Niche Dimensions and Sensory Functions
   in Amphibians 169
   R.G. Jaeger

Reptile Sensory Systems and the Electromagnetic
Spectrum 197
    W.R.A. Muntz

Sensory Ecology of Birds 217
    R.A. Suthers

Sensory Ecology of Mammals 253
    R.A. Suthers

3. *ADAPTIVE RADIATION OF SENSORY MODALITIES*

Functional Adaptations in Chemosensory Systems 291
    R.A. Gleeson

Adaptive Radiation of Mechanoreception 319
    H. Markl

Sound Reception in Different Environments 345
    A. Michelsen

Baroreception 375
    J.H.S. Blaxter

The Adaptive Radiation of Proprioceptors 411
    W. Wales

Peripheral Thermal Receptors 439
    R. Loftus

Photoreception 467
    M.A. Ali, M. Anctil, and L. Cervetto

A Survey of Vertebrate Strategies for Vision in
Air and Water 503
    J.G. Sivak

Ecological Aspects of Electroreception 521
    H.O. Schwassmann

4. *INDICES*
    Author Index 537
    Subject Index 569

1. GENERAL INTRODUCTION

GENERAL INTRODUCTION

M.A. ALI

Département de Biologie, Université de Montréal

Case Postale 6128, Montréal H3C 3J7, Canada

> Think of the beauty of a blank page
> before writing!
>
> Arab proverb

After admiring the beauty of a blank sheet for weeks, nay months, I made an outline of what is to be written and finally, with considerable trepidation, started desecrating the beauty of the sheet's blankness. Being a simplist by nature, it appeared to me that a general introduction should remain general, especially in a volume of this sort where specifics are being dealt with in the various chapters which follow. To write an historical general introduction, as has been done in quite a few instances which I have admired, would have been well nigh undesirable and indeed impossible due to my own shortcomings, the vastness of the subject and the fact that we like to believe that a new sub-discipline of Ecology is being launched. To cover, in an historical manner all the pioneering studies in Ecology and Sensory Physiology, which could be considered to have indicated the need for the term Sensory Ecology, would have been ideal but voluminous. It would have been incomplete and because of that, not truly useful. With these apologia in mind, I decided to write rather an essay outlining the concepts and reasons I had in mind when the Advanced Study Institute on this subject, a direct result of which this volume is, was planned. This is more easily said then done because it has been, and to a great extent will be, abstract. However, an attempt will be made with a great deal of confidence in the reader's indulgence.

The first point to be considered is: what is Sensory Ecology? Obviously it is a sub-discipline of Ecology as Population Ecology is and comparable to what Sensory Physiology is to Physiology. Ecology, originally termed Oekologie (Oikos-house; logia-speak, see) by Haeckel in 1866, is the study of the relations of the organisms to environing conditions, organic and inorganic. Haeckel originally viewed it from an evolutionary point of view. It is also used to denote the study of adaptations without special reference to their origin and even of the adjustment of individuals to their conditions, as in the distribution of plants. As a science or discipline then it is an epistemological concept. All biological relations boil down to:

a. perceptions - sensory events

b. responses - motor events

Thus, it is a twofold affair. Therefore, to speak of "Sensory Ecology" might appear to be partly pleonasmic. However, it is acceptable in so far as it specifies that the first (i.e. sensory) of the two basic aspects of Ecology is being considered. One may also hesitate to call an aspect or component *part* of a discipline or a sub-discipline. A sub-discipline in a sub-field which may or may not exist but the absence of which does not render the "discipline" non-existent. Thus, Ecology will exist without Sensory Ecology just as Morphology will exist without Anatomy or Histology. Also, Non-Sensory Ecology does not exist since sensation is essential to discern environmental factors. Nevertheless ecological relationships could also involve non-sensory controls or factors such as lack of food, over predation or pollution. In these cases the organism has no choice, just luck, as in bacteria and antibiotics; fish and pollution etc. Here we are dealing with situations wherein sensory modalities are brought into play to permit an animal to fully exploit its environment or evade or overcome the constraints.

An organism may be considered as a shell surrounded by the results of its evolutionary history or Phylogeny, which itself is mostly moulded by its physiological capacities to cope with, or its reactions to organic and inorganic environmental factors. Several sensory systems are combined to give a general impression of its surroundings. Noise and filtering - in the physiological sense - in the environment play an important rôle in the formation of this impression. For example, a persistent stimulus such a steady light, rustling of leaves, flow of water etc. are filtered by the sensory system. The problem of noise and filtering therefore is an important one in Sensory Ecology and Environmental Physiology.

Considering all these factors one may put forth a number of definitions of Sensory Ecology, all variations of essentially one theme.

INTRODUCTION

1. Sensory Ecology deals with the means by which the fitness of organisms of a species is optimised through adapting to the constraints of information input of both the physical and biotic environments.
2. Sensory Ecology deals with the sensory means by which the survival of organisms is rendered optimal in response to the constraints of the environment.
3. Sensory Ecology is the study of means by which the fitness of organisms is rendered optimal to cope with environmental pressures through adaptive radiation of sensory mechanisms.
4. Sensory Ecology deals with sensory strategies to cope with the environmental constraints.

Animal phyla appear at first glance to be a medley of unconnected groups but are in fact the products of a great number of evolutionary strategies appropriate to their ecological niches. The evolutionary process has developed sensory functions according to the needs of the organisms in the specific environments. Their diversity is the evolutionary answer to the existence of the multiplicity of habitats in nature.

Two types of evolutionary process produce the diversity.

1. The increasing general efficiency of a sensory function through the sub-cellular→cellular→tissue and organ grades. This process is a slow but continuous one that raises the upper limit of efficiency.

Examples may be provided from photo and sound reception. In protozoans such as *Amoeba* there is general photosensitivity; in the flagellate *Euglena* there is an intracellular eye spot; in the dinoflagellates *Proterythropsis* and *Erythropsis*, light receptors are even provided with lens-like devices. The situation is more advanced in planarians with eyes. After this one notices an increasing complexity in the polychaete annelids. It should be underlined that in many of these cases only anatomical evidence is available and conclusions are based on conjecture until physiological or behavioural evidence is provided. The photoreceptors in the higher forms have followed either the rhabdom or the ciliary lines. The former exemplified by crustaceans and insects and the latter by cephalopods and vertebrates. As far as sound reception is concerned one must assume, from general knowledge regarding Darwinian evolution, that sound perception in mammals is more efficient than in fishes. The word "efficient" remains to be specified. One probable progress evolved in this case is the replacement of *diffuse* reception (fishes) through the whole body or the forward (anterior) part of the body or the total head, by *spot* receptors (land vertebrates), i.e. typmanic and middle ear system. Then the substitution of a half coiled (birds) and then a fully coiled organ of Corti (mammals) for a short (amphibians) or elongate (reptiles) macula.

2. Adaptive radiation is a quicker process, involving the diversification at any particular level to populate the available niches with appropriately and adequately adapted sensory receptors. It is thus the emergence of *diverse* adaptive features within a relatively short time.

Here also examples may be provided from photo and sound reception. Within one family of fishes such as Percidae, while the ocular structure is similar, one species such as the perch (genus *Perca*) is an arhythmic form with a retina showing retinomotor responses of the epithelial pigment, rods and cones. There is also no reflecting material. On the other hand, in the pikeperches (genus *Stizostedion*) the cones are not capable of movement but serve as light guides and there is reflecting material, all adaptations for life in turbid waters or crepscular conditions. Even here, one species, the sauger (*S. canadense*), is more specialised for turbid water than the walleye (*S. vitreum*). Similar examples may be provided from other fish families, crustaceans, insects etc. In sound perception, in mammals it has apparently consisted in the diversified usage of specific *frequency ranges*. Some of those frequencies (and their harmonics) were united for *sonar* applications. Examples may also be readily provided from sound reception in insects, a number of fish families and birds.

Thus, *adaptive radiation* is a broadly described expression implying *comparison* because it encompasses *several* species, genera. families or orders.

Adaptation is a concept applicable to the particular problems of *one* species facing its *specific* set of relational problems. For example a tubular eye with a large lense and divided retinas in a deep-sea fish or the development of an association between bioluminescence and photoreception are visual *adaptations* for life in the depths. Similarly the development of a radar system in microchiroptera is an adaptation for feeding and survival in the dark. Pacific salmon (genus *Oncorhynchus*) developing an alarm reaction to minute quantities of mammalian (*e.g.* bear) odours in water is an adaptation for survival during upstream migration to spawn.

These two evolutionary processes have generated the staggering diversity of sensory structures. This diversity is thus understandable and meaningful in the light of the evolutionary process. All organisms have to preserve self and the race to survive. Success in this depends on efficient feeding and breeding, which leads to population growth with increasing competition and environmental resistance. Adaptive radiation of a successful group is an escape from the intensity of competition to varied niches. The role and efficiency of a sensory function are intelligible only in the light of the special habitats or niches with their special conditions demanding appropriate adaptations.

The basic problem is obvious *viz.* the survival of animals until they have successfully reproduced themselves. Our present business is to ascertain to what extent they depend on their sensory equipment to do this and, how the sensory functions have evolved adequately in relation to the requirements of the animal and the demands of the environment, which are often hostile. In order to survive and to reproduce, an animal has to feed, and in order to find food or to avoid becoming food itself and, to find a partner, it has to be able to move about, either as a larva or as an adult, actively or otherwise. To look at it in a slightly different manner: what are the environmental factors or conditions which have influenced the development and evolution of sensory structures and, how are the sensory functions modified to meet the requirements of the habitats, modes of life and locomotion?

One may also ask; did the sensory equipment evolve in response to the needs of the animal to function optimally vis-à-vis its environment or, did the animal choose and succeed in an environment best suited to the sensory functions which it possessed? Evidently, the answers to both these questions will be affirmative. These processes apply as well to the entire animal in general as to its sensory functions or organs in particular.

All too often sensory physiologists have looked at the response of this or that organ or tissue to one or other environmental factor without understanding the ecological implications. Ecologists, on the other hand, have investigated ecological relationships often without any reference to sensory modalities. Ethologists, who should know better, have studied behavioural reactions, given some attention to sensory modalities and often ignored environmental (ecological) parameters. These are the gaps we hope to fill with our contributions, which show that a bridge could indeed be built to span these disciplines and that one should encompass ecological parameters, sensory mechanisms and behavioural reactions. The reader may expect to have a number of surprises in the variety of examples given in the chapters that follow. They show how even the so called "lower animals", whose sensory modalities appear so different from ours at first glance, use them so effectively to survive, succeed and to reproduce in their environments. They also show how one should be careful in investigating these animals' reactions and interpreting them and above all understand that the world may not appear the same to them as it does to us, because we still do not possess, and may never do so, a thorough understanding of how they sense their environment. However, whatever understanding we have of their sense and their reactions to environmental parameters which we can measure, enables us to broaden our knowledge of our own senses, our own reactions and how we came to have them.

It seems appropriate to conclude this general introduction by quoting Bullock (Handbook of Sensory Physiology, Vol III/3, Springer-Verlag, N.Y.), "It would behoove us in humility, to learn....and to extend the range and depth of investigation of sensory systems on all fronts, lest we assume, as is so natural, that our own experience gives a reasonable approximation of the experience of other creatures".

ACKNOWLEDGEMENTS

A number of colleagues and friends helped by discussing the various concepts. I am grateful to them all but particularly to Pierre Couillard, Robert Jaeger, John Lythgoe, P.K. Menon and Paul Pirlot. However, I alone should be held responsible for the shortcomings.

2. ECOSENSORY FUNCTIONS IN ANIMALS

PHYLOGENETIC SURVEY OF SENSORY FUNCTIONS

M.A. ALI, R.P. CROLL & R. JAEGER

Dépt. Biol. Univ. Montréal, Dépt. Biol. McGill Univ. Montreal, Canada. & Dept. Zool. SUNY, Albany, U.S.A.

> Oh, for a life of sensations rather than of thoughts.
>
> John Keats: Letter to Benjamin Bailey
> 22 November 1817

The information given in this chapter was assembled with the purpose of giving a quick and convenient indication of the distribution of particular sensory functions among the different animal groups. The chapter is meant to complement the survey chapters in this volume which often focus on selected organisms as being illustrative of general mechanisms and which due to considerations of space were not able to give mention to all classes of organisms. For those desiring further information on certain sensory modalities in a class of animals references are given in the tables. The references selected were mostly of a review nature and tended to give the best overview of the general capabilities of the class. Further references to the original research papers are given in these reviews. On occasions when no recent review containing information on the sensory capabilities of class or order could be found, original research articles are referenced in the table. In addition to those sources specifically referenced in the tables the reader should be aware of the excellent reviews of comparative sensory physiology found in Prosser (1973), Bullock and Horridge (1965) and in the various volumes of the Handbook of Sensory Physiology (Springer-Verlag).

The particular sensory modalities included in the tables are, of course, not the only senses found in these animals but rather they were chosen for their wide phylogenetic distribution. For example, some evidence exists showing that gastropods, insects and birds respond to magnetic fields, but such studies are so scarce that a listing of the animals demonstrating a magnetic sense would not be warranted here. Similarly some fishes have well studied electroreceptors but this sense seems to be limited to only an exceedingly small fraction of the animal kingdom.

Even within this simplified categorization of sensory functions, problems of interpretation arise. Most organisms respond in some way to chemical stimulation. In the tables, those animals indicated as having well-established chemoreception have been shown to have low-threshold responses while animals indicated as having questionable evidence for chemoreception may respond only to large changes in pH or salt concentration. These responses to gross chemical stimulation may be mediated through cells not specialised for chemical detection. Metabolic changes may occur within these animals and these are then detected by neuronal elements thus producing a change in the animal's behaviour. Similarly, intense light and temperature may affect the nervous system only indirectly. In many cases this indirect monitoring of stimuli may be adequate for the animal's needs and no specialised sense organs have evolved. Snails may only need to be able to distinguish between an environment that is unfavourably hot or cold and one that is optimal for their metabolisms. Certain insects, however, use temperature cues for host detection and therefore have need of specialised, highly tuned temperature sense.

Another cautionary remark for those using these tables is that a lack of positive evidence does not constitute strong evidence against the existence of a certain sensory function. In order for an animal to respond to a stimulus it must have means of detecting it, and be motivated to respond to it. Escape responses to noxious stimuli are relatively easy to elicit. But elicitation of certain feeding behaviours by external stimuli is dependent upon intervening variables such as stomach distention and blood sugar concentration. If the external stimuli produce no response it could either be because the animal cannot sense the stimuli or because the animal's internal state is not conducive to feeding. Also many animals have not been closely studied regarding their sensory capabilities. Blank spaces in the table, then, must be interpreted only as a lack of positive evidence for the existence of a sensory modality. Whether this is due to an absence of sense organs or due to inadequate experimentation remains unresolved.

The distribution of sensory functions among the classes of invertebrates is shown in Table 1. The most striking feature of the table is the large number of gaps in our present knowledge of invertebrates. Hypotheses on the general trends in invertebrate sensory physiology must therefore be highly speculative. The world of most invertebrates is probably one mainly of tactile and chemical cues and of light intensity and perhaps directionality. Photoreception is widespread but acute vision is restricted mostly to cephalopod molluscs and various arthropods. Rudimentary form discrimination probably exists in certain annelids and gastropods. Most invertebrates adequately studied appear to respond quite well to low-threshold chemical stimulation. Generally the distant chemoreceptors are located externally either on the body wall or on specialised appendages like tentacles or antennae. This is in contrast to the olfactory nasal cavities of vertebrates. Proximal chemoreceptors are located where they would make contact with food, e.g., near the oral cavity and/or, as in the case of many insects, on the legs. Audition is not generally found except in insects but a vibration sense probably will prove to be widespread among the multicellular invertebrates. Proprioceptive sense and equilibrium sense serve as feedback controls in the orientation of the animal in its environment, and thus, are exceedingly important in the lives of many animals. Future research will, therefore, probably prove there to be a high selective advantage for having either one or both of these sensory modalities.

The distribution of sensory functions among the vertebrate orders are shown in Table 2. As can be readily seen we know much more about vertebrate sensory physiology than we do about that of invertebrates. Most vertebrates are highly motile to accommodate searching for food and suitable habitats and escaping from predators. Such mobility requires a fine sensing of information about changes (or steady-stage conditions) in the environment. Consequently, the sensory receptors listed in Table 2 seem to be nearly ubiquitous among vertebrates. Equilibrium is mediated through the inner ear common to all vertebrates. Proprioceptors scattered within the body allow controlled movement and internal homeostasis.

Among the fishes, vision occurs via the lateral eyes, although a few species are blind. Hearing, within the inner ears, has been well documented in a number of orders, but it probably is not very acute. Chemoreception occurs in the nasal organs, mouth and probably over the skin and is well developed, being a more important sensor for many species than either vision or hearing. Tactile receptors are scattered over the epidermal layer, and are often grouped on barbels or flaps; mechanoreceptors in the lateral line monitor water movement. Thermoregulation has been shown to occur in several species, and most species are probably sensitive to relatively small changes in water temperature.

Amphibians generally have well developed lateral eyes, although vision is more important for frogs than for salamanders and caecilians. Chemoreception is well developed in the latter two orders and is of lesser importance to frogs. Hearing is well developed in frogs, which possess external ears, but it is questionable how much information is detected by the inner ears of salamanders. Both orders are sensitive to touch and to changes in ambient temperature.

Reptiles have well developed lateral eyes, although a few squamates are blind. Snakes lack external ears, but other reptiles are quite sensitive to sound, particularly low frequencies. Chemoreception, especially via nasal organs and mouth, is acute in virtually all reptiles. Tactile receptors apparently occur over the body surface. Thermodetection allows many species to regulate body temperature through behavioural adjustments to ambient temperature.

Birds generally have excellent vision and hearing but relatively poor chemoreception. Mammals on the other hand are less stereotyped. Most orders have excellent chemoreception in the nasal organs and mouth. Vision varies among species from acute to poor. Hearing is generally well developed. Both birds and mammals are sensitive to touch over most of the body surface and homothermy requires sensing of ambient temperature as well as internal temperatures.

In conclusion, it is hoped that this assemblage of information in table form will facilitate the formation of newer, more sophisticated models of animal-environmental interaction that will better incorporate our knowledge of the sensory capabilities of the animals. This integration between sensory physiology and the ecological sciences will be beneficial in advancing our understanding of both areas of study.

Table 1. Distribution of sensory functions and environments among invertebrate classes. Number of species is given for each class. W, well established; P, probably present; Q, questionable presence; –, no reference found; F, freshwater; B, brackishwater; M, marine; S, subterranean (burrowing); L, on land; A, aerial; Ps, parasite.

Taxon	Photo	Thermo	Phono	Proprio	Tactile	Equilibrium	Chemo	Environments
Protozoa								
Mastigophora (3500)	W8	–	–	–	–	–	–	F,M,Ps
Sarcodina (8000)	–	–	–	–	–	–	–	F,M,Ps
Ciliata (6000)	–	–	–	–	W18	–	P26	F,M,Ps
Sporozoa (2000)	–	–	–	–	–	–	–	Ps
Porifera								
Calcarea (1000)	–	–	–	–	–	–	–	M
Hexactinellida (800)	–	–	–	–	–	–	–	M
Demospongiae (8000)	–	–	–	–	–	–	–	F,M,Ps
Scleropongiae (10)	–	–	–	–	–	–	–	M
Cnidaria								
Hydrozoa (1500)	W8	–	–	–	W8	W8	W8	F,M
Scyphozoa (200)	W8	–	–	–	–	W8	–	M
Anthozoa (6000)	W1	–	–	–	–	–	–	M
Ctenophora (90)								
Tentaculata	P8	–	–	–	W8	W33	P8	M
Nuda	–	–	–	–	–	–	–	M

Table 1: (continued)

Taxon	Photo	Thermo	Phono	Proprio	Tactile	Equilibrium	Chemo	Environments
Platyhelminthes								
Turbelleria (3000)	W8	–	–	–	Q8	W8	Q8	F,M
Trematoda (5100)	W45	–	–	–	P45	–	Q8	F,M,Ps
Cestoda (2000)	–	–	–	Q46	P8	–	–	F,M,Ps
Mesozoa (50)	–	–	–	–	–	–	–	M,Ps
Rhynchocoela (650)	W21	–	–	–	W8	W21	P21	F,M
Gastrotricha (400)	P8	–	–	–	P8	Q8	–	F,M
Rotifera (1500)	W17	–	–	–	–	Q8	–	F,M
Kinorhyncha (100)	W8	–	–	–	–	–	–	M
Nematoda (10000)	W13	P30	–	P32	W32	–	W32	F,M,S
Nematomorpha (230)	Q8	–	–	–	–	–	–	F,M,Ps
Acanthocephala (500)	–	–	–	–	–	–	–	Ps
Gnathostomulida (80)	–	–	–	–	P8	–	Q8	M
Annelida								
Polychaeta (5300)	W49	–	–	W25	W25	P33	W14	B,M
Oligochaeta (3100)	W31	–	–	–	W31	P33	W31	F,M,S
Hirudinea (500)	W19	P30	–	–	W38	–	–	F,M,L
Mollusca								
Gastropoda (35000)	W15	Q8	–	P8	W9	W33	W28	F,M,L
Amphineura (610)	W8	–	–	–	W8	–	W8	M

Table 1: (continued)

Taxon	Photo	Thermo	Phono	Proprio	Tactile	Equilibrium	Chemo	Environments
Aplacophora (610)	–	–	–	–	–	–	–	M
Bivalvia (20000)	W29,35	–	–	–	W39	–	W20	F,B,M
Scaphopoda (200)	–	–	–	–	–	–	–	M
Cephalopoda (650)	W53	–	–	W7	W52	W4	W51	M
Arthropoda								
Merostomata (5)	W22	–	Q8	W2	W27	W8	W2	M
Arachnida (36000)	W53	–	–	–	W8	W8	W16	L
Pycnogonida (600)	–	–	–	–	–	–	–	M
Crustacea (26000)	W44,50	P3	Q11	W40,11	W36	W11	W10,12	F,B,M,L
Insecta (750,000)	W33	W48	W42	W8	W8	W33	W6,24	F,S,L,A,Ps
Symphyla (120)	–	–	–	–	–	–	P8	L
Diplodopa (7500)	–	B8	–	–	P8	–	P23	S,L
Pauropoda (360)	–	–	–	–	–	–	–	L
Onychophora (65)	W8	–	–	–	P8	–	–	S,L
Pogonophora (80)	–	–	–	–	–	–	–	M
Sipuncula (330)	P8	–	–	–	P8	–	–	M
Echiura (100)	–	–	–	–	–	–	–	M
Priapulida (8)	P8	–	–	–	–	–	–	M
Tardigrada (350)	–	–	–	–	–	–	–	F,B,M
Pentastomida (70)	–	–	–	–	–	–	–	L

Table 1: (continued)

Taxon	Photo	Thermo	Phono	Proprio	Tactile	Equilibrium	Chemo	Environments
Phoronida (15)	–				–	–	–	M
Bryozoa (4000)	W43	–	–	–	W47	–	–	M
Entoprocta (60)	–	–	–	–	–	–	–	F,M
Brachiopoda (280)	P8	–	–	–	Q8	W8	Q8	M
Echinodermata								
Asteroidea (1600)	W5	Q5		–	W8	P5	W8	M
Ophiuroidae (2000)	P5	Q5		–	P5	Q5	W8	M
Echinoidae (800)	W8	Q5		–	W8	W5	W8	M
Holothuroidea (500)	W8	Q5		–	W8	W5	W5	M
Crinoidea (80)	Q5	Q5		–	W8	P5	Q5	M

REFERENCES FOR INTRODUCTION AND TABLE I

1. Anderson, P. & Mackie, G.O. (1977). Electrically coupled, photo-sensitive neurons control swimming in a jellyfish. Science 197: 186-188.
2. Barber, S.B. (1956). Chemoreception and proprioreception in *Limulus*. J. Exp. Biol. 131: 51-74.
3. Barber, S.B. (1961). Chemoreception and thermoreception. In: Physiology of Crustacea, Vol. II, Sense Organs, Integration and Behavior, ed. T.H. Waterman. Academic Press, New York.
4. Barber, W.C. (1968). The structure of mollusc statocysts with particular reference to cephalopods. Symp. Zool. Soc. (Lond.) 23: 37-62.
5. Binyon, J. (1972). Physiology of Echinoderms. Pergamon Press, Oxford, 264p.
6. Boeckh, J., Kaissling, K.E. and Schneider, D. (1965). Insect olfactory receptors. Cold Spring Harbor. Symp. Quant. Biol. 30: 263-280.
7. Boyle, P.R. (1976). Receptor units responding to movement in the octopus mantle. J. Exp. Biol. 1-9.
8. Bullock, T.H. & Horridge, G.A. (1965). Structure and Function in the Nervous System of Invertebrates. W.H. Freeman, San Francisco, 2 volumes.
9. Byrne, J., Castellucci, V. & Kandel, E.R. (1974). Receptive fields and response properties of mechanoreceptors innervating siphon skin and mantle shelf in *Aplysia*. J. Neurophysiol. 30: 1439-1465.
10. Case, J. (1964). Chemoreceptors on dactyl of *Cancer*. Biol. Bull. Woods Hole 127: 428-446.
11. Cohen, M.J. & Dykgraaf, S. (1961). Mecanoreception. In: Physiology of Crustacea, Vol. II, Sense Organs, Integration and Behavior. Ed. T.H. Waterman. Academic Press, New York.
12. Crisp, D.J. (1967). Chemoreception in barnacles. Biol. Bull. Woods Hole 133: 128-140.
13. Croll, N.A., Riding, J.L. & Smith, J.M. (1972). A nematode photoreceptor. Comp. Biochem. Physiol. 42A: 999-1009.
14. Davenport, D., Camorgis, G. & Hickok, J.F. (1960). Analysis of the behaviour of commensals in host-factor. I. A lesioned polychaete and a pinnotherid crab. Anim. Behav. 8: 209-218.
15. Dennis, M.J. (1967). Electrophysiology of the visual system of a nudibranch mollusc. J. Neurophysiol. 30: 1439-1465.
16. Drewes, C.D. & Bernard, R.A. (1976). Electrophysiological responses of chemosensitive sensilla in the wolf spider. J. Exp. Zool. 198: 423-435.
17. Donner, J. (1966). Rotifers. Frederick Warne, London, 80p.
18. Eckert, R., Nautch, Y. & Friedman, K. (1972). Sensory mechanisms in *Paramecium*. J. Exp. Biol. 56: 684-694.

19. Fioravanti, R. & Fuortes, M.G.F. (1972). Analysis of responses in visual cells of the leech. J. Physiol. (Lond.) 227: 173-194.
20. Franc, A. (1960). Classe des bivalves. In: Traité de Zoologie, Tome V, Fasc. II, p. 1845-2164. Ed. P.P. Grassé, Masson, Paris.
21. Gibson, R. (1972). Nemerteans, Hutchinson Univ. Lib. London, 224p.
22. Hartline, H.K. (1969). Visual receptors and retinal interaction. Science 164: 270-278.
23. Hodgson, E.S. (1958). Electrophysiological studies of arthropod chemoreception. III Chemoreceptors of terrestrial and freshwater arthropods. Biol. Bull. Woods Hole 115: 114-125.
24. Hodgson, E.S. & Roeder, K.D. (1956). Responses of chemoreceptor hairs in *Dipters*. J. Cell. Comp. Physiol. 48: 51-76.
25. Horridge, G.A. (1963). Proprioceptors and textile receptors in polychaete *Hormothoë*. Proc. R. Soc. Lond. Series B. 157: 199-222.
26. Jones, A.R. (1974). The Ciliates. Hutchinson Univ. Lib., London, 207p.
27. Kaplan, E., Barlow, R.B. Jr., Chamberlain, S.C. & Stelzner, D.J. (1976). Mechanoreceptors on the dorsal carapace of *Limulus*. Brain Res. 109: 615-622.
28. Kohn, A.J. (1961). Chemoreception in gastropod molluscs. Am. Zool. 51: 291-308.
29. Kennedy, D. (1960). Neural photoreception in a lanelli-branch mollusc. J. Gen. Physiol. 44: 277-299.
30. Laudien, H. (1974). Activity behavior, Chap. IV, p. 441-469. In Temperature and Life by H. Precht *et al.* Springer-Verlag, New York.
31. Laverack, M.S. (1963). The Physiology of Earthworms. Pergamon Press, London. 206 p.
32. Lee, D.L. (1965). The Physiology of Nematodes. Oliver and Boyd, Edinburgh, 154p.
33. Markl, H. (1974). The perception of gravity and of angular acceleration ininvertebrates. In: Handbook of Sensory Physiology 6 (1), Vestibular System, Part 1: Basic Mechanisms. Ed. H.H. Kornhuber. Springer-Verlag, Berlin.
34. Mazohin-Poršnjakov, G.A. (1969). Insect Vision. Plenum Press, New York, 306p.
35. McReynolds, J.S. & Gorman, A.L.F. (1970). Membrane conductances and spectral sensitivities of *Pecten* photoreceptors. J. Gen. Physiol. 56: 392-406.
36. Mellon, D. Jr. & Kennedy, D. (1964). Impulse origin and propagation in crayfish tactile receptors. J. Gen. Physiol. 47: 487-499.
37. Morgan, E. (1969). Responses of polychaete to hydrostatic pressure. J. Exp. Biol. 50: 501-513.

38. Nicholls, J.G. & Baylor, D.A. (1968). Specific modalities and receptive fields of sensory neurons in CNS of the leech. J. Neurophysiol. 31: 740-756.
39. Olivo, R.F. (1970). Mechanoreception function in the razor clam: sensory aspects of the post withdrawal reflex. Comp. Biochem. Physiol. 35: 761-786.
40. Paul, D.H. (1976). Role of proprioceptive feedback from nonspiking mechanosensory cells in the sand crab, *Emerita analoga*. J. Exp. Biol. 65: 243-258.
41. Prosser, C.L. (1973). Comparative Animal Physiology. 3rd. ed. Saunders, Philadelphia, 966p.
42. Roeder, K.D. (1966). Auditory system of noctuid moths. Science 154: 1515-1521.
    Michelsen, A. (1978). Sound reception in different environments. In: Sensory Ecology, Ed. M.A. Ali. Plenum Press, New York.
43. Ryland, J.S. (1970). Byozoans. Hutchinson Univ. Lib., London, 175p.
44. Shaw, S.R. (1972). Decremental conduction of the visual signal in barnacle lateral eye. J. Physiol. (Lond.) 220: 145-175.
45. Smyth, J.D. (1966). The Physiology of Nematodes. Oliver and Boyd, Edinburgh, 256p.
46. Smyth, J.D. (1969). The Physiology of Cestodes. W.H. Freeman, San Francisco.
47. Thorpe, J.P., Sheldon, G.A.B. & Laverack, M.S. (1975). Colonial nervous control of lophophore retraction in cheilostome Bryozoa. Science 189: 60-61.
48. Van Haga Reinauts, H.A. & Mitchell, R.K. (1975). Temperature receptors on tarsi of the tsetse fly, *Glossina morsitan West*. Nature 225: 225-226.
49. Wald, G. & Rayport, S. (1977). Vision in annelid worms. Science 196: 434-439.
50. Waterman, T.H. (1961). Light sensitivity and vision. In: The Physiology of Crustacea, Vol. II, Sense Organs, Integration and Behavior, ed. T.H. Waterman. Academic Press, New York.
51. Wells, M.L. (1964). Tactile discrimination of shape by *Octopus*. Quart. J. Exp. Psychol. 16: 156-162.
52. Wells, M.J. (1963). Taste responses in *Octopus*. J. Exp. Biol. 40: 187-193.
53. Wolken, J.J. (1975). Photoprocesses, Photoreceptors and Evolution. Academic Press, New York, 317p.

Table 2. Distribution of sensory functions and environments among vertebrate orders. Number of species is given for each order. W, well established; P, probably present; -, no reference found; F, freshwater; B, brackish water; M, marine; AE, amphibious, entire life; AP, amphibious, partial; S, subterranean; L, on land; A, aerial.

Taxon	Photo	Thermo	Phono	Proprio	Tactile	Equilibrium	Chemo	Environments
Cephalaspidomorphi								
Petromyzoniformes (31)	W18	P17	P25	W15	–	W15	W4	F,B,M
Pteraspidomorphi								
Myxiniformes (32)	P18	P17	P25	W15	–	W15	P9	M
Chondrichthyes								
Heterodontiformes (6)	W18	P17	P25	P15	P7	P15	P9	M
Hexanchiformes (6)	W18	P17	P25	P15	P7	P15	P9	M
Lamniformes (199)	W18	P17	W15	W15	P7	W15	W4	M
Squaliformes (76)	W18	P17	P25	P15	W12	P15	W4	M
Rajiformes (315)	W18	W17	W15	W15	P7	W15	W4	M
Chimaeriformes (25)	W18	P17	P25	P15	P7	P15	P9	M
Osteichthyes								
Ceratodiformes (1)	W18	P17	P25	P15	P7	P15	P9	F
Lepidosireniformes (5)	W1	P17	P25	P15	P7	P15	P9	F
Coelacanthiformes (1)	W18	P17	P25	P15	P7	P15	P9	M
Polypteriformes (11)	W1	P17	P25	P15	P7	P15	P9	F
Acipenseriformes (25)	W18	P17	P25	P15	P7	P15	P9	F,B,M

Table 2: (continued)

Taxon	Photo	Thermo	Phono	Proprio	Tactile	Equilibrium	Chemo	Environments
Semionotiformes (7)	W18	P17	P25	P15	P7	P15	W4	F,B
Amiiformes (1)	W1	P17	P25	P15	P7	P15	P9	F
Osteoglossiformes (15)	W18	P17	P25	P15	P7	P15	P9	F
Mormyriformes (101)	W18	P17	W15	P15	P7	P15	P9	F
Clupeiformes (292)	W1	P17	W30	P15	P7	P15	W4	F,B,M
Elopiformes (11)	W18	P17	P25	P15	P7	P15	P9	F,B,M
Anguilliformes (603)	W1	P17	W25	P15	W12	P15	W4	F,B,M
Notacanthiformes (24)	W15	P17	P25	P15	P7	P15	P9	M
Salmoniformes (508)	W1	P17	W15	W15	P7	W15	W4	F,B,M
Gonorynchiformes (16)	W18	P17	P25	P15	P7	P15	P9	F,B,M
Cypriniformes (3000)	W1	W17	W15	W15	P7	W15	W4	F
Siluriformes (2000)	W1	W17	W15	P15	W12	P15	W4	F,B,M
Myctophiformes (390)	W18	P17	P25	P15	P7	P15	P9	B,M
Polymixiiformes (3)	W18	P17	P25	P15	P7	P15	P9	M
Percopsiformes (8)	W18	P17	P25	P15	P7	P15	P9	F
Gadiformes (684)	W1	W17	W15	W15	W7	W15	W4	F,M
Batrachoidiformes (55)	W18	P17	P25	P15	P7	P15	P9	F,B,M
Lophiiformes (215)	W18	P17	P25	P15	P7	P15	P9	M
Indostomiformes (1)	W18	P17	P25	P15	P7	P15	W4	F
Atheriniformes (827)	W1	P17	W25	P15	P7	P15	W4	F,B,M

Table 2: (continued)

Taxon	Photo	Thermo	Phono	Proprio	Tactile	Equilibrium	Chemo	Environments
Lampridiformes (35)	W18	P17	P25	P15	P7	P15	P9	M
Beryciformes (143)	W18	P17	P25	P15	P7	P15	P9	M
Zeiformes (50)	W18	P17	P25	P15	P7	P15	P9	M
Syngnathiformes (200)	W18	W17	P25	P15	P7	P15	W4	F,M
Gasterosteiformes (10)	W1	W17	P25	P15	P7	P15	W4	F,B,M
Synbranchiformes (13)	W18	P17	P25	P15	P7	P15	P9	F,B,M
Scorpaeniformes (1000)	W1	W17	W15	P15	W12	P15	W4	F,B,M
Dactylopteriformes (4)	W18	P17	P25	P15	P7	P15	P9	M
Pegasiformes (5)	W18	P17	P25	P15	P7	P15	P9	M
Perciformes (6880)	W1	W17	W25	W15	P7	W15	W4	F,B,M
Gobiesociformes (144)	W18	P17	P25	P15	P7	P15	P9	F,M
Pleuronectiformes (520)	W1	W17	P25	W15	P7	W15	W4	F,B,M
Tetraodontiformes (320)	W18	P17	P25	P15	P7	P15	W4	F,B,M
Amphibia								
Gymnophiona (158)	P8	P22	P22	P19	P22	P22	P22	F,S,L
Caudata (300)	W8	W22	W14	W19	W8	W22	W11	F,AE,AP,S,L
Anura (1900)	W13	W22	W6	W19	W8	W22	W20	F,AE,AP,S,L
Reptilia								
Testudinata (320)	W27	P22	W2	—	P22	W22	W21	F,B,M,AE,AP,L

Table 2: (continued)

Taxon	Photo	Thermo	Phono	Proprio	Tactile	Equilibrium	Chemo	Environments
Rhynchocephalia (1)	W27	P22	W2	–	P22	W22	W21	L
Squamata (5840)	W27	P5	W2	–	P22	W22	W21	M,AE,AP,S,L
Crocodilia (21)	W27	P22	W2	–	P22	W22	W21	AP
Aves								
Sphenisciformes (17)	W24	W31	W10	P28	P28	W23	W3	M,AP
Struthioniformes (1)	W24	W31	W28	P28	P28	W23	P28	L
Rheiformes (2)	W24	W31	W28	P28	P28	W23	P28	L
Casuariiformes (8)	W24	W31	W28	P28	P28	W23	P28	L
Apterygiformes (3)	W24	W31	W28	P28	P28	W23	W3	L
Tinamiformes (32)	W24	W31	W28	P28	P28	W23	P28	A
Gaviiformes (4)	W24	W31	W28	P28	P28	W23	W3	A
Podicipediformes (20)	W24	W31	W28	P28	P28	W23	W3	A
Procellariiformes (93)	W24	W31	W28	P28	P28	W23	W3	A
Pelecaniformes (54)	W24	W31	W28	P28	P28	W23	W3	A
Ciconiiformes (109)	W24	W31	W28	W28	W3	W23	W3	A
Anseriformes (148)	W24	W31	W23	W28	W3	W23	W3	A
Falconiformes (271)	W24	W31	W23	P28	P28	W23	W3	A
Galliformes (244)	W24	W31	W10	W28	P28	W23	W3	A
Gruiformes (193)	W24	W31	W28	P28	P28	W23	W3	A

Table 2: (continued)

Taxon	Photo	Thermo	Phono	Proprio	Tactile	Equilibrium	Chemo	Environments
Charadriiformes (301)	W24	W31	W10	P28	P28	W23	W3	A
Columbiformes (305)	W24	W31	W23	P28	W3	W23	W3	A
Psittaciformes (315)	W24	W31	W23	P28	W3	W23	W3	A
Cuculiformes (146)	W24	W31	W28	P28	P28	W23	W3	A
Strigiformes (134)	W24	W31	W23	P28	P28	W23	W3	A
Caprimulgiformes (85)	W24	W31	W28	P28	P28	W23	W3	A
Apodiformes (395)	W24	W31	W28	P28	P28	W23	W3	A
Coliiformes (6)	W24	W31	W28	P28	P28	W23	P28	A
Trogoniformes (34)	W24	W31	W28	P28	P28	W23	P28	A
Coraciiformes (186)	W24	W31	W28	P28	P28	W23	W3	A
Piciformes (391)	W24	W31	W28	P28	W3	W23	W3	A
Passeriformes (4911)	W24	W31	W23	W28	W3	W23	W3	A
Mammalia								
Monotremata (6)	W29	W32	W32	W32	W32	W32	W26	AP,L
Marsupialia (250)	W29	W32	W16	W32	W32	W32	W26	AP,L
Insectivora (300)	W29	W32	W16	W32	W32	W32	W26	AP,S,L
Dermoptera (2)	W29	W32	W32	W32	W32	W32	W26	L
Chiroptera (800)	W29	W32	W30	W32	W32	W32	W26	A
Primates (195)	W29	W32	W30	W32	W32	W32	W26	L
Edentata (30)	W29	W32	W32	W32	W32	W32	W26	L

Table 2: (continued)

Taxon	Photo	Thermo	Phono	Proprio	Tactile	Equilibrium	Chemo	Environments
Pholidota (7)	W29	W32	W16	W32	W32	W32	W26	L
Lagomorpha (65)	W29	W32	W32	W32	W32	W32	W26	L
Rodentia (1800)	W29	W32	W30	W32	W32	W32	W26	AP,S,L
Cetacea (92)	W29	W32	W16	W32	W32	W32	W26	F,M,AE
Carnivora (252)	W29	W32	W30	W32	W32	W32	W26	AP,L
Pinnipedia (32)	W29	W32	W32	W32	W32	W32	W26	M,AP
Tubulidentata (1)	W29	W32	W32	W32	W32	W32	W26	L
Proboscidea (2)	W29	W32	W30	W32	W32	W32	W26	L
Hyracoidea (6)	W29	W32	W32	W32	W32	W32	W26	L
Sirenia (4)	W29	W32	W32	W32	W32	W32	W26	B,AE
Perrissodactyla (16)	W29	W32	W32	W32	W32	W32	W26	L
Artiodactyla (194)	W29	W32	W32	W32	W32	W32	W26	AP,L

REFERENCES FOR TABLE II

1. Ali, M.A. and Wagner, H.-J. (1975). Distribution and development of retinomotor responses. In M.A. Ali (ed.) Vision in Fishes: new approaches in research. Plenum Pr., New York; pp. 369-396.
2. Baird, I.L. (1970). The anatomy of the reptilian ear. In C. Gans (ed.) Biology of Reptilia 2: 193-275. Academic Pr., New York.
3. Bang, B.G. and Cobb, S. (1968). The size of the olfactory bulb in 108 species of birds. Auk 85: 55-61.
4. Bardach, J.E. and Villars, T. (1974). The chemical senses of fishes. In P.T. Grant and A.M. Mackie (eds.) Chemoreception in marine organisms. Academic Pr., New York; pp. 49-104.
5. Barrett, R. (1970). The pit organs of snakes. In C. Gans (ed.) Biology of the Reptilia 2: 277-300. Academic Pr., New York.
6. Capranica, R.R. (1976). Morphology and physiology of the auditory system. In R. Llinas and W. Precht (eds.) Frog neurobiology. Springer-Verlag, New York; pp. 551-575.
7. Flock, Ake (1971). The lateral line organ mechanoreceptors. In W.S. Hoar and D.J. Randall (eds.) Fish physiology V. Academic Pr., New York; pp. 241-263.
8. Goin, C.J. and Goin, O.B. (1971). Introduction to herpetology. W.H. Freeman and Co., San Francisco.
9. Hara, T.J. (1971). Chemoreception. In W.S. Hoar and D.J. Randall (eds.) Fish physiology V. Academic Pr., New York; pp. 79-120.
10. Kare, M.R. and Rogers, J.G. Jr. (1976). Sense organs. In P.D. Sturkie (ed.) Avian physiology. Springer-Verlag, New York; pp. 29-52.
11. Kauer, J.S. (1974). Response patterns of amphibian olfactory bulb neurones to odour stimulation. J. Physiol. 243:695-715.
12. Lagler, K.F., Bardach, J.E. and Miller, R.R. (1962). Ichthyology. University of Michigan Pr., Ann Arbor, Michigan.
13. Lettvin, J.Y., Maturana, H.R., McCulloch, W.S. and Pitts, W.H. (1959). What the frog's eye tells the frog's brain. Proc. Inst. Radio Eng. 47: 1940-1951.
14. Lombard, R.E. (1977). Comparative morphology of the inner ear in salamanders (Caudata: Amphibia). Contrib. Vertebrate Evol. 2 S. Karger Publ., New York.
15. Lowenstein, O. (1971). The labyrinth. In W.S. Hoar and D.J. Randall (eds.) Fish physiology V. Academic Pr., New York; pp. 207-240.
16. Masterton, B. and Diamond, I. (1973). Hearing: central neural mechanisms. In E.C. Carterette and M.P. Friedman (eds.) Handbook of perception III: biology of perceptual systems. Academic Pr., New York; pp. 407-448.
17. Murray, R.W. (1971). Temperature receptors. In W.S. Hoar and D.J. Randall (eds.) Fish physiology V. Academic Pr., New York; pp. 121-133.

18. Nelson, J.S. (1976). Fishes of the world. John Wiley & Sons, New York.
19. Noble, G.K. (1931). The biology of the Amphibia. McGraw-Hill Book Co., New York.
20. Ottoson, D. (1959). Comparison of slow potentials evoked in the frog's nasal mucosa and olfactory bulb by natural stimulation. Acta. Physiol. Scand. 47: 149-159.
21. Parsons, T.S. (1970). The nose and Jacobson's organ. In C. Gans (ed.) Biology of the Reptilia 2: 99-191. Academic Pr., New York.
22. Porter, K.R. (1972). Herpetology. W.B. Saunders Co., Philadelphia.
23. Schwartzkopff, J. (1968). Structure and function of the ear and of the auditory brain areas in birds. In A.V.S. DeReuck and J. Knight (eds.) Hearing mechanisms in vertebrates. Little, Brown and Co., Boston; pp. 41-59.
24. Sillman, A.J. (1973). Avian vision. In D.S. Farner and J.R. King (eds.) Avian biology 3: 349-387. Academic Pr., New York.
25. Tavolga, W.N. (1971). Sound production and detection. In W.S. Hoar and D.J. Randall (eds.) Fish physiology V. Academic Pr., New York; pp. 135-205.
26. Tucker, D. and Smith, J.C. (1976). Vertebrate olfaction. In R.B. Masterton, C.B.G. Campbell, M.E. Bitterman and N. Hotton (eds.) Evolution of brain and behavior in vertebrates. Lawrence Erlbaum Assoc., Hillsdale, N.J.; pp. 25-52.
27. Underwood, G. (1970). The eye. In C. Gans (ed.) Biology of the Reptilia 2: 1-97. Academic Pr., New York.
28. VanTyne, J. and Berger, A.J. (1976). Fundamentals of ornithology. John Wiley & Sons, New York.
29. Walker, E.P. (1975). Mammals of the world. John Hopkins University Pr., Baltimore, Maryland.
30. Webster, D.B. (1973). Audition. In E.C. Carterette and M.P. Friedman (eds.) Handbook of perception III: biology of perceptual systems. Academic Pr., New York; pp. 449-482.
31. Whittow, G.C. (1965). Regulation of body temperature. In P.D. Sturkie (ed.) Avian physiology. Cornell University Pr., Ithaca, New York; pp. 186-238.
32. Young, J.Z. (1966). The life of mammals. Oxford University Pr., London.

TAXES IN UNICELLS, ESPECIALLY PROTOZOA

P. COUILLARD

Département des Sciences Biologiques, Université
de Montréal, Montréal, Québec, H3C 3J7, Canada

> How now, Sir Proteus! Are you crept before us?
> The Two Gentlemen of Verona, Act IV, Sc. 2

1. INTRODUCTION

Protozoa form a group of unicellular, (acellular), prokaryotes at the animal end of a broad spectrum of organisms, the other extremity being occupied by unicellular algae such as Desmids and Diatoms. Borderline groups, such as the fungus-like Mycetozoa (Myxomycetes) or the green flagellates lie in the no-man's land, forever to be disputed by protozoologists and phycologists.

Protozoa are generally looked down as being ancient and primitive, terms which are mutually contradictory. Being ancient, they cannot be primitive since they have been improving themselves through evolution much longer than we have, as vertebrates or as mammals. It should not come as a surprise, therefore, to find in Protozoa highly complex and sophisticated mechanisms in such a crucial domain, evolution-wise as sensory ecology.

It is difficult to speak of protozoan behaviour without running into problems of semantics; most of the terms we must use have been borrowed from the vocabulary of animal or human ethology. Words like "cell, body, death, animal and plant" acquire a very special meaning when applied to unicells and concepts such as "memory, judgment, preference, choice or discrimination, avoiding or seeking", which are already difficult to apply to the higher invertebrates, are prone to lead us into inextricable pitfalls of teleology when used to describe or explain protozoan "behaviour".

Indeed, we must realise that, at the unicellular level, where nerves are replaced by molecular chains in membranes or by hypothetical microfibrillar networks, there are no such things as reflexion or reflexes, only what Jacques Loeb used to call "Forced movements" (Loeb, 1918). Thus I propose to show that, being built the way it is, the protozoan has no possibility but to act the way it does in answer to a given stimulus, a billion years of evolution insuring survival value to such a simple mechanism.

In discussing sensory adaptations of Protozoa to their environment, we shall be mostly concerned with taxes. Indeed, most protozoa are motile and they tend to express their moods and opinions by displacing themselves in the direction of greater confort or survival.

This account of protozoan taxes does not aim to be exhaustive, my choice of examples mostly reflects my own personal experience and preferences. I hope, by introducing the reader to a variety of ecophysiological situations, to stimulate interest in a biological group often neglected by the sensory ecologist.

For general considerations on protozoan behaviour, see: Jennings (1905), Loeb (1917), Fraenkel and Gunn (1961), Haldall, (1970), Perez-Miravete (1973), Carlile (1975) and Diehn *et al*. (1977).

2. PHOTOTAXIS

2.1 Generalities

For non-photosynthetic organisms, light is not an essential but an important ecological variable. It is generally tolerated, when present in moderate amounts, and avoided if excessive. Light is also the principal entraining factor (zeitgeber) for rhythmic activities of plants and animals.

For the cell biologist, light is too often a neglected experimental variable whenever living organisms are examined under the compound microscope. For example, Forget (1977), in my laboratory has shown that a phase contrast microscope, with quartz halogen illumination at 4.5 V., inflicts a radiant energy level of 10 300 W/m^2 to organisms being observed at 10 X 40. The use of suitable procedures, such as heat-absorbing filters, low voltage tungsten and dark-adapter observer can reduce this energy level to 0.8 W/m^2. In comparison, the solar constant at sea level (sun at zenith) is 900 W/m^2.

Thus, while looking at living cells, the microscopist should always think like the psychiatrist interviewing a patient: How would this subject behave if I were not observing him?

2.2 Species without characterized photopigments

2.2.1 *Amoeba proteus*

This common fresh-water amoeba contains no visible pigment save for the occasional green alga it may ingest, yet it has long been known to be negatively phototactic, especially to the short wavelengths of the visible spectrum. The effect of light on *Amoeba* was studied in detail by Mast (1911), (1932), who found that a light shock is only effective if applied to the so-called plasmagel sheet, the front of advancing granular endoplasm in a progressing pseudopodium. Such a precise localization of the blue-light sensitive site raises several interesting questions. Neither the membrane nor the main contracting gel tube seem to be involved. The specific effect of the light seems to be a gelation of the granular endoplasm which stops the advancing pseudopod, diverting it sideways, away from the illuminated area. As said before, no particular pigment accumulation can be found in any part of the cell and this particular reaction of *Amoeba* to blue light has yet to be integrated into any of the current theories of ameboid motion.

Forget (1977) has recently shown that panchromatic light intensities above 120 W/m^2 seriously disrupt the contractile vacuole cycle in *A. proteus*. Among observed effects, we note an increased variability in all parameters of the cycle, the appearance of plateau (vacuolar stasis), partial systoles and *pseudosystoles*, a new phenomenon in which the vacuole slowly decreases in volume, apparently losing its fluid into the surrounding cytoplasm, no pore to the outside being formed. Work in progress aims at obtaining an action spectrum for the effects of light on the contractile vacuole.

The ecological implications of our experiments are not immediate. Amoeba favours dimly lighted microhabitats, (laboratory strains are grown in the dark), and never encounters such severe light conditions in nature. In this perspective, we are planning a series of long-term experiments in which amebae will be submitted for days to continuous light regimes and examined periodically for cumulative effects.

More generally speaking, our so-called "unpigmented cells" are not completely devoid of light-absorbing molecules. Key compounds like flavins and cytochromes, strategically located in mitochondria, absorb in the blue (with maxima at 450 µm and 400-420 µm respectively). Disruption of mitochondrial function by light should not only affect energy metabolism but also accumulation by these organelles, of "control" ions such as calcium, and phosphate.

2.2.2. Stentor

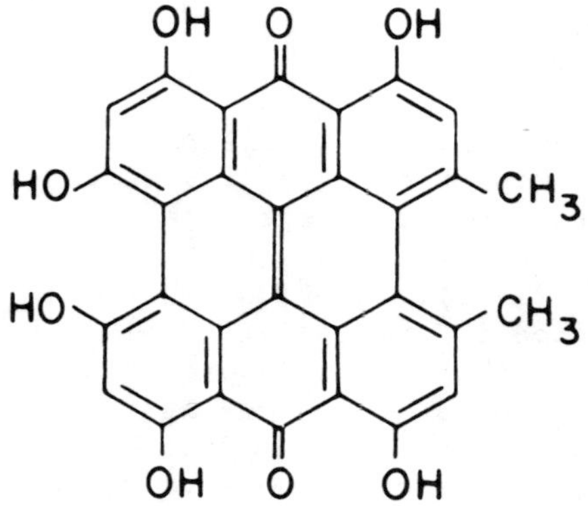

Fig. 1. Hypericin a mesonaphtho-dianthrone.

Members of this genus of trumpet-shaped freshwater spirotrich ciliates are remarkable by their contractility. Some species are pigmented and *S. polymorphus* contains symbiotic zoochlorellae but by far the most spectacular member of the genus is the blue *S. coeruleus*.* Its pigment, *Stentorin*, a mesonaphto dianthrone, is related chemically to blepharismin and hypericin (see section 2.2.3., this chapter) and is concentrated into globules which are found in cortical longitudinal stripes.

S. coeruleus can attach itself to substrates and feeds in an extended configuration. When stimulated mechanically or otherwise, it contracts to a rounded shape within a few tenths of a second. When swimming about, it assumes an intermediate pear-like form. These shape changes are made possible through the interplay of two distinct, antagonistic, mechano-effector systems, the microfibrillar myonemes and the microtubular postciliary fibers, respectively involved in body contraction and extension. (Huang & Pitelka, 1973).

Holt & Lee (1901) were first to report that *S. coeruleus* accumulates in shaded areas, the reaction was studied in detail by Mast (1906) and Tartar (1961) found that some strains of this species do not show this response to light. The mechanism of the light-avoiding reaction in *Stentor coeruleus* is now well known: positive illumination gradients induce *ciliary reversal,* an abrupt change in the polarity of the ciliary beat. As the stentor backs up, with some shift in orientation due to cell asymmetry, the effect of the stimulus wears off and forward movement is resumed most often in a different direction. Since negative light gradients do not have this reversing effect, the ciliate eventually ends up in the most obscure area of its accessible environment.

* For a general treatment of the genus, see Tartar (1961).

The electrophysiological correlates of light-induced ciliary reversal in *Stentor coeruleus* were elucidated by Wood (1976), using intracellular microelectrodes. He found that light can elicit three types of responses:

1- Graded photic receptor potentials, which fluctuate during illumination. These are attributed to an electrogenic pumping mechanism activated by light.

2- All-or-none action potentials, causing body contraction in extended animals.

3- In swimming stentors, regenerative responses (1-60 mV) directly associated with ciliary reversal and triggered by threshold receptor potentials. Furthermore, the action spectrum of the ciliary reversal response matches the absorption spectrum of stentorin. Finally, bleaching stentors in 0.3 M caffeine raises the stimulation threshold by a factor of 20.

According to Wood (1976) this makes stentorin a true photoreceptor pigment, not merely a photosensitizer.

2.2.3. *Blepharisma*

This close relative of *Stentor* is characterized by its pink colour due to the presence of the pigment blepharismin (chemically related to stentorin, see section 2.2.2., this chapter) in subpellicular granules.* *Blepharisma* species and strains can be found in all shades of colouration and albino mutants have also been reported. In addition to genetically-determined variation, cultures of *Blepharisma* grown in the light have a low pigment content.

Blepharismin is clearly a photosensitizing dye for its bearer. Well-pigmented individuals are killed by a brief exposure to visible or near-UV light (2,000 foot-candles) and for this, the presence of oxygen is necessary. Blepharismin can also act as an exogenous photosensitizer for non-pigmented species of Protozoa such as *Paramecium*. Strangely enough, from a teleological point of view, negative phototaxis has not been described in *Blepharisma* and unpigmented mutants do not seem to have any particular selective advantage in nature.

Naphtodianthrone pigments also occur in plants; hypericin, which is found in the genus *Hypericum* (St-John's Wort *Hypericaceae*), and Fagopyrin, in the genus *Fagopyrum* (Buckwheat, *Polygonaceae*).

We thus have a group of chemically-related pigments, two in closely-related protozoan genera, two in unrelated plants. Stentorin

---

* For a general treatment of the genus, see Giese, 1973.

acts as a photoreceptor and induces photoavoidance in its bearer while blepharismin is a photosensitizer, endogenous as well as exogenous. As for the plant pigments, while little is known of their raison d'être *in situ*, one, Hypericin, is dreaded by farmers for it induces an often fatal photosensitization in the cattle and sheep that eat the weed in question.

Clearly, much comparative work is needed to make some sense out of this rather unique ecological situation at least in the Protozoa. Can stentorin be shown to be a photosensitizer in *Stentor*? Does light induce some form of bioelectric response in the non-contractile *Blepharisma*? Have these pigment some evolutionary value for their respective bearers or must they be considered as unavoidable metabolic end-products to which *Stentor* and *Blepharisma* have had to adapt, each in its own way?

2.3.1. *Paramecium bursaria*

This paramecium* is easily recognised by the presence, in its cytoplasm, of several hundred zoochlorellae,** each within its own tiny vacuole (Karakashian *et al.* 1968). The association is mutually advantageous.

Paramecium obtains from its symbionts a number of products of photosynthesis, among which maltose and oxygen, and it probably also metabolizes degenerating algae. Thus, algae-containing paramecia can withstand starvation much better than those freed of their symbionts. As for the algae, in their homeostatic shelter, they obviously obtain from their host not only inorganic and possibly organic nutrients, but also abundant carbon dioxide for their photosynthetic needs. As an added benefit, they get a free ride to favourable light conditions. Indeed, *P. bursaria* is positively phototactic, other membres of the genus being rather indifferent to normal light intensities.

As recently shown by Saji and Oosawa (1974), the accumulation of *P. bursaria* in lighted areas is a case of photocapture. A quiescent *P. bursaria* under the microscope can be submitted to an increase of illumination without reacting. On the other hand, decreasing the light intensity induces a typical avoiding reaction: the paramecium backs up and spins about. Likewise, swimming *P. bursaria* move ahead in positive or neutral light gradients but they reverse and change direction in negative gradients. This leads necessarily to the accumulation of the ciliates in illuminated areas. This is clearly an "off" reaction, quite the reverse of what was seen earlier in *Stentor*. (see Section 2.2.2. this chapter).

* For a general treatment of the genus see van Wagtendonk (1974).
** Algal symbionts of invertebrates are discussed by Smith (1973).

We are tempted to correlate this particular phototactic capacity of *P. bursaria* with the presence of zoochlorellae in its cytoplasm. This could be substantiated by action spectrum determinations and verified by checking the light sensitivity of chlorella-free paramecia. If it is established that chlorellae, each within its vacuole, are the light-sensitive elements it remains to be seen how the message is passed on to the host cell and more precisely to the plasma membrane, which, in Paramecium, is the site of ionic events associated with ciliary reversal. (see section 4.1, this Chapter).

2.3.2. *Ophrydium*

Ophrydium versatile is a peritrich ciliate, a relative of the better-known *Vorticella*. It forms greenish, gelatinous, globular colonies up to 30 cm in size at the bottom of lakes and ponds. (Winkler and Corliss, 1965). Individuals are sessile, on thin anastamosing filaments, in tube-like cavities of the jelly mass. The cell is highly contractile and contains numerous symbiotic zoochlorellae.

If kept in continuous darkness, members of the colony respond, within 48 hours, by transforming into free-swimming, astomous, positively phototrophic *telotrochs*. These rapidly settle in well-illuminated areas, revert to the sessile form and start new colonies.

Teleologically speaking, this makes sense. Although *Ophrydium* has a functional cytostome and ciliary feeding apparatus, its algal symbionts need light for photosynthesis. Motile telotroch formation in abnormally-prolonged darkness turns what threatened to become a danger to survival into an ecological advantage, permitting the dissemination of new colonies into favourable niches.

Telotroch formation is a normal concomitant of binary fission and conjugation in ordinary Axenic Peritrichs. Its induction by a break in the daily light-dark cycle appears related to the presence of algal symbionts, just as the acquisition of phototaxis in *Paramecium bursaria*. In this case also, the intriguing problem of signal acquisition and cell-within-cell transfer of information remains unanswered. The mechanism of positive phototaxis in the telotroch has yet to be described.

2.3.3. *Strombidium*

Strombidium oculatum is a marine oligotrich ciliate whose behaviour was studied by Fauré-Frémiet (1948) in the tide pools of the Marine Station at Roscoff, France. It exhibits a remarkable adaptation to intertidal life: at high tide, when its tide-pool habitat is open to the sea, it becomes negatively phototactic and encysts, very likely in sheltered regions.

Fig. 2. *Strombidium oculatum* (modified from Fauré-Frémiet (1948) "s": "stigma" (See text.)

At low tide, it excysts and feeds, exhibiting a positive phototaxis. This cyclic phototaxism can persist, with tidal periodicity, under constant aquarium conditions.

The adaptative value of this intricate cyclic pattern of behaviour are evident. At low tide, *Strombidium* moves about and feeds when its native tidepool is isolated; on the rising tide, it seeks a sheltered niche and encysts upon the substratum thus avoiding being cast ashore or washed out in the open sea where, presumably, living conditions are not suitable.

Strombidium contains symbiotic algae, *not Chlorella*, since they are flagellated. (A freshwater species, *S. mirabile*, has symbionts of the genus *Chlamydomonas*). The phototaxis of Strombidium was attributed by Gruber (1844) to the presence of a green "eyespot" in the anterior portion of the cell. While common in green flagellates, stigmata have never been described in ciliates. As it turns out, the "eyespot" of Strombidium is, so to speak, a second-hand organelle! According to Fauré-Frémiet (1948) it consists of stigmata recuperated by the ciliate from degenerating algal symbionts!

This old investigation of Fauré-Frémiet lays the groundwork for a number of experimental enquiries:

1- Is phototaxis in *Strombidium* mediated by its borrowed stigma? We have seen earlier that Chlorella-containing Ciliates can exhibit a variety of phototactic responses without localized photoreceptors.

2- Is ciliary reversal involved, as in *Paramecium bursaria*?

3- Are there electrophysiological concomitants? How are they affected by cyclic changes?

2.4.1. *Euglena*

Most biologists remember *Euglena* from their elementary Biology course, as a green alga or protozoan (depending on whether they had

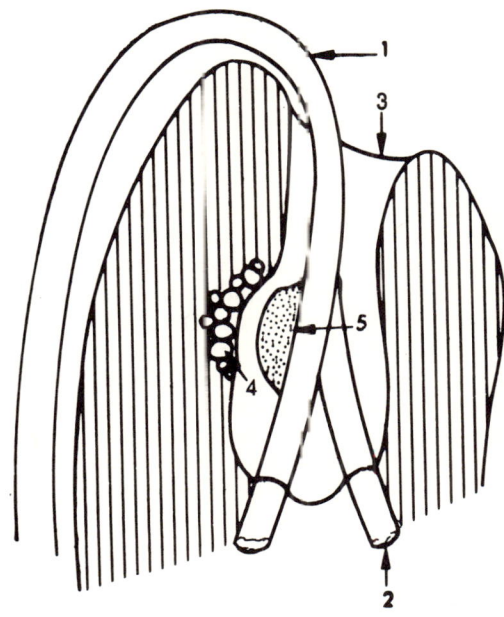

Fig. 3. *Euglena:* Disposition of locomotor and photoreceptor organelles.
1- Main flagellum
2- Vestigial flagellum
3- Mouth of reservoir
4- Stigma
5- Paraflagellar body

a botanist or a zoologist as a teacher), with a single anterior pulling flagellum and a light-sensitive orange eyespot. This is inaccurate in most respects: *Euglena* has *two* flagella, one of them vestigial, the flagellum *pushes* the cell and the stigma is not the light-sensitive organelle.

In normal movement, the main flagellum trails behind at a 30° angle with helicoidal waves traveling distally. The spirally-grooved body rotates on its long axis. The propulsive asymmetry of the pulsellum, compensated for by body rotation, results in a circular oscillation of the front end of the cell, the hind end more or less following the main path of movement.

Both flagella arise within the *reservoir*, a piriform invagination of the anterior tip of the cell. The long, functional, flagellum bears, near the base, a swelling, the *paraflagellar body*, (PFB). The stigma lies beneath the wall of the reservoir, next to the PFB. As confirmed by the electron microscope, there is no particular anatomical relationship between stigma and PFB.

Euglena is, of course, well known for its positive phototaxis to moderate light intensities. It reacts negatively to strong light. It also shows *photokinesis* a non-directional, light-induced acceleration of movement. (Wolken and Shih, 1958). Early observers of *Euglena* phototaxis describe the stigma as the site of photosensitivity in the cell, but as early as 1900, Wager proposed that the stigma is merely a shading device, acting in conjunction with the real photoreceptor, the paraflagellar body.

Ultrastructurally, the stigma is composed of some 40 electron-dense globules, 100-300 nm in diameter (Wolken 1956) while the PFB is a small (0.4 X 0.3 X 0.2 µm) slightly plano-convex structure located within the flagellar membrane. It shows a regular, quasicrystalline structure. (Kivic and Vesk, 1975. Wolken, 1977).

The mechanism of the phototactic response in *Euglena* has been reviewed by Checcuci (1976). When the euglena moves towards a light source, its PFB is never shaded by the stigma and the direction of movement is not affected. If illumination is from the side, body rotation causes the PFB to be shaded by the stigma, in an intermittent manner, for 2/3 to 3/4 of each spiral turn. At normal rotation frequency of 2 turns/sec., this gives a shaded/unshaded alternance of 350/150 msec. Whenever shading occurs, it triggers a temporary erection of the flagellum which shifts the cell axis sideways and reorients the path of the cell towards the light source. These minute, successive course corrections continue until the euglena points toward the light source and shading no longer occurs.

Simple experiments show that the effective stimulus for flagellar erection is the light-to-shade transition:

1- If a quiescent euglena is observed under the microscope, a decrease in illumination of 10-20% elicits spinning movements. (Flagellar erection).

2- In border-crossing experiments, a penetration of 10-15 µm into the dark precedes cell swerving. At a cell speed of 100-150 µm/sec, this represents the time constant of the normal shading stimulus.

Some hypotheses on the mechanism of flagellar erection have been put forth: A contraction of the walls of the reservoir (Piccinni and Omodeo, 1975) or an active straightening of the flagellum proper (Bovee and Jahn, 1972). Checcuci *et al.* (1974) have pointed to a possible involvement of the *paraflagellar rod*, an accessory structure in which ATPase activity was later demonstrated (Piccinni *et al.* 1975).

What are the pigments involved in *Euglena* phototaxis? The stigma, (shading device), is known to contain Beta-Carotene, Cryptoxanthin and Lutein (a carotenoid) (Ferrara and Blanchetti, 1976). It has, so far, proved impossible to characterize unambiguously the composition of the paraflagellar body. This organelle is too small and not absorbent enough to lend itself to microspectrophotometry and it has not yet been isolated in sufficient quantities for direct analysis. In the fluorescent microscope, it shows maximum excitation at 360 and 450 nm, wavelengths which are characteristic of flavins.

Action spectra of *Euglena* phototaxis have been, for a long time, ambiguous due to the interference of respective absorptions of stigma and PFB pigments in the visible. More recently, Checcuchi *et al* (1976) have obtained action spectra in the near UV and visible for three types of euglenas:

	Chloroplasts	Stigma	PFB	Phototaxis
Green	+	++	+	+
Etiolated. (dark-bleached)	(Proplastids)	+	+	+
Streptomycin-bleached	−	−	+	−

In all three types, an action spectrum characteristic of flavins has been obtained.

2.4.2. *Erythropsis*

Erythropsis pavillardi is a member of the Dinoflagellate family *Warnowiidae* (formerly *Pouchetiidae*), a group of rare and extremely fragile planktonic marine protozoa. Contrary to the typical dinoflagellate, *Erythropsis* does not use its flagella for locomotion but propels itself by the movement of a tentacle-like appendage, the piston. Hence, it does not rotate on its longitudinal axis when moving.

According to Grell (1973) *Warnowiidae* are heterotrophic and live in the deeper regions of the ocean. In these scotopic zones, light-sensitive organelles have to be quite efficient to serve either for orientation or for locating phototrophic prey. (See also Mornin and Francis, 1967).

Most members of the family have in common a large, prominent ocellum which has been described in detail by Greuet (1968) in *E. pavillardi* both at the photonic and at the electronic microscope levels. This large (40 μm) organelle represents, from the morphological point of view at least, a most extraordinary case of convergence with the vertebrate eye. It contains structures which most obviously invite comparison with cornea, lens. iris, vitreum (Hyalosome complex) and retina (melanosome), not to mention optic nerve, represented by a fibrillar bundle linking the melanosome with the body surface and the locomotor organ (piston).

Quite understandably, Greuet shows himself most reluctant in assigning functions to the various elements he describes, but, even at the submicroscopic level, one cannot help but draw homologies. Thus, an unlimited number of experimental opportunities present themselves with this unique material. The nature and extent of the phototactic behaviour, sensitivity and action spectrum of melanosome pigment, functional relationships between ocellum and locomotor organelle etc.

In another Dinoflagellate, of the genus *Glenodinium*, Dodge and Crawford (1969) have described a complex photoreceptor comprising an Euglena-type orange-coloured stigma coupled with a lamellar body, a stack of some 50 flattened vesicles, not unlike the outer segment of the vertebrate rod.

3. CHEMOTAXIS

3.1 *Escherichia coli*

It has been known since the early days of electron microscopy that Bacteria move by means of "flagella" and that they show chemotaxis, particularly towards food substances. (Adler, 1966). Early speculations about the mechanism of movement in bacterial flagella (Doetsch, 1966) have been confirmed by several techniques, since the first demonstartion of Mussill and Jarosch (1972). The flagellum of *E. coli* is a thin (200 Å), helical filament assembled out of subunits of a single protein, flagellin, of molecular weight 40,000. The flagellum acts by rotating on its long axis through the interaction of a "rotor", attached to its base, with a stator, this molecular motor being situated within the plasma membrane-cell wall complex of the bacterium. (de Pamphilis and Adler, 1971).

E. coli has about six flagella scattered on the cell surface (peritrichous). In normal locomotion the flagella trail behind, individually rotating counterclockwise and acting in coordination as a bundle. Occasionnally, the sense of flagellar rotation is abruptly reversed; the bundle of flagella flies apart, and the motion of the bacterium is suddenly perturbed while it undergoes "tumbling". (Larsen *et al.* 1974). Tumbling is always a transient phenomenon, followed by resumption of normal, more or less straight, movement, in a randomly different direction from before.

How such a simple response can lead to positive or negative chemotaxis can easily be realized: If *E. coli* moves within a gradient of an attractant, tumbling is minimum when the cell moves **up** the gradient. If the bacterium moves **down** the gradient, tumbling becomes more frequent. The reverse occurs within a gradient of repellent. Thus, *E. coli* can "find" food and "avoid" deleterious substances simply by controlling the frequency of its changes of direction. In a simplified fashion, this mechanism could be compared to the process of ciliary reversal which we have seen operating in Ciliate phototaxis and Thigmotaxis. (see sections 2.3.1. and 4.1 this chapter).

A biochemical mechanism for the tumbling response of bacteria to chemical gradients has been proposed by Ordal and Fields (1977). Intracellular free calcium levels exert a controlling influence on tumbling frequency. Attractants decrease free calcium through

binding to specific proteins. Calcium is released when these
receptor proteins become methylated. This is in accord with the
long-known fact that methionine is required for tumbling (Adler
and Dahl, 1967) and that low concentrations of external calcium
are also essential for this process.

Other questions arise when we consider the problem of
detecting chemical gradients for a cell the size of *E. coli*.
From actual performance, this bacterium can respond to gradients
leading to concentration differences as small as $1:10^4$ along its
own body length of 2 µm. This is clearly too much to expect; not
only are the concentration differences extremely small but the
time available for perception is extraordinarily short since
E. coli, in moving through such gradients, covers its own length
in 1/20 th of a second! We must rather assume that *E. coli* is
capable of temporal sensing, in other terms, that it posesses a
"memory" enabling it to store and compare chemical information
gathered along 20-100 body lengths. Not only does that increase
the gradient to be sensed to $1:10^3$ or $1:10^2$ but it also has the
advantage of leveling out statistical fluctuations.

3.2 *Amoeba proteus*

It is a cause of wonder, at first glance, that a large,
sluggish, bottombound organism such as the common freshwater ameba
should manage to capture such highly motile, free-swimming prey as
the small ciliate *Tetrahymena*, its favourite food in artificial
cultures. Indeed, a few minutes after adding prey to a dish of
hungry amebae, most cells are found with at least one, if not many
food vacuoles.

The events of phagocytosis are easy enough to observe; the mere
addition of Tetrahymena or even of the supernate from a Tetrahymena
suspension leads to a general speeding up of ameboid motion, thus
enhancing the probability and speed of capture. More specifically,
the presence of a tetrahymena within 30 µm of an ameba polarizes
the movement of an active pseudopod or induces the formation of a
new one in the region of the cell nearest to the prey. Upon reaching
the prey, this pseudopod invaginates into a *food cup*, the victim is
surrounded and rapidly enclosed within a *phagosome*. We shall not
concern ourselves here with the digestive processes that ensue. It
is notable that both food cup and phagosome are always much larger
than the prey, this used to be explained as resulting from the
violent movements of the prey during engulfment.

Pierre Marsot and I have attempted to throw some light upon
the mechanism of phagocytosis in *Amoeba proteus*. (Marsot and
Couillard 1972). We first studied the ingestion of gradually
simpler and simpler prey by *Amoeba*. These ranged from normal
tetrahymena, deciliated (non-motile) tetrahymena, dissociated cells

from freshwater Invertebrates (Hydra and Mussel) to semipermeable Nylon microcapsules (Chang, 1966) filled with *Hydra* tissue homogenate. All these preys are normally phagocytized, the usual oversize food cup and phagosome being formed. This implies that prey movement and surface characteristics are not necessary for its location and capture and that these factors are not essential for the large size of the phagocytic apparatus. Empty microcapsules are not phagocytized and boiling the homogenate or tightening the pores of the capsules greatly reduces their ingestion by *Amoeba*. This show that a chemotactic, diffusible product of prey metabolism is involved in capture.

Quantitative studies, in which a standard ameba population is presented with increasing non-motile cells or homogenate-microcapsules concentrations, show a marked inhibition of phagocytic activity at the higher prey concentrations. This is contrary to what one would expect if prey capture was a simple matter of probability. This, we explain by the hypothesis, also proposed by Jeon and Bell (1965) on other grounds, that chemotactic substances involved in phagocytosis have a biphasic effect: At low concentrations, they stimulate pseudopod formation and progression, while they inhibit it at high concentrations. This hypothesis can also account for food cup and phagosome formation: The prey, which emits chemotactic substance(s), finds itself at the center of a gradient which at first stimulates pseudopod progression. When the pseudopod nears the prey, it reaches high, (inhibitory) concentrations of the chemotactic substance. Its progression is locally inhibited, and food cup formation becomes automatic. Prey contact is not necessary nor is prey movement. Finally, the food cup and resulting phagosome are necessarily larger than the prey since they form at a distance from it.

This works fine for living prey, but *Amoeba* is known to occasionnally ingest inert particles such as glass, gold or quartz grains which are unlikely to emit chemotactic substances. This can be accounted for by a simple experiment: if clean glass microbeads or neutral resin particles are presented, they are mostly ignored by the amebae. If, on the other hand, the glass beads are previously coated with positively-charged Protamine Sulfate or if cation-exchange (positively-charged) resins are utilized, both types of inert particles are readily phagocityzed. In this case, however, the engulfment process is entirely different. No such thing as a food cup is formed, instead, randomly progressing pseudopods make snap contact with a target particle and then, the ameba litterally spreads itself in closest contact with the surface until the object is completely covered, finding itself, at the end, within the ameba. This, we interpret as resulting from the interplay of the negatively-charged ameba glycocalyx with a positively-charged surface.

If now, rate of phagocytosis is plotted as a function of concentration of positively-charged particles, a direct proportionality is found throughout the entire range, thus substantiating our impression of random predator-prey contact according to the target theory.

So, prey hunting, and capture by *Amoeba*, which appears at first like a highly purposeful behaviour, becomes reduced to the reaction of a pseudopod to a gradient of chemotactic substance with diphasic activity. Capture of an inert particle only requires that it should be "dirty" enough to carry a positive charge. All in all, being built the way it is, the ameba has no choice but to act the way it does!

Many questions remain open; what is the chemical identity of the chemotactic substances? What is the mechanism of its diphasic action on pseudopods? How does this integrate into current theories of amoeboid movement? From an evolutionary and ecological point of view, why has the prey not lost, through mutation-selection, the lethal aptitude of producing phagocytosis-inducing substances? My guess would be that it is an obligatory end-product of some essential metabolic reaction(s), a nice problem in comparative Biochemistry!

3.3. The gamones of *Blepharisma*

Conjugation is the sexual process characteristic of the Ciliates it amounts to a reciprocal fertilization involving cell-to-cell exchange of haploid nuclei between individuals of complementary mating types. Generally, the mutal recognition and subsequent union of conjugants involve specific macromolecules situated temporarily of permanently on the cell surface, as is the case for *Paramecium aurelia*.

In the large, pink spirotrich ciliate *Blepharisma intermedium*,* conjugation must be preceded by chemical communication between mating types (Miyake, 1974), leading to a complex sequence of interactions: mediated by *gamones* (hormones secreted by the gametes of one sex, acting on the gametes of the other sex) as follows:

1- Mating type I secretes G-1

2- G-1 reacts with Mating Type II cells (MT II) and transforms them for mating (MT II*)

3- G-1 also induces MTII cells to secrete their own gamone (G II).

* For a general treatment of the genus, see Giese (1973).

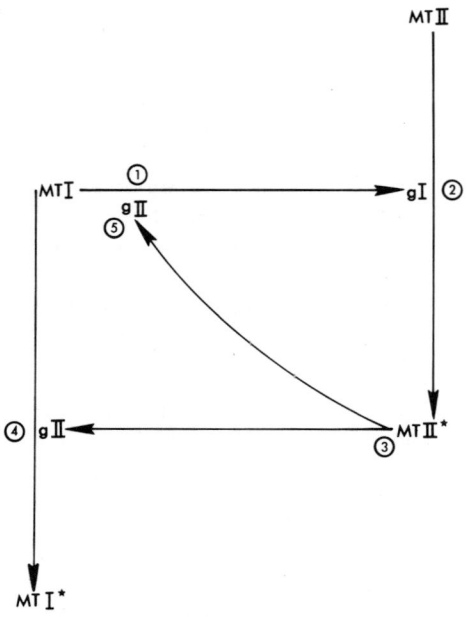

Fig. 4. Gamone secretion in *Blepharisma*. (Modified from Miyake 1974). See text for details.

4- G II, in turn, reacts with MT I cells and transforms them for mating (MT I*)

5- G II possibly reinforces G I secretion by MT I cells.

In *Paramecium*, only complementary mating types can unite; in *Blepharisma*, all mating combinations are possible, but only MT I X MT II unions lead to nuclear exchange.

Both gamones of *Blepharisma* have been characterized chemically; G I is a slightly basic glycoprotein of molecular weight 20,000, while G II is Calcium 3 (2' formyl amino 5 hydroxy benzoyl) lactate.

4. MECHANOTAXIS

4.1. The avoiding reaction in *Paramecium*

Upon striking an object, *Paramecium* undergoes the so-called <u>avoiding reaction</u>. It backs up, resumes forward movement, and repeats this trial-and-error sequence until the obstacle is bypassed. This has often been cited as a classical example or "purposeful" behaviour in a microorganism.

Modern work on this problem was initiated by Naitoh and Kaneko (1972) using biochemical models of *Paramecium caudatum*. In this technique, originally developed for muscle, motile cells are extracted at low ionic strength in the cold. The cells die and lose their ions and diffusible molecules. Only insoluble molecules remain in place within a completely permeable membrane remnant. Thus, the basic contractile machinery of the cell is kept intact, freely accessible to exogenous reagents. Arronet (1973).

Paramecium models, were subjected to various concentrations of calcium ions in the presence of ATP and magnesium. In calcium concentrations below 10^{-7} M, ciliary beat is reactivated and the models swim forward. Above 5×10^{-5} M Ca, they swim backwards. Intermediate concentrations gradually shift the direction of the effective stroke of the cilia. High calcium concentrations in the absence of Mg do not induce ciliary beat, but the non-motile cilia swing anteriorly to a position corresponding to their situation at the end of the reversed power stroke. Thus, magnesium and calcium have specific effects on the ciliary machinery of *Paramecium*; Magnesium is essential to the motion of these organelles, while calcium determines the polarity of the effective stroke.

We now turn to electrophysiological work by Eckert (1972) in collaboration with Naitoh (Eckert and Naitoh, 1972), using intracellular microelectrodes. They found that the plasma membrane of *Paramecium* responds differently to mechanical stimulation depending on which end of the cell is stimulated. Anterior stimulation induces a transient depolarization of the membrane while posterior stimulation induces a state of membrane hyperpolarization.

Taken as a whole, these findings of Naitoh and Eckert allow the avoiding reaction to be explained in biophysical terms (Naitoh, 1974). In the absence of mechanical stimuli, membrane pumps, or equivalent cytoplasmic segregative processes, hold intraciliary calcium concentration at low levels (below 10^{-7} M), and normal forward movement is maintained. Upon hitting an obstacle (forward mechanical stimulation), membrane depolarization allows sufficient external calcium to diffuse in to induce ciliary reversal and the paramecium backs off. The membrane repolarizes, calcium ions are pumped or segregated out and forward movement is resumed in a slightly different direction since some cell spinning has occurred when passing through intermediate internal calcium concentrations. Successive collisions will trigger the same sequence of reactions until the obstacle is avoided.

On the other hand, stimulation of the hind end of the paramecium will have opposite effects; transient membrane hyperpolarization causes an acceleration of forward ciliary beat, a typical and very useful escape reaction!

Thus, by the simple means of specific, localized electrophysiological responses, *Paramecium* has evolved the cellular equivalent of reflex behaviour. To quote Eckert: "While the metazoan nervous system converts analog signals (graded receptor potentials) to digital signals (all-or-none action potentials) and these back into graded analog signals (synaptic potentials), the bioelectric organization of the Ciliate is based entirely on analog-to analog transforms".

Similar, highly adaptive locomotory reflexes probably exist in other ciliates. We have already seen ciliary reversal in the phototactic *Paramecium bursaria* (see section 2.3.1. this chapter). In the larger spirotrich *Dileptus*, Doroszenski (1970) has been able to map precisely, by means of localized mechanical stimuli applied with a micromanipulator, a "backward response area" involving the front two thirds of the cell and a "forward response area" comprising the rearmost quarter of the ciliate, with an indifferent zone in between. It is highly probable that the ionic and electrophysiological correlates of both these systems should have much in common with what is already known in *Paramecium caudatum*.

It now seems that the calcium channels involved in ciliary reversal are located within ciliary membranes. In paramecia deciliated with chloral hydrate, the normal resting potential and the hyperpolarizing response to posterior stimulation are unaffected, but anterior stimulation fails to elicit ciliary reversal (depolarization and calcium inflow). If the paramecia are allowed to reciliate, the avoiding reaction is fully restored. (Ogura and Takahashi, 1976). Byrne and Byrne (1978) give scanning electron-microscope evidence identifying the ciliary membrane calcium channels with the so-called *intramembranous plaques* at the base of the cilium. "Paranoiac" mutants of *Paramecium*, characterized by long periods of continuous ciliary reversal, show altered plaque morphology.

Finally, the presence of specific calcium channels in ciliary membranes of *Paramecium* has led Duncan (1977) to an interesting speculation concerning the evolution of ciliary sensory cells, particularly the vertebrate retinal rod. It appears that photo-isomerization of rhodopsin eventually leads to changes in calcium permeability of outer segment discs and that the resulting calcium movements serve to amplify the photic stimulus and to transduce it to the plasma membrane. Since these discs are formed from infoldings of the outer segment plasma membrane, their calcium channels would be evolutionnary homologues of intramembranous plaques of protozoan cilia.

TAXES IN UNICELLS 49

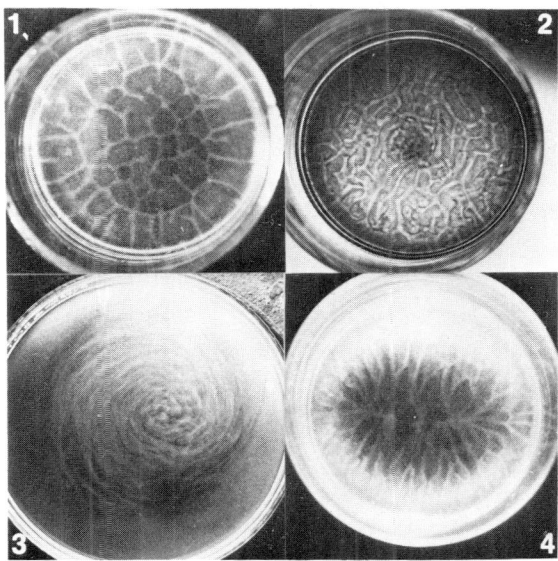

Fig. 5. Bioconvention patterns.
  Top: Normal patterns
    1: *Tetrahymena*
    2: *Euglena*
  Bottom: Disruption by light in *Tetrahymena*
    3: Tangential illumination
    4: Illumination from opposite sides.

5. GEOTAXIS

5.1. Bioconvection patterns

It is a matter of common observation that dense populations of small motile cells will spontaneously generate bioconvection patterns or "cells", analogus to the Bénard cells which are formed in heated fluids. (Whitehead, 1971).

Cultures of the small Ciliate *Tetrahymena* will readily form convection cells in flat-bottom dishes at depths greater than 2 mm (optimum, 1-2 cm) at cell densities higher than 150,000/ml (optimum, 1,300,000/ml). Time for the establishment of these patterns can be as short as 30 sec (Platt, 1961). Within a "cell", individuals swim down at the edges and up in the center. This is clearly a geotactic response, the patterns break down under weightless conditions, for example in aircraft flying paraboloid trajectories. (Winet, 1970). They are unaffected by factors which do not influence locomotion. In *Tetrahymena*, an unpigmented species, strong unilateral light will "push" the patterns away and cyclic ambient illumination can entrain daily rhythms, for example, under LD 12:12, pattern formation is slowest at dusk. The form of the "cells" is characteristic of the species, other variables being held constant. (Willee and Ehret, 1968).

We are still in the dark concerning the very basis of this phenomenon; what is the nature of the interaction between organisms which is responsible for such coordianted mass movements and what organizes these mass movements into Bénard-type cells?

5.2 Concrement vacuoles in Ciliates

According to Dogiel (1929), two families of Holotrich Ciliates, (*Butschliidae* and *Paraisotrichidae*), found in the intestine of the horse and related Perissodactyls, have, in the anterior part of the cell, a single, 10 - 18 µm, spherical vacuole containing a mass of small crystalline granules. Peripheral fibrils have also been described, surrounding the vacuole and leading towards the inner regions of the cell.

These so-called concrement vacuoles are interpreted by Dogiel strictly on morphological grounds, as organelles of equilibrium. If this should prove to be the case, it would represent a remarkable case of convergence with the statocysts of Metazoa, but clearly physiological research is needed before jumping to conclusions.

Muller's vesicles, of the Holotrich Ciliate *Loxodes*, have been interpreted by Pénard (1917) as being statocysts-like, but their close relationship with Golgi elements alternatively suggest an implication in hydric equilibrium.

REFERENCES

Adler, J. (1966). Chemotaxis in Bacteria. Science 153: 708-715.
Adler, J. (1976). The Sensing of chemicals by Bacteria. Scient. Am. 234 (4): 40-47.
Adler, J. and Dahl, M.M. (1967). A method for measuring the motility of bacteria and for comparing random and non-random motility. J. Gen. Microbiol. 46: 161-173.
Arronet, N.I. (1973). Motile Muscle and cell Models. Consultants Bureau, New York.
Berg, H.C. (1975). Bacterial behaviour. Nature 254: 389-392.
Berg, H.C. and Anderson, R.A. (1973). Bacteria swim by rotating their flagellar filaments. Nature 245: 380-382.
Bovee, E.C. and Jahn, T.L. (1972). A theory of Piezoelectric activity and Ion movements on the Relation of Flagellar Structures and their movements to the phototaxis of *Euglena*. J. Theor. Biol. 35: 259-276.

Buetow, D.E. (1968). The Biology of Euglena. Vol. 1. Academic Press, New York.
Carlile, M.J. (ed.) (1976). Primitive sensory and communicatives systems: The Taxes and Tropisms of Microorganisms and cells. Academic Press, New York.
Chang, T.M.S., Macintosh, F.C. and Mason, S.G. (1966). Semipermeable aqueous microcapsules I. Preparation and properties. Can. J. Physiol. Pharmacol. 44: 115-128.
Checcucci, A. (1976). Molecular Sensory Physiology of Euglena. Nature 63: 412-417.
Checcucci, A., Colombetti, G., Ferrara, R. and Lenci, F. (1976). Action Spectra for Photoaccumulation of green and colorless Euglena: Evidence for Identification of Receptor pigments. Photochem. Photobiol. 23: 51-54.
de Pamphilis, M.L. and Adler, J. (1971). Fine Structure and isolation of the hook-basal body complex of flagella from Escherichia coli and Bacillus subtilis. J. Bact. 105: 384-385.
Diehn, B., Feinleib, M., Haupt, W., Hildebrandt, E., Lenci, F. and Nultsch, W. (1977). Terminology of behavioural responses of motile microorganisms. Photochem. Photobiol. 26: 559-560.
Dodge, J.D. and Crawford, R.M. (1969). Observations on the fine structure of the eyespot and associated organelles in the dinoflagellate Glenodinium foliacum. J. Cell. Sci. 5: 479-493.
Doetsch, R.N. (1966). Some speculations accounting for the movement of bacterial flagella. G. Theor. Biol. 11: 411-417.
Dogiel, V.A. (1929). Die Sogenannte "Korkrement vacuole" des Infusorien als eine Statozyste betrachtet. Arch. Protistenk. 58: 319-348.
Doroszewski, M. (1970). Response of the ciliate Dileptus to mechanical stimuli. Acta. Protozool. 7: 353-362.
Duncan, C.J. (1977). A note on the evolution of the Transducer mechanism of the Vertebrate retinal rod. Experientia 33: 1310.
Eckert, R. (1972). Bioelectric control of ciliary activity. Science 176: 473-481.
Eckert, R. and Naitoh, Y. (1972). Bioelectric control of locomotion in the ciliates. J. Protozool. 19: 237-243.
Fauré-Frémiet, E. (1948). Le rythme de marée du Strombidium oculatium Gruber. Bull. Biol. Fr. Belg. 82: 3-23.
Ferrara, R. and Banchetti, R. (1976). Effect of Streptomycin on the Structure and Function of the Photoreceptor apparatus of Euglena gracilis. J. Exp. Zool. 198: 393-402.
Forget, J. (1977). La cinétique de la vacuole contractile chez Amoeba proteus: Effets de la lumière panchromatique à diverses intensités. Thèse M.Sc. Sciences Biologiques, Université de Montréal.
Fraenkel, G.S. and Gunn, D.L. (1961). The Orientation of Animals. Doner, New York.
Giese, A.C. (1967). Effects of Radiation upon Protozoa, in Chen, T.T. (ed.) Research in Protozoology, Vol. 2, pp. 256-267. Pergamon, New York.

Giese, A.C. (1973). Blepharisma. The biology of a Light-Sensitive Protozoan. Stanford University Press, Stanford, Calif.
Grell, K.F. (1973). Protozoology. Springer Verlag, Berlin.
Greuet, O. (1968). Organisation ultra-structurale de l'ocelle de deux Péridiniens Warnowiidae, *Erythropsis pavillardi* Kofoid et Swezy et *Warnowia pulchra* Schiller. Protistologica 4: 209-230.
Gruber, A. (1884). Die Protozoen des Hapens von Genus. Nova Acta Acad. Leopold 46: 475-539.
Halldal, P. (1970). Photobiology of Microorganisms. Wiley, London.
Holt, E.B. and Lee, F.S. (1901). The theory of phototactic response. Amer. J. Physiol. 4: 460-468.
Huang, B. and Pitelka, D.R. (1973). The Contractile process in the ciliate Stentor coeruleus I. The role of microtubules and Filaments. J. Cell. Biol. 57: 704-728.
Jahn, T.I. and Bovee, E.C. (1968). Locomotive and motile response in *Euglena*. Chap. 3, pp. 45-108 in Buetow, D.E. The Biology of *Euglena*, vol. 1. Acad. Press, New York.
Jennings, H.S. (1905). Behaviour of the lower Organisms. Indiana University Press, Bloomington, 1976. (Reprint).
Jeon, K.W. and Bell, L.G.E. (1965). Chemotaxis in large, free-living Amoeba. Exp. Cell. Res. 38: 536-555.
Kasakashian, S.J., Karakashian, M.W. and Rudzinska, M.A. (1968). Electron microscopic observations on the symbiosis of *Paramecium bursaria* and its intracellular algae. J. Protozool. 15: 113-128.
Koshland, D.E. (1974). The Chemotactic response in Bacteria. in Jaenicke, L. (ed.) Biochemistry of Sensory functions. Springer Verlag, New York.
Koshland, D.E. Jr. (1977). A response Regulator Model in a simple Sensory System. Science 196: 1055-1063.
Kung, C. and Eckert, R. (1972). Genetic modifications of Electric properties in an Excitable membrane. Proc. nat. Acad. Sci. U.S.A. 69: 93-97.
Larsen, S.H., Reader, R.W., Kort, E.N., Tso, W.W. and Adler, J. (1974). Change in direction of flagellar rotation is the base of the chemotactic response in *Escheridria coli*. Nature, Lond. 249: 74-77.
Leedale, G.F. (1967). Euglenoid Flagellates. Prentice Hall, Englewood clifs, N.J.
Loeb, J. (1918). Forced movements, Tropisms and Animal Conduct. Dover Publications, New York, (Reprint, 1973).
Marsot, P. and Couillard, P. (1972). La réaction phagocytaire chez *Amoeba proteus* I. Phagocytose de cellules dissociées d'Hydre d'eau douce. Can. J. Zool. 50: 745-749.
Mast, S.O. (1906). Light reactions in Lower organisms I. Stentor coeruleus. J. Exp. Zool. 3: 359-399.
Mast, S.O. (1932). Localized stimulation, transmission of impulses and the nature of response in Amoeba. Physiol. Zool. 5: 1-15.

Mast, S.O. (1911). Light and the behaviour of Organisms. Wiley, New York and London.

Miyake, A. (1974). Conjugations in the Ciliate *Blepharisma*: A possible Model System for Biochemistry of Sensory Mechanisms. in Jaenicke, L. (ed.) Biochemistry of Sensory Functions. Springer-Verlag, New York, pp. 299-305.

Mornin, L. and Francis, D. (1967). The Fine Structures of *Nematodinium armatum*, a naked dinoflagellate. J. Microscopie 6: 959-972.

Mussill, M. and Jarosch, R. (1972). Bacterial Flagella Rotate and do not Contract. Protoplasma 75: 465-469.

Naitoh, Y. (1974). Bioelectric basis of behaviour in Protozoa. Am. Zool. 14: 885-893.

Naitoh, Y. and Kaneko, H. (1972). Reactivated Triton-Extracted models of Paramecium: Modification of ciliary movement by Ca. ions. Science 176: 523-524.

Ogura, A. and Takahashi, K. (1976). Artificial deciliation causes base of calcium, dependent responses in *Paramecium*. Nature, Lond. 264: 170-172.

Ordal, G.W. and Fields, R.B. (1977). A Biochemical mechanisms for Bacterial Chemotaxis. J. Theor. Biol. 68: 491-500.

Penard, E. (1917). Le genre *Loxodes*. Revue Suisse Zool. 25: 453-489.

Perez-Miravete, A. (1973). Behaviour of Microorganisms. Plenum, New York.

Piccinni, E. and Omodeo, P. (1975). Photoreceptors and Phototactic programs in Protista. Boll. Ital. Zool. 42: 57-79.

Piccinni, E., Albergoni, V. and Coppellotti, O. (1975). ATPase Activity in Flagella from *Euglena gracilis*. Localization of the Enzymes and Effects of Detergents. J. Protozool. 22: 331-335.

Platt, J.R. (1961). Bioconvention patterns in cultures of Free-swimming organisms. Science 133: 1766-1767.

Prescott, D.M. (1968). Biology of a new species of *Ophrydium* from Amazon Rio Negro river. J. Protozool. 15(suppl.): 7.

Saji, M. and Oosawa, F. (1974). Mechanism of photoaccumulation in *Paramecium bursaria*. J. Protozool. 21: 556-561.

Smith, D.C. (1973). Symbiosis of Algae with Invertebrates. Oxford Univ. Press, London.

Tartar, V. (1961). The Biology of *Stentor*. Pergamon Press, New York.

Van Wagtendonk, W.J. (ed.) (1974). *Paramecium*, a current survey. Elsevier, New York.

Whitehead, J.A. (1971). Cellular Convection. Am. Scient. 59: 444-451.

Willie, J.J. and Ehret, C.F. (1968). Circadian Rhythm of Pattern formation in Populations of a Free-Swimming Organism *Tetrahymena*. J. Protozool. 15: 789-792.

Winet, H. (1970). The Influence of gravity and origin of Bioconvention in *Tetrahymena pyriformis* cultures. Ph.D. Thesis UCLA 1969. Diss. Abstr. Int. B 30: 5303.

Winkler, R.N. and Corliss, J.D. (1965). Notes on the rarely described green colonial protozoan *Ophrydium versatile*. Trans. Am. Microsc. Soc. 84: 127-137.

Wolken, J.J. (1956). A Molecular Morphology of *Euglena gracilis* var. *bacillaris*. J. Protozool. 3: 211-221.

Wolken, J.J. (1977). *Euglena*: The photoreceptor system for phototaxis. J. Protozool. 24: 518-522.

Wolken, J.J. and Shin, E. (1958). Photomotion in *Euglena gracilis*. I. Photokinesis. II. Phototaxis. J. Protozool. 5: 39-46.

Wood, D.C. (1976). Action Spectrum and Electrophysiological responses correlated with the photophobic response of *Stentor coeruleus*. Photochem. Photobiol. 24: 261-266.

ECOSENSORY FUNCTIONS IN LOWER INVERTEBRATES

Enrico A. FERRERO

Istituto di Zoologia e Anatomia Comparata,

Università di Trieste, I 34137 TRIESTE - ITALIA

> "Nec aliter, si parva licet componere magnis"
> Vergil, Georg. IV,176

The animals ranging from Porifera to Pseudocoelomata, which are generally classified as lower invertebrates, represent a variety of diversified groups. Some are important for the number of species, for the efficiency in spreading into many ecological niches, or for the interaction with human life (when parasites or as part of ecosystem processes) while others are less important because of the restricted number of species and/or stenoecious habits.

Such a variety makes the study of sensory adaptations worthwhile on the assumption that the sensory systems have reached different degrees of evolution depending on the level of structural complexity achieved by the animal, and that the different and sometimes restricted distribution of sensory modalities is traceable to the environmental characteristics of the habitat.

However, the study of sensory ecology of lower intertebrates suffers from the dearth of information on their sensory systems. The dimensions of the sensory receptors and the relationships with the neighbouring tissues restrict physiological investigations; and a knowledge integrating morphological, physiological and behavioural data is far from being complete. Sometimes anatomical evidence is available but the actual link with the observed behaviour is missing. Sometimes behavioural responses suggest the existence of sensory capacities, but their morphological bases and physiological characteristics are unknown.

The aim of the present chapter is to organise some of the existing data in a few representative examples showing how and to what extent the sensory capacity is used in different environmental conditions ranging from more uniform to more complex habitats.

The heterogeneity of life styles, even within a single group, and the heterogeneity of the experimental approaches used in the investigations led me to favour some groups and sources of information. My personal bias towards the biology of Turbellaria and the ultrastructural characterisation of the sensory receptors is evident, as it is difficult to disregard one's own interest and experience in getting acquainted with the complete collection of the literature. Recent review articles and books, mainly on coelenterate and nematode ecology and behaviour have been quoted; they are excellent sources for further detailed references.

Taxonomy of the lower invertebrates is continuously evolving. These animals show the first evolutionary attempts which were carried out at a multicellular and organ level, before establishing any effective morphological and developmental patterns. Although their phylogenetic relationships can be reasonably inferred, their systematic arrangement is still under question (Dougherty, 1963). Since a detailed analysis of the taxa is beyond the purpose of this paper, reference has been made to general textbooks in Zoology. Table I summarises how the taxa are distributed in different habitats, but it does not pretend to be authoritative in the classification followed, or more precise and exhaustive than a general outlook.

If we overlook systematic and phylogenetic problems and consider the lower invertebrates as a whole, we can recognize the following major trends:

1) Increasing complexity of internal organization
2) Increasing ability to perceive and integrate outer stimuli, and to respond with elaborated and flexible behavioural patterns.
3) Development of sensory cells and organs, as a function of the "richness" of environmental stimuli which they are confronted with, both in space and in time.
4) Formation of associations enabling the animal to lead more "advanced" life than the structural level achieved.
5) "Regressions" to more steady environmental conditions than those compatible with the achieved level of organisation, as for example:

    a) cave-life conditions

    b) various levels of parasitic interactions.

LEVELS OF ORGANIZATION

Within the lower Metazoa Meglitsch (1972) ranges four large groups called Parazoa, Radiata, Acoelomate Bilateria and Pseudocoelomata. Lumping different phyla together may be a questionable procedure depending much on individual views; however, it helps to clarify the different organisational levels stressing the most peculiar features.

Parazoa have achieved the cell level. Cellular differentiation in sponges is established, but a high degree of totipotentiality is still kept. A certain degree of integration yet occurs as the ultrastructural studies by Pavans de Ceccatty et al. (1970) point out; but the occurrence of a distinct "nervous system" is much under question.

Radiata have developed tissue layers, each with a definite function. The nervous system, the synapses and the electric action potential made their first appearance; though some indistinct border lies between nervous, muscular and epithelial conduction properties (neuroid conduction, see Mackie, 1970, for a review).

Acoelomata and *Pseudocoelomata* with a defined mesodermic layer and mesodermal derivatives like muscles, mesenchymal cells, reproductive organs, excretory and circulatory systems, have reached an organ level. Such an organization is more demanding, in terms of definition of an outer limiting barrier and of internal coordination, including transport problems and homeostasis. The appearance of a pseudocoelome is partly a response to these requirements. On the other hand the complexity of an organ level brings problems of surface metabolic exchanges, which a flattening of the body shape may improve, and of mechanical support which is met by the development of a strong outer cuticle, like that in some pseudocoelomates, nematodes in particular. A thick cuticle impairs contact of the superficial receptors with the outside environment, so that sensory organs are housed in sensory pits (nematode amphidia, phasmidia etc.) or invested by protective cuticular layers (antennae, setae etc.).

PRINCIPAL STEPS OF THE EVOLUTION OF THE NERVOUS SYSTEM

The use of sensory information also is subject to different levels of integration, and a general trend is shown of increasing development of communication and motor reactions. Horridge (1968) sketches a tentative history of the evolution of the nervous system.

Independent effectors regarded as the first step of reaction are the testimony of a general property of living systems, irritability. In sponges closure of the oscula by means of a sphincter-like ring

TABLE I - DISTRIBUTION OF PHYLETIC GROUPS IN ENVIRONMENTS

Phylum (n° of species)	Groups	Aquatic				Terrestrial		Parasitic	
		Sea	Brackish	Freshwater	Interstitial	Cave	Land	Ecto-	Endo-
PORIFERA (3000 sp.) sponges	Class Calcarea	X							
	Hexactinellida	X							
	Demospongiae	X		X					
CNIDARIA (9000 sp.) corals, jellyfish sea-anemones	Class Hydrozoa	X		X	X				
	Scyphozoa	X							
	Anthozoa	X							
CTENOPHORA (90 sp.)	Class Tentaculata	X							
	Nuda	X							
PLATYHELMINTHES (13000 sp.) flatworms	Class Turbellaria	X		X	X	X	X	X	X
	Gnasthostomulida	+			X				
	Monogenea							X	X
	Trematoda								X
	Cestoidea								X
MESOZOA (50 sp.)	Ord. Dicyemida							X	X
	Orthonectida							X	X
NEMERTINEA (600 sp.)		X		X	X		X		

Groups	Aquatic				Terrestrial		Parasitic	
Phylum (n° of species)	Sea	Brackish	Freshwater	Interstitial	Cave	Land	Ecto-	Endo-
ACANTHOCEPHALA (800 sp.) Thorny headed worms								X
ROTIFERA (2000 sp.) Class Seisonacea	X							
Bdelloidea			X̲		X			
Monogononta		X	X̲	X				
GASTROTRICHA (150 sp.) Ord. Macrodasyoidea	X̲		X	X				
Chaetonotoidea	X		X̲	X				
KINORHYNCHA (100 sp.)	+			X				
NEMATODA (10000 sp.) roundworms Subcl. Aphasmidea	X		X	X		X		X
Phasmidea	X		X			X		X̲
NEMATOMORPHA (230 sp.) horse-hair worms Ord. Gordioidea			X					X juv.
Nectonematoidea	X							X
ENTOPROCTA (100 sp.)	X̲		X					

o An underlined mark (X) means the main habitat occupied by the group, + further specifies the exclusive habitat, Juv.= during the juvenile stage.

of elongated cells is reported as dependant on the administration of strong, possibly noxious, stimuli. Temporal and spatial summation are reported as well as spread of excitation over several centimetres (Bullock and Horridge, 1965). Cnidarian nematocysts, for long considered as typical independent effectors, are now found (Mariscal, 1974, for a review) to be controlled by the physiological state of the animal, the nervous system and the neurosensory cells (Westfall, 1973). *Local reflexes* do not involve travelling of sensory stimulation over long distances, however a limited degree of integration is allowed either through direct sensory-motor synapses, or local pacemaker systems. Cnidaria and Turbellaria show examples of such behaviour, especially when the periphery is set free from higher hierarchical control.

Cephalization is related to the development of the bilateral symmetry, muscle development and locomotion on a surface. Grouping of sensory receptors on the anterior end helps both scanning of the environmental conditions as soon as they are met during locomotion, and centralization of the nervous system. Within the Turbellaria such a process can be seen as successive morphological and functional steps: the superficial nerve net becomes more and more reduced, direct connexions with the cerebral ganglion are established and the behaviour patterns are encoded as rigid sequences within the cerebral ganglion (Bullock and Horridge, 1965; Koopowitz, 1970; Koopowitz et al., 1976). Finally the rigidity of fixed behavioural sequences, which is advantageous for adding hierarchically superimposed steps, restricts plasticity and adaptation of responses. Planarians, which as far as we know are the most advanced from this point of view and are the only ones extensively studied, display *conditioning* abilities and *learning*. They became a very popular subject for experimental psychophysiologists, and the possible molecular encoding of *memory* has raised diverging opinions. Corning and Kelly (1973) have summarised the data and the discussion on the subject; I suggest their work to interested students.

DISTRIBUTION OF THE SENSE ORGANS IN RELATION TO HABITATS

Regarding natural selection as a process promoting and acting upon diversification, and species preservation as a conservative process resting on uniformity, we may sensibly infer that the "richer" a habitat is of environmental stimuli the more heterogeneous will be the available ecological niches, and the more variable the sense organs which will be used in them. In this case we may call "progressive" the sensory controlled behaviour that enables life habits which are ahead of the structural level reached, and "regressive" those life habits that draw back to environmental conditions which are more uniform than can be stood by the acquired

level of complexity. It is interesting to note that uniform
conditions are usually related to species preservation processes,
whereas meeting heterogeneous environments is associated with
the dispersal phase. In Table II is a proposed distribution of
ecosystems and life habits as a function of increasing complexity
of the environmental stimuli. In addition the stimuli can be
spatially heterogeneous and/or temporally inconstant; the former
condition gives rise to oriented behaviour and localised
distribution, while the latter gives rise to transient response
or cyclic activity.

CONSTRAINTS POSED BY THE MECHANICAL PROPERTIES OF THE MEDIUM

Mechanical isotropy of a medium is compatible with a radial
symmetry and a ciliary beating from the "anterior" to the "posterior"
end. Planktonic larval stages of lower invertebrates are, as a
rule, ciliated. Symmetrical ciliary locomotion is retained in
Ctenophora and increased in complexity by the comb-plates of the
adult. On the contrary, movement along the interfaces of two
media with different densities, like the sea-bottom, adds to the
antero-posterior another functional asymmetry, the "superior-
inferior" one. The ciliary movement is superseded by creeping
movement which is effectively performed by a better developed
musculature. The need of a wide surface for increasing contact
and metabolic exchanges leads to the morphological differentiation
of a dorso-ventral flattening. These adaptive steps can be
exemplified by the free-swimming ciliated larvae of Demospongiae
(Bergquist et al., 1970 & Table III) which switch from ciliary
swimming to creeping habits before settling on the hard substrate.
Within Cnidaria, benthic *Halcampa* larvae (Nyholm, 1949) are
unciliated and disperse by creeping, as do *Craspedacusta* asexual
buds by means of muscular activity; dispersal of *Clava* planulae
by crawling is more extensive over smooth surfaces than over rough
(Williams, 1965). The Platyctenea, aberrant flat creeping Ctenophora,
show a reduction of the comb-plates and the development of a
"creeping sole", which actually is an everted pharynx with a ciliary
surface and with increased musculature. The trend of the reduction
of cilia and increase in musculature is found also in Turbellaria.
The fully ciliated catenulid *Stenostomum* strongly depends on cilia
for its swimming, the stop, start and reversal of ciliary beat being
possibly under nervous control (Rampitsch, 1941). With the pond-
living triclads, which glide with the ventral surface on the
substrate, the ventral nerve cords become prominent and are
sometimes the only nerve cords. Furthermore, thick muscle layers
concomitant with a distinct ventral nerve plate occur in terrestrial
triclads (Hyman, 1951), where locomotion can only be accomplished
by muscular creeping, and no buoyancy lift helps to counteract
gravity.

TABLE II - PROPOSED DISTRIBUTION OF ECOSYSTEMS AND LIFE HABITS ARRANGED WITH RESPECT TO THE INCREASING COMPLEXITY OF THE ENVIRONMENTAL STIMULI

↑ increasing complexity of the environmental stimuli

Parasitic	Free-living			
Endo-	Seawater	Freshwater		Terrestrial
Ecto-		Standing waters	Moving waters	
	Pelagic Benthic			
	infauna epifauna	pelagic benthic		hypogeal
	aphotic ──hadal──	infauna epifauna		
	──abyssal──	──profundal──		interstitial
	──bathyl──	limnetic infralittoral	rivers	
	photic neritic littoral	eulittoral	streams	
	estuarine brackish water	pools and swamps		

→ increasing complexity of stimuli

TABLE III - BEHAVIOUR OF LARVAE AND THE ECOLOGICAL SITUATION OF THE ADULT SPONGE °

Species	Swimming pattern	Length of free life	Light response	Ecological situation of adult sponge
Haliclona sp.	1) Directional swimming with constant rotation 2) Corkscrew	9-10 h	Positively phototactic moderated when creeping	Upper and middle mid-littoral, around and under stones
Mycale macilenta	Corkscrew	over 48 h	negatively phototactic	Littoral fringe to 4 m; cryptic
Halichondria moorei	Creep along substrate, rarely swim, move upward	20-60 h	no response	Fringing middle and upper mid-littoral pools and rocks in standing pools

° (modified after Bergquist et al. 1970)

Thigmotaxis, considered as the tendency to place the largest possible area of body surface in contact with the substrate, has a survival value for animals subject to strong currents (streams) or wave action. It becomes a spatially orienting response too; and touch- and contact chemoreceptors, probably evolved as prey detectors, can then be used for habitat selection and settlement. When animals crawling on unequal surfaces, show a high degree of complexity of the nervous system and centralised patterns in locomotor coordination, like the ditaxic muscular movement in polyclads, proprioception may be assumed, though this has not yet been identified in lower invertebrates on a physiological basis.

If the requirements of benthic creeping animals versus the pelagic ones can be understood, less can be said of the interstitial animals (infauna). This habitat has only recently been recognised as a distinct one, characterised by narrow sand interstices, liability to drying, low oxygen conditions and frequent rearrangement of surface layers by mechanical disturbances. It is also considered a conservative environment where smaller groups, likely to be superseded in open spaces, found their ecological niche and survived (Delamare Deboutteville, 1960). Many groups (see Table I) occupy this habitat. They tend in general to have an elongated body form. According to their dimensions and the interstices they occupy there is a full range of locomotory strategies: cilio-muscular (Turbellaria), gliding (Gastrotricha), hitching and pulling (Kinorhyncha), helicoidal (Nematoda). According to Swedmark (1964) the diameter of interstices between sand grains is the most important factor in the distribution, along with temperature, salinity and oxygen tension. Wallace (1958) finds indeed a correlation between granulometry, water film thickness and movement of the nematode *Heterodera schachtii*; however our knowledge is still incomplete.

Thigmotaxis and the frequent occurrence of adhesive glands (for instance Turbellaria, see Tyler, 1976, for a review; and Gastrotricha) associated with presumptive tangoreceptors, suggest that loss of contact with the walls of the interstices or possibly strong abrupt mechanical stimulation, brings these anchoring devices into action. The trend shown by interstitial animals to keep statocysts (usually considered a primitive feature) and to dispose of the eyes (Delamare Deboutteville, 1960) is consistent with the environment, in particular for the deeper living species. Wieser (1959) correlated the distribution of ocelli and ocellar pigment with the habitat of marine nematodes and found that even in closely related genera the ones living on littoral or sublittoral algae had ocelli in a much larger percentage than the ones living in sands (Table IV). Otoplanid Turbellaria possess cerebral photoreceptors, "Sehkolben", (Bedini and Lanfranchi, 1974) together with ciliary presumptive photoreceptors (Ehlers and Ehlers, 1977), and Acoela possess reddish pigment spots, all of them devoid of a

TABLE IV - OCCURRENCE OF MARINE NEMATODES WITH OCELLI AND OCELLAR PIGMENT WITH RESPECT TO THEIR HABITAT°

	LITTORAL	LITTORAL	SUBLITTORAL	LITTORAL	LITTORAL	SUBLITTORAL	
	Algae exposed	Algae sheltered	Secondary substrates	Sand exposed	Sand sheltered	Coarse bottom	

% of specimens in samples collected in different habitats, with ocelli and ocellar pigment

Ocelli	14.8	7.9	11.8	–	0.05	1.0	
Ocellar pigment	48.7	22.4	7.7	0.2	–	0.5	

% of the total number of specimens with ocelli and ocellar pigment, living in a given habitat

							Total number
Ocelli	44.9	33.6	20.3	–	0.4	0.8	256
Ocellar pigment	57.4	36.9	5.2	0.3	–	0.2	658

°(modified after Wieser, 1959)

screening pigment cup. Non-directional photosensitivity or klinophototaxis could still be retained and used during vertical migrations (see p. 13), whereas gravireception may take over when, or at depth where, the downwelling light is no longer an orientational cue (light is virtually zero below 10 cm of sand).

Adaptation to a burrowing life brings back some of the problems of locomotion confronted in a mechanically isotropic medium, though friction makes it a much more energy consuming habitat. On the one hand we find a resumed round or possibly streamlined profile of the body, and on the other a muscular mechanism imposed by the magnitude of the reacting forces, to overcome and to withstand them when the body is anchored inside a hole. Sensory adaptations to burrowing are those connected with monitoring tension of muscles and direction of movement. In fact in the burrowing anemone *Ceriantheopsis*, the morphological evidence of a putative proprioceptor has been recorded (Peteya, 1973), and in the holdfast of the hydroid *Corymorpha*, whose implantation is considered geotropic, an endodermal unciliated statocyst has been reported (Campbell, 1972). This information parallels the more frequent occurrence of statocysts in the burrowing polychaetes (Fauvel, 1907) or the occurrence of statocysts in the burrowing sea-cucumbers (synaptulids).

MORE OR LESS "STIMULATING" HABITATS

In addition to the mechanical properties of the medium, where movement takes place, enabling us to distinguish between swimming, creeping, burrowing, or more or less sedentary life habits, the habitats can be further categorized as more or less "stimulating".

1. Aphotic habitats

The aphotic pelagic and benthic districts of both sea and lakes, as well as caves and cave pools, can be certainly ranked as the poorest in respect of physical stimuli. Variations in light, temperature, oxygen, salinity and even pressure can be ruled out as physical stimuli for abyssal species; and if existing they must have a biological origin like bioluminescence, chemical communication or sources of vibration. Wolff (1970) in his list of abyssal groups reports only a few lower invertebrates adapted to such an environment, though one actinian and one nematode species have been found below 10,000 m, the sensory equipment of which, however, is little known. Intracellular bioluminescent mechanisms in the ctenophore *Mnemiopsis* are inhibited by hydrostatic pressures exceeding 200 Atm (Chang and Johnson, 1959), whereas light is reported (see discussion in Morin, 1974) to inhibit bioluminescence in Coelenterata; that restricts the depth range in which bioluminescence can be used.

Continuity of the interstitial environment through the continental freshwaters and ground-waters complicates the definition of a cave aquatic habitat (Delamare Deboutteville, 1960; Vandel, 1964) but true cave-dwelling species have been reported within Turbellaria, Nemertinea, and Nematoda. Stating that they show reduction or loss of photoreceptors along with pigmentation is a platitude (Hyman, 1951; Durand and Gourbault, 1975) but, as common to lower invertebrates, a general aversion to light, especially at short wavelengths and high intensity, is retained. Mitchell (1974) studying the behaviour of the blind cave-dwelling triclad *Sphalloplana zeschi* in experimental conditions, finds a thermopreferendum around $20^{\circ}C$, close to the cave temperature, and he suggests that mainly chemical cues are used for food detection.

2. Spatially heterogeneous stimuli in the photic pelagic district

The behaviour of animals living in this district is related to the abundance of primary producers and food supply in general and to the more varied spatial distribution of physical stimuli. Planktonic animals are usually photopositive, and include the photopositive meroplanktonic larval stages produced by benthic forms to gather food and increase dispersal. Photopositive behaviour is not necessarily related to the possession of specialised sense organs, as a "dermal light sense" (Steven, 1963) can be invoked. The concentration at the surface of the sea of the jellyfish *Tiaropsis* which have ocelli and display photokinetic behaviour, is according to Zelickman et al. (1969) connected with the distribution of their phytoplanktonic prey; (more discussion is delayed to the section on vertical migrations). Recent studies on the morphology of the eyes (Singla, 1974) and on the interactions of pacemakers (Passano, 1976; Anderson and Mackie, 1977) in the hydromedusae contribute to our knowledge of their behaviour upon steady or abruptly changing illumination. Light can be used for absolute orientation in space, as Mackie and Singla (1975) point out correlating the occurrence of righting and steering reflexes in ocellate hydromedusae, in contrast to the "blind" *Stomotoca*. Gravity too plays a spatial orienting role in the open sea. Singla (1975) proposes a simple model relating structure of the ciliary statocyst and the righting behaviour in *Aequorea*. Evolution of closed ciliary statocysts from vibration receptors has been suggested by Horridge (1969); in Ctenophora geotactic orientation set by the balancer cilia of the apical organ through the direction of beating of the comb-plates, is under control of vibration receptors either on the tentacles or in the apical organ itself (Horridge, 1965a; Krisch, 1973). In the mechanically homogeneous pelagic habitat vibrational sources are either small planktonic preys, for the detection and capture of which vibration receptors are actually used (Horridge, 1965b; Hernandez-Nicaise, 1974), or ripples of the sea surface in rough weather conditions or close to the sea-shore. Therefore, reversal of the usually negative geotaxis by the vibration receptors is interpreted by Horridge (1965a) as an adaptive mechanism to bring the animal into waters drifting from the seashore and back to the open sea.

MacDonald (1975) reports evidence of an increase in activity of Cnidaria at moderate (<200 Atm) pressures and pressure dependent responses are listed by Knight-Jones and Morgan (1966) in pelagic lower invertebrates. The lack of positively identified receptors as opposed to what Aronova (1974) found in Ctenophora, either morphologically or physiologically, casts some doubt on listing pressure reception as a sensory modality in lower groups. As MacDonald (1975) points out, pressure as well as temperature and in some respects light, interfere with basic metabolic processes; therefore an alteration of animal function, if existing, can be considered as a "general response". In addition gravity, added to light, temperature (to some extent) and pressure gradients (in opposing directions) can all be integrated as a vertically orienting cue in pelagic animals. Some of these aspects will be discussed again in connection with vertical migrations.

Currents, temperature and salinity may form boundaries of discontinuity both between adjacent superficial and vertical layers. That would certainly affect the distribution and dispersal in poor swimming, stenothermal or stenohaline forms, especially in coastal waters where extensive mixing of freshwater and seawater occurs. Holoplanktonic stenoecious species restricted to particular water masses can be used as indicators (Vinogradov, 1970), others can be more tolerant and wide-spread. Aggregation at salinity and temperature discontinuity layers is reported by Arai (1973, 1976) for *Sarsia* (Cnidaria) and *Pleurobrachia* (Ctenophora). It is however doubtful whether there are distinct sensory organs for these sensory modalities or, as Arai (1976) suggests, sensitiveness is mediated through the common effect of temperature and salinity on density and ultimately on the rate of increase of pressure. A change in the rate of flow of the water masses in two contiguous layers can be a significant stimulus eliciting a behavioral adaptive response, but it is unlikely to happen alone, not associated with discontinuities of temperature and salinity, on whose effect I have commented above, or, in case of vertical currents, with light and pressure changes. A rheoreception, opposed to vibration and touch reception, is not a well established sensory modality in pelagic lower invertebrates though cnidarian statocyst (Singla, 1975) and ctenophore apical organ (Horridge, 1974) may act as an acceleration detecting device too.

3. The intertidal zone

We regard the intertidal zone as the most heterogeneous environment (Newell, 1970; Vernberg and Vernberg, 1972; Meadows and

Campbell, 1972) easily split up into a variety of microhabitats. All the physical factors are at work here and some biological ones too. Light is beyond any doubt the most prominent stimulus controlling the population density of the coastal phyto- and zooplankton, the distribution of littoral photosynthetic algae and the release and settlement of larvae of sessile intertidal animals. Correspondence of larvae light response and habitat preferences of the adult has been paradigmatically demonstrated in two species of intertidal sponges (Table III, Bergquist et al., 1970). In general after a positive phototactic young stage, preferring diffuse, blue-green (450-510 nm) light, a photonegative settling stage follows. Sessile adult organisms, even if devoid of known photoreceptors, show a phototropic response, like *Metridium, Calliactis & Eudendrium*.

Factors other than light may influence larval settling (Crisp, 1974; Mackie, 1974; Brewer, 1976; Müller et al., 1976). Some factors like O_2 and CO_2 concentrations, may be related rather to general metabolic needs of the animal, as already pointed out, than to the occurrence of specific receptor cells; others like adhesion and effect of currents could be connected with ciliary receptors morphologically identified by Lyons (1973a). Chemical recognition of the substrate is likely in the case of the larvae of *Sertularella miurensis*, which settle preferentially on *Sargassum* and show limited dispersal (Nishihira, 1967a,b), as well as in *Actinia equina*, where the planulae reenter the adult after a period of planktonic life and before metamorphosis (Chia and Rostron, 1970); differential settling around male or female *Allopora* isolates is reported by Ostarello (1976). Chemoreception is used by the adults for food and substrate selection, intraspecific aggregations, sex attraction and mating behaviour, and finally for interspecific associations. Amino acids and polypeptides of low mol.wt. usually elicit chemotactic or chemotropic responses and even fully integrated behavioral sequences. Lenhoff (1974, 1976) has recently reviewed the subject as regards Cnidaria, whereas scattered reports are available of free-living Platyhelminthes (Coward and Johannes, 1969; Ash et al., 1973; Koopowitz et al., 1976) and Nematoda (Croll, 1970; Ward, 1973, 1976; McLaren, 1976, for references). In the polyclad turbellarian *Stylochus mediterraneus* food and habitat selection do coincide. The worm is an active predator on mussel and oyster beds within which it lives (see Gelleni et al., 1977, for a review). Water conditioned with the mussel, (*Mytilus gallo-provincialis*) tested in a Y maze, acts as an attractant, whereas that conditioned with the oyster (*Crassostrea angulata*), does not. Associations may be species-specific (Meadows and Campbell, 1972, for a review). In polyxenous species an "ingestive conditioning" (Landers and Rhodes, 1970) may take place at an early stage fixing the specificity of food or host according to the environmental conditions and laying the bases for divergent evolution of populations.

The effect of waves in the littoral zone has certainly contributed to the evolution of thigmotactic and rheotactic responses as well as of adhesive organs. Negative rheotaxis has an adaptive value in preventing dispersal both by the currents drifting from the shore in rough sea conditions and by the tides. In the interstitial turbellarian *Convoluta psammophila*, morphological evidence for such a sensory modality has been presented (Bedini et al., 1973).

4. Freshwater habitats

Standing freshwater ecosystems, but for the periodical effect of tides, should not differ too much from the sea in the constraints put on the inhabiting fauna. In ponds, swamps and temporary bodies of water changes in chemicals, humidity, and temperature conditions are as much extreme as in littoral rockpools, and are likely to have affected the sensory equipment and the behaviour of the animals in much the same way. Unfortunately no detailed studies are available on sensory adaptations in these habitats for lower invertebrates. It has been already stated that large variations in pH, O_2, temperature, humidity and salinity are withstood by means of mechanisms other than capacity for behavioural adaptations, often involving evolution of forms of resistance (suspended life in Rotifera, "dauer" eggs in various groups) or of complex life cycles.

Moving freshwater ecosystems present zonation patterns dependent upon velocity of current and temperature, the latter usually increasing from well to estuary. Distribution of triclad turbellarians in streams has been studied both in the field and experimentally (Lock, 1972, 1975; Reynoldson, 1974; Kawakatsu, 1974) and it seems highly consistent. According to Lock and Reynoldson, *Crenobia alpina*, a stenothermal ($4°$ -$20°C$) planarian that can withstand currents up to 14 cm/sec, is found in the stream sections with swiftly running waters, while *Polycelis felina* is found in the slower flowing sections, even though the temperature regime is suitable for both species. *Polycelis nigra* occupies the warmer, stiller region on the coastal plain. The slope of the terrain is the cause of spatial separation. Kawakatsu (1974) on the contrary, reviewing data on the planarians of the Japanese Islands, stresses the importance of thermal factors possibly associated with food availability. No investigations have been carried out on the differential sensory equipment among these species. The fact that "collar cell" receptors, whose ultrastructure is described in the freshwater species, *Polycelis tenuis* (Bowen and Ryder, 1974), are not mentioned in the terrestrial triclad *Bipalium kewense* by Storch and Abraham (1972) may suggest that this cell type is responsible for rheosensitivity.

5. Other spatially orienting cues

Many experimental observations have been carried out on the behavioural responses to sensory modalities whose morphological bases and ecological value are presently little understood. Galvanotaxis, usually directed to the cathode, is discussed by Viaud (1954) and Brown (1971) in planarians, and by Croll (1970) in nematodes. In the latter the matching of the threshold electric potentials for the response with the range of negative potentials developed by root hairs in plants, may account for a host finding function in plant parasites; however the meaning of galvanotaxis in other animals is difficult to explain, unless as a general property of the living systems. Brown (1971), investigating the problem of the biological clock-compass, shows in planarians a complex action of electrical gradients superimposed on the terrestrial magnetic field and time. Sensitivity to natural and artificial magnetic fields, relevant to the problem of geographical orientation and to gamma radiations, is claimed by the same author.

TEMPORALLY INCONSTANT STIMULI

Stating that the environmental stimuli, whose effects due to differential spatial distribution have been discussed above, are not constant even in a given place, is a truism. Moreover, spatially heterogeneous stimuli may be analysed by the animal's sensors as a temporal sequence so that a klinotactic orientation in the sense of Fraenkel and Gunn (1961), ensues. As we know from the information theory, random signals have little information content, so that they can be dealt with by the animal with a fast response or in a purely statistical way. On the contrary a constant temporal succession of events "makes sense" to the animal and brings about adaptations to those sensory inputs.

The tidal rhythm and the diel light are short period changes cycle; long period changes are the seasonal cycles and, to some extent, the lunar cycles (Brown, 1971). Since the beginning of the century (Gamble and Keeble, 1904) the behaviour of the interstitial acoel *Convoluta roscoffensis* has been known. It migrates to the moist sand surface when exposed by the tide and forms mucous films on which the worms aggregate (Fraenkel, 1961). In undisturbed sands the animal shows a negative geotaxis, whereas it quickly swims back into the sand at any vibration. The stronger the mechanical stimulus the deeper the animal buries itself (Fraenkel, 1929). The worm carries symbiotic zoochlorellae and is also positively phototactic if undisturbed. Gamble and Keeble (1904) have reported that the tidal rhythm of vertical migration is kept up to one week in the laboratory. The behaviour displayed is biologically advantageous because it brings the worms to superficial layers where, because of exposure to light, photosynthesis can be more effectively accomplished, while avoiding the effects of the tides and wave action. The allied species *Convoluta psammophila* shares with the previous species the symbiosis with the zoochlorellae and the pattern of life habits. The ultrastructure of its unciliated statocyst (Ferrero, 1973) shows that muscle fibres are inserted onto the statocyst capsule and the position of the inserted muscles is consistent with the rotational movements observed in the living animal. Although the receptors responsible for the upwards or downwards migrations geared to the tidal rhythm are unknown, the persistence of the periodical movements may be explained by a partly endogenous control. A tentative hypothesis, integrating the available data, suggests that an endogenous rhythm, together with light, affects the taxis by rotating the statocyst according to the upwards orientation. Sudden vibrations or those due to the incoming tide cause the rotation of the statocyst by $180°$, so that the geotactic "mood" is inverted; a condition which persists longer the stronger the vibrational stimulus delivered. A parallel can be drawn with the changes in geotactic sensitivity displayed by the ctenophores (Horridge, 1974), though the underlying structures are different.

Circadian cycles are mainly regulated by light. They may appear as vertical migrations like in the free-living animals, or cycles of expansion and contraction like in sedentary forms. Medusae and syphonophores reach or are closer to the surface during the night. Boundaries of discontinuity may restrict the extent of the migration and confine the animals to a given range of depth. Therefore vertical migrations can not be considered an adaptation for reaching the top layers which are richer in planktonic food. Of course no single factor can explain the vertical migrations (see discussion in Vinogradov, 1970); however, the interpretation that each species follows an optimal light level is commonly agreed upon by the majority of the authors. The ocellate *Leuckartiara* shows a distinct pattern of high densities in the top layers during the night and in the deep layers in daylight (Russell, 1925) but species lacking known photoreceptors equally perform vertical migrations. The occurrence of photosensitive neurons in medusan pacemaker systems may account for such a behaviour (Anderson and Mackie, 1977). The larvae of the trichostrongyle nematode show vertical migrations in the field (Rogers, 1940) and are mainly recovered from grass at dawn and dusk; i.e. at intermediate illumination conditions. Croll (1965) proved experimentally that at comparable light intensities the worms followed the straighter path between two points and consequently travelled further than at extreme illuminations.

Sedentary animals also present behaviours entrained by the periodical light patterns. Illumination elicits a phototropic response in hydropolyps and actinians; it controls the cycles of expansion/contraction of the column (Tardent et al. 1976) and induces phototactic locomotion. According to Passano and McCoullogh (1964) the rhythm of activity imposed by the natural illumination cycle has an adaptive value. For example, it synchronises the activity of *Hydra* to that of its prey, usually small crustacea, which in turn are dependent upon the phytoplanktonic food source. Indeed starvation increases the positive phototaxis in *Hydra littoralis* (Feldman and Lenhoff, 1960).

Light-dark adaptation and retinomotor (or photomechanical) responses are present in some lower invertebrate groups. They tend to increase sensitivity to light after a dark period and to decrease it in bright light conditions. Transients in illumination are as a general rule connected to increased activity in circadian rhythms, where peaks are associated to dawn and dusk. In addition to the extensively studied example of *Hydra*, photoactivation occurs in dark-adapted nematodes. It declines within 4-6h in continuous light and is resumed again to the maximal level after 3 h of dark adaptation. In turbellarian triclads dark-adaptation, measured as the amplitude of the ocellar potential to a test flash after increasing periods of dark, is completed in about 4 min, but the intensity and duration of the adapting light is critical in determining the time course of the recovery (Brown and Ogden, 1968). In the turbellarian rhabdocoel, *Dalyellia*, retinomotor responses are demonstrated morphologically after a 12 h D- 12 h L illumination regime (Bedini et al., 1977). Light-adaptation, consisting in the screening of the photoreceptor processes by the pigmented eye cup is fast (10 min); dark-adaptation takes about 9 h. However no systematic sampling in the field has been carried out to test the vertical distribution of the worms during the 24 h period.

Light conditions influence and time the reproduction in these animals. Light inhibits gemmulation in sponges (Rasmont, 1970) whereas in the hydroid *Clava* it elicits the release of the positive photoklinotactic larvae (Williams, 1965). That would assist the movement of the larvae up to the littoral zone enabling them to spot the algal fronds on which they settle eventually. An interesting series of observations made by Clément and Pourriot (1972) show how the alternate generation cycle of the rotifer, *Notommata*, is regulated by the photoperiod. Long photoperiods induce the switching from parthenogenesis to heterogamy and production of fertilised winter eggs. The action spectrum of this phenomenon is compared by Pourriot and Clément (1973) with that reported for the phototropic (Viaud, 1940, 1943 a,b) and phototactic (Menzel and Roth, 1972) responses in other rotifers; Clément (1975) concludes that the cerebral eyes are possibly the receptors involved in the detection of the photoperiods and in the control of the reproductive cycle.

THE SOCIAL LIFE OF LOWER INVERTEBRATES

1. Intraspecific aggregations

Aggregations may be set up either casually because of sharing a restricted habitat and/or a common behaviour towards a physical stimulus, or actively by interindividual "recognition". In coelenterates we have seen examples of "passive" aggregations at discontinuity layers (Arai, 1973, 1976), of limited dispersal of the larvae in some bottom living forms like *Halcampa, Sertularella, Clava,* or of an "active recognition" like in *Actinia equina* and *Allopora* larvae (see pp. 59,64). It certainly helps to exploit an area suitable for settlement and increases the (chances) of sexual reproduction.

The species-specific aggregation of planarians which was tought to occur mechanically by means of an ortho- or klinokinetic response (Fraenkel and Gunn, 1961) to heterogeneous illumination, is according to Reynierse et al. (1969) a complex behaviour making use of different cues: photokinesis, possibly, for distant stimuli and distinctive morphology and chemical secretions as proximal stimuli which orient the worms towards the aggregations. In nematodes chemical factors and thigmokinetic responses play a role in bringing the animals together and in mass co-ordinated movements referred to as "swarming" (Croll, 1970). The adaptive value of such phenomena has been interpretated by Wallace (1963) as increasing survival in dry conditions or on exposure to ultraviolet radiation.

In both triclads (Best et al., 1969, 1974; Pigon et al., 1974) and rhabdocoels (Fiore, 1971; Fiore and Joalé, 1973; Heitkamp, 1972) a mechanism for self-inhibition of population growth in crowded conditions has been observed. In planarians the control is exerted on fissioning; ablation of the cephalic margins releases fissioning in grouped animals at the same rate as in the isolated intact ones. Since vision and discharge of pheromones into the habitat had been experimentally ruled out as sensory cues, chemical as well as tactile stimuli may mediate the social control response. When contact between animals is partially prevented, stimulus reception should be attenuated; indeed a partial release of fissioning in grouped animals ensues. On these bases the authors suggest that contact chemoreception is involved. In rhabdocoels the switching from delivery of subitaneous (thin shelled) to dormant (thick shelled) eggs is a socially controlled event. Here a specific water-soluble pheromone inhibits the production of subitaneous eggs in young worms but no evidence is presented that the response is mediated by a sensory receptor.

2. Interspecific non parasitic associations

Two questions, relevant to the subject of this chapter arise:

a) which sense-detected signals are used in establishing an association between two, sometimes distantly related species?

b) how does the association affect the sensory capacities of one or both the partners?

Thorough investigations carried out mainly in cnidarians show that often complex, integrated behavioural patterns have evolved. Ross (1974) recently reviewed this subject and little more can be added. In the well known symbiosis between sea anemones and hermit crabs, mechanical and chemical stimuli are both used in recognition and transfer of the polyps to their hosts. *Hydractinia echinata* is usually found on gastropod shells inhabited by hermit crabs. Müller et al. (1976) believe that velocity gradients, set up by objects moving close to the planula poised on the substrate, are the stimuli for its attachment and transfer. If the movement stops, the planula resumes crawling about. Initiation of metamorphosis is due to chemical factors released by certain marine bacteria, forming a film on the shell, and to vibrations produced by the crab. However, the planulae attach themselves and metamorphosie on sterilized shells provided they are occupied by the hermit crab.

Transfer of the actinian *Calliactis parasitica* on *Pagurus bernhardus* is triggered, according to Ross and Sutton (1961a), by a substance of molluscan origin contained in shells and periostracum; the stepwise behavioural pattern is released as soon as the tentacles come in contact with this material. Active behaviour from the host crab can promote and help the transfer. Gentle, rhythmical tactile stimuli, and even experimental electrical stimulation at low frequency, applied to the column, make the anemone, *Calliactis*, relax and detach. Slow electrical activity in a special excitable system in the ectoderm of *Calliactis parasitica* precedes detachment and settlement, and it can be related to this behaviour (McFarlane, 1969a, b). Such a behavioural response is not developed in the non-commensal species and even in commensal ones the effect is not always predictable (Ross, 1974). Forcible transfer by the crab only is described for some other species but in any case the final settlement on the shell depends on the chemical recognition of the substrate by the anemone (Ross and Sutton, 1961 b).

In associations a mutual tolerance between the partners is usually developed and the behaviour modified so that one or both species are better adapted than in the solitary condition. Some fish and crustaceans dwell among the tentacles of cnidarians but are not stung and killed by the nematocysts. The control appears to be connected, at least for the fish, with the mucous layers developed (Mariscal, 1974). Schlichter (1976) gives some evidence that the substances released by the sea anemones act as inhibitors of discharge of their own cnidocytes. Such molecules are released in the environment and possibly absorbed on the mucous epidermal layer of the fish. So the fish acquires a specific protection which lasts as long as its relationship with the anemone is maintained.

Symbiosis of cnidarians with algae is likely to affect the response of the animals to light. However, conflicting evidence has been reported. According to Pearse (1974a,b) the pattern of expansion and contraction and the occurrence of phototactic behaviour displayed by the sea anemone *Anthopleura*, is modified by symbiotic zooxanthellae. Anemones without symbionts expand and contract irregularly under changes of light and do not show phototactic responses. Anemones with zooxanthellae expand in moderate light and contract under intense light or in darkness and they also display positive or negative phototaxis depending on the experimental light conditions. Aposymbiotic anemones, i.e. anemones from which the symbiotic zooxanthellae have been removed, show the same pattern of expansion and contraction as the actual symbionts but never show a phototactic behaviour. The flexible phototactic behaviour and the expansion/contraction pattern retained may play an important role in favourably regulating the amount of light to which the symbiotic algae are exposed. On the other hand Pardy (1976) denies any difference in phototactic response shown by symbiotic and aposymbiotic *Hydra viridis* as well as by animals treated with an inhibitor of the photosynthesis. No data are available on the effects on behaviour of the symbiosis between zoochlorellae and various acoel and rhabdocoel turbellarians, and it should be worth investigating.

INTERSPECIFIC PARASITICAL ASSOCIATIONS

Parasitism represents a widespread life habit in the phyla Platyhelminthes and Nematoda and the rule for Acanthocephala and Mesozoa. These phyla constitute about a half of any text-book on parasitology and we can say that metazoan parasites fall mostly within the lower invertebrates (see Table I). Parasitology, like protozoology, is a subject in itself and even though the stimuli and the general reactions involved are the same as in free-living species, the biology of the parasite-host association has to be seen as a single system in which one component is strictly adapted and geared in a well balanced way to the other.

A complete, effective parasitical relationship requires the maximum of specificity, i.e. strict conditions of host and host organ selection, of timing of the life cycle and of the releasing stimuli; little flexibility is possible, and meeting unfavourable conditions is heavily paid in terms of reproductive efficiency and survival. A "perfect" parasite is also adapted to live in a uniform environment. Usually it takes advantage of the homeostatic capabilities of the host for reducing its own energy expenditure for maintaining life, and devotes its efforts to the reproduction. Reproduction and dispersal are in fact the weak points of the "perfect" parasite. But for the ones which are transmitted directly from host to host, and even here some resting form exists, the endoparasite must face the outer environment for self-propagation and dispersal. Parasites show various strategies for acquiring "mobility". They may just emit well protected products of reproduction (capsulated eggs, resting larval stages etc.) into the environment and rely on the chance that they get picked up by the suitable host or by an intermediate host. So the amount of reproductive products becomes the limiting factor. Parasites may use vectors to whose life habits and specificity is committed the task of spreading the infection to the final host. They may evolve free-living stages which actively move seeking the host, in which case their behaviour must be selective in spotting the environmental conditions where the chances of meeting the host are increased. Finally the number of active stages can be increased by adding a reproductive step in an intermediate host so that the requirements of specificity and dispersal are met twice in the life cycle of the parasite. We may conclude that in parasitic life strictness of habits and adaptation to dispersal are the two aspects of the dialectics of nature and when one increases the patterns found in the other become more complex.

Unfortunately this is the way the parasitologist thinks; his model is the "perfect" parasite and all variations are seen as tending to improve its parasitic efficiency. The zoologist cannot believe in the sudden appearance of a perfectly adapted parasite, but sees it as having arisen in a series of steps by which a certain stage of an animal life cycle has progressively taken advantages of a more uniform environment for its own biological purposes, reproduction being only one of them. In Cnidaria for example, the early hydroid phase of *Polypodium* is endoparasitic in the ovary of sturgeon (Raikova, 1963); *Peachia* larvae ingested by medusae, develop further in the gastrovascular cavity and then move out to feed on the gonads (Spaulding, 1972). In this case the specificity of parasitism is related more to dispersal than to reproductive needs. In Nematomorpha, only the growth of the juvenile stage is accomplished in the insect host, sex recognition, reproduction and hatching occurring in fresh- or seawater.

Considering the independent but similar evolutionary patterns found in Platyhelminthes and Nematoda a general trend can be formulated. After an active larval stage the animal becomes more and more strictly associated, possibly through ectoparasitic stages, to the physiological characteristics of a certain class of hosts. Specificity builds up both by genetically acquired immunity in the host, and adaptation to restricted ecological niches within the host organs by the parasite. It is of some interest that the intestinal parasites are the most common; for, the alimentary tract is indeed in continuity with the outer environment and enables an easy penetration of the parasite through food, and an easy dispersal of reproductive products through faeces. Some species of the blood fluke *Schistosoma* still use the host's digestive tract for emission of the eggs.

SENSE ORGANS IN PARASITES

The introductory discussion given above implies critical differences in the behaviour of parasites at various stages. The problem of how this can be related to the ecology of the parasites and their hosts has been recently focussed by Fallis (1971), by Canning and Wright (1972), and by Kennedy (1975). The interest in the role played by the sense organs has prompted the reviews by Lyons (1973b) and McLaren (1976). In parasitic life sensory information is certainly more needed in the free-living stages where the spectrum of different stimuli is more complex to analyse, than in the restricted hemeostatic milieu of the sedentary parasitic stages; it is also more needed when the life cycles are geared to the life habits of the hosts and of the vectors, rather than when occurring in a single organ of a single host all the time. Finally when reproduction depends on the encounter of dioecious animals sensory information is more necessary than in parthenogenetic or hermaphroditic forms.

Adopting the view of the parasitologist we can start considering the "perfect" endoparasite settled in its specific niche. Usually it needs some contact with the surrounding tissues if it dwells in a cavity within which a fluid flows, like parasites of the intestine or blood vessels, or as an ectoparasite in an environment of flowing water. Hooks and sucker-like adhesive organs are developed at the anterior and/or posterior end. Like in the higher invertebrates with similar needs (octopus sucker; Graziadei and Gagne, 1975) adhesive organs are equipped with receptors for testing the substrate and for monitoring the degree of adhesion. Lyons (1973c) describes beneath the tegument of the papillae bulging out of the surface of the *Entobdella soleae* opisthaptor, nerve endings piled stack-like. Compression of the papillae, when the monogenean is attached on the epidermis of its host, is recorded as a distortion of the nerve endings, much in the same way as the organ of Bayer in the leech (Damas, 1973) and the tactile bodies in vertebrates work. In the newly excysted *Fasciola hepatica* (Bennett, 1975) the oral sucker is surrounded by sensory pits, possibly with a chemosensory function, whereas the domed aciliated nerve endings are considered contact pressure receptors. Tapeworms, which used to be considered as the most "degenerate" parasites completely adapted to cutaneous absorption and to reproduction, present typical ciliated endings (Blitz and Smyth, 1973; Jones, 1975) probably linked with chemoreceptive functions. Since the alimentary canal of the host presents a variety of pH and nutrients, intestinal parasites have to locate the suitable conditions; these may change during the developement of the worm and give rise to oriented migrations. *Hymenolepis diminuta* attaches initially in the middle portion of the gut, then moves forwards toward the duodenum. In hosts with resected bile ducts however, it moves to a more posterior location (Braten and Hopkins, 1969). Therefore it is likely that bile secretions control migration, but no direct proof exists that chemoreceptors are involved. Nematodes have papillae associated with the lips and they are quite prominent in the intestinal ascarids.

Tactile papillae are necessary also for mating in dioecious species where meeting the partner may be a matter of chance and close contact of the copulatory organs ensures the efficacy of fertilisation. Caudal papillae are more developed in the male nematodes and in hookworms the posterior end expands in a copulatory bursa. Nematomorpha possess similar adaptations, and in addition they can recognise the females at a distance and respond to touch, but the sensory organs involved are still unknown. Sensory structures should be assumed in the dioecious digenean *Schistosoma*; the female resides in a ventral longitudinal groove of the male and contact may be monitored by the tegumental tubercules. If reproduction must be secured, overcrowding and massive infections must be prevented. Intraspecific competition acts primarily in this sense; reduced establishment, growth rate and egg production may be due to metabolic factors (Roberts, 1966). However since a social regulation of population growth is active in free-living forms it would be worth investigating if social communication takes place. Cyclic activity patterns and migrations within the host body are often connected with reproduction. Release of eggs may occur at times when the chances of meeting the next host are high; in *Enterobius vermicularis* it is more frequent during the night and is accompanied by a migration of the gravid females from the ileo-caecal region to the anus. In the blood fluke *Schistosoma haemotobium* there is a peak of release of eggs around mid-day when the probability of the humans entering the water in tropical regions is higher, whereas strains of *Schistosoma japonicum*, parasitic on rats, release their cercariae at dusk when rodent activity is greatest. The case of the microfilariae is paradigmatic. They appear in the peripheral blood during the night when the primary vectors, blood feeding mosquitoes, are more active. Fluctuations in the arterial oxygen tension and other physiological factors of the host may be the responsible stimuli. Reversal of the sleep schedule reverses the periodicity of the filariae; however, transfusion to a host whose circadian rhythm was phase-shifted affords evidence for an endogenous rhythm which is kept for 8 h (Hawking, 1968). Different strains of the filaria, *Loa Loa*, show a periodicity which is directly correlated to that of the vector. Nocturnal strains in monkeys, which aggregate at night, use a nocturnal insect vector; diurnal strains in man use vectors with diurnal biting activity (Duke, 1972). The periodical light stimulation may be the actual "Zeitgeber" even in strict endoparasites. Although no distinct photoreceptors are described in endoparasitic forms, filarial worms, living in the heart of dogs, show *in vitro* a sensitivity to visible and ultra-violet light (Earl, 1959).

Seasonal cycles synchronise parasite and host activity. In the monogeneans, ectoparasitic on fish, the peak of reproduction coincides with the spawning season of their host when many adults and large shoals of susceptible juveniles are available. *Mazocraes alosae* produces eggs which develop and hatch rapidly within the very month in which its host, the shad (an anadromous migratory fish), leaves the estuaries in masses. The mechanism of co-ordination is supposed to be a common response of both host and parasite to cyclic environmental stimuli like temperature and the photoperiod (Kennedy, 1975). Cyclic physiological conditions of the host may control the outbursts of reinfection in the nematodes, *Haemonchus contortus* and *Toxocara canis*, parasites of sheeps and dogs; the declining of the immune response during pregnancy or variations in the hormonal balance activate the dormant larvae. In *Haemonchus* the generation time of the parasites becomes shorter with the increasing temperature, so that increasing reinfection takes place in the newborn susceptible progeny during the summer.

The coordination of the behaviour of the parasite in the host with that of the host or vector enhances the success of the reproductory products in finding another suitable host, and bridges the gap between the need of uniformity and the need of dispersal. As the parasites may undergo complex migrations in the host before establishing in the target organ (nematodes for instance), they evolved elaborate life cycles and behavioural patterns for increasing dispersal, like in Digenea. In these, two free-swimming larval stages alternate with two reproductive phases. The first larval stage, the miracidium, is infective to aquatic snails which enter the life cycle as intermediate hosts. The rule that free-living forms are equipped with a variety of sense organs, as opposed to the endoparasitic stages, is fully confirmed here. The miracidium possesses eyes, gravity receptors, superficial receptors and a very active ciliary movement (Brooker, 1972). Since the time of survival as free-swimming organisms is restricted, the behaviour must be directed to finding the host quickly in its habitat. Therefore the miracidia show specific orientation responses to the physical characteristics of the host habitat (Wright, 1971; Cable, 1972). The first step involves positive phototaxis and negative geotaxis; it takes the animal to the water surface where snails are more frequent. Finding that *Schistosoma japonicum* reverses its taxis in dependence on temperature, becoming photonegative below 15°C, clearly demonstrates the adaptation to the habits of its hosts which at that temperature congregate at the bottom. The second step is the dispersal of the miracidia by random movements which bring the larvae into the range of the host. In the third step the response is directed to chemical or other stimuli produced by the host. The mucus as well as extracts of parts of the snail and conditioned water are effective in eliciting an oriented behaviour. Amino acids and short chain fatty acids act as attractants (MacInnis, 1965). The animal moves up a gradient though not following a straight line (Wright, 1959; Ulmer, 1971); this is probably due to a chemokinesis more than to a true taxis (Wilson and Denison, 1970). Reproduction in the snail, the intermediate host, compensates for the high mortality of miracidia. Eventually the following free-swimming stage, the cercaria, is released. It is less specific in its behaviour and though most show positive phototaxis and negative geotaxis, the presence of the eyes is not a rule. Wright (1971) reports specific attraction to green surfaces in cercariae encysting on the grass around ponds; whereas others, parasites of fishes, exhibit a shadow response; the activity increases and they swim upwards to contact the host. Cercariae parasitising benthic arthropods are photonegative. Rheoreception can be postulated in cercariae that cease swimming when in a current; such a response increases the probability of being drawn over the gills of a crustacean host they can attach and penetrate. Finally contact chemoreceptors play a role in *Schistosoma* cercariae; they recognise the presence of skin lipids which stimulate penetration.

PERSPECTIVES

The purpose for which this review was undertaken is to demonstrate that groups highly diversified in structural complexity and in life habits exhibit common features when the same general constraints of the environment are acting. Because of the heterogeneity, no more than general trends can be identified. However, they are stimulating more detailed research within smaller systematic groups adapted to different environmental conditions. The work by Wieser (1959) on the striking difference in occurrence of nematode ocelli in exposed or sheltered littoral species, is exemplary. Parallel work in the field on quantitative samples and in the literature on the systematic description and distribution of the species might bring out valuable and more clear cut results.

Missing information about the physiology of many receptors which are described morphologically, or whose existence is inferred on behavioural bases, prevents definite statements on the function of the sensory equipment on the whole in lower invertebrates. Only for photoreception is it possible to notice a comprehensive approach; however, the observations are still scattered over so few and sometimes far related species that conclusions, at the moment, can be drawn only from the comparison of general categories of life habits, and on the assumption that the meagre data may be representative of the whole group or habitat.

The difference in the sensory endowment between free-living and parasitic forms is evident; the comparative evolution of sensory adaptation to parasitism can be traced in groups presenting a full transition from the independent life to the parasitic one. Nematodes again would be a suitable material; however they show a limited tendency in reducing and simplifying their sense organs also in completely endoparasitic forms. The fact that free-living nematodes also can withstand very unfavourable and unusual habitats suggests that parasitism in nematodes may have arisen in easy transition. An animal with a thick cuticle and well developed internal homeostasis occupies a very peculiar ecological niche.

In conclusion the circumstantial evidence about receptor functions obtained from different sources, i.e. morphological, functional and behavioral, fits in consistently with the main trends established here, so that sensory ecology too helps in the functional interpretation of those sensory capacities which are assumed on morphological descriptions or inferred on behavioural bases.

ACKNOWLEDGEMENTS

I am indebted to Prof. M.A. Ali, and to Prof. E. Ghirardelli for stimulating discussions and critical revision of the manuscript. Conversations with Dr. W. Wales and Dr. C. Bedini helped in bringing out valuable suggestions which are gratefully acknowledged.
Prof. P.K. Menon was kind enough to improve the language.

REFERENCES

Anderson, P.A.V. and Mackie, G.O. (1977). Electrically coupled, photosensitive neurons control swimming in a jellyfish. Science 197: 186-187.
Arai, M.N. (1973). Behaviour of the planktonic coelenterates, *Sarsia tubulosa, Phialidium gregarium* and *Pleurobrachia pileus* in salinity discontinuity layers. J. Fish. Res. Board Canada 30: 1105-1110.
Arai, M.N. (1976). Behaviour of planktonic coelenterates in temperature and salinity discontinuity. In: Coelenterate Ecology and Behavior. Ed. G.O. Mackie, Plenum Press, New York, p. 211-218.
Aronova, M. (1974). Electron microscopic observation on the aboral organ of Ctenophora. I. The gravity receptor. Z. mikrosk. anat. Forsch. (Leipz.) 88: 401-412.

Ash, J.F., McClure, W.O. and Hirsch, J. (1973). Chemical studies of a factor which elicits feeding behaviour in *Dugesia dorotocephala*. Anim. Behav. 21: 796-800.

Bedini, C., Ferrero, E. and Lanfranchi, A. (1973). The ultrastructure of ciliary sensory cells in two Turbellaria Acoela. Tissue & Cell 5: 359-372.

Bedini, C., Ferrero, E. and Lanfranchi, A. (1977). Fine structural changes induced by circadian light-dark cycles in photoreceptors of Dalyelliidae (Turbellaria, Rhabdocoela) J. Ultrastr. Res. 58: 66-77.

Bedini, C. and Lanfranchi, A. (1974). The fine structure of photoreceptors in two otoplanid species (Turbellaria, Proseriata). Z. Morph. Tiere 77: 175-186.

Bennett, C.E. (1975). Surface features, sensory structures, and movement of the newly excysted juvenile *Fasciola hepatica* L.J. Parasitol. 61: 886-891.

Bergquist, P.R., Sinclair, M.E. and Hogg, J.J. (1970). Adaptation to intertidal existence: reproductive cycles and larval behaviour in Demospongiae. In: The Biology of the Porifera. Ed. W.B. Fry, Symp. Zool. Soc. Lond. 25: 247-271.

Best, J.B., Goodman, A.B. and Pigon, A. (1969). Fissioning in planarians: control by the brain. Science 164: 565-566.

Best, J.B., Howell, W., Riegel, V. and Abelein, M. (1974). Cephalic mechanism for social control of fissioning in planarians. I. Feed-back cue and switching characteristics. J. Neurobiol. 5: 421-442.

Blitz, N.M. and Smyth, J.D. (1973). Tegumental ultrastructure of *Raillietina cesticillus* during the larval-adult transformation, with emphasis on the rostellum. Int. J. Parasitol. 3: 561-570.

Bowen, I.D. and Ryder, T.A. (1974). The fine structure of the planarian *Polycelis tenuis* (Iijima) III. The epidermis and external features. Protoplasma 80: 381-392.

Braten, T. and Hopkins, C.A. (1969). The migration of *Hymenolepis diminuta* in the rat's intestine during normal development and following surgical implantation. Parasitol. 59: 891-905.

Brewer, R.H. (1976). Some microenvironmental influences on attachment behavior of the planula of *Cyanea capillata* (Cnidaria: Scyphozoa). In: Coelenterate Ecology and Behavior. Ed. G.O. Mackie, Plenum Press, New York, p. 347-354.

Brooker, B.E. (1972). The sense organs of trematode miracidia In: "Behavioural aspects of parasite transmission", edited by Canning, E.U. and Wright, C.A. Zool. J. Linn. Soc. 51 Suppl. 1: 171-180.

Brown, F.A. (1971). Some orientational influences of non visual terrestrial electromagnetic fields. Ann. N.Y. Acad. Sci. 188: 224-241.

Brown, H.M. and Ogden, T.E. (1968). The electrical response of the planarian ocellus. J. gen. Physiol. 51: 237-253.

Bullock, T.H. and Horridge, G.A. (1965). Structure and function in the nervous systems of invertebrates. W.H. Freeman & Co., San Francisco.

Cable, R.M. (1972). Behaviour of digenetic trematodes. In: "Behavioural aspects of parasite transmission", edited by Canning E.U. and Wright, C.A. Zool. J. Linn. Soc. 51 Suppl 1: 1-18.

Campbell, R.D. (1972). Statocyst lacking cilia in the coelenterate *Corymorpha palma*. Nature 238: 49-51.

Canning, E.U. and Wright, C.A. (Eds.) (1972). Behavioural aspects of parasite transmission. Zool. J. Linn. Soc. 51 Suppl. 1.

Chang, J.J. and Johnson, F.H. (1959). The influence of pressure, temperature and urethane on the luminescent flash of *Mnemiopsis leidyi*. Biol. Bull. (Woods Hole) 116: 1-14.

Chia, F.S. and Rostron, M.A. (1970). Some aspects of the reproductive biology of *Actinia equina* (Cnidaria: Anthozoa). J. Mar. Biol. Ass. U.K. 50: 253-264.

Clément, P. (1975). Ultrastructure de l'oeil cérébrale d'un rotifère *Trichocerca rattus*. J. Microsc. (Paris) 22: 69-86.

Clément, P. and Pourriot, R. (1972). Photopériodisme et cycle hétérogonique chez certains Rotifères Monogonontes. I. Observations préliminaires chez *Notommata copeus*. Arch. Zool. Exp. Gen. 113: 41-50.

Corning, W.C. and Kelly, S. (1973). Platyhelminthes: the Turbellarians. In: "Invertebrate learning". Ed. W.C. Corning, J.A. Dyal and A.O.D. Willows, Plenum Press, New York vol. 1.

Coward, S.J. and Johannes, K.B. (1969). Amino acid chemoreception by the planarian *Dugesia dorotocephala*. Comp. Biochem. Physiol. 29: 475-478.

Crisp, D.J. (1974). Factors influencing the settlement of marine invertebrate larvae. In: Chemoreception in Marine Organisms. Ed. P.T. Grant and A.M. Mackie, Academic Press, London p. 105-141.

Croll, N.A. (1965). The klinokinetic behaviour of infective *Trichonema* larvae in light. Parasitol. 55: 579-582.

Croll, N.A. (1970). Behaviour of Nematodes. Edward Arnold, London.

Damas, D. (1973). Etude ultrastructurale des organes tégumentaires de Bayer (complex épithélio-musculaire) chez l'hirudinée *Glossiphonia complanata* L. C.R. Séances Acad. Sci. Ser.D Sci. Nat. 276: 2545-2548.

Delamare Deboutteville, C. (1960). Biologie des Eaux Souterraines Littorales et Continentales. Hermann, Paris.

Dougherty, E.C. (Ed.) (1963). The lower Metazoa: comparative biology and phylogeny. University of California Press, Berkeley.

Duke, B.O.L. (1972). Behavioural aspects of the life cycle of *Loa*. In: "Behavioural Aspects of Parasite Transmission", Ed. E.U. Canning and G.A. Wright, Zool. J. Linn. Soc. 51 Suppl. 1: 97-107.

Durand, J.P. and Gourbault, N. (1975). Etude cytologique des organes photorécepteurs de *Dendrocoelopsis (Amyadenium) chattoni*

(Triclade paludicole hypogé). Ann. Spéléol. 30: 129-135.
Earl, P.R. (1959). Filariae from the dog *in vitro*. Ann. N.Y. Acad. Sci. 77: 163-175.
Ehlers, B. and Ehlers, U. (1977). Die Feinstruktur eines ciliären Lamellarkörpers bei *Parotoplanina geminoducta* Ax (Turbellaria, Proseriata). Zoomorphologie 87: 65-72.
Fallis, A.M. (Ed.) (1971). Ecology and physiology of parasites. Adam Hilger, London.
Fauvel, P. (1907). Recherches sur les otocystes des annélides polychètes. Ann. Sci. Nat. (Zool.) 6: 1-149.
Feldman, M. and Lenhoff, H.M. (1960). Phototaxis in *Hydra littoralis*: rate studies and localization of the "photoreceptor". Anat. Rec. 137: 354-355.
Ferrero, E. (1973). A fine structural analysis of the statocyst in Turbellaria Acoela. Zoologica Scripta 2: 5-16.
Fiore, L. (1971). A mechanism for self-inhibition of the population growth in the flat-worm *Mesostoma ehrenbergi* (Focke). Oecologia (Berl.) 7: 356-360.
Fiore, L. and Ioalé, P. (1973). Regulation of the production of subitaneous and dormant eggs in the turbellarian *Mesostoma ehrenbergi* (Focke). Monitore Zool. Ital. (N.S.) 7: 203-224.
Fraenkel, G. (1929). Über die Geotaxis von *Convoluta roscoffensis*. Z. vergl. Physiol. 10: 237-247.
Fraenkel, G. (1961). Quelques observations sur le comportement de *Convoluta roscoffensis*. Cah. Biol. Mar. 2: 155-160.
Fraenkel, G. and Gunn, D.L. (1961). The Orientation of Animals. Dover Publication, New York.
Galleni, L., Tongiorgi, P., Ferrero, E. and Salghetti, U. (1977). *Stylochus mediterraneus* Galleni (Turbellaria, Polycladida) predator of the mussel *Mytilus gallo-provincialis* Lmk. Mar. Biol. (in press).
Gamble, F.W. and Keeble, F. (1904). The bionomics of *Convoluta roscoffensis* with special reference to its green cells. Q. J. Micr. Sci. 47: 363-431.
Graziadei, P.P.C. and Gagne, H.T. (1975). Sensory innervation in the rim of the octopus sucker. J. Morphol. 150: 639-680.
Hawking, F. (1968). The 24-hour periodicity of microfilariae: biological mechanisms responsible for its production and control. Proc. Roy. Soc. Lond. B Biol. Sci. 169: 59-76.
Heitkamp, U. (1972). Die Mechanismen der Subitan- und Dauereibildung bei *Mesostoma lingua* (Abilgaard, 1789) (Turbellaria, Neorhabdocoela). Z. Morphol. Tiere 71: 203-289.
Hernandez-Nicaise, M.L. (1974). Ultrastructural evidence for a sensory-motor neuron in Ctenophora. Tiss. Cell 6: 43-47.
Horridge, G.A. (1965a). Relations between nerves and cilia in Ctenophores. Am. Zool. 5: 357-375.
Horridge, G.A. (1965b). Non-motile sensory cilia and neuromuscular junctions in a ctenophore independent effector organ. Proc. Roy. Soc. Lond. B Biol. Sci. 162: 333-350.

Horridge, G.A. (1968). The origin of the nervous system. In: "Structure and function of the nervous tissue". Ed. G.H. Bourne, Academic Press, New York p. 1-31.

Horridge, G.A. (1969). Statocysts of medusae and evolution of stereocilia. Tiss. Cell. 1: 341-353.

Horridge, G.A. (1974). Recent studies on the Ctenophora. In: "Coelenterate Biology", Ed. L. Muscatine and H.M. Lenhoff, Academic Press, New York p. 439-468.

Hyman, L.H. (1951). The Invertebrates: Platyhelminthes and Rhynchocoela the Acoelomate Bilateria. McGraw-Hill, New York.

Jones, A. (1975). The morphology of *Bothriocephalus scorpii* (Müller) (Pseudophyllidea, Bothriocephalidae) from littoral fishes in Britain. J. Helminthol. 49: 251-261.

Kawakatsu, M. (1974). Further studies on the vertical distribution of freshwater planarians in the Japanese Islands. In: "Biology of the Turbellaria". Ed. by M.W. Riser and M.P. Morse, McGraw-Hill, New York p. 291-338.

Kennedy, C.R. (1975). Ecological Animal Parasitology. John Wiley, New York.

Knight-Jones, E.W. and Morgan, E. (1966). Responses of marine animals to changes in hydrostatic pressure. Oceanogr. Mar. Biol. Ann. Rev. 4: 267-299.

Koopowitz, H. (1970). Feeding behaviour and the role of the brain in the polyclad flatworm, *Planocera gilchristi*. Anim. Behav. 18: 31-35.

Koopowitz, H., Silver, D. and Rose, G. (1976). Primitive nervous systems. Control and recovery of feeding behavior in the polyclad flatworm, *Notoplana acticola*. Biol. Bull. (Woods Hole) 150: 411-425.

Krisch, B. (1973). Über das Apikalorgan (Statocyste) der Ctenophore *Pleurobrachia pileus*. Z. Zellforsch. mikrosk. Anat. 142: 241-262.

Landers, W.S. and Rhodes, E.W. (1970). Some factors influencing predation by the flat-worm, *Stylochus ellipticus* (Girard) on oysters. Chesapeake Sci. 11: 55-60.

Lenhoff, H.M. (1974). On the mechanism of action and evolution of receptors associated with feeding and digestion. In: "Coelenterate Biology". Ed. by L. Muscatine and H.M. Lenhoff, Academic Press, New York p. 211-243.

Lenhoff, H.M., Heagy, W. and Danner, J. (1976). A view of the evolution of chemoreceptors based on research with cnidarians. In: Coelenterate Ecology and Behavior. Ed. G.O. Mackie, Plenum Press, New York, p. 571-579.

Lock, M.A. (1972). The responses to current flow of two stream-dwelling triclads, *Crenobia alpina* (Dana) and *Polycelis felina* (Dalyell). Oecologia (Berl.) 10: 313-320.

Lock, M.A. (1975). An experimental study of the role of gradient and substratum in the distribution of two stream-dwelling triclads, *Crenobia alpina* (Dana) and *Polycelis felina* (Dalyell) in North Wales. Freshwat. Biol. 5: 211-226.

Lyons, K.M. (1973a). Collar cells in planula and adult tentacle ectoderm of the solitary coral *Balanophyllia regia* (Anthozoa, Eupsammiidae). Z. Zellforsch. Mikrosk. Anat. 145: 57-74.

Lyons, K.M. (1973b). The epidermis and sense organs of the Monogenea and some related groups. Adv. Parasitol. 11: 193-232.

Lyons, K.M. (1973c). Scanning and transmission electron microscope studies on the sensory sucker papillae of the fish parasite *Entobdella soleae* (Monogenea). Z. Zellforsch. Mikrosk. Anat. 137: 471-480.

MacDonald, A.G. (1975). Physiological Aspects of Deep Sea Biology. Cambridge University Press, London, 450 p.

MacInnis, A.J. (1965). Responses of *Schistosoma mansoni* miracidia to chemical attractants. J. Parasitol. 51: 731-746.

Mackie, G.O. (1970). Neuroid conduction and the evolution of conducting tissues. Q. Rev. Biol. 45: 319-332.

Mackie, G.O. (1974). Locomotion, flotation and dispersal. In: Coelenterate Biology. Ed. L. Muscatine and H.M. Lenhoff, Academic Press, New York p. 313-357.

Mackie, G.O. and Singla, C.L. (1975). Neurobiology of *Stomotoca*. I. Action systems. J. Neurobiol. 6: 339-356.

Mariscal, R.N. (1974). Nematocysts. In: Coelenterate Biology. Ed. L. Muscatine and H.M. Lenhoff, Academic Press, New York p. 129-210.

McFarlane, I.D. (1969a). Two slow conduction systems in the sea anemone *Calliactis parasitica*. J. Exp. Biol. 51: 377-385.

McFarlane, I.D. (1969b). Co-ordination of pedal-disk detachment in the sea anemone *Calliactis parasitica*. J. Exp. Biol. 51: 387-396.

McLaren, D.J. (1976). Nematode sense organs. Adv. Parasitol. 14: 195-265.

Meadows, P.S. and Campbell, J.I. (1972). Habitat selection by aquatic invertebrates. Adv. Mar. Biol. 10: 271-382.

Meglitsch, P.A. (1972). Invertebrate Zoology. Oxford University Press, London.

Menzel, R. and Roth, F. (1972). Spektrale Phototaxis von Planktonrotatorien. Experientia (Basel) 28: 356-357.

Mitchell, R.W. (1974). The cave-adapted flatworms of Texas; systematics, natural history and responses to light and temperature. In: Biology of the Turbellaria. Ed. N.W. Riser and M.P. Morse, McGraw-Hill., New York.

Müller, W.A., Wicker, F. and Eiben, R. (1976). Larval adhesion, releasing stimuli and metamorphosis In: Coelenterate Ecology and Behavior. Ed. G.O. Mackie, Plenum Press, New York, p. 339-346.

Newell, R.C. (1970). Biology of intertidal animals. Logos Press Ltd., London.

Nishihira, M. (1967a). Observation on the selection of algal substrate by hydrozoan larvae, *Sertularella miurensis* in nature. Bull. Mar. Biol. Stn. Asamushi 13: 35-48.

Nishihira, M. (1967b). Dispersal of the larvae of a hydroid, *Sertularella miurensis*. Bull. Mar. Biol. Stn. Asamushi 13:49-56.

Nyholm, K.-G. (1949). On the development and dispersal of athenaria actinia with special reference to *Halcampa duodecimcirrata* M. Sars. Zool. Bidr. Upps. 27: 467-505.

Ostarello, G.L. (1976). Larval dispersal in the subtidal hydrocoral *Allopora californica* Verrill (1866). In: Coelenterate Ecology and Behavior. Ed. G.O. Mackie, Plenum Press, New York, p. 331-337.

Pardy, R.L. (1976). Aspects of light in the biology of green hydra. In: Coelenterate Ecology and Behavior. Ed. G.O. Mackie, Plenum Press, New York, p. 401-407.

Passano, L.M. (1976). Strategies for the study of the coelenterate brain. In: Coelenterate Ecology and Behavior. Ed. G.O. Mackie, Plenum Press, New York, p. 639-645.

Passano, L.M. and McCoullogh, C.B. (1964). Co-ordinating systems and behaviour in *Hydra* I. Pacemaker system of the periodic contractions. J. Exp. Biol. 41: 643-664.

Pavans de Ceccatty, M., Thiney, Y. and Garrone, R. (1970). Les bases ultrastructurales des communications intercellulaires dans les oscules de quelques éponges. In: "The Biology of the Porifera". Ed. W.G. Fry, Symp. Zool. Soc. Lond. 25: 449-466.

Pearse, V.B. (1974a). Modification of sea anemone behavior by symbiotic zooxanthellae: expansion and contraction. Biol. Bull. (Woods Hole) 147: 630-640.

Pearse, V.B. (1974b). Modification of sea anemone behavior by symbiotic zooxanthellae: phototaxis. Biol. Bull. (Woods Hole) 147-641-651.

Peteya, D.J. (1973). A possible proprioceptor in *Ceriantheopsis americanus* (Cnidaria Ceriantharia). Z. Zellforsch. Mikrosk. Anat. 144: 1-10.

Pigon, A., Morita, M. and Best, J.B. (1974). Cephalic mechanism for social control of fissioning in planarians. II. Localization and identification of the sensory receptors by electron micrographic and ablation studies. J. Neurobiol. 5: 443-462.

Pourriot, R. and Clément, P. (1973). Photopériodisme et cycle hétérogonique chez *Notommata copeus* (Rotifère Monogononte): influence de la qualité de la lumière. Spectres d'action. Arch. Zool. Exp. Gen. 114: 277-300.

Raikova, E.V. (1963). An early parasitic stage in the life-cycle of *Polypodium hydriforme* Ussov (Coelenterata). Dokl. Akad. Nauk. SSSR 154: 742-743 (in Russian).

Rampitsch, J. (1941). Versuche über die cilioregulatorische Fortbewegung des Turbellars *Stenostomum leucops*. Zool. Anz. 133: 253-258.

Rasmont, R. (1970). Some new aspects of the physiology of freshwater sponges. In: The Biology of the Porifera. Ed. W.G. Fry Symp. Zool. Soc. Lond. 25: 415-422.

Reynierse, J.H., Gleason, K.K. and Ottemann, R. (1969). Mechanisms producing aggregations in planaria. Anim. Behav. 17: 47-63.

Reynoldson, T.B. (1974). Ecological separation in British triclads (Turbellaria) with a comment on two american species. In: "Biology of the Turbellaria", edited by N.W. Riser and M.P. Morse, McGraw-Hill, New York p. 213-228.

Roberts, L.S. (1966). Developmental physiology of cestodes I. Host dietary carbohydrate and the crowding effect in *Hymenolepis diminuta*. Exp. Parasitol. 18: 305-310.

Rogers, W.P. (1940). The effect of environmental conditions on the accessibility of third stage trichostrongyle larvae to grazing animals. Parasitol. 32: 208-226.

Ross, D.M. (1974). Behaviour patterns in associations and interactions with other animals. In: "Coelenterate Biology", Ed. L. Muscatine and H.M. Lenhoff, Academic Press, New York p. 281-312.

Ross, D.M. and Sutton, L. (1961a). The response of the sea anemone *Calliactis parasitica* to shells of the hermit crab *Pagurus bernhardus*. Proc. Roy. Soc. Lond. B Biol. Sci. 155: 266-281.

Ross, D.M. and Sutton, L. (1961b). The association between the hermit crab *Dardanus arrosor* (Herbst) and the sea anemone *Calliactis parasitica* (Couch). Proc. Roy. Soc. Lond. B Biol. Sci. 155: 282-291.

Russell, F.S. (1925). The vertical distribution of marine macroplankton. An observation on diurnal changes. J. Mar. Biol. Ass. U.K. 13: 769-809.

Schlichter, D. (1976). Macromolecular mimicry: substances released by sea anemones and their role in the protection of anemone fishes. In: Coelenterate Ecology and Behavior. Ed. G.O. Mackie, Plenum Press, New York, p. 433-441.

Singla, C.L. (1974). Ocelli of Hydromedusae. Cell Tiss. Res. 149: 413-429.

Singla, C.L. (1975). Statocysts of Hydromedusae. Cell Tiss. Res. 158: 391-407.

Spaulding, J.G. (1972). The life cycle of *Peachia quinquecapitata*, an anemone parasitic on medusae during its larval development. Biol. Bull. (Woods Hole) 143: 440-453.

Steven, D.M. (1963). The dermal light sense. Biol. Rev. (Camb.) 38: 206-240.

Storch, V. and Abraham, R. (1972). Elektronmikroskopische Untersuchungen über die Sinneskante des terricolen Turbellars *Bipalium kewense* Moseley (Tricladida). Z. Zellforsch. mikrosk. Anat. 133: 267-275.

Swedmark, B. (1964). The interstitial fauna of marine sand. Biol. Rev. (Camb.) 39: 1-42.

Tardent, P. (1976). The reactions of *hydra attenuata* Pall, to various photic stimuli. In: Coelenterate Ecology and Behavior. Ed. G.O. Mackie, Plenum Press, New York, p. 671-683.

Tyler, S. (1976). Comparative ultrastructure of adhesive systems in the Turbellaria, Zoomorphologie 84: 1-76.

Ulmer, M.J. (1971). Site-finding behaviour in helminths in intermediate and definitive hosts. In: "Ecology and Physiology of Parasites", Ed. A.M. Fallis, Adam Hilger, London p. 123-159.
Vandel, A. (1964). Biospéologie. Gauthier-Villars, Paris.
Vernberg, W.B. and Vernberg, F.J. (1972). Environmental Physiology of Marine Animals. Springer Verlag, Berlin.
Viaud, G. (1940). Recherches expérimentales sur le phototropisme des Rotifères. Bull. Biol. Fr. Belg. 74: 249-308.
Viaud, G. (1943a). Recherches expérimentales sur le phototropisme des Rotifères. II. Bull. Biol. Fr. Belg. 77: 68-93.
Viaud, G. (1943b). Recherches expérimentales sur le phototropisme des Rotifères. III. Bull. Biol. Fr. Belg. 77: 224-242.
Viaud, G. (1954). Conception nouvelle du galvanotropisme animal. Expérience sur les planaires. Experientia (Basel) 10: 233-242.
Vinogradov, M.E. (1970). Vertical distribution of the oceanic zooplankton. Israel Program for Scientific Translations, Jerusalem.
Wallace, H.R. (1958). Movement of eelworms. I. The influence of pore size and moisture content of the soil on the migration of larvae of the beet eelworm *Heterodera schachtii*, Schmidt. Ann. Appl. Biol. 46: 74-85.
Wallace, H.R. (1963). The biology of plant parasitic nematodes. Edward Arnold, London.
Ward, S.N. (1973). Chemotaxis by the nematode *Coenorhabditis elegans*: identification of attractants and analysis of the response by use of mutants. Proc. Nat. Acad. Sci. 70: 817-821.
Ward, S.N. (1976). The use of mutants to analyse the sensory nervous system of *Coenorhabditis elegans*. In: The Organization of Nematodes, Ed. by N.A. Croll, Academic Press, London p. 365-382.
Westfall, J.A. (1973). Ultrastructural evidence for a granule-containing sensory-motor interneuron in *Hydra littoralis*. J. Ultrastr. Res. 42: 268-282.
Wieser, W. (1959). Free-living marine nematodes IV. General part. Report of Lund University Chile Expedition, 1948-49. K. Fysiogr. Sallsk. Lund Handl. N.F. 70
Williams, G.B. (1965). Observations on the behavior of *Clava squamata*. J. Mar. Biol. Ass. U.K. 45: 257-273.
Wilson, R.A. and Denison, J. (1970). Studies on the activity of the miracidium of the common liver fluke, *Fasciola hepatica*. Comp. Biochem. Physiol. 32: 301-313.
Wolff, T. (1970). The concept of the hadal or ultral abyssal fauna. Deep Sea Res. 6: 95-124.
Wright, C.A. (1959). Host location by trematode miracidia. Ann. Trop. Med. Parasitol. 53: 288-292.
Wright, C.A. (1971). Flukes and Snails. George Allen & Unwin, London
Zelickman, E.A., Gelfand, V.I. and Shifrin, M.A. (1969). Growth, reproduction and nutrition of some Barents Sea hydromedusae in natural aggregations. Mar. Biol. 4: 167-173.

COELOMATE INVERTEBRATES (except Crustacea, Arachnida and Insecta)

M.A. ALI

Département de Biologie, Université de Montréal

Case Postale 6128, Montréal H3C 3J7, Canada

> Look you, the worm is not to be trusted but in the keeping of wise people;
>
> Antony and Cleopatra, Act V, Sc. 2.

    The groups to be considered in this chapter (Table 1) form a vast and very heterogeneous assemblage. This renders the task of doing a detailed survey of their ecosensory functions very difficult but challenging and interesting. The main purpose is to assemble the information available mostly in appropriate textbooks and review articles, and present it with a view to pointing out relationships among habitats, sensory functions, modes of life, feeding and locomotion. One can readily see that at times the distinctions are not always clear. Parasitism is a mode of life as well as feeding. The same group of cells could carry out more than one sensory function e.g. tactile and chemoreceptive. Or, a function could be carried out by a group of cells which do not form an organ such as in the case of the sensory cells which form clusters and are photoreceptive.

    Four tables (see also Ali, Croll and Jaeger, this volume) were prepared which form the nucleus of this chapter. Since most readers will not be specialists in invertebrate biology but will be interested in general interrelationships, this presentation intends merely to point out general principles of broad significance. Also, this assemblage hopefully will eliminate the need, for almost all users, to look through an extensive literature. To make it easier for the author and to avoid overloading the tables and text, we have

TABLE 1. HABITATS OF GROUPS CONSIDERED

Phylum	Groups / Class	Aquatic			Amphibious		Terrestrial		
		Fresh-water	Brackish	Marine	Entire life	Partial	Subterranean	Land	Aerial
Annelida >8,700 spp.	Polychaeta (bristle worms) >5,300 spp.			X					
	Oligochaeta (earthworms & aquatic worms) 3,100 spp.	X		X	X		X		
	Hirudinea (leeches) > 500 spp.	X		X				X	
Arthropoda >750,000 spp.	Merostomata (horseshoe crab) 5 spp.			X					
	Chilopoda (centipedes) 3,000 spp.							X	
	Symphyla (centipede-like) 120 spp.							X	
	Diplopoda (millipedes) 7,500 spp.							X	
	Pauropoda (grub like) 360 spp.							X	
Onychophora (caterpillar like) 65 spp.							X	X	
Pogonophora (beard worms) 80 spp.				X					
Sipuncula (peanut worms) 330 spp.				X					
Echiura (worm like) 100 spp.				X					

COELOMATE INVERTEBRATES   93

HABITATS OF GROUPS CONSIDERED

		Aquatic			Amphibious		Terrestrial		
Phylum	Class	Fresh-water	Brackish	Marine	Entire life	Partial	Subter-ranean	Land	Aerial
Priapulida (cucumber shaped) 8 spp.				X					
Tardigrada (water bears) 350 spp.		X	X						
Pentastomida (tongue worms) 70 spp.		Parasites of reptiles, birds and mammals.						X	X
Phronida (worm like) 15 spp.				X					
Bryozoa 4,000 spp.	Stenolaemata			X					
	Gymnolaemata (most spp.)			X					
	Phylectolaemata 50 spp.	X							
Entoprocta 60 spp.				X					
Mollusca >80,000 spp.	Gastropoda (conches, snails, slugs) >35,000 spp.	X		X					
	Monoplacophora (chiton like) 10 spp.			X					
	Polyplacophora (chitons) 600 spp.			X					
	Aplacophora (worm like) 130 spp.			X					

HABITATS OF GROUPS CONSIDERED

Phylum	Groups Class	Aquatic Fresh-water	Brackish	Marine	Amphibious Entire life	Partial	Terrestrial Subter-ranean	Land	Aerial
	Bivalvia (clams, oysters, mussels) 20,000 spp.	X	X	X					
	Scaphopoda (tusk shells) 200 spp.			X					
	Cephalopoda (nautilus, cuttlefish, squid, octopus) 650 spp.			X					
Brachiopoda (lamp shells) 280 spp.				X					
Echinodermata 5,300 spp.	Asteroidea (starfishes) 1,600 spp.			X					
	Ophiuroidea (sea stars) 2,000 spp.			X					
	Echinoidea (sea urchins) 800 spp.			X					
	Holothuroidea (sea cucumbers) 500 spp.			X					
	Crinoidea (sea lilies) 80 spp.			X					
Hemichordata 100 spp.	Enteropneusta (acorn worms)			X					
	Pterobranchia			X					
Chordata (Urochordata) 2,100 spp.	Ascidiacea (sea-squirts)			X					

HABITATS OF GROUPS CONSIDERED

Phylum	Class	Aquatic			Amphibious		Terrestrial		
		Fresh-water	Brackish	Marine	Entire life	Partial	Subter-ranean	Land	Aerial
	Thaliacea			X					
	Larvacea			X					
Chaetognatha 65 spp.	(arrow-worms)			X					

abstained from citing all the references in the text but have given a bibliography containing the references which we have consulted, cited or not.

In the preparation of the tables, the protostomes were divided arbitrarily into two groups, one composed of Annelida, Arthropoda and the minor phyla and the other of Mollusca and Brachiopoda. The third part includes the deuterostomes. The Annelid superphylum (arachnids crustacea and insects not included in this chapter; see Gogala, this volume) is almost entirely restricted to the marine environment, except most of the Oligochaeta and Hirudinea and the non-merostomate chelicerates, which have colonised land and freshwater also. The Mollusca and the Brachiopoda are almost entirely marine except for some Gastropoda and Bioalvia while all the deuterostome invertebrates are marine (Table 1). The Pentastomida, as parasites of reptiles, birds and mammals may be considered to be terrestrial. Thus, one may safely say that the aquatic milieu in general, and the marine one in particular, are the natural habitats of these invertebrate groups, except for the small number stated above which have been driven to the terrestrial environment by competition, to succeed as either free-living forms e.g. earthworms, centipedes, ectoparasites (leeches) or endoparasites (Pentastomida).

When one surveys the known sensory functions of these animals (Table 2; see also Ali, Croll and Jaeger, this volume) it seems that photoreception is by far the most common, followed by mechanoreception and chemoreception. However, sensory modes have not been tested adequately in many forms, and this impression might be misleading. Nevertheless all of these groups, except the crinoids, inhabit waters shallow enough to permit light penetration. The photic quality of their environments vary greatly but unfortunately very little is known about the photopigments or photophysiology in these groups except in a few polycaetes (Yingst, Fernandez and Bishop; Wald and Rayport, 1977) gastropods, cephalopods and, of course *Limulus*. It would appear that the photopigments will be well adapted to function optimally in the given environments. This as well as the study of the operation of the photoreceptors in most of these groups is still virgin field. The wide distribution of chemo-/ and mechanoreceptor functions is to be expected in view of the feeding and locomotory modes of these animals (Tables 3 and 4). The interrelationships among them are intricate and call for means to detect prey, predator and partner, often using chemical and vibratory clues as shown in Mollusca (Wood, 1968; Field and MacMillan, 1973) and annelida (Daly, 1973) for example. It is interesting that even here the sea lilies are an exception. Being mostly deep-sea benthic forms it should be less critical in their case not to possess these senses. On the other hand, they are a poorly studied group from the point of view of physiology and even ecology, as are also most of the others.

TABLE 2. THE MAIN SENSE ORGANS IN THE ANIMAL GROUPS EXAMINED.

Annelida
- Polychaeta — Eyes (2-4 prs); epidermal photoreceptor cells; nuchal organs (ciliated pits or slits; chemoreceptive); ciliated sense organs; statocysts; tactile cells;
- Oligochaeta — Pigment cup ocelli; sensory cells with free nerve endings;
- Hirudinea — Free nerve endings; sensory cells with terminal bristles; eyes (2-10) sensory papillae; respond to water pressure vibrations; moving shadows; attracted by temperature.

Mollusca
- Gastropoda — Eyes; tentacles; osphradia; statocysts; chamoreceptors (?);
- Monoplacophora — Not known (Post-onal tentacles and velum may be sensory)
- Polyplacophora — No cephalic eyes or tentacles; subradular organ; aesthetes: sometimes made up of bundles of sensory cells forming ocellus; tactile and photoreceptor cells; sensory epithelium patches; (litre osphradia?)

Aplacophora — Not known

Bivalvia — Pallial tentacles (tactile and chemoreceptor cells); statocysts; ocelli; sensory epithelium patch (osphradium?);

Scaphopoda — Tentacles (captacula);

Cephalopoda — Eyes; osphradia; statocysts; tactile cells; chemoreceptor cells;

Arthropoda
- Merostomata — Eyes (lateral, median); frontal organ (chemoreceptor); spines of gnathobases (chemo).
- Chiliopoda — Ocelli (few to many); organs of Tömösvary (vibrations; audition); last pair of legs (?);
- Symphyla — Stylus (?); organs (?) of Tömösvary;
- Diplopoda — Ocelli (2-80); tactile hair and peg and conelike projections (chemoreceptor) on antennae; organs of Tömösvary;

Pauropoda	Disclike sensory organs (Tömösvary?); antennae (with club shaped sensory structure);
Onychophora	Eye (small); tubercles and other areas of integument with sensory cells;
Pogonophora	Tentacles (sensory?);
Sipuncula	Sensory cells; nuchal organs (ciliated pits; chemoreceptive); pigment cup ocelli;
Echiura	No specialised sense organs; proboscis very sensitive to touch and seems to be chemosensory too.
Priapulida	Papillae on proboscis and trunk;
Tardigrada	Bristles and spines; simple eye spots (red or black pigmented cell);
Pentastomida	No specialised sense organs;
Phoronida	Sensory cells;
Bryozoa	No specialised sense organs; tentacles sensitive and carry sensory cells, tactile, chemosensory and sensitive to water current;
Entoprocta	Sensory cells (with projecting bristles on body surface especially outer side of tentacles and along calyx margin)?
Brachiopoda	Statocysts (1 pair); mantle margin; mantle setae (tactile)
Echinodermata	
Asteroidea	Eye spots at the tips of arms (80–200) sensory cells (light, contact, chemical); podia (righting response);
Ophiuroidea	General epithelial sensory cells; podia (chemoreception);
Echinoidea	Sensory cells in epithelium, especially on spines, pedicellariae and podia; spheridia (statocysts); photoreceptor cells on tube feet;
Holothuroidea	Sensory cells in epidermis; warts and tubercles on body surface with cluster of sensory cells; statocysts; photoreceptor cells at base of tentacles;
Crinoidea	No specialised sensory organs; sensitive to touch and shows righting reaction, photosensitive and prefers darker places.

Hemichordata

    Enteropneusta   Preoral ciliary organ; neurosensory cells on surface epithelium;

    Pterobranchia   Tentacles; neurosensory cells;

Chordata (S. Phyl. Urochordata)

    Ascidiacea   No specialised sense organs; sensory cells abundant on siphons, buccal tentacles; atrium (tactile), chemoreceptive; pigmented cups in cylindrical cells; Larvae have ocellus and statocyst.

    Thaliacea   Not known;

    Larvacea   Not known;

Chaetognatha   Eyes; sensory bristles; head organ (ciliary loop) function not known.

TABLE 3. MODES OF FEEDING / NUTRITION

Phylum	Class	Raptorial	Herbivorous	Omnivorous	Scavengers	Browsers	Direct deposit feeders	Indirect deposit feeders	Filter feeders	Ectoparasites	Endoparasites
Annelida	Polychaeta	X	X	X	X	X	X	X	X		
	Oligochaeta	X	X		X	X	X	X	X		
	Hirudinea	X								X	
Arthropoda	Merostomata				X						
	Chilopoda	X									
	Symphyla		X								
	Diplopoda		X								
	Pauropoda			X	X						
Onychophora		X									
Pogonophora							?				
Sipuncula							X				
Echiura							?				
Priapulida		X									
Tardigrada			X								

TABLE 3. MODES OF FEEDING / NUTRITION

Phylum	Class	Raptorial	Herbivorous	Omnivorous	Scavengers	Browsers	Direct deposit feeders	Indirect deposit feeders	Filter feeders	Ectoparasites	Endoparasites
Pentastomida											X
Phoronida									X		
Bryozoa									X		
Entoprocta									X		
Echinodermata	Asteroidea	X									
	Ophiuroidea	X		X							
	Echinoidea	X	X	X	X	X	?				
	Holothuroidea			X	X	X	X		?		
	Crinoidea								X		
Hemichordata	Enteropneusta			X		X					
	Pterobranchia								X		
Chordata (Urochordata)	Ascidiacea								X		
	Thaliacea								X		
	Larvacea								X		
Chaetognatha		X									

TABLE 3. MODES OF FEEDING / NUTRITION

Groups Phylum	Class	Raptorial	Herbivorous	Omnivorous	Scavengers	Browsers	Direct deposit feeders	Indirect deposit feeders	Filter feeders	Ectoparasites	Endoparasites
Mollusca	Gastropoda	X	X	X	X	X	X	X	X	X	X
	Monoplacophora		X	X							
	Polyplacophora		X								
	Aplacophora	X			X					X	
	Bivalvia								X		
	Scaphopoda	X					X		?		
	Cephalopoda										
Brachiopoda									X		

TABLE 4. MODES OF LIFE & LOCOMOTION

Phylum	Class	Sedentary: sessile	Sedentary: non-sessile	Tubicolous	Boring	Burrowing	Crawling	Walking-running	Swimming	Commensal	Parasitic
Annelida	Polychaeta		X	X		X	X		X	X	
	Oligochaeta			X		X	X			X	
	Hirudinea						X		X		X
Arthropoda	Merostomata					X			X		
	Chilopoda						X	X			
	Symphyla							X		X	
	Diplopoda					X		X		X	
	Pauropoda					X					
Onychophora							X	X			
Pogonophora				X							
Sipuncula					X	X					
Echiura						X					
Priapulida				X		X					
Tardigrada							X				
Pentastomida											X

TABLE 4. MODES OF LIFE & LOCOMOTION

Phylum	Class	Sedentary: sessile	Sedentary: non-sessile	Tubicolous	Boring	Burrowing	Crawling	Walking-running	Swimming	Commensal	Parasitic
Phoronida				X							
Bryozoa		X		X						X	
Entoprocta		X								X	
Echinodermata	Asteroidea						X				
	Ophiuroidea		X				X	?			
	Echinoidea				X	X	X	X			
	Holothuroidea		X			X	X				
	Crinoidea	X									
Hemichordata	Enteropneusta					X					
	Pterobranchia			X							
Chordata (Urochordata)	Ascidiacea	X									
	Thaliacea								X Planctonic		
	Larvacea								X Planctonic		
Chaetognatha									X		

TABLE 4. MODES OF LIFE & LOCOMOTION

Groups		Sedentary: sessile	Sedentary: non-sessile	Tubicolous	Boring	Burrowing	Crawling	Walking-running	Swimming	Commensal	Parasitic
Phylum	Class										
Mollusca	Gastropoda						X		X		X
	Monoplacophora						X				
	Polyplacophora		X				X				
	Aplacophora						X				
	Bivalvia	X	X			X	X		X	X	X
	Scaphopoda					X					
	Cephalopoda							X	X		
Brachiopoda		X	X			X					

When we examine modes of feeding (Table 3), two aspects are striking. First, there are versatile groups which are successful, with members capable of various modes of feeding. The Polychaeta, Gastropoda and Echinoidea fall under this grouping. Second, there are successful groups which have specialised modes of feeding, such as Hirudinea, Bivalvia, Cephalopoda and Asteroidea. The first group may be considered to represent success due to the use of adaptive radiation as an evolutionary process while the second symbolise success due to increase in efficiency and an evolutionary process. This kind of parallelism may also be drawn with reference to the sensory functions associated with feeding habits (e.g. Bivalvia, Cephalopoda as two representative groups).

Modes of life and locomotion (Table 4) are more complex and hence more difficult to interpret. However, it is evident that classes such as Polychaeta, Gastropoda and Echinoidea are versatile. In general, the versatile groups are more successful, adaptive radiation having played an important role. The Cephalopoda form an exception, its success having depended on a mode of locomotion, one of the most efficient in the animal kingdom, with concomittant development and efficiency of their eyes and tentacles (Wells, 1966).

Thus, it is self-evident that the maintenance of animal life demands continuous adjustment to changing conditions in the environment. Some of these changes will have no significant effect upon an animal's life. Others may influence its capacity to survive and reproduce, or at least to carry out its normal activities efficiently. To these it must adapt, which means that it must be able to respond to them in such a way as to promote its survival. An animal must therefore possess sensitive structures, receptors, which can be excited by an appropriate range of environmental stimuli to enable it to respond adaptively. The pattern of stimulus and response must have existed from the beginning, for living forms could not have survived without it. Presumably, it depended at first upon the disturbance by the environment of some basic property of photoplasmic organisation. This property is perhaps to be found in the nature of the membrane that forms the cell's surface. The significance of the membrane is seen readily in two types of cell that are intimately associated with animal behaviour - the nerve cell and the muscle cell. Their composition differs from that of the surrounding medium in having a higher concentration of potassium and lower concentration of sodium. In association with this, the membrane carries an electric charge and is polarised, the charge being the resting potential. An active control of potassium is a general feature of living systems and it is likely that polarisation of the surface membrane is a common, even universal, property of living matter. Thus, from an early stage of evolution the primary effect of environmental change appears to have been the creation of localised states of instability in surface membranes, involving changes in their tonic permeability. This would have brought about a flow of ions resulting in some measure of depolarisation. Membrane disturbances could certainly be conducted over cell surfaces, but the conduction effect is decremental with distance from the point of initial disturbance. This is a rudimentary mode of conveying information and is useful only in the acellular animals, the Protozoa, although the ciliate *Paremecium* apparently produces unequivocal action potentials (Eckert, 1972). In some protozoans, kinetodesmata and other intra-cytoplasmic fibrils, may play a part in conduction and coordination. If this is so, it is difficult to see how the conducting mechanism could be similar to polarisation changes at the cell surface. The fibrils could have evolved independently as a conducting mechanism, different from those of multicellular animals. The situation is not clear in the Parazoa (sponges) either. In them, some form of signal, but not action potentials, can evidently be transmitted from cell to cell as mechanical stimulation results in localised contractions of the body (Pavans de Ceccaty, 1974). Something more is needed to provide long distance signalling that is essential for the organised responses of the more organised Enterozoa and it is the special province of the nervous system and, to an extent, of the epithelial conduction systems (Mackie, 1965). This is believed to be absent from Parazoa, well established in Cnidaria and is a characteristic and familiar feature of the coelomates.

The nervous system has resulted from an important cellular specialisation, the neurones. It is able to translate excitation into coded signals (impulses), which it can propagate over long distances with decrement. Although neurones appear highly diversified, the common principle of their organisation is the possession of three components: a cell body, a dendritic zone, and an axon. Like other types of cell, the neurone must carry out vegetative functions also, these are mainly located in the cell body, which is responsible for the maintenance of the dendritic field and axon, and for the production of certain secretions, the neurohumours and the neurohormones. The dendritic zone is the receptor region of the cell. It may be specialised as a receptor structure, sensitive to environmental stimuli; or it may be part of a motor neurone or internuncial neurone. The axon, or nerve fibre, conducts nerve impulses away from the cell body. It is typically so specialised that when an impulse reaches it, it transmits a chemical signal across a barrier, the synapse. This signal may excite the dendrites of other neurones or effector structures, such as muscle or glands, that is, the agents of response. The axon endings of the neurosecretory neurons may be associated with blood vessels to form neurohaemal organs, which release the secretion into the blood. In the light of all this information, the role of the nervous system in the animal's behaviour and consequent reaction to its environment will be discussed. Emphasis is placed particularly on secretion since it plays a very important role in the lives of the animals dealt with here.

Among the groups considered, one may distinguish two types of nervous system: the "primitive" one which lacks an organised central nervous system (CNS) and the more advanced which has it. This shows that the CNS, which interconnects different pathways, is not an essential feature of neural organisation. A primitive nervous system exists in three groups of invertebrates: the Coelenterata (Cnidaria and Ctenophora), the Echinodermata, and the Hemichordata. A dominant feature of the primitive is that the nerve cells are arranged in one layer and form an irregular nerve net or plexus. Neural pathways become differentiated within this net, to varying degrees in the three groups. This may be regarded as foreshadowing the association of fibres into the macroscopically visible tracts that form the nerves in the most advanced systems. As this association develops, the net part of the system diminishes in importance, although it may still have some significance. Where a central nervous system is present, the neural pathways constitute the familiar reflex arcs of classical physiology. These pathways may be laid down during development, in which case they are genetically determined and are therefore independent of the experience of the animal. These form the neural basis of inborn reflex actions. They are characterised in general by the simplicity of their neural pathways. The inborn reflex responses are not restricted to animals with a CNS as shown in cnidarians (see Ferrero, this volume). Hence, we may extend the concept of reflex response to include the behaviour of animals with a primitive nervous system.

Reflex responses may be considered to be the unit components of behaviour. However, animal behaviour is commonly of much greater complexity. When inborn, it is expressed as instinctive behaviour, which consists of reflex actions that are unified or integrated into patterns of activity involving the whole organism. In contrast to simple reflex responses, instinctive behaviour is usually excited by a complex pattern of stimulation, termed a releaser. It is commonly associated with an internal drive which is manifested in seeking, or appetitive behaviour, in which the animal actively explores the potentialities of its environment instead of passively waiting, as it were, to be stimulated. Although the effect of drive is often clear, its physiological basis is not well understood and is probably intimately associated with sensory cues. It perhaps results from characteristic patterns of activity in the nervous system, and more particularly in the CNS, in the advanced systems.

How adaptive behaviour can be effectively organised out of a relatively small repertoire of reflex responses is illustrated in the behaviour of the polyplacophoran mollusc, *Lepidochiton cinereus*, as analysed by Evans (1951). Its response to the turning of a stone on the undersurface of which it is found is an adaptive behaviour and depends upon light, gravity and humidity. When the tide is out the animal is usually on the under-surface of the stone and when the stone is turned upside down in bright light, it creeps over to the surface which is now lowermost. This protects the animal from exposure. When exposed to bright light, the animals move at random, the rate of movement being more or less dependent upon the intensity of illumination. A locomotory reaction of this type, in which the stimulus brings about a variation in linear velocity, is called an orthokinesis (orthos, straight). *Lepidochiton* shows negative orthokinesis. As for humidity, chitons will not move across a dry surface but are active on a moist surface, much more than when completely submerged in water. It appears then that they are stimulated to move when they become exposed as the tide falls, so that they are likely to find protection before the substratum dries. In this they are aided by gravity, which varies according to whether they are immersed or exposed. A response like this in which movement is oriented with reference to the source of the stimulus, is termed a taxis; *Lepidochiton* is negatively geotactic. Survival in a constantly fluctuating environment would be almost impossible if behaviour depended solely upon rigidly determined patterns of activity, so that some degree of flexibility of response is also necessary. This is achieved by efferent pathways being linked with a wide range of incoming information. To this may be added the properties of synapses and other factors which also promote flexibility of neural action. Responses depend upon the internal state of the animal; thus the reactions of a hungry animal will differ from those of a well-fed one. Reactions will also vary with age and state of development; in particular, the state of sexual maturity. Here, the secretion of hormones is important, influencing the excitability of the nervous system.

Behaviour can be modified in the light of the previous experience of the animal. This is learning, which may be defined, following Thorpe, (1963), as that process which manifests itself by adaptive changes in behaviour as a result of experience. It is beyond the scope of this chapter to discuss its details but suffice it to say that learning takes many forms, which are inevitably limited by the structure and complexity of the nervous system. There is, however, much variation and flexibility in the behaviour patterns of animals and these reflect differences in the level of organisation of the nervous system. We shall briefly examine how the requirements of adaptive behaviour are met in the main groups mentioned above. There are two modes of approach. One may study one topic (responses to light or temperature) in a group of animals or, one may examine the range of reactions throughout one species alone. Both modes of approach are required and they need to be coordinated.

THE PRIMITIVE NERVOUS SYSTEM

As mentioned above, a primitive nervous system exists in the Coelenterata, the Echinodermata, and the Hemichordata. It is beyond the scope of the present chapter to discuss the first group (see Ferrero, this volume).

The movements of Asteroidea depend upon the properties of a hydro-static skeleton, and these properties are expressed at a high level of morphological differentiation and a more complex pattern of behaviour. When a starfish is crawling, most of its thousand or so tube-feet are so engaged, and the impression we get is one of considerable disarray, for there is no discernible phase relationship coordinating their movement into a common rhythmic pattern. Nevertheless, there is evidence of neural control. The feet show a common direction of movement, for all the feet, regardless of the arm to which they belong, point and step in a general direction. The direction of movement varies from time to time but one arm will always be the leading one. Variations in the general pattern of movement are brought about in response to stimulation. These variations may result in exchanges in the speed of stepping, from 3 to 10 steps per minute in *Asterias rubens* (Smith, 1950). They could result in changes in connexion with feeding or in direction. This could be related to neural organisation. The nervous system of the starfish is remarkable not only because it retains the form of a nerve net, but also because a major part of it remains in its primitive position within the ectoderm. The ectoneural nervous system as a whole includes the ectodermal receptor cells, together with the association pathways that are formed by the nerve cells in the plexus, and into which sensory information is conveyed. It is, therefore, a sensory system. In the Asteroidea the motor nervous system is partially represented by the hyponeural nervous system. Although these two systems are morphologically separable from each other, they come into close conjunction at certain points, notably in the region of the radial nerve cord and at the origins of the lateral motor nerves. Since radial symmetry has been secondarily adapted in the Asteroidea, their nerve net has either carried much farther, or regressed to the tendencies seen in the Coelenterata for the establishment of through-conduction pathways.

The combination of diffuse conduction and through conduction can be demonstrated by experimental manipulation of the animal (Smith, 1950). If a starfish is inverted over a glass cylinder, and the dorsal surface is stimulated with a probe, four patterns of response, resulting from the progressive spread of impulses, could be seen. These are: 1. movements of the pedicellariae and spines, the former opening and closing their valves. The maximum response is shown only in the vicinity of the stimulus, being absent even beyond a distance of 5 mm. This is reminescent of the response in some Cnidaria and appears to be the action of the ectoneural nerve net. 2. The extension and bending of the feet, that lie immediately underneath the site of the stimulus. If a short incision is made through the epidermis, and the stimulus applied immediately above it, the response is abolished. This indicates that the excitation from the nerve cord is being propagated along through-conduction pathways that run transversely in the deeper part of the ectoneural plexus. The failure of the excitation to circumvent the small incision shows that there is no spread along the length of the arm, so that the superficial plexus is in this case not involved. 3. Lateral protraction of the feet that lie distally and proximally to the point of stimulation. This response depends upon the propagation of impulses through the nerve cord, for if part of this is removed the response is abolished. 4. All the stepping feet throughout all the arms show an increased rate of stepping. This response is not limited to the site of the stimulus and further it is accompanied by a change in the direction of movement. If the axial nerve cord is cut in any one arm the feet distal to the cut no longer respond, while cutting of other parts of the nervous system has no effect. The importance of central control becomes apparent when the starfish is moving normally. If a single arm is isolated from the body and it has no connexion at all with any part of the nerve ring it will move predominantly with the base foremost. If, however, it retains even a small piece of the ring, then it will move with its tip foremost. The balance of central and peripheral control on the behaviour of the starfish as shown by these and other observations, is obviously of great adaptive value. The stimuli resulting from the normal contact of the feet with the substratum evoke no localised reflex responses, but favour continued movement. A brief dorsal stimulus affects only the immediately neighbouring foot, causing a temporary retraction if it is already protracted, probably saving it from any immediate harm. With prolonged stimulation the central control of movement comes into action in such a way as to bring about the removal of the animal from the source of possible injury. Thus, these responses fulfil the primary requirement of adaptation: the survival of the individual.

Although we know very little about the hemichordate nervous system it is certain that it is similar indeed to the ectoneural component of the echinoderm system. A peculiarity of the hemichordates is that their nerve cells lie entirely outside and above the plexus, with their nuclei at a lower level than those of the epidermal cells. The receptor elements of this system are primary sensory cells, with their distal fibres passing into the plexus. An intriguing feature of the hemichordates, and one that labels their nervous system as more primitive than that of echinoderms, is the absence of any clearly defined motor system. This may, however, be correlated with their relatively inactive, burrowing mode of life

(Table 4). Burrowing relies upon the passage of peristaltic waves backwards over the body surface, and particularly over the proboscis (Knight-Jones, 1952). These waves are effected by the propagation of excitation down the dorsal nerve cord, from the extreme anterior end. If the proboscis is cut so that only the dorsal nerve cord is left undamaged, the peristaltic waves continue. If the dorsal nerve cord is cut out at one point, but the whole of the rest of the proboscis is left intact, the waves are disrupted at the point of the cut. The ventral cord of the trunk, not the dorsal, is responsible for the retreating movement into its burrow.

The regular arrangement of the fibres of the plexus, and the collection of some of them into tracts or cords, is an expression of the establishment of definitely oriented pathways. Also, this system, despite its apparent simplicity, has powers of integration which ensure that the animal reacts with the behaviour pattern of a whole organism, and that responses are total and integrated. If an animal is stimulated halfway along its body, it will burrow or retreat but not both simultaneously. The neurocord does not appear to function as a CNS so that the problem is wide open for further study.

ANNELID NERVOUS SYSTEM

Since these animals move in one direction, and consequently are bilaterally symmetrical with the major receptor systems at the anterior end, there is a marked specialisation of the CNS at the forward end leading to (or resulting from) the process of cephalisation and the appearance of that morphologically and physiologically complex structure, the brain. This trend is already present in the Platyhelminthes. The further development of neural organisation can be well studied in Annelida, particularly in relation to metamerism which is the foundation of their locomotory mechanism (Dales, 1967). The CNS is compact in these animals with two ventral nerve cords, often fused. The receptors are usually bipolar sense cells, lying peripherally. They are particularly numerous on the parapodia of the nereid worms.

The mode of action of the metameric nervous system is illustrated in locomotion which depends on the integration action of a specialised nervous system. The integration of the segmental reflexes into the behavioural pattern of the worm is the function of the CNS. If the nerve cord is cut, leaving the animal otherwise intact, the continuity of movement of the whole body is interrupted but the regions anterior and posterior to the cut show normal locomotor pattern. This shows that the integration of the segments depends upon the propagation of impulses from segment to segment along the length of the CNS. If an earthworm is cut into two but left connected by the ventral nerve cord, peristaltic movements

continue in both portions in a coordinated manner. This can only
be due to the conduction of excitation through the nerve cord. This
can also be shown by suspending isolated pieces in a saline bath.
Tension or touch evoke peristalsis only if the nerve cord has been
left intact. Gray and Lissmann (1938) have shown that the segmental
locomotor reflexes are also dependent upon peripheral excitation
evoked by the stimulation of segmental receptors. A decapitated
earthworm suspended in water shows no peristaltic movements but
they appear if the worm is removed from the support of the water
into the air. The same happens if the worm is subjected to tension
while in water. In both situations movement is evoked by the
stimulus of stretching. Another important factor is tactile
stimulation applied to the ventral surface. A suspended worm will
show peristalsis when its body is in contact with the substratum,
but may cease to do so when it is removed from that contact. A
situation similar but complicated by a more specialised body with
suckers is seen in the leech. In the intact leech these movements
are regulated by excitation arising in the suckers by contact with
a substratum. Fixation of the anterior sucker to the substratum
is followed by a wave of contraction of the longitudinal muscles,
while activity of the circular muscles follows stimulation of the
posterior suckers. Still the suckers are not the only factors as
can be shown by their removal or denervation. Locomotor motions
continue as long as the leech is in contact with the ground. The
movements cease if the worm is lifted off the substratum by
passing threads underneath it. Here also, as in the earthworm,
tactile stimulation of the ventral surface is required if it is to
move. However, in the intact leech the suckers provide time signals
for the commencement of peristalsis. The leech also differs from
the earthworm in being adapted for both swimming and crawling. The
former calls for dorso-ventral undulations. The slow rhythms of
terrestrial locomotion are only shown when there is ventral tactile
stimulation. When this is removed the rapid rhythms of swimming
occur. In the earthworm stretch reflexes do not operate when the
worm is moving over a smooth surface but do so when it is on a
rough surface which offers resistance. In these circumstances both
tactile and stretch reflexes cooperate with CNS.

The cerebral (suprapharyngeal) ganglion is another puzzling
feature. An earthworm without it can eat, crawl and copulate but
is restless, overactive and burrows less efficiently. A polychaete
without a supraoesophageal ganglion is unduly active but does not
feed, burrow, and is insensitive to light and chemical stimuli.
The nervous connexions of these ganglia seem to make them important
sensory centres but we do not know how. In the leech also the
removal of the ganglion results in loss of muscular tone and a lack
of crawling. There is good evidence that the cerebral ganglion of
annelids secretes hormones, so our interpretation of the results
mentioned above must be cautious.

The study of learning capacity in annelids has shown that the cerebral ganglion is not as effective as the brain of higher animals in this respect. Acquired behaviour patterns persist in nereids and earthworms even after the severence of the cerebral ganglion from the rest of the CNS. One must also realise the difficulty of interpreting these results due to the effect the removal, or severance, of the cerebral ganglion may have on the inflow of information from cephalic sense organs. This could be lost to the animal. However, the peristomal cirri are connected to the ventral nerve cord and provide an exception. In prospective, the brain may be considered as primarily a sensory and integrative centre for reflex activities, whereas the acquisition of learned responses occurs throughout [the CNS in annelids as well as in Horridge's (1962) cockroach].

MOLLUSCAN NERVOUS SYSTEM

Adaptive radiation of molluscs is exemplified dramatically by their nervous system which ranges from what may appear to be a diffuse, decentralised system in the chitons to the highly evolved sensory and central nervous mechanisms in the cephalopods which are comparable with those of the arthropods and vertebrates. The absence of metamerism is an obvious factor in their neural organisation. However, there is a similar trend in the establishment of a ganglionic system with tendency to centralise. The condition of the nervous system reflects the modes of life of the main molluscan groups. A well-organised CNS may appear when its evolution is promoted by its high selective value. This condition does not occur in the Gastropoda and Bivalvia, which are essentially inactive or sedentary, e.g. deprived of any wide range of environmental stimulation. Further, the evolution of the protective shell has created a behavioural response of retreat rather than exploration when exposed to such stimulation. Their ganglia are centainly centres for reflective responses, involving mostly restricted sensory and motor activities, and yet, learning ability has been demonstrated in slugs (Gelperin, 1974) and *Aplysia* (Kendel, 1976). While the relationships of cephalisation to the life habits of gastropod is obvious, it is more striking in the case of the lamellibranchs which have no head, as their mode of life is sedentary and withdrawn. In correlation with this their nervous system is very simple. Both in Gastropoda and Bivalvia the fields of action of the ganglia are limited as demonstrated by experiments. If the pedal ganglia of *Aplysia* are removed there is increased tonus and contraction. If on the other hand, the cerebral ganglia are removed there is an increase in motor activity. This suggests that the pedal ganglia inhibit tonic contractions while the cerebral ganglia, like the brain of Annelida, inhibit locomotor activity. In Bivalvia the palps, otocysts and osphradia are controlled by the cerebropleural ganglia. The visceral ganglion innervates siphons, pallial sense organs and much of the mantle. The reactions of *Mytilus* to changes in the condition of water could be abolished by the removal of the visceral ganglion. In *Pecten* there is a wide departure from the typical bivalve mode of life, as it is capable of jet propulsion by rapid contractions of the so-called fast muscles. Here the visceral ganglia have come to form the largest single component of the nervous system. In the absence of the head, the visceral ganglia assume the dominant role while the cerebropleural ganglia are small and moved backwards.

The evolution of cephalopods has followed lines very different from those of other molluscs, the fundamental difference lying in their pelagic and predatory mode of life. Obviously, they must have started from a common ancestry with the more sluggish and more typical molluscs. The link is *Nautilus* which is quite different from the other living cephalopods. It has an external shell and its eyes are simple, lacking lens and iris, probably used for light detection only. Chemoreception would appear to be more important. Although, its CNS shows considerable cephalisation as in other cephalopods, it is comparatively simple with only three pairs of lobes connected by commissures to form a ring-shaped complex. Functional analysis of these will be interesting in relation to their behaviour. The nautilus and squid have received little attention by comparison with the octopus. The modern forms can be held easily in laboratories and have been better studied. While the reduction of the shell is a commitment to pelagic life the secretive and retiring habits of octopus make it an exception. Associated with the speed and freedom of movement of these is their well-developed vision and the dramatic ability to change colours. The complex organisation of the CNS is remarkable. It forms a concentration around the oesophagus referred to as the brain. Removal of the basal lobes hinders integrated movements. There are certain areas of the supra-oesophageal region that are called "silent areas". These are concerned with memory and learning (Young, 1961; Wells, 1962). Octopus is very capable of acute visual and tactile discrimination as seen by classical conditioning experiments. It has a good memory but for all its acuity, the visual discrimination has marked limitations as regards shapes. This may be related to the structural peculiarity of the eye. Its tactile sense has also been studied and found unable to distinguish surface irregularities. The octopus cannot learn to distinguish objects by their weights not because it lacks proprioceptors but because the movement of tentacles is regulated by axial ganglia in the arms. It cannot judge the position of its mechanoreceptors and consequently is unable to estimate the surface patterning of an object. Despite its intelligence therefore, it is unable to use its tentacles in new manipulations skills. This peripheral restriction of proprioceptive information is a consequence of soft bodies. Animals with exoskeletons can readily provide for central monitoring of the positions of body parts.

GIANT NERVE FIBRES

These were first described in 1836 in the CNS of Crustacea. They have since been found to be of widespread occurrence being found in Cestoda, Nemertea, Annelida, Arthropoda, Mollusca, Hemichordata and Pisces (Bullock and Horridge, 1965). They may reach a diameter of 700 µm as compared with 20 µm for a typical vertebrate nerve fibre, and some are myelinated. Since these have arisen independently on a number of occasions, they must have a considerable adaptive value since they permit rapid propagation of impulses. They are particularly concerned in reactions such as instant retreat from danger of a harmful environment. This has been shown in the earthworm in which cutting the giant fibres abolishes the rapid end-to-end contractions that constitute its protective response.

Giant fibres are well developed in cephalopods. Those in squid have been well studied; the contractile response that results from stimulation of the giant fibres causes water to be forced out of the mantle cavity propelling the animal backwards. Thus, they are of importance in preserving the life of the animal by their speed of conduction. The represent an extreme demonstration of the selection pressure which has favoured increased efficiency of nerve conduction. For comparisons, fine fibres conduct at about 0.025 m/sec; lateral giant fibres at 7 m/sec, and the median giant fibres at 17 m/sec to 25 m/sec. The difference is partly due to the elimination of synaptic barriers and partly due to the increase in diameters. Conduction speed is improved by the presence of myelin sheath around the fibre as shown in the vertebrates, most of which do not have giant fibres.

NEUROHUMOURS

As mentioned earlier, the nerve tissue has two distinct yet interrelated functions: propagation of nerve impulses and the synthesis and discharge of secretions. The latter function is of an equal importance in the groups we are dealing with. The neurohumours or chemical transmitter substances bring about the conduction of nerve impulses across synaptic junctions or muscle end-plates. These are formed in the nerve cell body and are passed down the axon and stored at its terminal. Neurohumoral function has been attributed to a host of substances: acetylcholine, noradrenaline, adrenaline, dopamine, octopamine, hydroxytryptamine, aputamic acid and GABA (Gerschenfeld, 1973; Walker, 1977). Acetylcholine provides the most convincing evidence for neurohumoral action in the invertebrates. It is known to be present in the Protozoa and in all of the main invertebrate groups except the Porifera, Coelenterata, and Urochordata. Whether it is physiologically active in a wide range of coelomates is still very uncertain.

The best evidence comes from work with Bivalvia. It exerts an inhibitory effect on the heart of *Venus mercenaria*. However, much work needs to be done. Information regarding the distribution of the catecholamines, noradrenaline, adrenaline and dopamine in the invertebrates is so scarce that it is difficult to evaluate their possible physiological role. Dopamine, however, appears to be present in significant amounts in a variety of invertebrates, including coelenterates, and may thus be considered a physiologically active transmitter in the sensory systems and CNS of many forms. 5-hydroxytryptamine, an indole alkylamine, is widely distributed in the nervous systems of Annelida, Crustacea, and certain Mollusca. It can influence the heart beat and other activities of these animals, and influence the proprioceptors of the legs of Crustacea. In the crustaceans the identity of this substance is doubtful but it is certain that they possess amines. It is clear that chemical transmission is an aspect of neural organisation in invertebrates that needs much further investigation.

NEUROHORMONES

In addition to the "ordinary" nerve cells there is a second type that differs from them in containing a secretory product that is readily stainable and could be observed by light microscopy. This product is known as a neurosecretion. This also is elaborated in the cell body and passed down the axon and released from the axon endings. The basic difference is that it is not restricted to local and transitory action but instead passes into the blood stream and circulates in the body, producing specific physiological effects at points even remote from its origin. The cells producing it are neurosecretory cells and their function is to secrete neurohormones which form part of the endocrine secretions of the body. The significance of this is that it extends endocrine activity to the nervous system, thus introducing a mechanism that converts a neural signal into a chemical one. Neurohumours are different from neurohormones because they act locally. Substances such as noradrenaline may act as both in vertebrates but this is perhaps an exception. The neurohormones have been extensively studied in the invertebrates since they are the predominant hormones in them (Highman and Hill, 1977). In the gastropod *Aplysia*, an egg-laying hormone of protein nature is present in the so-called neurosecretory "by cells". Presumably, sensory stimuli associated with copulation are transmitted to the cerebral and pleural ganglia which in turn activate the bag cells. The latter respond by repetitive spike activity and hormone release so that shedding of oocytes is synchronised with sperm transfer from the mating partner (Kupperman and Kandel, 1974; Highman and Hill, 1977). In the annelids regulation by neurohormones has been reasonably well studied thus permitting some generalisations. They have neurosecretory cells in their cerebral ganglia. Four types of such cells exist in

Nereis, three in *Lumbricus*, and the leech *Theromyzon rude*. Evidence for this is not unequivocal except for one type in *Nereis*. Part of the information has come from studies of growth and regeneration in nereids which showed that these depend upon a neurohormone produced by the cerebral ganglion and that their decline results from a reduction in this secretion. Sexual maturation is also under hormonal control, as shown by the phenomenon of epitoky (Clark and Olive, 1973). Fertilisation may be carried out individually as by the non-epitokous *Nereis diversicolor* in which the male enters the burrow of the female or it may spawn at the surface. Epitoky is a sophisticated device for securing the same end. A further condition for success is that large numbers of individuals shall become sexually mature simultaneously as in eunicid, nereid and syllid polychaetes. The three groups show how natural selection, using common potentialities, leads to parallel development of adaptive mechanisms. Here the unifying influence is the capacity for neurosecretory regulation of sexual maturation.

It is known now that this regulation operates in conjunction with environmental cues. Not surprisingly, in view of the lunar periodicity which is a feature of this process, this factor proves to be the intensity of moonlight. It is correlated with the neurosecretory activity of cerebral ganglia of members of a population. Evidence comes from studies with *Eunice fucata* which can respond to weak moonlight. Immature worms are photonegative when light intensity is higher than 0.1 lux and are confined to burrows at full moon. When mature, their epitokal regions are photopositive at intensities above 0.05 lux stimulating them to swarm at quarter moon. In *Platynereis dumerilii* the spawning maximum is around new moon. Sexual maturation can be artificially induced by exposing individuals to varying photoperiods corresponding to those of the lunar cycle after a critical stage of development has been reached. It seems also possible to imprint worms to cycles of lunar photoperiodicity so that the effect of the cycles is manifested after an interval of time. In this way individuals of *P. dumerilii* can be induced to spawn synchronously by artificial exposure to appropriate cycles of photoperiod. This response may be shown even if the treatment is terminated before maturity, and constant illumination substituted for the photoperiod. The effect of extrainment can last up to three months. In all this the cerebral hormone appears to be the regulator. Here also considerably more work needs to be done not only with the other groups but even the annelids, as evidence is still fragmentary. Alleged neurosecretory cells have been described in a wide range of other invertebrates. It cannot also be assumed that invertebrate hormonal mechanisms must be exclusively or mostly neurosecretory ones. For example in the octopus, control of the maturation of the gonads operates through the optic receptors and neural pathways, with no participation of neurosecretory mechanisms (Wells, 1976). From the available evidence it appears that the early evolution of endocrine systems depended on the secretory capacity of the nerve cell, and this may have antedated the development of epithelial endocrine cells which are still unknown in annelids and are of minor importance even in crustaceans and insects. Neurosecretory phenomena are a feature of the annelid-arthropod branch; there is also acceptable evidence of neurosecretion in arachnids and myriapods. It is also intriguing that in this respect the annelid-arthropods resemble the vertebrates more than their fellow deuterostomes do. Although more evidence is necessary, one may speculate that the neurosecretory systems in the vertebrates and annelid-arthropods must be the result of convergent evolution.

Chemical coordination systems play also an intraspecific role. This could have been the earliest form of chemical communication. The metabolic products released from the body must have affected neighbouring individuals. If this were beneficial, it would have entailed selective advantage, leading to further elaboration of those products and enhancement of sensitivity to them. We saw examples of this in the regulation of spawning where in addition to the factors mentioned, chemical signals may have been exchanged between the animals. One example is *Nereis succinea* in which a secretion from the mature eggs and gravid females induces the males to spawn, the presence of sperm then inducing the females to spawn. Another case is that of the female oysters, in which spawning is evoked by some substance present in the testes and sperm. This leads to a chain reaction of spawning of the females, inducing other males to do so and so on (Galtsoff, 1961). These substances are known as gamones. Our knowledge of this type of communication is very scarce except for the sex attractants of insects.

There may also be substances which promote interspecific communication or relationship. This is particularly so in the aquatic environment which permits more easily the exchange of materials. An extreme case of such a relationship is the stimulation of spawning in sea urchins by the spring phytoplankton bloom, the cue for spawning being apparently some substance associated with, or released by the phytoplankton (Himmelman, 1975). The adaptive advantage here is to obtain optimal food availability and temperature conditions for the planktotropic larvae. These substances may also be metabolites and may be critical in maintaining inter-specific relationships such as symbiosis and parasitism, where chemical means are one of the devices by which partners or hosts and parasites are bought together. Very little is known about this but it may have commonly resulted from one species having sensed and used for its benefit the metabolite of another species, which then became its partner or host.

ACKNOWLEDGEMENTS

I thank Dr. Michel Anctil, Professor J.F. Case, Mr. Roger Croll and Professor P.K. Menon for criticising an earlier version of this chapter and offering improvements. I alone am, however, responsible for its present shortcomings. I also thank Miss Margaret Pertwee for assistance in the preparation of the final version.

REFERENCES

Barnes, R.D. (1974). Invertebrate Zoology. 3rd. ed. Saunders Philadelphia, 870 p.
Barrington, E.J.W. (1967). Invertebrate Structure and Function. Nelson, London, 549 p.
Boolootian, R.A. (ed.) (1966). Physiology of Echinodermata. Interscience Publishers, New York. 822 p.
Bullock, T.H. and Horridge, G.A. (1965). Structure and Function of the Nervous Systems of Invertebrates. Vols 1 and 2. Freeman, San Francisco, 1719 p.
Clark, R.B. and Olive, P.J.W. (1973). Recent advances in polychaete endocrinology and reproductive biology. Oceanogr. Mar. Biol. Ann. Rev. 11 (Cited in Highman and Hill, 1977).
Dales, R.P. (1967). Annelids. Hutchinson Univ. Lib., London, 200 p.
Daley, J.M. (1973). The ability to locate a source of vibrations as a prey-capture mechanism in *Haemothoë inbricata* (Annelida Polychaeta). Mar. Behav. Physiol. 1: 305-322.
Eckert, R. (1972). Bioelectric control of ciliary activity. Science 176: 473-481.
Edwards, C.A. and Lofty, J.R. (1972). Biology of the Earthworms. Chapman and Hall, London, 283 p.
Evans, F.G.C. (1951). An analysis of the behaviour of *Lepidochitona cinereus* in response to certain physical features of the environment. J. Anim. Ecol. 20: 1-10.
Field, L.H. and MacMillan, D.L. (1973). An electrophysiological and behavioral study of sensory responses in *Tritonia* (Gastropoda, Nudibranchia). Mar. Behav. Physiol. 2: 171-185.
Galtsoff, P.S. (1961). Physiology of reproduction in molluscs. Am. Zool. 1: 273-289.
Gardiner, M.S. (1972). The Biology of Invertebrates. McGraw-Hill, New York, 954 p.
Gelperin, A. (1974). Olfactory basis in homing behavior in the giant garden slug, *Limax maximus*. Proc. Nat. Acad. Sci. U.S.A. 71: 966-970.
Gerschenfeld, H.M. (1973). Chemical transmission in invertebrate nervous systems and neuromuscular junctions. Physiol. Rev. 53: 1-119.
Gray, J. and Lissmann, H.W. (1938). Studies in animal locomotion. VII. Locomotory reflexes in the earthworms. J. Exp. Biol. 15: 506-517.

Highman, K.C. and Hill, L. (1977). The Comparative Endocrinology of the Invertebrates. 2nd. ed. Edward Arnold, London, 357 p.
Himmelman, J.H. (1975). Phytoplankton as a stimulus for spawning in three marine invertebrates. J. Exp. Biol. Ecol. 20: 199-214.
Horridge, G.A. (1962). Learning of leg position of the ventral nerve cord in headless insects. Proc. R. Soc. Lond. B157: 33-52.
Jahan-Perwar, B. and Fredman, S.M. (1976). Chemoreception in *Aplysia*. In Neurobiology of Invertebrates, Gastropoda Brain. Ed. J. Salanki. Akad. Kiadó, Budapest, p. 511-524.
Kandel, E.R. (1976). Invertebrate nervous systems and the mechanisms of behavior. In The Nervous System, Vol. 1, The basic neuro-sciences. Ed. R.O. Brady. Raven Press, New York p. 663-669.
Knight-Jones, E.W. (1952). On the nervous system of *Saccoglossus cambrensis*. Trans. R. Soc. Lond. B236: 315-354.
Kohn, A.J. (1961). Chemoreception in gastropod molluscs. Am. Zool. 1: 291-308.
Kupfermann, I. and Kandel, E.R. (1970). Electrophysiological properties and functional interconnections of two symmetrical neurosecretory clusters (by cells) in abdominal ganglion of *Aplysia*. J. Neurophysiol. 33: 865-876.
Mackie, G.O. (1965). Conduction in the nerve-free epithelia of siphonophores. Am. Zool. 5: 439-453.
Pavans de Ceccaty, M. (1974). Coordination in sponges. The foundations of integration. Am. Zool. 14: 895-903.
Smith, J.E. (1950). Some observations on the nervous mechanisms underlying the behavior of starfishes. Symp. Soc. Exp. Biol. 4: 196-220.
Thorpe, W.H. (1963). Learning and Instinct in Animals. 2nd ed. Methuen, London, 558 p.
Wald, G. and Rayport, S. (1977). Vision in annelid worms. Science 196: 1434-1439.
Walker, R.J. (1977). Putative transmitters in invertebrates. Biochem. Soc. Trans. 5: 841-844.
Wells, M.J. (1962). Brain and Behavior of Cephalopods. Heineman, London, 171 p.
Wells, M.J. (1966). Cephalopod sense organs. In Physiology of Mollusca. Eds. K.M. Wilbur and C.M. Yonge, Vol. 2. Academic Press, New York, p. 523-545.
Wells, M.J. (1976). Hormonal control of reproduction in cephalopods. In Perspectives in Experimental Biology. Ed. P. Spencer Davies, Vol. 1: Zoology. Pergamon Press, Oxford, p. 157-166.
Wilbur, K.M. and Yonge, C.M. (Eds.) (1964). Physiology of Mollusca. Vol. II. Academic Press, New York, 645 p.
Wood, L. (1968). Physiological and ecological aspects of prey selection by the marine gastropod *Urosalpinx cinerea* (Prosobranchia: Muricidae). Malacologia 36: 267-320.

Yingst, D.R., Fernandez, H.R. and Bishop, I.G. (1972). The spectral sensitivity of a littoral annelid, *Nereis mediator*. J. Comp. Physiol. 77: 225-232.

Young, J.Z. (1961). Learning and discrimination in the octopus. Biol. Rev. 36: 32-96.

ECOSENSORY FUNCTIONS IN INSECTS

(WITH REMARKS ON ARACHNIDA)

M. GOGALA

Department of Biology, Biotechnical Faculty and

Institute of Biology, University of Ljubljana, Yugoslavia

INTRODUCTION

The large number of insect species and their adaptations to the diversity of habitats and modes of life present an inexhaustable source for the study of ecophysiological adaptations in general and ecosensory functions in particular. According to the data from Borror and DeLong (1971) given in Romoser's (1973) handbook there are over 700,000 insect species described, with the biggest systematic groups being Coleoptera (290,000), Lepidoptera (110,000), Hymenoptera (100,000), Diptera (90,000), Hemiptera (55,000) and Orthoptera (20,000). All the other groups do not exceed 5,000 known species. Even the evaluation of ecosensory functions of bigger groups of insects such as Coleoptera, Hymenoptera or Diptera would require a lot of work and perhaps all the pages in this book. So the only possibility to remain in the reasonable limits of a chapter is to restrict ourselves to certain modalities and certain cases of sensory adaptations which seem to be typical of insects. Because of many similarities between the two groups there are some remarks added also for Arachnida, another important group of Arthropoda, which was almost as successful in the evolution as Insecta (estimated number of species 30,000).

GENERAL CHARACTERISTICS

In the vast number of animals belonging to different groups of insects we still can find some common properties, which hold for most if not for all insects and which are of great importance for their sensory relationship to the environment. Such common properties are from my point of view the following:

a) chitinous exoskeleton
b) relatively small physical dimensions
c) aerial locomotion
d) relation to the evolution of the higher plants

The first two properties are common also to the arachnids while the latter two are not common even for all of the insects, but yet typical for most of them.

Chitinous exoskeleton determines the general form and organisation of sensory organs, sensillas being in principle the "channels" through the cuticle with auxiliary chitinous structures, which are important in the transduction of stimuli. Characteristics of the cuticular layer with extremely wide range of mechanical properties (e.g.: possible extension 1,3 - 1500% - Neville, 1975) were studied by Barth (1969;1970 a) also in Arachnida. These properties are especially important for the reception of mechanical stimuli and for the production of vibrational communicative signals (see below). Chitinous material can be important also in photoreception and in other relationships to the photic environment, because it can be highly transparent (Neville, 1975) with suitable refractive properties to be a functional part in photoreceptive organs (Carricaburu, 1967; Seitz, 1969) or can have extremely high optical density because of the pigment deposits in the cuticle. Good mechanical properties enable also the evolution of very elaborate structures, which give rise to the intense structural colours in many Lepidoptera, Coleoptera and others and which have important etho- and ecological functions (Fig. 5) (see below: u. v. patterns! Neville, 1975; Fox and Vevers, 1960). The electrical properties of the insect exoskeleton, also important for the sensory relationship to the environment, will be covered in the last part of this chapter.

Most of the adult insects range in size between 1 mm and 1 cm, some bigger ones reaching the next size class till 1 dm with only few being bigger than 1 dm or smaller than 1 mm. This range of physical dimensions which holds also for Arachnida, enables the insects to intrude habitats not available to bigger animals, but it has also some interesting effects on the sensory relationships, like the development of the near field sound communication in some groups, e. g.: Diptera (Bennet-Clark, 1971; 1975; see also Michelsen, this volume).

Another common property of the pterygote insects is flight. In this connection again some special adaptations of sensory organs have evolved, important for successful orientation (mechanoreceptors, photoreceptors) and for the regulation of motor activity during flight (Gewecke, 1974). Arachnida are in general not airborne, but some of them use the air streams for a special kind of locomotion (small spiders), or to catch prey (net building spiders). Both Chelicerata and Insecta use the anemomenotactic orientation as an important mechanism in their life (Linsenmair, 1968; 1973). Anemoreceptors and vibroreceptors are therefore important class of sensory organs in Arachnida and Insecta.

The evolution of many insects was closely related to the evolution of the higher plants. This factor has been very important in the development of the sensory organs, which are involved in this relationships and often tuned to the signals from the plants. Among the best examples are again the u.v. signals from flowers and u.v. sensitivity of pollinators (Daumer, 1958; Frisch, 1965; Mazokhin-Porshnyakov, 1969; Eisner et al., 1969). Different chemical substances from the plants can serve as attractants or repellents for the insects or even as a source of their own pheromones (Priesner, 1973).

Before we pay attention to the different groups of insects, their sensory abilities and some cases of interesting ecosensory relations among insects and their environment, we should briefly review the types of sensory organs present in a generalised insect.

There is a characteristical set of different cuticular sensilla, which can be scattered on the body surface or grouped to form more or less compact sensory organs. They can be all probably derived from primitive hair or thorn like sensilla. Examples of such sensilla are sensilla trichodea, s. campaniformia, s. placodea etc.

Other ubiquitous types of sensory organ in insects are scolopophorous or chordotonal organs, built up from specialised sensory and supporting cells. Those units can be stretched between two parts of the exoskeleton (amphinematic) or attached to the cuticle only on one side (mononematic). Scolopophorous sensilla usually occure in bundles, forming with accessory structures the complex organs like Johnston's organ, subgenual organs etc.

The third type of sensory organs are the photoreceptive organs, in a typical case organised as compound eyes and ocelli (imaginal and larval - stemmata).

In table 1, the types of sensory organs are listed according to the modalities they deal with. More information on the structure, function and distribution of different sensory organs in insects (and Arachnida) can be found in specialised entomological literature (Romoser, 1973; Eidmann, 1970; Kaestner, 1956; 1972; Rockstein, 1974; Dethier, 1963; Wigglesworth, 1965; Mazokhin-Porshnyakov, 1977).

Table 1. Sense organs in insects (I) and arachnids (A). Based on information obtained mostly from Kaestner (1956; 1972), Eidmann (1970), Romoser (1972).

Modalities	Types of receptor organs

Chemoreception:
  contact (taste)
  distant (olfaction)
	I.) sensilla trichodea, s. styloconica, ...
	s. trichodea, s coeloconica, s. ampullacea, s. basiconica, s. placodea, ...
	A.) s. trichodea, s. ampullacea,...

Hygroreception:	I.) s. basiconica, coeloconica, s. trichodea (?), ...
Thermoreception:	I.) bimodale sens. organs, cuticular patches (IR detectors ?)
Mechanoreception: contact (tactile sense, statorec., rheorec., propriorec., vibrorec.), distant (hearing)	I.) s. trichodea, trichobothria, s. chaetica, s. squamiformia, s. campaniformia, scolopidia, stretch rec. compound organs: statocysts, hair plates, Johnston's organ, subgenual, tympanal o, ... A.) s. trichodea, s. ampullacea, trichobothria, (pseudostigmal o.), slit sensilla → lyriform o., comb like o. (Scorp.), malleoli (Solif.)
Photoreception:	I.) compound eyes, stemmata, ocelli A.) ocelli (median, lateral)
Electroreception, Magnetoreception:	unknown
Unknown:	Haller's o. (Ixodides)

Before we proceed with the closer description of some examples of ecosensory relationships we should overview the habitats in which the members of insect orders live. Such a review gives of course a very superficial information but still enables us to make some general statements on the ecology of different insect groups which is of interest in the evaluation of their sensory capabilities. From

Table 2. Habitats of larval (l) and adult (a) insect and arachnid stages, the systematic order followed is that of Grzimek (1969; 1971) and information is from various sources.

	Aquatic			Terrestrial			
	Freshw.	Marine	Amph.	Land	Subterranean	Aerial	Paras.
	A	B	C	D	E	F	G
INSECTA							
Protura							
Collembola		(+)	+	+	+		
Diplura							
Thysanura							
Pterygota							
Ephemeroptera	+l			+a		+a	
Plecoptera	+l			+a		+a	
Odonata	+l			+a		+a	
Orthopteria:			+	+		(+a)	+
(Grylloblattaria)							
(Saltatoria)							
(Phasmida)							
(Dermaptera)							
(Diploglossata)							
Blattia:			+	+	+	+a	
(Blattaria)							
(Mantodea)							
(Isoptera)							
(Zoroptera)							
Embioptera				+		(+a)	
(Psocia:)							
Psocoptera				+		(+a)	
Phthiraptera				(+)			+
(Mallophaga)							
(Anoplura)							
Thysanoptera				+			
(Hemipteria:)							
Heteroptera	+	(+)	+	+	(+)	+a	+
Homoptera				+		+a	
(Coleopteria:)							
Coleoptera	+	+	+	+	+	+a	
Strepsiptera				(+)		(+a)	+
(Neuropteria:)							
Megaloptera	+l			+a		+a	
Raphidides				+		+a	
Planipennia	+l			+		+a	

Table 2 cont.

| | Aquatic | | | Terrestrial | | | |
	Freshw.	Marine	Amph.	Land	Subterranean	Aerial	Paras.
	A	B	C	D	E	F	G
(Mecopteria:)							
Mecoptera				+		+a	
Trichoptera	+1			+		+a	
Lepidoptera	+1			+	(+)	+a	
Diptera	+1	+1	+	+	+	+a	+
Siphonaptera				(+)			+
Hymenoptera	(+)			+		+a	+
ARACHNIDA							
Scorpiones				+			
Pedipalpi				+	+		
Palpigradi				+	+		
Araneae	+	(+)	+	+	+	(+)	
Ricinulei				+	+		
Pseudoscorpiones				+	+		
Solifuga				+			
Phalangidae				+			
Acarinae	+		+	+	+		+

table 2 we can see that the only major habitat which has not been occupied by the insects is the marine one, with the exceptions like some Coleoptera and Diptera in the upper littoral and the marine water strider *Halobates*. Unfortunatelly almost nothing is known about sensory specialisation of these animals, but the critical physiological and ecological factors for the survival of marine forms were probably other than perceptual (respiration, osmoregulation). There are some extreme habitats, which are not separately presented in the table, like high altitude environment, tissue of the living plants, which are simply covered under terrestrial environment in this review. Also in these cases little is known about sensory adaptations of the animals, specialised to such extreme environments and so it would be worth to investigate them.

One fact, which can be seen from these data (table 2) is the ecological diversity or unity of different insect orders. There are typical polyvalent groups like Coleoptera, Diptera and Heteroptera and other more uniform specialised groups like Psocoptera, Thysanoptera or even bigger groups as Homoptera, Neuroptera or Lepidoptera. This superficial uniformity of this group hides a vast variety in the mode of life, feeding behaviour and daily period of activity, which can have a pronounced influence on the sensory adaptations of those insects.

SPECIAL ADAPTATIONS

In the following part of this chapter, I would like to present some examples of ecosensory adaptations typical for the insects and their relation to the environment. It should be by no means taken as a complete coverage but rather as an individual choice of examples!

Photoreception

First we should discusse the photoreceptors and their abilities in relation to the light stimuli available in the environment and important for the insects. As we know from the numerous investigations from last decades one characteristic feature of insect spectral sensitivity is a high sensitivity for the u.v. light (Autrum and Thomas, 1973; Rockstein, 1974; Mazokhin-Porshnyakov, 1969). Spectral sensitivities, measured electrophysiologically by means of mass (ERG) or single cell recordings show often the double peaked curves with a main peak near 500 nm and the second peak near 350 nm (Fig. 1 b). Such a spectral sensitivity can be due to the presence of single (Autrum and Zwehl, 1964) or double peaked (Burkhardt, 1962) receptors in the eyes. It means that the insects must have some profit from high u.v. sensitivity.

If we compare the spectral sensitivity curves with the daylight spectra, we should first mention Dartnall's criticism on the use of $dN/d\lambda$ spectra or even $dE/d\lambda$ spectra, available in different handbooks of physics or meteorology (Dartnall, 1975). According to the techniques, used in electrophysiological determination of spectral sensitivity using filters with more or less constant width of transmission bands we prefer for this comparison the $dN/d\lambda$ light emission curves. Because of the topic organisation of the insect eyes the insects are unlike other animals (Dartnall, 1975) usually looking at the sun and the sky. So there are good reasons to compare the spectral curves of sensitivity with the spectral curves of the direct sunlight and the emission spectra of the sky (Fig. 1).

The u.v. part of the sunlight spectrum, measured at the sea level represent a relativelly weak band of total irradiancy. On the contrary, the secondary emission of the "blue" sky shows pronounced peaks in blue and u.v. part of a spectrum because of a scattering effect in the atmosphere (Henderson, 1970). And to this source of light are adapted photoreceptors of most insects and many arachnids at least in the dorso-frontal part of the eyes. The highest sensi-

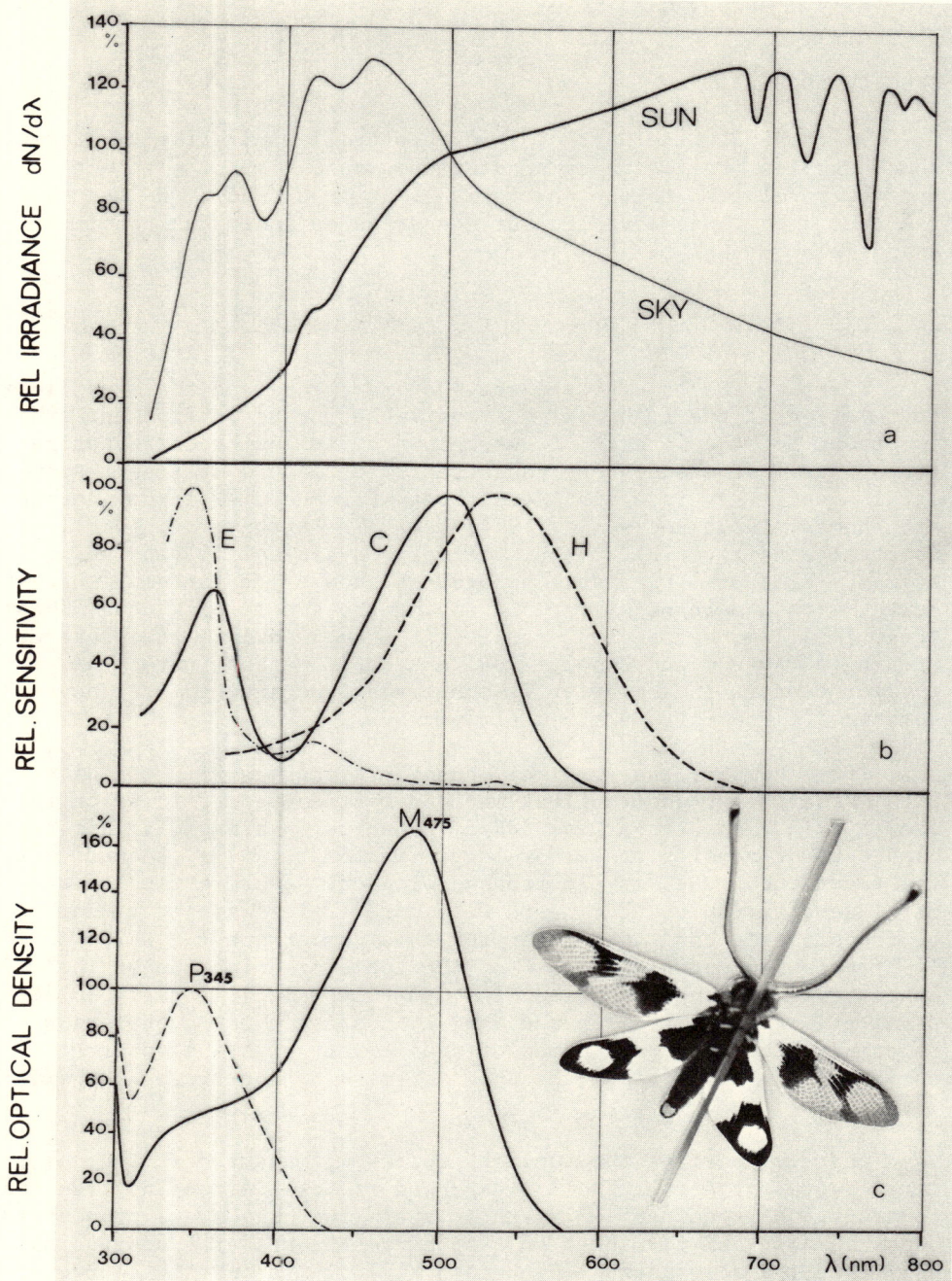

Figure 1

tivity to u.v. or blue light is usually found in this part of the eye, which is usually exposed to the direct illumination of the sky-light. The best examples for this statement are the insects with the so called double or divided compound eyes like *Ascalaphus* (Gogala, 1967), *Bibio* (Burkhardt and Motte, 1972), *Ephemerella* (Fig.2) and others. One reason for such distribution of the u.v. receptors is in the importance of the u.v. rays in the orientation of insects by the so called sky compass (Frisch, 1965; Wehner, 1976). The detection of the polarised light patterns on the sky, so important in successfull orientation of insects is for different reasons confined to the specialised u.v. receptors (Wehner, 1976). Of course, there must be some other reasons for such exclusive u.v. sensitivity in the dorso-frontal part of the eye as in the case of *Ascalaphus*. The possible explanation, proposed by Mazokhin-Porshnyakov (1959) and Gogala (1967) is in the uniform background for detection of nearby objects, caused by strong scattering of this wavelength of light. In predatory animals like Odonata or Ascalaphidae this can be an important priority. Another factor, which also improves the detection of prey e.g. small insects flying in the air, is visual acuity, it being higher with the same optical organization of ommatidia with light of shorter wavelengths.(For a discussion on the factors influencing the detection of moving targets see Horridge, 1977).

Is it the explanation for the distribution of u.v. receptors in *Ephemerella* males in the detection and recognition of sexual partners? They dance in the twilight, where the relative intensity of the short wavelength is high, but the absolute intensity of the light is low (Henderson, 1970). There are many other examples of divided compound eyes in insects (Simuliidae, Blepharoceridae and others) but also in the cases where morphologically a clear division does not exist between two parts of the eye there is at least a

Fig. 1. Comparison of the spectral irradiances of the daylight (a) with the spectral sensitivities of the eyes (b) and with the visual pigment absorbance curves (c). a) Spectral irradiances of the sun and the sky, presented as relative photon flux per wavelength interval. Curves normalised at 500 nm. Data derived from Moon (1940) and Hess (1939) in Henderson (1970). Sun: h = 30°, sky: 90° from the sun; sea level. b) Relative spectral sensitivity curves of the mayfly frontal compound eyes (*Ephemerella sp.*, male) (E), compound eyes of the blowfly (*Calliphora erythrocephala*) (C), and of a man (photopic spectral sensitivity curve) (H). c) Spectral absorbance curves of the ultraviolet visual pigment P_{345} of the owlfly (*Ascalaphus macaronius*) and its photoproduct M_{475}, which is stable and photoreconvertable.

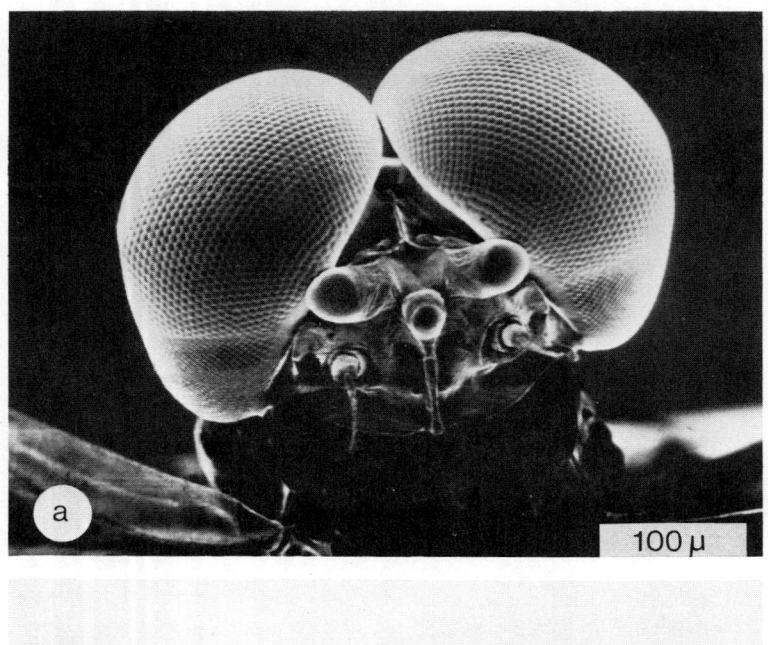
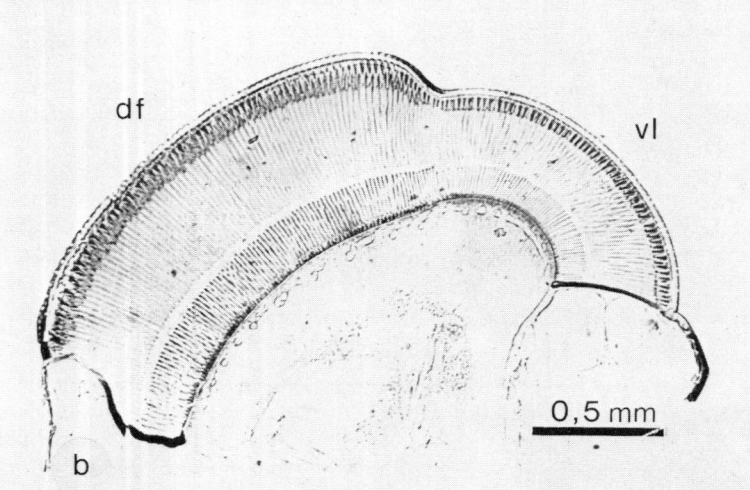

Fig. 2. Insect divided eyes: a) scanning electron micrograph of the head of *Ephemerella sp.* (male) with the big divided compound eyes and the three ocelli. b) Micrograph of a section through the divided compound eyes of *Ascalaphus macaronius*. df. - u.v. sensitive dorso-frontal part, vl - ventro-lateral part of the eye with the main peak of the spectral sensitivity in u.v. and the secondary peak near 520 nm (Orig.).

Fig. 3. U.v. patterns of flowers and insects. a,b) *Hieracyum silvaticum*; c,d) *Gonepteryx rhamni*. a,c) "normal" photography (Agfa Ortho film, filter GG-400, Schott, Mainz). b,d) u.v. photography (Agfa Ortho film, filter UG-1, Schott, Mainz; electronic flash) (Orig.).

functional specialisation of different parts of the compound eye, which is often reflected also in the overall spectral sensitivity of this parts of the eyes.

In the environment there are many other relevant u.v. stimuli for the insects, which are not restricted to the sky. Since the work from Daumer (1958), Mazokhin-Porshnyakov (1969), Obara (1970) and later Nekrutenko (1964) and Eisner et al. (1969) we know about u.v. "hidden" patterns on the flowers and on the insects themselves (Fig. 3, 5), which are biologically important as guide-signals for pollinators (Frisch, 1965) or as secret advertising and recognition signals for the mating partners (Obara, 1970; Eisner et al., 1969). A comparison of reflected light from the wings of a butterfly *Gonepteryx rhamni* with the overall spectral sensitivity of their eyes

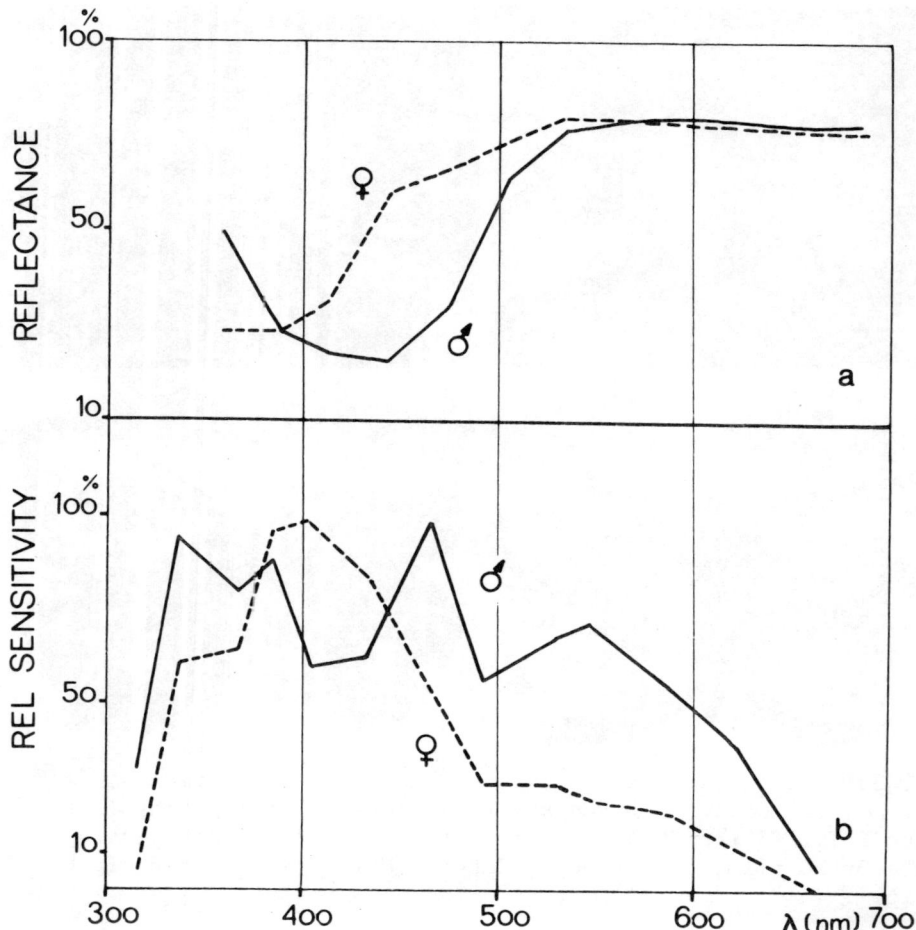

Fig. 4. A comparison of the reflectance spectra from the upper side of the front wings (a) to the spectral sensitivity curves (b) of the males and females of the butterfly *Gonepteryx rhamni*. Sensitivity peaks of the males coincide with the characteristic differences in reflectances, which are typical for sex and species (Smerdu and Gogala, unpublished data).

shows interesting differences between both sexes, which can be explained by the tuning of light receptor organs to the biologically important signals from their sexual partners and food sources on the other side (Fig. 4). In most cases u.v. receptors of insects are a part of their colour vision system as was shown first in the classical experiments with the honey bees (Frisch, 1965).

Fig. 5. Scanning electron micrographs of the scales of the butterfly *Gonepteryx cleopatra*. U.v reflecting iridescent scales are shown in micrographs a,b (right side), c and d. Thin film interference colours are due to the cuticular layers spaced $\lambda/2$ apart, which can be clearly seen in micrographs d and c (damaged part of a scale). U.v. absorbing (e) and pigment bearing scales (b - left side, f) of the upper side of the front wings are also shown. Magnification in c, e and f is the same (Orig.).

Colour vision, which is typical of most insects and arachnids investigated till now (Michieli, 1959; Mazokhin-Porshnyakov, 1969), seems to be very easily adapted to the way of life of these animals. One good example seems to be the spectral response of the photoreceptors of the stalked-eye flies Diopsidae to the light available in the tropical forest (Burkhardt, 1972). Better knowledge of the ethology, ecology and sensory physiology of single arthropod species will provide many other examples of ecophysiological adaptations of photoreceptors to the biologically important light stimuli.

There is an interesting question concerning the colour vision and the previously mentioned polarised light detection in the eyes of insects and arachnids. According to the degree of polarisation, plane of polarisation and orientation of molecules of the photopigment in the single sensory cell, the excitation level in this cell can vary pronouncedly despite the wavelength and the intensity of the light stimulus being constant. The same variation of the excitation of this cell can be reached with the constant polarisation parameters but with the change of the intensity and/or the wavelength. It is evident that the animals should clearly discriminate between those parameters if they should use them as biologically important clues. How this discrimination is secured at least in some insects is explained in a comprehensive paper by Wehner (1976). It cannot be excluded, that the polarisation parameters of the light stimuli, at least in some cases, cause the enhancement of some colours or the opposite, that the colour contrast enhances the detection of the plane of polarisation in non twisted rhabdoms. One of the best animals for such experiments would be the water striders (Gerridae, Heteroptera) with their specialised ommatidia, oriented to the water horizont (Bohn and Täuber, 1971).

The investigation of u.v. receptors and their visual pigment in the case of *Ascalaphus* and other insects prooved the existence of stable meta-products and a reversible photoreconversion dynamics according to the photon catching probability of both (or all) stable forms of visual pigment (Gogala, Hamdorf, Schwemer, 1970; Schwemer, Gogala, Hamdorf, 1971; Hamdorf, Gogala, 1973; Hamdorf, Schwemer, 1975). In the case of the u.v. visual pigment of *Ascalaphus* P_{345} peaks the meta-form at 475 nm (M_{475}) (Fig. 1 c). In *in vitro* experiments with illumination of visual pigment extracts, rhabdom or whole eye preparation with monochromatic light, different ratios between visual pigment and its photoproduct are established according to the wavelength of the light (Gogala, Hamdorf, Schwemer, 1971). Similar results are obtained with the di- or poly-chromatic light illumination, the P/M ratio being dependent on a spectral distribution of the available light energy and on the absorptive properties of the visual pigment (for a review see Hamdorf and Schwemer, 1975).

The position of the P and M absorption peaks is so very important in the automatic regulation of proportion of P to M form of the pigment, which has an important influence on the sensitivity of an insect eye. The dependency of the electrically measured sensitivity of the eye and the P/M ratio of the visual pigment is not a simple one and is still not completely understood (Hamdorf and Schwemer, 1975; Stark, 1977), but in general there is no doubt that the higher concentration of visual pigment in the P state enables the arthropod eye for the higher absolute sensitivity (Razmjoo and Hamdorf, 1976; Barns and Goldsmith, 1977).

Suitable position of λ_{max} of photoproducts with usually higher specific absorption can under certain environmental light conditions automatically reestablish sufficient P/M ratios for effective vision (Fig. 1). This photoregeneration process can be supported by the thermodynamic dark regeneration process, as it seems to exist in some insect eyes (Stavenga, 1975).

It should be mentioned that in the insects, with colour vision the process of photoregeneration provides an automatic system for the adjustment of sensitivity of single types of colour receptors to preserve the correct and more or less constant colour perception under illumination with different colour temperatures (Höglund, Hamdorf and Rosner, 1973).

At the end of this part of the chapter we could say that in the case of insects also we can see great adaptive possibilities of visual pigments to environmental stimuli as is well known in the case of fishes (compare the chapter of Lythgoe; Ali, Anctil and Cervetto, this volume).

Chemoreception

In accordance with the restrictions mentioned in the introduction to this chapter I do not intend to cover this part of the sensory interactions between the environment and arthropods despite their importance. Nevertheless, I would like to mention at least a few studies of chemosensory interactions. Good physiological and ethological data are available from the works of Dethier (1963), Kaissling (1971), Priesner (1973) and Hodgson (1974). In this connection also should be mentioned the chapter on chemoreception in the book of Mazckhin-Porshnyakov (1977), written by Elizarov and his previous texts cited in this book. Because of the practical importance of the chemosensory interactions (attractants, repellents, pheromones!) this is a promising and exciting field for further study, also from an ecophysiological point of view.

Hygroreception

It is generally accepted, that the insects and other animals covered with the chitinous exoskeleton are effectively protected against water loss through the cuticle. Still there is a great variability in the preferential relative humidities (RH), in the effects of RH on the reproduction and development and on survival of these animals (Bursell, 1974 b). Humidity receptors are especially important for some animals to keep up their water balance and to find the suitable habitat and sources of food and water. Around the fresh transpiring leaves there is a steep humidity gradient (Waterhouse, 1950; Flitters, 1968; after Bursell, 1974 b) which can be detected by humidity receptors of the insects, which use the plants as a source of food for example. The blood sucking insects (mosquitoes) orient themselves also by the detection of water vapours from homoeothermic animals in addition to other sensory clues (Wright, 1975).

The hygroreceptors are usually specialised sensilla trichodea, which seem typically to detect as mechanoreceptors the deformation of the chitinous structures, caused by the loss or gain of water at different relative humidities (e.g. sensillum capitulum, Yokohari and Tateda, 1976; Pinet and Bernard, 1972). Another example supporting this idea is found among typical cave animals (troglobionts), which are usually very sensitive to RH drop in the environment.

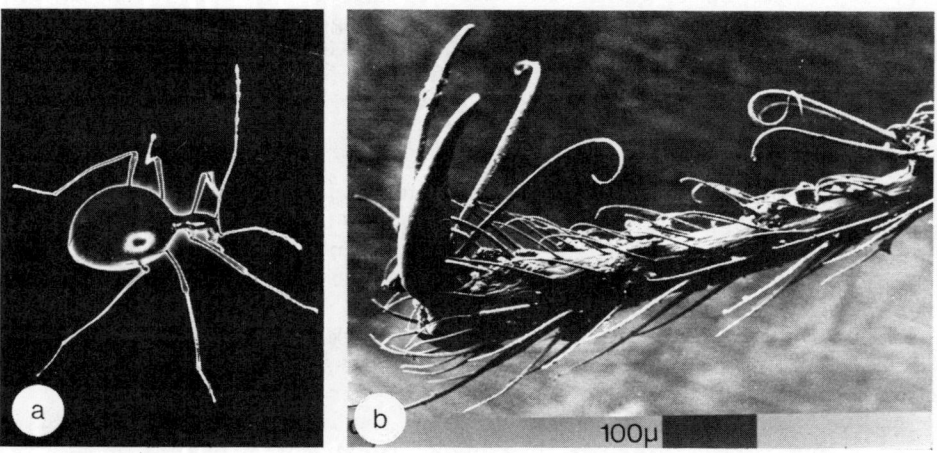

Fig. 6. *Leptodirus hochenwarti*, a typical blind cave dwelling species of beetles (a) from Yugoslavia. b) Scanning electron micrograph of a hind leg tarsus with the special flattened hairs with presumably hygroreceptive function (Orig.).

On the underside of the tarsi of *Leptodirus hochenwarti* there are
special flattened hairs or setae which fold and unfold very quickly
under different humidity conditions (Fig. 6). One can observe the
bending and straightening of these structures under the microscope
while approaching and removing a wet brush or something similar to
the tip of tarsi, without touching them. These animals are reacting
behaviouraly very sensitively to the change of humidity and do not
spread beyond the limits of moist or wet substrate (Michieli and
Gogala, unpublished data). It is interesting, that these hairs are
located on the underside of the tarsi so as to be in close contact
or in the vicinity of the substrate they are walking on. Unfortune-
ately the direct proof of the sensory finction of these hair-like
structures is still missing. Similar structures have been described
in other insects as adhesive structures, important in the locomotion
(empodium and euplantulae or pulvilli). It would be worthwhile to
analyse the different types of cuticular sensilla in other insect
species which are specialised to live in such habitats with extremely
high and stable RH (e.g.: other cave animals, soil animals, insects
living inside of plants etc.). Moist and dry hygroreceptors are for
such animals of high importance for the survival and successful
reproduction, because the eggs and larval stages are often directly
or indirectly (infection) more sensitive to changes in the humidity
(Bursell, 1974 b).

Thermoreception

Another environmental factor which is usually related to humidity is temperature. Sensory information about the temperature,
absolute and relative and their gradients is important for the survival of all animals in general and the insects in particular. Insects are of course poikilotherms, but at least some species are
able to regulate their body temperature to a certain extent (e.g.:
bees, moths, etc.; see Bursell, 1974 a). The effective behavioural
regulation of body temperature is one of the reasons for the presence of sensitive and reliable thermoreceptors. The other reasons
are food finding in some animals (e.g.: parasites of homoiotherms)
or more commonly finding micro habitats which are necessary for the
effective search and occupation of specialised ecological niches.
Such an example is the case of the buprestide beetle (*Melanophila
acuminata*) which is able to locate forest fires at great distances
(Evans, 1964; 1966 after Neville, 1977) (see also Loftus, this
volume).

Typically, only the so called cold-receptors were found in the
insects, the rare exeption being the bug *Oncopeltus fasciatus* (Čokl,
1972). These animals react behaviorally very clearly to temperature
gradients (Fig. 7). It has been shown that the antennal thermoreceptors respond by an increase of the spike frequency to the in-

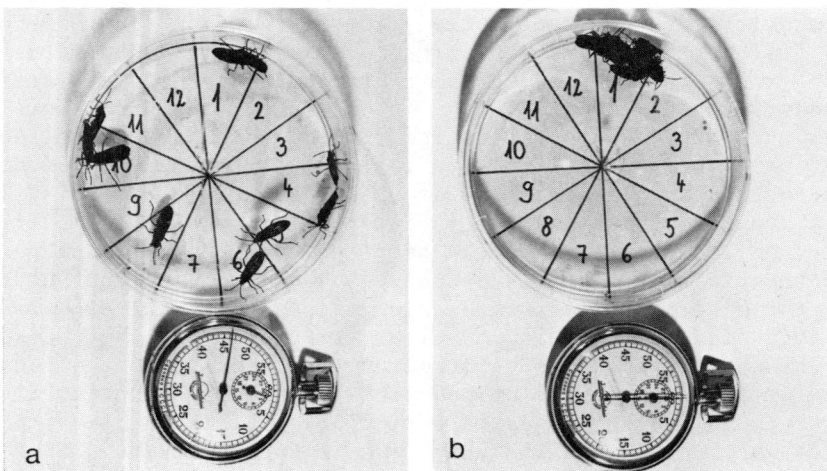

Fig. 7. Distribution of the bugs *Oncopeltus fasciatus* in the petri dish during IR illumination of the sector 1. a) Situation after 1 min 46 s and b) situation after 30 min of IR illumination (Tungsten lamp 25 W, 30 cm distance, filter UG-8, Zeiss, Jena).

crease of temperature. Still we do not know the details of the receptors involved, nor do we understand the ecological and behavioral background of their pronounced thermophilic reaction. The possible explanation could be in the importance of ecological thermoregulation for the small animals like insects. If this is true, a reliable and sensitive thermoreceptor as that in *Oncopeltus* can be expected in many other insects, if not in all of them. With the aid of such receptors the insects could find the temperature optima for their activity and thereby increase their chances of success. Actual measurements on some dipteran species have shown, that the temperature excess, gained from a sun radiation of 1.10^3 W/m² can in larger animals like *Phormia* reach values of 15°C (Bursell, 1974 a). This is one good reason for the existance of sensitive thermoreceptors in such animals but other possibilites should not be neglected.

Mechanoreception

One of the most important categories of sensory organs are mechanoreceptors, which show among insects a vast variety of specialisation from mechanoreceptive proprioreceptors (see Markl; Wales, this volume) over stato, anemo, vibro to the phonoreceptors. Good reviews are available from Schwartzkopff (1974), Michelsen (1974), Michelsen and Nocke (1974). In this chapter I would like to emphasise some questions of related environmental stimuli, for which the insects have evolved distinct specialised receptors. This is the complex of the substrate vibrations, the near field air or water particle movements (sound wind) and the sound in a far field. For analysis insects use their pressure or pressure gradient receivers, displacement receptors and specialised vibroreceptors. There are many examples among insects which make use of all four principles to detect vibrations in the environment. Many Orthoptera possess tympanal organs which can act as pressure or pressure gradient receivers (Michelsen, 1971), according to the frequency range. In the legs they have the subgenual organs for the detection of substrate vibrations and they have hair sensilla (e.g.: on the cerci), which can act as the air particle movement receptors for the sound in the near field conditions (Petrovskaya, 1969). New investigations on the nerve pathways from the tympanal and subgenual organs in *Locusta* and *Decticus* (Čokl, Kalmring and Wittig, 1977; Čokl and Serša, unpublished data) and the electrophysiological investigations on their interneurons show that the information from both types of organs often influence the activity of the same neuron in the ventral nerve cord.

When we consider the biological meaning of the acoustical and vibrational signals in insects we can generally say that most of the signals play a part in their premating and mating behaviour. They are important for finding sexual partners (phonotactic response: Regen, 1913; 1923; Murphey and Zaretzky, 1972) for their recognition and stimulation.

What are the advantages of having many different inputs for the vibrational stimuli? There is no simple answer but we can make more or less probable speculations about this. Advertising their presence to the sexual partner or to the rival can be hazardous if the predators can locate precisely the site where the sound emitting animal is sitting. With the use of different channels for vibrational stimuli the location of the partner by another insect can be more precise that it is for some predators (e.g. birds) which use mainly sound pressure receivers. Insects communicating in the near field can exactly exchange the essential informations about each other and still remain hardly audible for the potential predators, which do not have air movement receptors (e.g. *Drosophila*, Bennet-Clark, 1971; 1975). The intensity levels measured with the pressure sensitive or movement sensitive transducer in the near field can be as much apart as 35 and 70 dB respectively. So the signal to noise level is high enough for the insects and the privacy of the channel against sound pressure sensitive predators is maintained. Such receptors can enable the animal also with the information about the approaching predator, as in the case of the caterpillars of *Barathra brassicae*, which can detect the predatory wasps *Dolichovespula media* with their specialised hair sensilla and with appropriate behaviour raise their chance of survival (Tautz, 1977 and personal communication).

Mutatis mutandis the advantages of the communication channel through the substrate vibrations can be understood. In the example of singing bugs from the families Cydnidae and Pentatomidae, we can also observe a very low sound pressure level in the near field (Haskell, 1957; Gogala, 1969; 1970; Gogala, Čokl, Drašlar, Blažević, 1974) and on the other hands the sufficient amplitudes of the substrate vibrations for successful communication within a few centimetres distance. Some results on the bug *Canthophorus* suggest a very interesting possibility of the tuning of the acoustical signals to the resonance properties of the plant stem on which they live as monofags (Gogala, et al. 1974). The bugs in the separate chambers responded to each other with rival alternation only when the transmission of the signals was provided through the stem of their feeding plant *Thesium*. Similar examples should be investigated in other insects which do use the substrate vibration as the main communicative channel and are more or less strictly bound to one feeding plant or other substrate. Physical measurements of the spread of acoustic vibrations over such plant organs and tissues and their resonance frequencies would be highly desirable.

In the case of Perlidae and Sialidae rhythmic modulations is the only important parameter in communication and, the resonance characteristics of the substrate are negligible (Rupprecht, 1968; 1969; 1975). The explanation would be that these insects are as imagos not bound specifically to any particular substrate and in their songs frequency modulation is practically absent. This is not true for the bugs mentioned above in many species of which pronounced and constant frequency modulation or frequency band modulation occurs (Fig. 8) (Gogala, 1969; Gogala and Razpotnik, 1974).

Chitinous exoskeleton is very suitable for the production of sound in a broad sense (Leston, 1957; Haskell, 1974, see also Michelsen this volume) and on the other hand it is similarly adaptable for the "construction" of sensory organs, which are tuned to all possible sources of biologically relevant acoustical signals. Very good examples for this are the night active flying insects like moths and lacewings, which are chased by echolocating bats. The specialised sensory tympanal organs have evolved on different parts of the body, with different anatomy but with the same common property i.e. they are sensitive to the ultrasound emitted by bats and usually are able to recognise the presence and the direction of the ultrasound source, which is important for survival during their

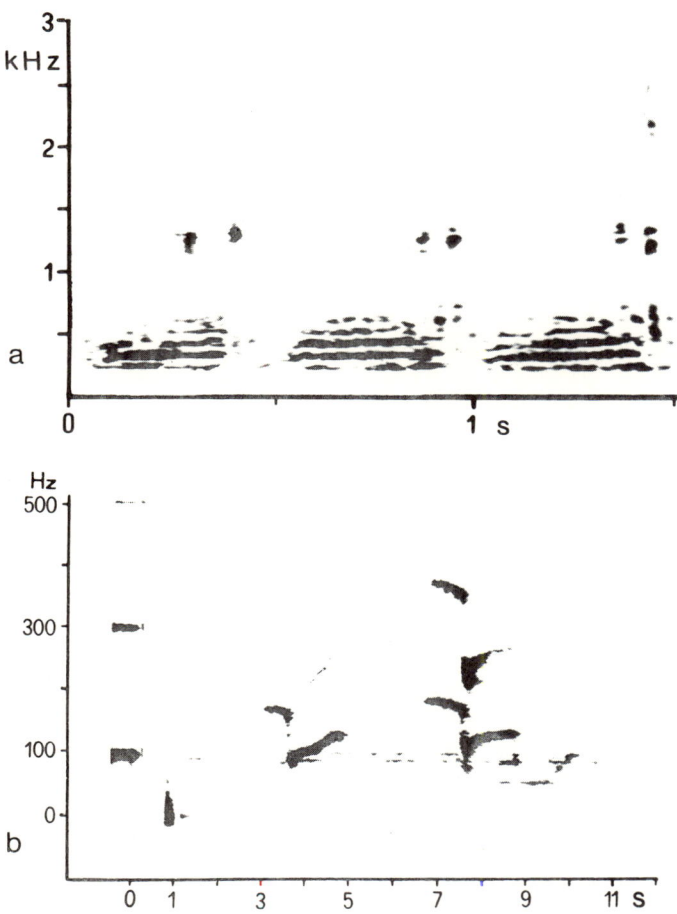

Fig. 8. Sonagrams of the vibrational signals of the male bugs *Tritomegas bicolor* (a) and *Piezodorus lituratus* (b) as examples of frequency (b) rsp. frequency band modulated signals (a) in insects, communicating by substrate borne sound. b) 100 Hz calibration signal at time 0. (Modified from Gogala et al. (1974) (a) and Gogala and Razpotnik (1974) (b)).

nocturnal activity (see a review-by Sales and Pye (1974)).

At the end of this part of the chapter, I would like to return to the hair sensilla of the trichobotrium type, which are in general accepted as a prototype for an air particle movement receptor (Fig. 9). Among the insects and arachnids they are widely spread and I have already mentioned their possible role as air particle movement detectors in the acoustic near field. But, on the other hand, we have to remember, that such function as a kind of hearing organ has been in many instances denied or at least not proven (Markl, 1973). The extirpation of trichobothria in Cydnidae does not prevent the acoustic communication and alternation between animals (Gogala, Čokl, Drašlar and Blažević, 1974). Hoffmann (1967) could not record the responses of trichobothria in scorpions to the biologically relevant levels of sound. There are many other possibilities for the main function of such organs in insects and arachnids, especially anemoreception (Linsenmair, 1968; 1973) or tangoreception, but the extremely high sensitivity of this mechanoreceptors to different kinds of mechanical and other stimuli leaves open the door for other speculations. Many properties of trichobothria remind us of the acoustico-lateralis system of lower vertebrates, with the single units responding in definite direction with excitation or inhibition, and these units being organised in a well defined system to cover all possible directions of stimulation (Fig. 9) (Drašlar pers. comm.). There are also other similarities; possibility of the existance of the so called "Ferntastsinn", role in acoustic reception and, possibility of the detection of the electrical fields, which will be discussed in the last part of this chapter.

Fig. 9. Trichobothria of the bug *Pyrrhocoris apterus*. a) The position of the trichobothria and the directional sensitivity of the lateral organs. Arrows indicate the direction of the movement of the hair, which causes the excitation of the sensory cell. Circles indicate the nondirectional units, responding to the displacement of the hair in any direction from the resting position. Numbers indicate abdominal sternites. b,c) Scanning electron micrographs of the trichobothria in the field of trichomes (macula opaca) Drašlar, unpublished data).

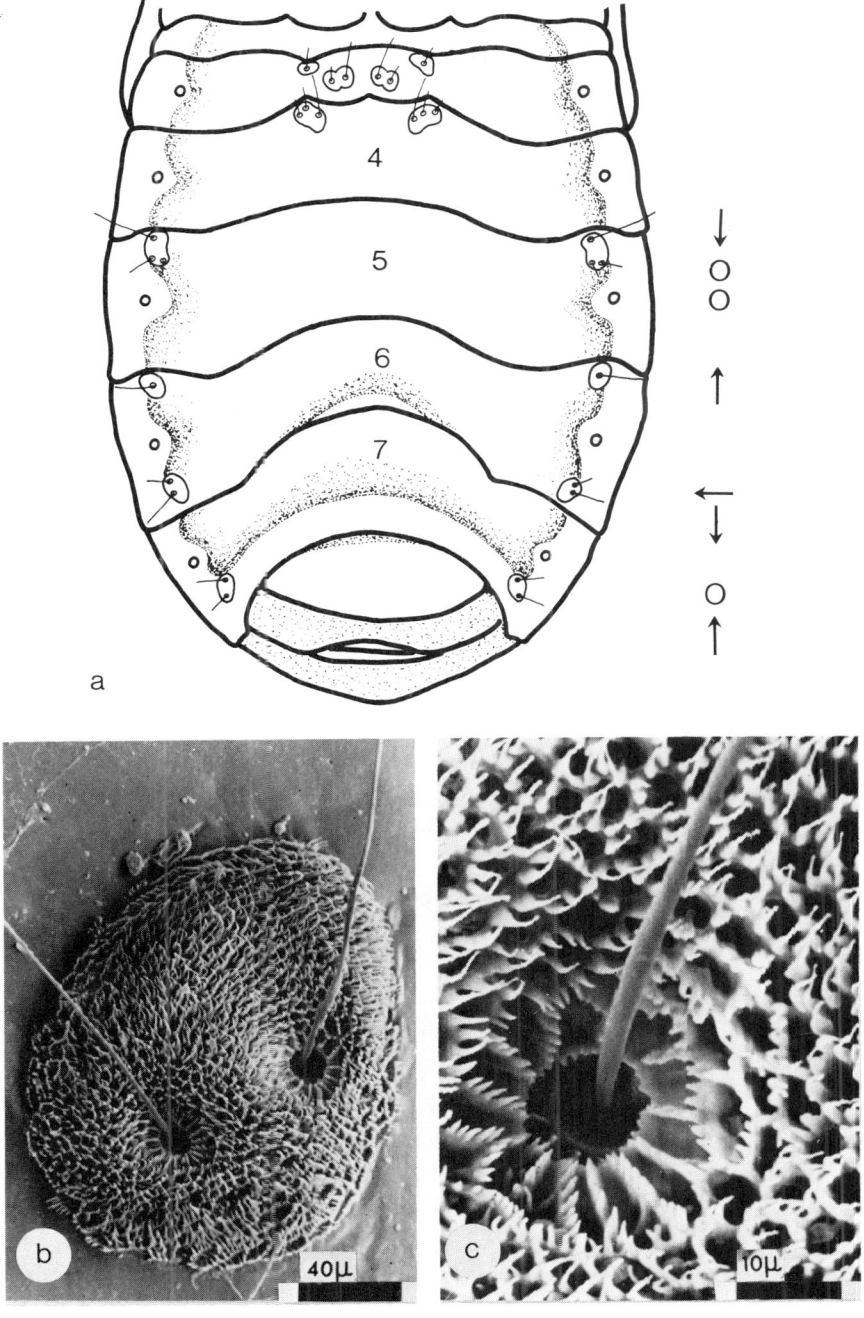

Figure 9

Electric and magnetic stimuli

Our knowledge of the electric fields in the environment of insects and around the insects themselves is still very fragmentary (for reviews see Warnke, 1973; 1978; Mazokhin-Porshnyakov, 1977). The effects of electric charges upon the behaviour of insects are known (Warnke, 1976).

Different sources of electric fields were investigated by Altman and Warnke (1973), Warnke (1973; 1975) and Becker (1976). One of the most interesting phenomena was established by Warnke. He measured the potentials on the different parts of the body of the bees with impendance transducers. The potentials measured were of the class of several tens of volts on some parts of the exoskeleton. Highest potentials were measured on membranous and glandular parts of the cuticle (including antennae), other parts

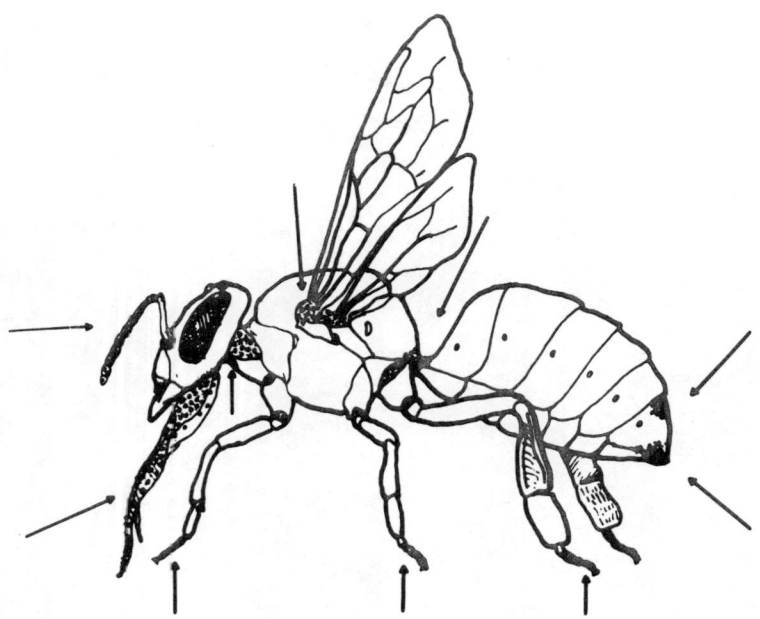

Fig. 10. The body surface of a bee can be divided roughly into two areas as far as its electrical behaviour is concerned. All the membranous and glandular surfaces of the cuticula show great variations in potential (stippled areas), whereas the rest of the surface is at a low potential, up to +1v. Over fairly short distances, there may be fairly high intensities of electric field, 25v/0.2cm (after Warnke, 1973; 1976).

being at the same time almost at the zero level potential to the earth (Fig. 10). The charging is the highest during the flight of animals through ionised air. In standing or walking animals the charging and discharging is through the arolia of the tarsi effective. According to the posture of the legs the arolia can make better or weaker contact with the substratum so that the animals can regulate the discharging rate of their body surface relative to the substratum (Warnke, 1973). Big differences in the conductance of the insect cuticle were found by other workers, according to different samples and moulting cycles (for references see Neville, 1975).

Very interesting suggestion was made also by Warnke (1975) and Becker (1976) namely that at least some insects use their electric fields in communication. Warnke claims this for the bees and Becker for termites which, according to his data are able to coordinate the building of their galleries in separate boxes until the electrostatic isolation or strong electric field interference prevents it. In the later case the communicative signals should be in the range of the low frequency ac fields which have in general similar but stronger effects than the dc fields, also on the behaviour of *Drosophila*, bees and wasps (Mazokhin-Porshnyakov, 1977).

As stated previously, some types of mechanoreceptors can function also as detectors for electric fields around the insects or arachnids and other sources in their habitat. It is well known, that it is possible to move the trichobothrial hairs electrostatically with the aid of small capacitors, the plates of which are charged to a level of a few tens to approximatelly one hundred Volts. (Hoffmann, 1967; Drašlar, pers. comm.). Such mechanoreceptors can really be responsible for the detection of communicative electric signals since the electric gradients around the insects or arachnids are sufficient for eliciting the bioelectric response in the sensory cells of these organs. This is a rich and interesting field for further study, but many data from the literature should be carefully reexamined.

Since the electric and magnetic events are interrelated, it is not surprising that many authors try to explain the effects of magnetic fields upon the insects (and other animals) with the effects on electric charges on and around them. The ability of the insects to orient themselves in the earth's magnetic field has been proven many times (for review see Martin and Lindauer, 1973) but the receptor mechanism is still unknown. One possible explanation could be the effect of the Lorentz forces. They act upon charged surfaces, moving in the magnetic field, and in accordance with the Hall effect give rise to the changes of this electric charge (Warnke, 1975).

The whole complex of orientation and communication in insects and other arthropods using electric and magnetic fields deserves further attention from physical, physiological and ecological points of view.

CONCLUSION

Few years ago only a few scientists believed the early papers of F. Schneider (1960, 1975 a) on magnetic orientation of may-beetles and flies, but now there is not much doubt about it. Recently the same author claims that insects are able to detect gravitational waves (Schneider, 1975 b). It will indeed be interesting to examine the situation after another 10 or 20 years.

One may end by quoting Sir Vincent Wigglesworth:"Insects ... are so varied in form, so rich in species and adapted to such diverse conditions of life, that they afford unrivalled opportunities for physiological study" and by adding that this could also be extended to sensory ecology.

REFERENCES

Altmann, G. and Warnke, U. (1973). Registrierung von Tierbewegungen mit Hilfe der Körperoberflächenladungen. Experientia,29: 1044.
Autrum, H. and Thomas, I. (1973). Comparative physiology of color vision in animals . In: Handbook of sensory physiology. Central processing of visual information A, edited by R. Jung, Springer-Verlag, Berlin, New York.
Autrum, H. and Zwehl von V. (1964). Die spektrale Empfindlichkeit einzelner Sehzellen des Bienenauges. Z. vergl. Physiol. 48: 357-384.
Barns, S.N. and Goldsmith, T.H. (1977). Dark adaptation, sensitivity, and rhodopsin level in the eye of the lobster, *Homarus*. J. comp. Physiol. 120: 143-159.
Barth, F.G. (1969). Die Feinstruktur des Spinnenintegumentes. I. Die Cuticula des Laufbeins adulter häutungsferner Tiere. Z. Zellforsch. 97: 137-159.
Barth, F.G. (1970). Die Feinstruktur des Spinnenintegumentes. II. Die räumliche Anordnung der Mikrofasern in der lamellierten Cuticula und ihre Beziehung zur Gestalt der Porenkanäle. Z. Zellforsch. 104: 87-106.
Becker, G. (1976). Reaction of termites to weak alternating magnetic fields. Naturwissenschaften, 63: 201-202.
Bennet-Clark, H.C. (1971). Acoustics of insect song. Nature (London), 234: 255-259.
Bennet-Clark, H.C. (1975). Acoustics and the behaviour of *Drosophila*. Verh. Dtsch. Zool. Ges. 68: 18-28.
Bohn, H. and Täuber, U. (1971). Beziehungen zwischen der Wirkung

polarisierten Lichtes auf das Elektroretinogramm und der Ultra-
struktur des Auges von *Gerris lacustris*, L. Z. vergl. Physiol.
72: 32-53.
Borror, D.J. and De Long, D.M. (1971). An Introduction to the study
of Insects. Holt, Rinehart and Winston Inc. New York, 3 rd ed.
Burkhardt, D. (1962). Spectral sensitivity and other response cha-
racteristics of single visual cells in the arthropod eye.
Symp. Soc. Exp. Biol., 16: Biol. receptor mechanisms, 86-109.
Burkhardt, D. (1972). Electrophysiological studies on the compound
eye of a stalked-eye fly *Cyrtodiopsis dalmanni* (Diopsidae,
Diptera). J. comp. Physiol. 81: 203-214.
Burkhardt, D. and De la Motte, I. (1972). 1. Electrophysiological
studies on the eyes of Diptera, Mecoptera and Hymenoptera.
In: Inform. Proc. in the Visual Syst. Arthropods, edited by
R. Wehner., Springer Verlag, Berlin, Heidelberg, New York.
Burkhardt, D., De la Motte, I. and Seitz, G. (1966). Physiological
optics of the compound eye of the blow fly. In: The functional
organization of the compound eye, edited by C.G. Bernhard,
Pergamon Press, Oxford, New York.
Bursell, E. (1974 a). Environmental aspects - temperature. In:
The physiology of Insecta, 2. edition, edited by M. Rockstein,
Academic Press, New York, London.
Bursell, E. (1974 b). Environmental aspects - humidity. In: The
physiology of Insecta, 2. edition, edited by M. Rockstein,
Academic Press, New York, London.
Carricaburu, P. (1969). Catadioptrique de l'oeil composé. Vision
Res. 9: 1523-1536.
Carricaburu, P. and Chardenot, P. (1967). Spectres d'absorption de
la corneé de quelques Arthropodes. Vision Res. 7: 43-50.
Čokl, A. (1972). Thermoreception of the bug *Oncopeltus fasciatus*
(slov., summary engl.). Biol. Vestn. 20: 39-45.
Čokl, A., Kalmring, K. and Wittig, H. (1977). The responses of au-
ditory ventral-cord neurons of *Locusta migratoria* to vibration
stimuli. J. comp. Physiol. 120: 161-172.
Dartnall, H.J.A. (1975). Assessing the fitness of visual pigments
for their photic environments. In: Vision in Fishes, New
approaches in research, edited by M.A. Ali, Plenum Press,
New York, London.
Daumer, K. (1958). Blumenfarben wie sie die Bienen sehen. Z. vergl.
Physiol. 41: 49-110.
Dethier, V.G. (1963). The Physiology of Insect Senses. Methuen,
London, John Wiley et Sons, New York.
Eidmann, H. and Kühlhorn, F. (1970). Lehrbuch der Entomologie. 2.
Auflage. Verlag Paul Parey, Hamburg, Berlin.
Eisner, T., Silberglied, R.E., Aneshansley, D., Carrel, D. and
Howland, H.C. (1969). Ultraviolet Video-Viewing: The television
camera as an insect eye. Science, 166: 1172-1174.
Fox, H. and Vevers, G. (1960). The nature of animal colours. Sidg-
wick and Jackson, London.

Frisch, von K. (1965). Tanzsprache und Orientierung der Bienen. Springer Verlag, Berlin, Heidelberg, New York.

Gewecke, M. (1974). The antennae of insects as air-current sense organs and their relationship to the control of flight .In: Exp. Analysis of Insect Behaviour, edited by L. Barton Brown, Springer Verlag, Berlin, Heidelberg, New York.

Gogala, M. (1967). Die spektrale Empfindlichkeit der Doppelaugen von *Ascalaphus macaronius* Scop. (Neuroptera, Ascalaphidae). Z. vergl. Physiol. 57: 232-243.

Gogala, M. (1969). Die akustische Kommunikation bei der Wanze *Tritomegas bicolor* L. (Heteroptera, Cydnidae). Z. vergl. Physiol. 63: 379-391.

Gogala, M. (1970). Artspezifität der Lautäusserungen bei Erdwanzen (Heteroptera, Cydnidae). Z. vergl. Physiol. 70: 20-28.

Gogala, M., Čokl, A., Drašlar, K. and Blažević, A. (1974). Substrate-borne sound communication in Cydnidae (Heteroptera). J. comp. Physiol. 94: 25-31.

Gogala, M., Hamdorf, K. and Schwemer, J. (1970). Der UV-Sehfarbstoff bei Insekten. Z. vergl. Physiol. 70: 410-413.

Gogala, M. and Razpotnik, R. (1974). An oscillographic sonagraphic method in bioacoustical research (slov., summary engl.) Biol. Vestn. 22: 209-216.

Goldsmith, T.H. and Bernard, G.D. (1974). The visual system of insects . In: The physiology of Insecta, 2. edition, edited by M. Rockstein, Academic Press, New York, London.

Grzimek, B. (ed.) (1969). Grzimeks Tierleben, Enzyklopädie des Tierreichs,2: Insekten. Kindler Verlag, Zürich.

Grzimek, B. (ed.) (1971). Grzimeks Tierleben, Enzyklopädie des Tierreichs,1: Niedere Tiere. Kindler Verlag, Zürich.

Hamdorf, K. and Gogala, M. (1973). Photoregeneration und Bereichseinstellung der Empfindlichkeit beim UV-Rezeptor. J. comp. Physiol. 86: 231-245.

Hamdorf, K. and Schwemer, J. (1975). Photoregeneration and the adaptation process in insect photoreceptors . In: Photoreceptor optics, edited by A.W. Snyder and R. Menzel, Springer Verlag, Berlin, Heidelberg, New York.

Haskell, P.T. (1957). Stridulation and its analysis in certain Geocorisae (Hemiptera, Heteroptera). Proc. Zool. Soc. (Lond.) 129: 351-358.

Haskell, P.T. (1974). Sound production . In: The Physiology of Insecta, 2. edition, edited by M. Rockstein, Academic Press, New York, London.

Henderson, S.T. (1970). Daylight and its spectrum. Adam Hilger Ltd., London.

Hodgson, E.S. (1974). Chemoreception . In: The physiology of Insecta, 2. edition, edited by M. Rockstein, Academic Press, New York, London.

Hoffmann, C. (1964). Zur Funktion der kammförmigen Organe von Skorpionen. Naturwissenschaften, 51: 172.

Hoffmann, C. (1967). Bau und Funktion der Trichobothrien von *Euscorpius carpathicus* L. Z. vergl. Physiol. 54: 290-352.
Höglund, G., Hamdorf, K. and Rosner, G. (1973). Trichromatic visual system in an insect and its sensitivity control by blue light. J. comp. Physiol. 86: 265-279.
Horridge, G.A. (1977). The compound eye of the insects. Scientific American, 237 (1.): 108-121.
Kaestner, A. (1956). Lehrbuch der speziellen Zoologie. Band I.: Wirbellose, 3. Lieferung. Gustav Fischer Verlag, Jena.
Kaestner, A. (1972). Lehrbuch der speziellen Zoologie. Band I.: Wirbellose, 3. Teil: Insecta. Gustav Fischer Verlag, Stuttgart.
Kaissling, K.E. (1971). Insect Olfaction. In: Handbook of Sensory Physiology, vol. IV., edited by L.M. Beidler, Springer Verlag, Berlin, Heidelberg, New York.
Leston, D. (1957). Stridulatory mechanisms in Terrestrial species of Hemiptera, Heteroptera. Proc. Zool. Soc. (Lond.), 128: 369-386.
Linsenmair, E.K. (1968). Anemomenotaktische Orientierung bei Skorpionen (Chelicerata, Scorpiones). Z. vergl. Physiol. 60: 445-449.
Linsenmair, K.E. (1973). Die Windorientierung laufender Insekten. In: Orientierung der Tiere im Raum, edited by M. Lindauer, Fortschritte Zool. 21: (2/3), 59-79.
Markl, H. (1973). Leistungen des Vibrationssinnes bei wirbellosen Tieren. In: Orientierung der Tiere im Raum, edited by M. Lindauer, Fortschritte Zool. 21: (2/3), 100-120.
Martin, H. and Lindauer, M. (1973). Orientierung im Erdmagnetfeld. In: Orientierung der Tiere im Raum, edited by M. Lindauer, Fortschritte Zool. 21: (2/3), 211-228.
Mazokhin-Porshnyakov, G.A. (1959). Colorimetric study of colour vision in the dragonfly (in russ.). Biofizika, 4: 427-436.
Mazokhin-Porshnyakov, G.A. (1969). Insect vision. Plenum Press, New York.
Mazokhin-Porshnyakov, G.A. (1977). Guide-book in the sensory physiology of insects (in russ.). Moscow Univ. Press.
Michelsen, A. (1971). The Physiology of the locust ear: I., II., III. Z. vergl. Physiol. 71: 49-128.
Michelsen, A. (1974). Hearing in Invertebrates. In: Handbook of sensory physiology, vol. V/1, edited by W.D. Keidel and W.D. Neff, Springer Verlag, Berlin, Heidelberg, New York.
Michelsen, A. and Nocke, H. (1974). Biophysical aspects of sound communication in insects. In: Advances in insect physiology, vol. X., edited by J.E. Treherne, Academic Press, New York, London.
Michieli, S. (1959). Analysis of scototactic (perigrammotactic) reactions in arthropods (in slov., german summary). Razprave SAZU, Cl. IV. 5: 237-286.
Murphey, R.K. and Zaretsky, M.D. (1972). Orientation to the calling song by female crickets, *Scapsipedus marginatus* (Gryllidae).

J. exp. Biol. 56: 335-352.
Nekrutenko, Y.P. (1964). The hidden wing-pattern of some paleartic species of *Gonepteryx* and its taxonomic value. J. Res. Lepid. 3: 65-68.
Neville, A.C. (1975). Biology of the Arthropod Cuticle. Springer Verlag, Berlin, Heidelberg, New York.
Obara, Y. (1970). Studies on the mating behavior of the white cabbage butterfly, *Pieris rapae crucivora* Boisduval. III. Near-ultra-violet reflection as the signal of intraspecific communication. Z. vergl. Physiol. 69: 99-116.
Petrovskaya, E.D. (1969). On frequency selectivity of cercal receptors in the cricket, *Gryllus domesticus* (russ., engl. summ.). J. evolutionary Biochem. Physiol. 5: 337-338.
Pinet, J.M. and Bernard, J. (1972). Essai d'interpretation du mode d'action de la vapeur d'eau et de la temperature sur un recepteur d'insecte. Ann. zool. ecol. anim. 4:
Priesner, E. (1973). Artspezifität und Funktion einiger Insektenpheromone. Fortschritte Zool. 22: 49-135.
Razmjoo, S. and Hamdorf, K. (1976). Visual sensitivity and the variation of total photopigment content in the blowfly photoreceptor membrane. J. comp. Physiol. 105: 279-286.
Regen, J. (1913). Über die Anlockung des Weibchens von *Gryllus campestris* L. durch telephonisch übertragene Stridulationslaute der Männchens. Pflügers Archiv f. Physiol. 155: 1-10.
Regen, J. (1923). Über die Orientierung des Weibchens von *Liogryllus campestris* L. nach dem Stridulationsschall des Männchens. Sitz. Ber. d. Akad. d. Wiss. Matem. naturw. Kl. 1. 132: 81-88.
Rockstein, M. (1974). The physiology of Insecta (2. edition). Academic Press, New York, London.
Romoser, W.S. (1973). The Science of Entomology. Macmillan, New York.
Rupprecht, R. (1968). Das Trommeln der Plecopteren. Z. vergl. Physiol. 59: 38-71.
Rupprecht, R. (1969). Zur Artspezifität der Trommelsignale der Plecopteren (Insecta). Oikos 20: 26-33.
Rupprecht, R. (1975). Die Kommunikation von *Sialis* (Megaloptera) durch Vibrationssignale. J. Insect. Physiol. 21: 305-320.
Sales, G. and Pye, D. (1974). Ultrasonic communication by animals. Chapman and Hall, London.
Schneider, F. (1960). Der experimentelle Nachweis einer magnetischen und elektrischen Orientierung des Maikäfers. Verh. d. Schweiz. Naturforsch. Ges. im Kanton Aargau 132-134.
Schneider, F. (1975 a) Der experimentelle Nachweis magnetischer, elektrischer und anderer ultraoptischer Informationen. Z. ang. Ent. 77: 225-236.
Schneider, F. (1975 b). Gibt es sinnesphysiologisch wirksame Gravitations-wellen? Ein Problem der ultraoptischen Orientierung. Vierteljahrsschrift der Naturforsch. Gesell. in Zürich, 120: 33-79.

Schwartzkopff, J. (1974). Mechanoreception . In: The physiology of Insecta, 2. edition, edited by M. Rockstein, Academic Press, New York, London.

Schwemer, J., Gogala, M. and Hamdorf, K. (1971). Der UV-Sehfarbstoff der Insekten: Photochemie in vitro und in vivo. Z. vergl. Physiol. 75: 174-188.

Seitz, G. (1969). Untersuchungen am dioptrischen Apparat des Leuchtkäferauges. Z. vergl. Physiol. 62: 61-74.

Stavenga, D.G. (1975). Dark regeneration of invertebrate visual pigments . In: Photoreceptor optics, edited by A.W. Snyder and R. Menzel, Springer Verlag, Berlin, Heidelberg, New York.

Tautz, J. (1977). Reception of medium vibration by thoracal hairs of caterpilars of *Barathra brassicae* L. (Lepidoptera, Noctuidae). J. comp. Physiol. 118: 13-31.

Warnke, U. (1973). Physikalisch-physiologische Grundlagen zur luftelektrisch bedingten Wetterfühligkeit der Honigbiene (*Apis mellifica*). Diss. Math. Nat. Fak. Univ. Saarland, Saarbrücken.

Warnke, U. (1975). Insekten und Vögel erzeugen elektrische Felder. Umschau, 75: 479.

Warnke, U. (1976). Effects of electric charges on honeybees. Bee World 57: 50-56.

Warnke, U. (1978). Information - transmission by electrical biofields . In: Communication of biological systems by means of electromagnetic fields, edited by F. Popp, Urban-Schwarzenberg, in press.

Wehner, R. (1976). Porarized-light navigation by insects. Scientific American 235: (1.) 106-115.

Wigglesworth, V.B. (1972). The principles of insect physiology. 7 th edition. Chapman and Hall Ltd., London.

Wright, R.H. (1975). Way mosquito repellents repel. Scientific American 233: (1.) 104-111.

FISHES: VISION IN DIM LIGHT AND SURROGATE SENSES

J.N. LYTHGOE

MRC Vision Unit, University of Sussex

Falmer, Brighton, BN1 9QG, U.K.

INTRODUCTION

The amount of information that the eye can gain from a light source ultimately depends upon the number of photons that the light carries and in dim light it is a famine of photons rather than inadequate sensitivity of the eye that limits the ability of the eye to detect fine detail, rapid movement, colour and contrast. The absorption of a single photon by a visual pigment molecule is sufficient to isomerise the chromophoric group of the molecule and thus to set in train the events that ultimately lead to the sensation of vision (for a review see Knowles and Dartnall, 1977). In man, at least, it requires the absorption of only 5 - 10 photons in an area covered by 500 rods to initiate a sensation of vision (Hecht, Schlaer and Pirenne, 1942). Ripps and Weale (1976) consider that the absorption of 1, 2 or 3 photons is probably sufficient to impart visual information although at this low level conclusions based on such information are not reliable.

The number of photons falling on a surface is expressed as number per unit area per unit time per unit frequency (or wavelength interval). It is not possible to predict exactly where a photon will fall and it is only possible to quote a statistical probability for the number that will arrive within a defined area. One of the most important visual tasks is to distinguish between two areas of an image on the basis of brightness, which involves comparing the number of photons that fall within the two areas. If the light is dim the actual number of photons captured will be few and the chance fluctuation in photon number over a short period of time may swamp the difference between the two areas measured over a longer period of time.

There are several ways of increasing the number of photons sampled and hence reduce error due to chance fluctuations. But all carry penalties which may decisively degrade the visual efficiency of the animal in other ways or make it physically so ungainly that it cannot survive.

THE GEOMETRY OF THE EYE

In dim light it is important to have the brightest image that is practicable on the retina. It is relevant whether the image consists of small points of light or whether the source is an extended one such as an ordinary landscape in daylight. The optics of the situation are analysed and compared by Rodiek (1973), Kirschfeldt (1974), and Land (1977).

For a point source, the brightness of the image is proportional to the square of the pupil aperture. In other words, a large eye is necessary to see points of light such as the stars in the sky or points of bioluminescence in the dark sea. For extended light sources it is the ratio of the aperture of the eye to the focal length that is important. This is the "f" number used by photographers and describes the acceptance angle of the lens. Thus for extended light sources, a small eye can give a bright image (but at the expense of acuity) provided the focal length is proportionally short.

There are other adaptations to the geometry of the eye and head that increase the brightness of the retinal image without inordinately increasing the size of the eye. Many dim-light animals have a tapetum behind the retina that reflects light that has already passed unabsorbed through the retina, back through it a second time (Nicol, Arnott and Best, 1973, for a review). A rostral aphakic space coupled with sighting grooves in the snout along the direction of view increases the effective aperture of the eye in the forward direction (Munk and Frederiksen, 1974). The retina of deep-living or strictly nocturnal fishes are usually rod-dominated and the rods may be arranged in layers to catch the greatest possible number of photons (Walls, 1942; Locket, 1970, for reviews). Photon capture is further made more efficient in deep sea fishes by the presence of rhodopsins in the rods that have their wavelength of maximum absorption (λ_{max}) at the 470-480 nm region of the spectrum and are thus tuned to absorb most efficiently those wavelengths of light to which the deep oceans are most transparent (Lythgoe, 1972; Muntz and McFarland, 1977, for reviews).

SUMMATION AREA AND TIME

It is a commonplace of statistics that the larger the samples the finer the difference that can be revealed. At the retinal level the number of photons counted can be increased by sampling over a wider area. This improves the perception of contrast between large areas, but necessarily worsens the perception of fine detail. Integration over a wide area is achieved either by increasing the cross-sectional area of individual visual cells, or by bunching them together into discrete units or by routing the signals from many rods through a single neurone so that the signals from many rods are integrated together (for reviews see Walls, 1942; Rodiek, 1973).

The number of photons that fall within a given area depends upon the time period of exposure. The sampling time of the retina, which is variously known as the memory time, integration time, or summation time, is estimated by finding the longest time that a decrease in photon flux can be compensated for by an increase in exposure time (Ripps and Weale, 1976). For human scotopic vision the sampling time is about 0.1 secs. The frequency of a flashing light that just becomes indistinguishable from a constant light (the flicker fusion frequency) is related to the retinal integration time. Man has a flicker fusion frequency of 50 - 60 cycles/seconds in bright light but less than 10 at scotopic levels. Very similar values have been found in fishes such as the sea breams (Sparidae) and the anchovy (*Engraulis mordax*) (Blaxter, 1970, ex Protasov *et al.*, 1960).

A direct effect of longer integration times in the retina is that images moving across the retina, are simply not seen if their presentation time at any point on the retina is short compared to the integration time. Photographers are familiar with a similar effect. A time exposure of a few seconds of a busy street scene will show stationary cars and fixed buildings. It may even show standing pedestrians, but the scene will be apparently devoid of all moving cars and walking pedestrians.

An animal adapted to dim light conditions, therefore, has two built-in disabilities that are a necessary consequence of increased sensitivity. Its visual acuity is poor and it will fail to see objects moving fast relative to itself, either because they are moving, or because the observer is. The detection of objects that are both small and fast is doubly difficult for a nocturnal animal.

COLOUR VISION

A receptor that responds to light from only a narrow frequency interval will obviously be less sensitive than one that is sensitive to the full spectral range of available light. By its very nature, colour vision involves comparing the signals from two or more receptor types, each mostly sampling at a different spectral band. Colour vision, therefore, involves a loss of sensitivity that may be decisive at low light intensities and it is unlikely that colour vision is often a useful sense in dim light. The only exception could be in the detection of bioluminescence, for here the individual points of light may themselves be bright although the overall irradiance of the visual scene is low.

A particular problem in underwater vision is the reduction of visual contrasts. This comes about through the addition of non-image-forming diffuse light to the visual image arising from light scattered by suspended particles in the water and by the water molecules themselves (Duntley 1963; Lythgoe 1972). This diffuse light is one source of visual noise. It means that more photons must be sampled before differences in the number incident on different retinal areas become detectable. Contrast detection is always difficult at night. Under water the contrasts are themselves reduced by light scatter and vision becomes even more difficult.

THE QUIT POINT

As the conditions for vision become more and more difficult, the visual information that can be got from ever larger and more specialised eyes is no longer worth the effort in obtaining it. At this point other sensory modalities, such as the various lateral line senses, are developed instead and the eye is reduced in size and complexity and, presumably, in importance.

Only in the extreme cases of cave-dwelling species such as the Iranian blind fish, *Noemacheilus smithi* (Greenwood, 1976), is the eye completely absent. In most cases it seems that a residual capacity to sense light makes it worth retaining the eye in some sort of "caretaker" capacity. There is an elegant example of this in the bats studied by Chase (1972) and described by Suthers in this volume. Some microchiroptera feed on small flying insects by night. To do this visually would be an impossibly exacting task for the insects are small and fast-moving. Instead, the echo-location sense is greatly developed and the eyes are very small, probably being used only for such crude tasks as detecting light at the mouth of the roosting cave or perhaps for setting diel rhythms. By comparison the microchiroptera that feed on larger and more slow-moving objects such as nectar-bearing flowers, fruit and on vertebrate blood have retained quite reasonably functional eyes.

THE FAUNAL BREAK

Whereas inshore waters show enormous variations in colour and clarity, the deep oceans show less variability (see Jerlov, 1976, for a review of light penetration in the oceans). Thus it is possible to estimate at what depth the daylight will have fallen to some fraction of its intensity at the surface.

FISHES: VISION IN DIM LIGHT AND SURROGATE SENSES

Clarke and Denton (1962) have calculated that the deepest mesopelagic fishes can detect noon daylight in the clearest ocean water is around 1000 m. In somewhat less clear ocean water (Jerlov type II as compared to type I) Dartnall (1975) calculates that the equivalent depth for just perceptible daylight is 250 m. In the clear waters of the Indian Ocean, Clarke and Kelly (1964) concluded that the lower limit for daylight vision varies from 700 - 1300 m.

The open ocean is conventionally divided into three horizontal zones. The uppermost zone occupies the top 100 m where there is sufficient light for both photosynthesis and vision. This is the epipelagic zone. The middle zone, the mesopelagic zone, extends from 100 m to about 1000 m and here there is sufficient light for vision but not for photosynthesis. The bathypelagic zone extends from 1000 m to the bottom and there is not enough daylight for either photosynthesis or vision. There is little mixing between the bathypelagic and mesopelagic faunas. The mesopelagic animals, both fish and invertebrates, generally have large, well-developed eyes, photophores, and often silvery, transparent or countershaded bodies that serve as camouflage in the dim light. Bathypelagic animals generally have small eyes, poorly-developed photophores and black, red or brown non-reflecting bodies (Marshall, 1971).

The boundary between the bathypelagic and mesopelagic faunas is known as the faunal break. There is now strong evidence for Marshall's view that the faunal break corresponds to the deepest depth that animals can see daylight. During the 1965 SOND cruise to the Eastern North Atlantic, there were both careful light measurements (Kampa, 1970) and animal collections that included decapods (Foxton, 1970), amphipods (Thurston, 1976) and fishes (Badcock, 1970). The faunal break at the clearest station occurs at 650 - 700 m where the energy of the downwelling diffuse light is about 10^{-6} µw/cm^2/nm at 480 nm. This corresponds to a light intensity in photons of 5.27 photons/cm^2/sec/10^{13} Hertz (see Dartnall, 1975). This is near the scotopic broad field threshold for man and is about 1 log unit brighter than Clarke and Denton's (1962) estimate for the visual threshold of deep-sea fishes.

There is, of course, a rôle for vision below the depth that daylight can penetrate due to the presence of bioluminescence. However, as Marshall points out, bioluminescence becomes less common in deeper living animals. Perhaps this is because one of the prime uses for bioluminescence is to obliterate the silhouette of animals seen from below against the downwelling daylight (Denton, Gilpin-Brown and Wright, 1972). With no daylight, such camouflage is no longer needed. Bioluminescence notwithstanding, the value of vision to fishes living in deep water is obviously diminished, for fishes living deeper than 1000 m or so often have small or regressed eyes

(Munk, 1964, 1966; Marshall, 1971) although exceptions are many (Locket, 1971).

SENSES THAT SUBSTITUTE FOR VISION

The "decision" to develop other sensory systems when the information provided by the eye is not worth the investment put into it depends largely on whether there are other sensory modalities already present that are potentially able to substitute for vision. Walls (1942) uses the concept of pre-adaptation to explain how fishes such as the brotulids and catfishes have chemical and tactile senses particularly well-developed that suit them to a nocturnal bottom-grubbing life and life in very deep water. In particular, the brotulids and catfishes, have often been successful colonisers of caves where the sense of vision is absolutely useless.

THE ACOUSTICO-LATERALIS SYSTEM

On the basis of morphology, embryology and perhaps of physiology as well, (Russell and Sellick, 1976) the cochlear, vestibular organs and lateral line of lower vertebrates, together form a single organ system collectively known as the acoustico-lateralis. However, the acoustico-lateralis has the ability to monitor a wide range of environmental variables that can conveniently be classed as either mechanical, such as water displacement, vibration and the direction of gravity; or electromechanical, such as the form of electrical fields in the water and the earth's electromagnetic field. We may imagine that the perception of gravity and relatively high frequency "sound" vibrations in the water are of a kind familiar to us through our own senses, but the sensitivity residing in the lateral line to transient displacements in the water and to perturbation in the electrical field are of a kind foreign to our senses. Many deep-living fishes have particularly well-developed acoustico-lateralis systems. It is reasonable to suppose that the acoustico-lateralis is well-developed as a substitution for the diminished information provided by the eyes.

THE WATER DISPLACEMENT SENSE

The organ that monitors water displacement is the neuromast. Neuromasts are found in cyclostomes, fishes, larval amphibia and aquatic adult urodeles and some aquatic ranales (Russell, 1976). In all these groups, the neuromasts have a very similar shape consisting of a bundle of sensory cells, from which sprout fine hairs embedded in a gelatinous cupular. The neuromasts are found on the head and body, generally in lines, and this tendency towards linear array is best seen in the lateral line that runs along the flanks of many teleosts fishes. In addition to this major line, there are also subsidiary branches that run around the eyes and over the head.

In most teleost fishes the neuromast organs are sunk in mucus-filled canals that communicate to the exterior periodically through pores and it is these pores that are visible at the surface. In species that live in the calm of very deep water and caves, the neuromasts are not protected in canals or grooves but protrude directly into the water, (Marshall, 1971). Indeed, in some deep-living ceriatoid angler fishes and true cave fishes the neuromasts are actually stalked.

There has been much discussion about whether the lateral line organs respond to pressure differences or to water currents (see Russell 1976, for a recent review). The present consensus appears to be that the neuromast organs themselves respond only to the sheering force between cupular and cell body set up by water displacements. They do not respond directly to pressure changes. Water displacements are set up by water currents which may result from, amongst others, the movements of neighbouring fishes, convection movements in the water, tidal currents, the motion of the fish itself, or by the near field currents set up in the immediate vicinity of oscillating sound-producing bodies (Pumphrey, 1950; Harris and Bergeijk, 1962). In parenthesis, it should be noted that the pressure waves of true "far field" sound also cause displacements of water particles, but Harris and Bergeijk remark that a by no means inconsiderable sound having an acoustic pressure of 1 µbar causes a negligible 1 $\overset{o}{A}$ngstrom displacement of the water particles.

Although the lateral line organs themselves may not respond to transient pressure changes, Russell (1976) points out that the gas-filled cavity of the swimbladder could transduce pressure changes into displacement movements in the lateral line. Denton and Blaxter (1976), in a study of the mechanical relationship between the clupeid inner ear, swimbladder and lateral line, show that the swimbladder acts merely as a gas reservoir for "topping up" the gas in the auditory bullae. The pro-otic membrane in the bullae divides the gas filled part of the bullae from the liquid filled part that is continuous with the liquid in the utricular macula and lateral line. The presence of a compressible element, namely, the gas in the bullae, ensures that pressure differences will cause displacements in the non-compressible liquids of the lateral line and utricular macula.

The acoustico-lateralis system, therefore, is able to monitor both pressure changes and water displacements. In addition, the neuromast organs are directionally sensitive, an attribute that seems to be related to the arrangement of the stereocillia protruding into the cupular from the sensory cells. Information on direction and amplitude of the displacement is presumably computed more centrally by comparing the output of more than one neuromast organ (Russell, 1976).

In considering the sensory capacity of the lateral line, it is necessary to separate gravity waves travelling along the surface (and perhaps waves on the density discontinuity of the seasonal thermocline) (Woods, 1971), from sound or direct water displacements caused by currents in a homogeneous volume of water. Unlike sound waves, the velocity of surface waves is a function of frequency, the velocity reaching a minimum at 15.4 Hz and increasing at higher and lower wavelengths. (Schwartz, 1971) showed that in several species of surface-feeding fishes the threshold amplitude of a wave varies with frequency and is lowest at 10 - 40 Hz, thus making the fishes most sensitive to the waves that travel slowest. Although this means that the animal is relatively insensitive to the waves that reach it first, it does mean that the ability to deduce the position of direction of a localised disturbance at the water surface from the time of arrival of the surface waves it produces will be the more precise. Russell (1976) estimates that a fish should be able to measure a just perceptible difference in the time of arrival of surface waves, which are about 1 mm apart, when the amplitude of the waves are about 50 times threshold and their frequency about 13.5 Hz. Schwartz (1974) estimates that a fish is able to locate a surface disturbance, such as those set up by a floundering insect trapped in the surface film, at a distance of 3 - 3.5 body lengths away.

The localisation of the source of water disturbances is an important property of the lateral line organs. There are good circumstantial reasons for believing that a long linear array of receptor organs carried along an unflexing body is particularly useful for gaining spatial information about both electrical and mechanical disturbances. Thus Lissman (1958) has shown that in evolution electrically-sensitive fishes tend to have elongate bodies with a long dorsal fin that propels the animal by undulations that travel along its length. Fishes that depend heavily upon their lateral line mechanical sense also have elongate bodies. Many lateral line fishes have two modes of swimming (Bone, 1971). In the fast mode an elongate fish such as the scabbard fish, *Acanopus carbo*, and the pike, *Esox lucius* (Marshall, 1971) swim with powerful flexions of the body and caudal fins. In the fast "escape" mode, the afferent output from the dogfish lateral line is inhibited (Roberts and Russell, 1972). However, in the slow mode, which is presumably when prey is detected, the body is kept straight and is propelled either by the dorsal fine alone, or by gentle sculling with the caudal fin. Within the scabbard fish family (Trichiuridae) as a whole Bone (1971) recognises two types of slow swimming. The first, shown by five genera, involves gentle sculling with a specially modified caudal fin. The second, shown by three genera, involves the enlarged dorsal fin and the reduction of the caudal and anal fin. In the sculling type, the lateral line is carried along the midline of the flank; but in the dorsal fin swimmers the lateral line runs along the ventral surface, presumably to remove it as far as possible from the disturbance set up by the dorsal fin.

The lateral line system that requires the body to be kept straight presumably has superior spatial acuity and because unwanted "noise" from the fish's own swimming movements is kept to a minimum, sensitivity to distant sources should be good. Indeed, Pumphrey (1950) estimates that a scabbard fish ought to be able to detect a mackerel-like fish at a range of 16 - 32 m. In actively swimming schooling species, such as the saith, *Pollachius virens*, the lateral line plays a role in the maintenance of schooling distance. Pitcher, Partridge and Wardle (1976) have shown that temporarily blinded fish are able to maintain regular schooling distance from their nearest two neighbours and, indeed, maintain their distance more closely than do normally-sighted fish.

THE LABYRINTH SENSE

Like the lateral line organ the labyrinth in vertebrates is derived embryologically from the auditory placode. The basic sensory cell, the hair cell, is common to both and they are both most likely stimulated by sheering forces set up by fluid displacements. In evolutionary terms, there are good reasons for thinking that the labyrinth is homologous in evolution to part of the lateral line system (Bergeijk, 1967).

Fishes are often denied visual clues in the water either because it is dark or the water is too turbid for vision, or being in midwater there are no visible "landmarks". Gravity and acceleration forces may be strongly developed to give the fish information about its position in the water. According to Bergeijk, the essential evolutionary step leading to an awareness of gravity and acceleration forces was the acretion of mineral deposits in contact with the hair cells of the acoustico-lateralis. These deposits have a greater inertia than the sensory hair cells. Thus an angular acceleration will tend to set up a sheer force between the kinocillium in contact with the mineral deposit and the cell itself. The hair cell is sensitive to sheer forces, which is a function of the water displacement sense, but the presence of a heavy crystalline deposit in contact with the hair cell might also have given awareness of angular acceleration and possibly to gravity also.

The hypothetical lateral line hair cells with their crystalline deposits (Pumphrey, 1950) would be unable to distinguish between angular acceleration and to water displacement and the one modality could interfere with the other. To overcome the difficulty, the lateral line might be sunk beneath the surface and to overcome unwanted signals due to changes in angular velocity set up by flexions of the body itself, the new acceleration sense is located near the centre of movement in the rear part of the head. Lastly, the acceleration sense needs to be sensitive to changes in all three spatial dimensions and the three orthogonally arranged semi-circular canals are an adaptation to this need.

These ideas are difficult to test phylogenetically because the primatine chordate, Amphioxus, has no labyrinth, but the sea lamprey, *Myxine*, already has a fairly well-developed labyrinth, albeit with only one semi-circular canal. The homologies of the labyrinth have been extensively discussed by de Burlet (1934). During the entire subsequent evolutionary history of the vertebrates after the agnatha, the labyrinth remained essentially unchanged with three semi-circular canals containing crystalline otolithes.

HEARING

It is convenient to define hearing as sensitivity to compression waves in the far field of the sound source. Since particle displacements in water are exceedingly minute, of the order of 1 ångstrom, it is unlikely that pressure waves can set up sheering forces across the hair cells sufficient to stimulate them. Nevertheless, it is perfectly evident that fishes are sensitive to sound (see, for instance, Popper and Fay, 1973). Indeed, Myrberg et al. (1972) have shown that sharks can locate a sound source 125 - 400 m away. One of the most sensitive fishes to sound is the blind cave fish, *Astyanax jordani*, which has a threshold down to nearly - 50 decibels (re 1 microbar) at a frequency of 1000 Hz (Popper, 1970).

THE ELECTRIC SENSE

Humans have nothing similar to the electric sense of fishes and we have very little idea of the extent of the electric sense amongst them. It is known, however, that the electric fishes studied by Lissman (1958) are species that live in turbid water, are nocturnal, and have poorly-developed eyes. In extreme cases, like *Typhlonarke*, the animal is almost or entirely blind.

Physically the difference between mechanical and electrical stimuli are profound, but the specialised electroreceptors of the lateral line organs are homologous with the neuromast organs: Electroreceptors are quite variable in form, but share the feature that the receptor cells are housed at the inner end of a tube containing a jelly of low electrical resistance, but whose walls have electrical insulators. The voltage gradient between the interior of the fish and the external water is thus compressed to the small distance across the sensitive cells, which act as a voltmeter (Kalmijn, 1974). The electric organs are thus directionally sensitive according to the orientation of the canal relative to the electric gradient. In common with the wavelength-sensitive neuromasts of the water displacement sense, some more central comparison of the signals from differently-orientated electroreceptors is necessary to get information about the orientation of the external electric fields.

A body of different ionic composition from the surrounding water will set up an electric field around it provided that the body is not surrounded by an insulating layer. Fishes lying buried from sight in a soft substratum set up such fields, especially around the respiratory and buccal membranes that are periodically exposed during respiratory movements. The detection of these electric fields must be a potent aid in the search for food since Kalmijn (1974) has shown that the dogfish, *Scyliorhinus*, and rays, are able to detect plaice, *Pleuronectes platessa*, at a range of 10 cm by this means.

There is a possibility that the electric sense may be sufficiently sensitive to detect the electro-magnetic field of the earth and to orientate to it as a navigational aid (Akoev, Ilyinski and Zadan, 1976). The dorsal ampullae of lorenzini in *Trygon pastinaca* and *Raja clavata* are thought to be sensitive enough to detect the electric currents induced in them by the earth's geomagnetic field.

ACTIVELY-PRODUCED FIELDS

Many species of fish possess electric organs that the pioneer work of Lissman (1958) and Lissman and Machin (1958) has shown to be an aid to both navigation and communication. The electric discharges produced for these purposes are much less than those employed by some species such as the electric ray, *Torpedo marmorata* to stun their prey. The former are therefore known as weakly electric fish.

The electric discharge for either navigation or recognition falls into one of two groups. In the first group, "hummers", the discharge is a continuous sinusoidal wave of rather high frequency which remains remarkably constant, changing only at the approach of another species having a similar discharge rate. The two signals would jam each other and one of the fish changes the frequency of its discharge to avoid this jamming.

"Buzzers" discharge at a much lower frequency which is variable, the frequency increasing when some interesting event is detected. One family may include both buzzing and humming species. For instances within the Gymnotidae, *Sternopygus*, *Eigenmannia*, and *Apteronotus* are hummers, and *Electrophorus*, *Gymnotus* and *Hypopomus* are buzzers.

There is only hypothesis at the moment concerning the different capabilities of wave and pulse species. Scheich and Bullock (1974) suggest that hummers are particularly good for long distance social communication and have a good temporal acuity. If the fish is itself moving rapidly or is surrounded by objects that are moving rapidly past it, as in swiftly running water, detection is easier using these high-frequency discharges. Buzzers may be better for the assessment of slow-moving or stationary objects, but may suffer from the disadvantage that they are easily jammed, not only by other buzzing species (to which they respond by a change of discharge frequency) but also by such natural events as lightning discharge.

REFERENCES

Akoev, G.N., Ilyinski, & Zadan, P.M. (1976). Responses of Electroreceptors (Ampullae of Lorenzini) of skates to electric and magnetic fields. J. Comp. Physiol. 106: 127-136.

Badcock, J. (1970). The vertical distribution of mesopelagic fishes collected on the SOND Cruise. J. mar. biol. Assoc. U.K. 50: 1001-1044.

Bergeijk, W.A. von (1967). The evolution of vertebrate hearing. In: Contributions to sensory physiology, 2: 1-49. Neff, W.D. (Ed.), N.Y., Academic Press.

Blaxter, J.H.S. (1970). Light: Fishes. In: Marine ecology, Vol. 1, Part 1 (Environmental factors) Edit. by Kinne, O., Wiley Interscience, Lond. pp. 213-320.

Blest, A.D. & Land, M.F. (1977). The physiological optics of *Dinopis subrufus* L. Koch. A fish-lens in a spider. Proc. Roy. Soc. B. 196: 197-222.

Bone, Q. (1971). On the scabbard fish *Acanopus carbo*. J. mar. biol. Assoc. U.K. 51: 219-226.

Chase, J. (1972). The role of vision in echo-locating bats. Ph.D. Thesis, Indiana University.

Clarke, G.L. & Denton, E.J. (1962). Light and Animal Life. In: The Sea, Vol. I. ed. by M.N. Hill, pp. 456-468. N.Y., London, Interscience.

Clarke, G.L. & Kelly, M.G. (1964). Variation in transparency and in bioluminescence on longitudinal transects in the western Indian Ocean. Bull. Inst. Monaco 64: 20 pp.

Dartnall, H.J.A. (1975). "Assessing the fitness of visual pigments for their photic environment". In: Vision in Fishes, (ed. by M.A. Ali). New York & London, Plenum Press.

de Burlet, H.M. (1934). Vergleichende Anatomie des statoacustischen Organs. In: "Handbuck der vergleichenden Anatomie der Wirbeltiere". (Bolk *et al.*, Eds.), Vol. II, pp. 1293-1432. Munich, Urban & Schwarzenberg.

Denton, E.J., Gilpin Brown, J.B. & Wright, P.G. (1972). The angular distribution of the light produced by some mesopelagic fish in relation to camouflage. Proc. roy. Soc. (B) 182: 145-158.

Denton, E.J. & Blaxter, J.H.S. (1976). The mechanical relationship between the clupeid swimbladder, inner ear and lateral line. J. mar. biol. Assoc. U.K. 56: 787-807.

Duntley, S.Q. (1963). Light in the Sea. J. opt. Soc. Amer. 53: 214-233.
Foxton, P. (1970). The vertical distribution of pelagic decapods (Crustacea: Natantia) collected on the SOND Cruise, 1965. II. The Penaeidea and general discussion. J. mar. biol. Assoc. U.K. 50: 961-1000.
Greenwood, P.H. (1976). A new and eyeless cobitid fish (Pisces, Cypriniformes) from the Zagros Mountains, Iran. J. Zool. Soc. Lond. 180: 129-137.
Harris, G.G. & Bergeijk, W.A. van (1962). Evidence that the lateral line organ responds to water displacements. J. acoust. Soc. Amer. 34: 1831-1841.
Hecht, S., Shlaer, S. & Pirenne, M.H. (1942). Energy, Quanta and Vision. J. gen. Physiol. 25: 819-840.
Jerlov, N.G. (1976). Marine Optics. Amsterdam, Oxford, N.Y., Elsevier.
Kalmijn, A.J. (1974). The detection of electric fields from manmade and animate sources other than electric organs. In: Handbook of Sensory Physiology, III/3 ed. by A. Fessard, pp. 147-200. Springer-Verlag, Berlin.
Kirschfeld, K. (1974). The absolute sensitivity of Lens and Compound Eyes. Z. Naturforsch. 29c: 592-596.
Knowles, A. & Dartnall, H.J.A. (1977). Photobiology of Vision. In: The Eye, Vol. IIB., Ed. by H. Davson., London & N.Y., Academic Press.
Lissmann, H.W. (1958). On the function and evolution of electric organs in fish. J. exp. Biol. 35: 156-191.
Lissmann, H.W. & Machin, K.E. (1958). The mechanism of object location in *Gymnarchus niloticus* and similar fish. J. exp. Biol. 35: 451-486.
Locket, N.A. (1970). Deep sea fish retinas. British Med. Bull. 26: 107-111.
Locket, N.A. (1971). Retinal structure in *Platytroctes apus*, a deep-sea fish with a pure rod fovea. J. mar. biol. Assoc. U.K. 51: 79-91.
Lythgoe, J.N. (1972). The adaptation of visual pigments to their photic environment. In: Handbook of Sensory Physiology, Vol. VII/1, pp. 566-603. Ed. by H.J.A. Dartnall, Berlin, Heidelberg, N.Y., Springer-Verlag.
Marshall, N.B. (1971). Explorations in the life of fishes. Cambridge, Mass., Harvard.
Munk, O. (1964). Ocular degeneration in deep-sea fishes. Galathea Rep. 8: 21.
Munk, O. (1966). Ocular anatomy of some deep-sea teleosts. Dana Rep. 70: 1-62
Munk, O. & Frederiksen, R.D. (1974). On the function of aphakik apertures in teleosts. Videnskabelige meddelelser fra Dansk Naturhistorisk forening 137: 65-94.

Myrberg, A.A., Ha, S.A.J., Walewski, S. & Banbury, J.C. (1972). Effectiveness of acoustic signals in attracting epipelagic sharks to an underwater sound source. Bull. mar. Sci. 22: 926-949.

Nicol, J.A.C., Arnott, H.J. & Best, C.G. (1973). Tapeta lucida in bony fishes (Actinopterygii): a survey. Can. J. zool. 51: 69-81.

Pitcher, T.J., Partridge, B.L. & Wardle, C.S. (1976). A blind fish can school. Science 194: 963-965.

Popper, A.N. (1970). Auditory capacities of the Mexican Blind cavefish (*Astyanax jordani*) and its Eyed Ancestor (*Astyanax mexicanus*). Animal behaviour 18: 552.

Popper, A.N. & Fay, R.R. (1973). Sound detection and processing by teleost fishes: a critical review. J. Acoust. Soc. Amer. 53: 1515-1529.

Pumphrey, R.J. (1950). Hearing. Symp. Soc. exp. Biol. 4: 3-18.

Ripps, H. & Weale, R.A. (1976). The visual stimulus. In: The Eye, Vol. IIa., pp. 43-99, Ed. by H. Davson, N.Y. Academic Press.

Roberts, B.L. & Russell, I.J. (1972). The activity of lateral line efferent neurones in stationary and swimming dogfish. J. exp. Biol. 57: 435-448.

Rodieck, R.W. (1973). The Vertebrate Retina. San Francisco, W.H. Freeman & Co.

Russell, I.J. (1976). Amphibian lateral line receptors. In: Frog Neurobiology, pp. 513-550. Ed. by Llinas R. & Precht, W. Berlin, Heidelberg, N.Y., Springer.

Russell, I.J. & Sellick, P.M. (1976). Measurement of potassium and chloride ion concentrations in the cupulae of the lateral lines of *Xenopus laevis*. J. Physiol. 257: 245-255.

Scheich, H. & Bullock, T.H. (1974). The detection of electric fields from electric organs. In: Handbook of Sensory Physiology Vol. III/3. Ed. by A. Fessard, pp. 201-256. Springer-Verlag, Berlin.

Schwartz, E. (1971). Die ortung von wasserarellen durch oberflächenfische. Z. vergl. Physiol. 74: 64-80.

Schwartz, E. (1974). Lateral-line mechanoreceptors in fishes and amphibians. In: Handbook of Sensory Physiology, Vol. III/3. Ed. by A. Fessard, pp. 257-278. Berlin, Heidelberg, N.Y., Springer-Verlag.

Thurston, M.H. (1976). The vertical distribution and diurnal migration of the crustacea Amphipoda collected during the SOND Cruise, 1965. II. The Hyperiidea and general discussion. J. mar. biol. Assoc. U.K. 56: 383-470.

Walls, G.L. (1942). The Vertebrate Eye and its adaptive radiation. N.Y., London, Hafner.

Woods, J.D. (1971). Micro-oceanography. In: Underwater Science, Ed. by Woods, J.D. & Lythgoe, J.N. pp. 291-317, Oxford.

ECOLOGICAL NICHE DIMENSIONS AND SENSORY FUNCTIONS IN AMPHIBIANS

ROBERT G. JAEGER

Dept. of Biol. Sciences, State Univ. of N.Y. at Albany

Albany, New York   12222 U.S.A.

INTRODUCTION

One of the goals of sensory ecology is to understand how taxonomic groups of animals converge or diverge in sensory functions to solve similar perceptual problems. Several chapters in this book point out how parameters of the physical environment influence sensory modes: i.e., in particularly murky waters, fishes may emphasize electroreception rather than photoreception for communication (H. Schwassmann this volume); in waters of varying quality for light transmission, different species of fishes may have eyes adapted morphologically and physiologically to perceive maximally in different parts of the visible spectrum (J. Lythgoe, this volume). This chapter on amphibians will compare and contrast the primary sensory modes (chemoreception, sound reception, photoreception) utilized by anurans and caudates. The modes emphasized by each Order are viewed as adaptations to several ecological constraints on information transmission. The constraints considered here are imposed by both the physical environment and by the number of species packed into a community.

AMPHIBIANS

The class Amphibia is composed of about 3100 species subdivided into three orders (Porter, 1972). The caecilians (Gymnophiona) are poorly known ecologically, behaviorally and taxonomically and consequently will not be considered farther here. The salamanders (Caudata) are basically circumpolar with a recent invasion and extensive species radiation of one family (Plethodontidae) in Central America and Northern South America (Wake, 1970; Wake & Lynch, 1976). Frogs and toads (Anura), on the other hand, reach their greatest species diversity in the pan-tropical areas, with as many as 81 sympatric species in one community in Ecuador (Crump, 1974), with diversity decreasing into the temperate regions (Savage, 1973). Although the two orders are basically different in centers of geographic distribution, species in both groups have evolved similar modes of habitat utilization: obligatorily aquatic, semi-aquatic, terrestrial, fossorial and arborial species. Both orders have many species that lay eggs in or near water, with the young passing through an aquatic larval stage, while other species lay eggs on land, which hatch directly into subadults. Yet anurans and caudates have generally different ecological adaptations. Caudates have short legs and long tails, the latter used for swimming by aquatic species, storage of adipose tissue and anti-predator defense, and their mobility is relatively limited. Anurans, however, are tailless and possess long hind legs used in rapid and extensive mobility (Zug, 1972). Caudates, in general, also tend to be confined to more moist and secretive microhabitats than are anurans, perhaps due to the former group's reduced mobility.

THE PRIMARY MODES OF PERCEPTION

In this section, I shall review a number of studies concerning sensory functions in amphibians which I believe to be relevant to amphibian ecology. The information developed here will then be used in later sections of this chapter to build several conceptual models concerning amphibian sensory ecology.

1. Chemoreception

Chemoreception is well developed in salamanders and probably serves as the primary sensory mode for detecting information about the environment. On the other hand, chemoreception appears to be only weakly developed in frogs and toads and thus plays a minor role in environmental sensing. Madison (1977) extensively reviewed the literature on chemical communication in amphibians.

There is scant evidence of chemical communication in anurans. A study of Rabb & Rabb (1963) indicated that the aquatic frog *Pipa pipa* (family Pipidae) uses chemical cues in courtship behavior. Of more ecological importance, a series of studies by Grubb (1970, 1973a, 1973b, 1975) showed that a number of species of anurans are able to use environmental odors to locate breeding ponds. Given a choice of water from the individual's own pond versus water from another pond used by the same species, the individual tended to choose the water from the home pond. However, not all species tested responded to such odor cues.

Compared to anurans, the literature on caudates is relatively rich in examples of chemical communication. Chemoreception is par-

ticularly important in courtship behavior. Twitty (1955) showed
that secretions produced by females of the red-bellied newt, *Taricha
rivularis* (Salamandridae), act as a signal toward which males orient
in water. Such "tracking" of conspecifics of the opposite sex is
probably common among both terrestrial and aquatic species of salamander, although rigorous experiments are lacking. Early studies by
Organ (1958, 1960a, 1960b) and the more recent and excellent ethological studies of Arnold (1972, 1976) strongly suggest that olfactory
communication is intimately involved in the complex courtship rituals
performed by many salamanders. Salamanders of the family Plethodontidae possess nasolabial grooves (Sever, 1975), running from either
external naris to the upper lip, which serve to transport odoriferous
particles from a substrate, touched by the nasolabial cirri, into the
internal nares (Brown, 1968). Arnold (1972) suggested that "tapping"
behavior, which is systematically performed as the animals move about
and touch the cirri to the substrate, is the primary means of chemoreception in plethodontid salamanders. Tapping behavior apparently
allows the male to determine the species, sex and reproductive condition of females (Arnold, 1976). Mental hedonic glands (Sever,
1976), differentiated mucous glands on the chin of male plethodontid
salamanders, are used in a "slapping" behavior toward females, and
they seem to have an aphrodisiac effect upon the female during courtship (Arnold, 1976). Arnold (1972, 1976) has provided the best information to date as to the uses of olfactory communication in salamander courtship.

There is considerable evidence that chemoreception plays a major
role in homing behavior by salamanders. The aquatic newt, *Taricha
rivularis*, has been studied extensively (Twitty, 1959, 1966; Grant,
Anderson & Twitty, 1968). When displaced over various distances
from their home streams, individuals readily homed following extirpation of the lateral eyes but generally did not home after severence
of the olfactory nerves. Another salamander, which inhabits ground
litter in forests, *Plethodon jordani* (Plethodontidae), also exhibits
homing behavior which is highly oriented, direction-independent and
distance-dependent (Madison, 1969; Madison & Shoop, 1970). Although
Madison (1972) could not totally rule out other environmental cues,
he concluded that displaced salamanders tend to orient into the wind
and respond to air-borne familiar odors from the home area; those
familiar odors could come from conspecifics living in and around the
displaced animal's home area.

Recent research indicates that chemical cues are used in relatively complex communication systems among salamanders that are not
directly related to courtship. Madison (1975), using air-borne odors
in a two-choice olfactometer, found that the terrestrial salamander
Plethodon jordani could distinguish between conspecifics of the same
sex who had previously occupied adjacent versus more distant home
ranges in relation to the test-animal's home range. During the non-
breeding season, the experimental animals preferred (moved toward)

the odors of neighboring conspecifics, in preference to non-neighbors. During the breeding season, though, neither sex demonstrated a significant distinguishing preference. Madison (1975) interpreted these data as suggesting discrimination of odors based on kinship relationships. Work by two persons in my laboratory shows an equally sophisticated communication system in the terrestrial salamander *Plethodon cinereus*. Tristram (1977) showed that these salamanders can use substrate olfactory cues to differentiate between their own markings and those of conspecifics. "Tapping" (Arnold, 1972, 1976) was used as a behavioral indicator, and the salamanders tapped their nasolabial cirri significantly more frequently on substrates containing their own odors than on substrates marked by conspecifics. However, the salamanders exhibited significantly more "escape behavior" when exposed to conspecific-marked substrates as compared to own-marked substrates. McGavin (1978) extended this research and, again using tapping behavior as a behavioral indicator of olfactory "interest", found that individuals of *P. cinereus* can learn to distinguish among the substrate odors of conspecifics; i.e., they tapped significantly less frequently on substrates marked by individuals with which they had prior olfactory contact than on substrates marked by unfamiliar conspecifics. These data are similar to those obtained by Madison (1975) for responses of *P. jordani* to odors of neighboring (familiar) and non-neighboring (unfamiliar) conspecifics. Individual recognition of conspecifics could be accomplished by markings from skin and/or cloacal glands. Tristram (1977) noted that male and female *P. cinereus* frequently touch their chins to the substrate as if marking, and Sever (in press) has discovered a diversity of glands in the cloacae of several species of male plethodontid salamanders. A preliminary study by Ritterman & Jaeger (unpublished data) suggested that *P. cinereus* can distinguish between substrates marked with their own fecal material and those marked with feces of conspecifics.

Finally, Jaeger & Gergits (MS) found that substrate signal markers can be used in both intra- and interspecific communication. Given a choice, in the laboratory, of own-marked substrate or a substrate marked by a conspecific, both male and female *Plethodon cinereus* significantly preferred (moved toward) their own substrates to those of conspecific males but made no such choice when paired against substrates of conspecific females. Thus, the movements and consequent substrate choices of the salamanders are influenced by the identity of the salamanders previously occupying the substrate, and the information upon which the choice is made apparently is transmitted by a pheromone which contains sexual information. Substrate choice is made by a withdrawal from the appropriate conspecific pheromone rather than an approach toward the individual's own peromone; i.e., males withdrew from conspecific male-marked substrates but not from conspecific-female marked substrates. Interspecific information is also transmitted between *P. cinereus* and a sibling species *P. nettingi shenandoah* (Jaeger & Gergits, op. cit.), as determined by laboratory experiments. In almost every sexual pairing, individuals of *P. cinereus* tended to withdraw from substrates previously marked by congeners in favor of their own-marked substrates and *P. nettingi shenandoah* did the same in every sexual pairing with *P. cinereus*. Thus, the skin and/or cloacal chemical cues placed on substrates by these salamanders contain species-identification as well as sex-identification information. These data on sex, individual and species recognition via chemical cues will be used later in this chapter to construct models on competition in salamander communities.

One further, possible use of chemoreception in salamanders should be mentioned here. A preliminary experiment by Jaeger (unpublished data) indicated that *P. cinereus* may be able to track certain prey items via chemical cues left by the prey on substrates. Salamanders were presented with two substrates: one covered with water in which a potential prey item had been soaked, and the other a water control. The salamanders preferred the prey-marked substrate when the potential prey was a slow moving animal (snail) but not when it was a fast moving animal (centipede). More detailed studies are needed to clarify whether salamanders can actually follow chemical trails left by natural prey; such experiments could prove interesting in light of findings that some newborn snakes and lizards can differentiate potential prey on the basis of chemical cues alone (Burghardt, 1969, 1970, 1973; Burghardt & Abeshaheen, 1971).

The area of chemoreception in salamanders needs a great deal more exploration. Two distinct approaches have been attempted so far. First, the behavioral approaches discussed here merely infer that chemoreception occurs when an animal responds differently to two different but largely unspecified cues, such as "substrate markings" or "air-borne odors". There is no independent check that such markings or odors actually do exist and no precise information as to their sources. The second approach is recording from olfactory neurons when known chemicals (of types normally not experienced by salamanders) are applied to the nasal receptors (Kauer, 1974; Kauer & Shepherd, 1975). A pairing of these methods could yield much information. For example, extracts from various glands in the skin or cloacae of salamanders could be made. These extracts could then by used to determine olfactory neuronal responses as well as whole-animal behavior as a function of the species (prey, conspecific, congener, etc.), sex, reproductive condition, social status, etc. of both the animal transmitting and the one receiving the chemical signal. In short, there is an exciting future for chemoreceptive studies in salamanders.

2. Sound Reception

While chemoreception plays a strong role in environmental sensing for caudates and a weak role for anurans, the opposite is true for sound reception. All salamanders lack the tympanic membrane and middle ear cavity, although there is a well developed inner ear (Lombard, 1977); there is some evidence that the inner ear may perceive some information via vibrations from the environment, but the significance of this is poorly understood. Except for an occasional squeek, salamanders are also voiceless.

Sound reception and photoreception are the primary sensory channels for adult frogs and toads. There is an enormous literature on sound reception, which I shall merely touch upon.

The morphology and function of the anuran inner ear were described in some detail by Lombard & Straughan (1974), and the morphology and physiology of the entire auditory system was reviewed by Capranica (1976) for adults and tadpoles. Almost all species of frogs have tympanic membranes and use some type of vocal communication. Such communication by sound has at least three functions: reproductive isolation, courtship in the broadest sense (attraction of females by males, release calls, etc.) and territorial advertisement.

Schiøtz (1973) stated that the overwhelming function of the mating call for most anuran species is to serve as a premating isolating mechanism. This is accomplished by each species having a relatively stereotyped call that is different from the calls of other sympatric species. Indeed, some sympatric species of anurans are extremely similar morphologically but can easily be distinguished by humans on the basis of their calls alone; *Hyla versicolor* and *H. chrysoscelis* (Hylidae) are good examples of this (Zweifel, 1970). Schiøtz's (1973) conclusion is based on a considerable amount of research by several workers. The evidence largely hinges on character displacement of calls. For a diversity of species (Bufonidae, Leptodactylidae, Hylidae, Microhylidae), closely related species-pairs that are sympatric have greater differences in voice than those same species-pairs from allopatric populations (Blair, 1955, 1958, 1962; Littlejohn, 1959, 1965, 1968). Other evidence for calls as premating isolating signals is covered by Schiøtz (1973). Apparently the species-specific information in the mating calls comes from the temporal pattern, or pulse repetition rate, of the calls (Straughan, 1973). Capranica (1976) demonstrated that both males and females are selectively tuned, to some extent, to the frequencies of the call of their own species. That is, their hearing is most sensitive to the species' own call, and the upper cut-off frequency is somewhat species specific. This, of course, could provide an effective means by which premating isolation could occur when a number of sympatric species are breeding, and thus calling, simultaneously. In *Acris* (Hylidae), there are geographic dialects in mating calls which are matched by the frequency sensitivity of the auditory nervous system (Capranica, Frishkoph & Nevo, 1973).

Straughan (1973, 1975) and Straughan & Heyer (1976) proposed that the frequency component of the anuran call is directed toward "channelization of species specific sounds" to avoid interspecific interference and confusion with other environmental sounds.

Calls by male frogs attract females to the places where the males are stationed (e.g., Straughan & Heyer, 1976), and such calls may be structurally so designed as to allow transmission of information over either long or short distances. Gerhardt (1976), working with the green treefrog *Hyla cinerea*, suggested that the low frequency components of the male's call can attract females from a distance, while high frequency components become important for species identification as the female gets closer to the male or chorus. The type of habitat and locations of the frogs affect the distance at which the high energy components have enough energy to influence the female's behavior; this illustrates the effect of environmental filtering on information transmission in frog calls. Gerhardt (1975) also found that some species of frogs approximate non-directional sound sources while other species are directional.

Calls also act as a reference point for migrating frogs of both sexes (Landreth & Ferguson, 1966). Many species, such as *Rana catesbeiana*, (Ranidae), utilize release calls; unreceptive females or males that are amplexed by another male emit this call and are usually released (Capranica, 1968).

Of particular ecological interest are the territorial calls of anurans. Emlen (1968) observed that male bullfrogs, *Rana catesbeiana*, establish territories during the breeding season against conspecific males. The territories are defended by stereotyped postures, approaches and combats, with intense intermale competition for females. Wiewandt (1969), using playback recordings of calls, showed that the "mating" call of the bullfrog can release an attack by a territorial male. The territorial male responds to the call with a specific "bonk" vocalization. Capranica (1968) demonstrated that the territorial calls consist of three types: one made by males only, one only by females and a third by both sexes. Crump (1972) reported that the male poison-arrow frog, *Dendrobates granuliferus* (Dendrobatidae), also exhibits territorial behavior involving vocalization, posture and combat. Males of the neotropical treefrog, *Eleutherodactylus coqui* (Leptodactylidae), have a two-note call; males and females of the species respond to different notes in the call (Narins & Capranica 1976). In this way, a male can apparently partition information concerning territorial defense against other males and courtship cues for females.

Some anurans may use sound reception for locating prey. Jaeger (1976) reported that the marine toad, *Bufo marinus* (Bufonidae), appears to use the calls of another species of frog to locate members of that species, which are then preyed upon by the toad. Using sound to locate prey may be useful to nocturnal frogs encountering poor visual perception.

Much information has recently been acquired concerning the auditory system and sound communication in anurans, especially from the studies of R. Capranica and his associates. However, research on anuran behavioral interactions, both intra- and interspecific, is just beginning, and sound reception undoubtably plays a major role in such interactions. Also, little is known about how sounds are used by frogs and toads in detecting potential predators and prey.

3. Photoreception

Photoreception plays an important role in environmental sensing by both frogs and salamanders. Vision, by means of the lateral eyes, is intimately involved in prey-catching behavior and probably in predator avoidance. Also, the lateral eyes and extraoptic receptors apparently are used in orientation, homing and habitat selection.

A great deal of research has been performed on amphibian visual neurophysiology, but only a fraction of this is applicable to perceptual ecology. One of the most interesting papers, by Lettvin, Maturana, McCulloch & Pitts (1959), showed that the output of the anuran retina is composed of four facets of the visual image: (1) local sharp edges and contrast, (2) movement of edges, (3) local dimmings produced by movement or rapid general darkening, and (4) the curvature of edge of a dark object. The last one of these they called the "bug perceivers" because it is ideal for detecting small insect prey. The fibers involved respond best when a small object that is smaller than a receptive field enters that field, stops and then moves about. The response is not affected by changes in lighting of by a moving background, and the response is not stimulated by the background *per se*, whether it is moving or still. In short, the anuran eye has an excellent system for isolating potential prey of a correct size from the noise of the background. Ingle (1968, 1970, 1971, 1973, 1975) has performed excellent studies correlating visual releasers with prey-catching behavior in anurans.

In the toad *Bufo bufo*, the first part of the prey-catching behavior is turning toward the prey, and this behavior is determined by the angular velocity, stimulus size and direction of the contrast presented by the prey item (Ewert, 1968a). There is an optimum velocity of prey movement (Ewert, 1968b), related to position in the visual field (Ewert & Ingle, 1971), which presumably prevents a toad from wasting energy striking at a prey that is moving too fast to be hit with the tongue. The components of feeding behavior (turning movement, binocular fixation and snapping at the prey) are also controlled in part by the contrast of the prey with its background (Ewert, 1969) and the number of moving prey items grouped together

(Ewert & Härter, 1969). While small objects in the visual field elicit the components of feeding behavior, large objects induce avoidance behavior (ducking, turning away or jumping), but both types of behavior depend on the degree of brightness contrast and velocity of movement of the object (Ewert, 1970). The above studies, which interface visual perception with feeding behavior, provide an excellent starting point for ecologists who are interested in food niche partitioning by sympatric anurans in various communities; this will be considered later in this chapter.

The use of vision by salamanders in processing information about food and predators is less well understood. Most salamanders, except for some cave-dwelling species, possess lateral eyes and the eyes are generally large relative to body size. My observations on a number of terrestrial species of salamanders indicate that visual input is needed for short-distance location of prey objects. When a potential prey, of a suitable ingestable size, is nearby, the salamander turns its head toward the prey, moves in a straight line toward the prey, usually stops if the prey ceases movement, then strikes with the tongue when the prey moves. However, since some salamanders can apparently track prey by odor (Jaeger, unpublished data) and can detect airborne odors over a considerable distance (Madison, 1972, 1975), it is difficult to separate visual from olfactory cues in foraging behavior.

Studies interfacing visual physiology and feeding behavior, such as those given above for anurans, are needed for salamanders. My general impression is that several cues used in the feeding behavior of anurans are also used by salamanders: i.e., size and movement of prey. There is little evidence on visually cued predator avoidance in salamanders, and I have seen no such responses when terrestrial salamanders are confronted with natural predators (such as snakes and shrews) in the laboratory.

Anurans that lay eggs in water tend to return each year to the same body of water for breeding (e.g., Heusser, 1969). Grubb (1970, 1973a, 1973b, 1975) suggested that chemoreception may play a role in such orientation for some species, but there is a large body of evidence that photoreception provides primary cues in this type of orientation in anurans. When frogs are released in outdoor test enclosures, from which only the sky is visible, they tend to orient along a line (Y-axis) perpendicular to the land-water interface of breeding sites. Such visually cued orientation, based on celestial cues, has been demonstrated for a large number of species (Dole, 1972, 1973; Ferguson, Landreth & McKeown, 1967; Ferguson, McKeown, Bosarge & Landreth, 1968; Gorman & Ferguson, 1970; Jordan, Byrd & Ferguson, 1968; Landreth & Ferguson, 1966, 1967, 1968; Taylor & Ferguson, 1969). Y-axis orientation requires information about the position of the sun for many species, but some species may use stars or stars and moon (Ferguson, 1971). The lateral eyes are apparently involved in Y-axis orientation, but extraoptic photoreceptors are also used in celestial orientation and setting of the biological clock (Adler, 1970, 1971, 1976; Justis & Taylor, 1976; Taylor & Ferguson, 1970).

Although Madison (1972) stressed chemoreception as a primary factor in salamander orientation and homing, recent evidence indicates that photoreception plays a role also. Both sighted and eyeless tiger salamanders, *Ambystoma tigrinum* (Ambystomatidae), can use linearly polarized light in spatial orientation, indicating that extraoptic photoreceptors are functional (Adler & Taylor, 1973; Taylor & Adler, 1973). These authors suggested that plane-polarized light can provide continuous knowledge of the position of the sun to migrating salamanders even when the sun is not visible, such as in wooded areas. The types of extraoptic photoreceptors used by caudates were reviewed by Adler (1976).

From an ecological perspective, one of the most interesting facets of photoreception in amphibians comes from studies of phototaxis. Although research on phototactic behavior has been conducted for a long time (Loeb, 1918), only recently have attempts been made to specify adaptive functions for phototaxis. Torelle (1903) showed, in rather crude studies, that two species of *Rana* frogs tended to respond photopositively to white light intensities and, when presented with combinations of "monochromatic" light, tended to move toward stimuli from that area of the visible spectrum which humans perceive as blue. W.R.A. Muntz and his associates continued this study and through a series of experiments derived much of the present knowledge on what may be called the "blue response" of anurans. Similar to Capranica's studies relating sound reception and behavior, Muntz performed experiments which interfaced phototactic behavior and visual physiology. Muntz (1962a) confirmed the behavioral "blue response" in *Rana temporaria* and stated that it is not a function of spectral sensitivity but is independent of both intensity and saturation. The on-fibers to the diencephalon were shown to respond more strongly to blue light than to any other color (Muntz, 1962b). Muntz (1962a) attempted an ecological hypothesis for the blue response: the escape-to-water hypothesis. Light from ponds or open sky over ponds should appear bluer to a semi-aquatic frog, such as *Rana*, than light from surrounding vegetation which is dominated by middle wavelengths (green). By jumping toward blue, the frog would land in water and effect an escape from an approaching terrestrial predator. Tadpoles of *Rana*, on the other hand, exhibit a "green response" (Muntz, 1963a, 1964), approaching green stimuli more strongly than any other colors, and this response gradually shifts to the blue response at metamorphosis. Looking at salamanders of the family Salamandridae, Muntz (1963b) found that one species *(Triturus cristatus)* is photopositive to white light intensities and demonstrates the blue response; another species *(Salamandra salamandra)* is photonegative to white light intensities and responds most strongly to the two ends of the visible spectrum (a U-shaped spectral response peaking in red and violet).

As a continuation of Muntz's studies, Jaeger & Hailman (1971) reported that both the blue-mode and the U-shaped spectral responses could be found in different species of frogs, and that they correspond to photopostitive and photonegative behavior, respectively, toward white light intensities. Experiments on 121 species of anurans (Jaeger & Hailman, 1973) demonstrated that each species has a preferred white light intensity (optimum ambient illumination, or O.A.I.) to which it will respond phototactically; most species prefer illuminance above 90 lux, a few less than 0.01 lux and the rest somewhere between these values. Most species with low O.A.I.'s are tropical forms and tend to be either nocturnal or active in very dimly lit microhabitats or both, and they tend to forage on relatively slowly moving organisms. Hailman & Jaeger (1974) and Jaeger & Hailman (1976a) proposed that the blue response is based on true color vision but the U-shaped spectral response is probably a function of spectral sensitivity. They proposed the following relationship between responses by anurans to intensity and spectral cues (Fig. 1): (1) when the ambient illumination is brighter than the species' O.A.I., the frog responds photonegatively and exhibits the U-shaped spectral response, since the ends of the visible spectrum appear dimmest to its eyes due to spectral sensitivity; (2) when ambient light is dimmer than the O.A.I., the frog responds photopositively and gives the blue response based on true color vision; (3) when ambient light is at the species' O.A.I., the animal attempts to maintain that intensity and is indifferent to colors. This model, then, proposes that phototaxis is merely a means by which frogs of a given species seek the ambient illumination which is optimal for their visual system; moving toward blue light coming from the sky allows the frog to move up a light gradient (the blue response, Fig. 1), until it reaches optimum conditions, in a habitat that may be patchy in terms of light intensities (such as forest with breaks in the canopy). Once the frog achieves or exceeds its O.A.I., there is no consistent color in the environment which can cue a down-gradient of intensity, and the frog responds by intensity cues alone (thus the spectral sensitivity function of the U-shaped spectral response in Fig. 1). The model was tested experimentally (Hailman & Jaeger, 1976) and was found to be valid; a given frog could be made to exhibit either the blue or U-shaped response depending on the intensity of the light presented. Hailman & Jaeger (1974) suggested that by placing itself at the O.A.I., a frog is able to experience the maximum amount of information from the environment, particularly concerning prey and predators. Field studies in the neotropics (Jaeger & Hailman, MS) suggested that different species of frogs forage under conditions of light that roughly approximate their O.A.I. values, although there are complications due to pupillary dynamics, retinal light- and dark-adaptation and circadian rhythms. Nevertheless, species of frogs seem to forage along time and habitat niche dimen-

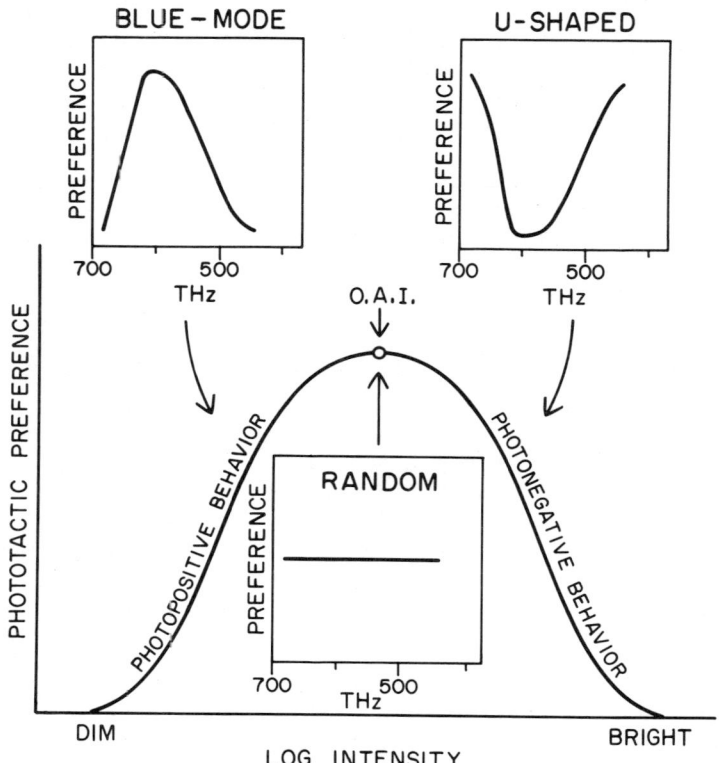

Fig. 1. The model of phototaxis. When ambient light intensity is below the frog's O.A.I., the frog is photopositive and expresses a blue-mode response (upper left insert). When intensity is above the O.A.I., the frog is photonegative and expresses a U-shaped spectral response (upper right insert). When intensity is at the O.A.I., the response to spectral stimuli becomes random (lower center insert). Adapted from Fig. 3 in Hailman & Jaeger (1976).

sions such that ambient light matches the O.A.I. In sum, we proposed (Hailman & Jaeger, 1974, 1976) that circadian rhythms first set the time of day when foraging occurs; then, when pupillary and retinal adaptational adjustments fail to compensate for extremes in ambient light, the frog makes such adjustments bodily through phototaxis until it reaches optimal ambient conditions. Hailman & Jaeger (1974) showed that Muntz's (1962a) escape-to-water hypothesis is incorrect, since species that never escape to water, or never go to water even to breed, (i.e., *Eleutherodactylus*, Leptodactylidae), also give the blue response.

Jaeger & Hailman (1976b) confirmed Muntz's (1963a, 1964) green response for tadpoles of a number of families. They proposed that by swimming toward green, a tadpole is able to find patches of vegetation which serve as food and shelter from predation, and several experiments supported the hypothesis.

Phototactic behavior, then, may serve as a means by which different species of adult anurans living sympatrically may reduce interspecific competition through habitat partitioning. A model of this will be presented later in the chapter.

4. Other types of reception

Of course amphibians have other types of receptors for sensing environmental information, but the sensory ecology of these is not well understood. Tadpoles and aquatic frogs have lateral lines (Murray & Capranica, 1973), which are excellent for deriving information under aquatic conditions. Tactile reception is undoubtably important during courtship behavior of salamanders, which perform stereotyped contact movements (Arnold 1972, 1976), and frogs, which mate via amplexus. However, no further consideration will be given to these.

PERCEPTION AS SHAPED BY ENVIRONMENTAL DIMENSIONS

Anurans and caudates have somewhat different means for gathering information from the environment. Anurans primarily utilize sound and photoreception while caudates rely primarily on olfaction and, to an unknown degree, vision. Even within these taxonomic groups, sensory modes are often partitioned. For example, different species of frogs have different O.A.I.'s and utilize different frequency ranges of sounds in their calls, while salamanders apparently transmit species specific information via chemical cues. I shall now discuss some aspects of amphibian ecology and attempt to relate sensory functions to environmental constraints and ecological niche dimensions. Some of the arguments are, at this point, quite speculative.

ECOLOGICAL ENVIRONMENT OF SALAMANDERS

1. Constraints of the Physical Environment

Salamanders tend to be either aquatic or to lead secretive lives in terrestrial habitats, such as in fossorial burrows, forest-litter, arboreal bromeliads or under rocks and logs. In both aquatic and heterogeneous terrestrial habitats, chemical cues are excellent for transmitting long-term information concerning spatial defense, courtship, etc. by often relatively sedentary animals. Water carries chemical cues over long distances; terrestrial substrates can be marked and odors carried by air. Because of the great diversity of skin and cloacal glands in salamanders, an enormous amount of information can be transmitted by chemical signals alone. Even where long distance information is necessary, such as migration to bodies of water and homing, air-borne chemical information seems to be adequate. The ecological adaptiveness of photic cues in orientation of salamanders is still poorly understood, although such photic cues apparently are used. Since chemical cues function for both short and long distance information, salamanders do not greatly utilize vision (as do some mammals, for example) or hearing (as do frogs) for long distance sensing of the environment. However, vision probably plays an important role in the fine tuning of short distance feeding and courtship behavior. Sound is not effectively utilized even though there is a well developed inner ear.

2. Intraspecific Competition: A Model

Food is periodically a limited resource for the forest-litter salamander *Plethodon cinereus* (Jaeger, 1972, MS). When the forest litter is wet, the salamanders forage in and on the litter and, depending on ambient temperature and thus the salamander's metabolic rate, food may or may not be limited. However, during dry periods, the salamanders are prevented from foraging in all but the deepest patches of litter due to the threat of desiccation, and food then becomes limited relatively independent of temperature. I previously proposed (Jaeger, 1972) that competition for food occurs through exploitation; i.e., one salamander is merely better at obtaining food items than another. It now appears that interference competition plays a role, in which salamanders may aggressively defend foraging territories. Although territoriality *per se* (i.e., a defended space) has not been convincingly demonstrated thus far, there is increasing evidence that it exists. Both Grant (1955) and Thurow (1976) observed what appears to be territorial behavior in various species of plethodontid salamanders. Certainly intraspecific aggression does occur (e.g. Organ, 1960a, 1960b; Thurow, 1976; Gergits, unpublished data on *P. cinereus*). Of interest here is the evidence for olfactory advertisement of substrates. Tristram (1977) showed that *P. cinereus* behaves differently, via tapping behavior, toward own-marked and conspecific-marked substrates, and Jaeger & Gergits (MS) showed that members of this species often choose their own substrates in preference to conspecific-marked substrates (depending on the combination of sexes involved). McGavin (1978) proposed that individual recognition of conspecifics by means of substrate odors could be used in territorial defense; some birds respond aggressively toward the calls of unfamiliar conspecifics but exhibit reduced aggression toward the calls of familiar neighboring conspecifics. McGavin proposed that *P. cinereus* could use olfactory cues in the same way; an individual could increase its fitness by reducing "needless" aggression toward neighboring territorial salamanders that are not infringing on the individual's own territory.

Considerable research is still required to understand whether territorialism is common among terrestrial salamanders. Competition for food in animals often takes the form of a defended foraging space, which is more efficient than purely exploitative competition (Miller, 1969), since with interference the resolution of competition is relatively rapid. Also, the well developed olfactory communication system in salamanders would be excellent for producing long term signal markers. Finally, terrestrial salamanders seem to have a long search time during foraging (in relation to pursuit and handling times) (Fraser, 1976), and this search time could be used simultaneously in defending a foraging site, as in Schoener's (1971) Type I predator. In sum, the model here predicts that one way a terrestrial salamander can maximize fitness is to locate an optimum foraging patch in the forest-litter (or elsewhere) where prey availability is relatively predictable, to mark that patch with an odoriferous advertisement, and to defend the patch against invasion by conspecifics. Further studies on chemoreception in salamanders may reveal that its primary function, aside from involvement in courtship, is to allow individuals to partition space, and thus food, in a patchy environment.

3. Homing Behavior

It is easily understandable why salamanders that breed in water have evolved the ability to home. Merely because a salamander has survived the larval stage in a pond or stream and reached adulthood conveys "knowledge" on that individual that that body of water is suitable for its own young, in terms of predictability of predator load, predictability of moisture level, etc. In short, an individual might inhance its fitness by returning to a body of water that conveys some predictability to the individual rather than going to an unknown body of water that has no *a priori* predictability.

However, it has not been clear why terrestrial salamanders home to the spot from which they were removed. If moved even a short distance, individuals tend to home (Madison, 1972). Given the above model of territorial partitioning of habitats, though, homing does make sense. If suitable habitat is partitioned by conspecifics, a displaced salamander might find itself in another individual's territory. As a general rule-of-thumb for territorial vertebrates, the territory holder has the advantage over the invader in aggressive encounters. Consequently, unless an individual has been displaced from an extremely food-poor territory, that individual's fitness may best be served by rapidly returning to the home area. Where studied, such returns do tend to be rapid (Madison, 1972). A salamander returning "home" could use directional cues provided by either its own substrate markings on the home site or by the odors of familiar neighboring conspecifics (see Madison, 1975).

4. Courtship Behavior

Courtship behavior involves a complexity of olfactory, visual and tactile cues (Arnold 1972, 1976). As with aquatic salamanders (Twitty, 1955), males of terrestrial species probably track females (and perhaps vice versa) along olfactory trails or by air-borne odors. Courtship in terrestrial salamanders usually occurs during discrete periods of the year. Such movements in search of mates may enforce the necessity of later homing to the individual's territorial site, hypothesized above. It is unlikely that a terrestrial salamander would be forcibly "displaced" from its home site except by some human researcher. However, after finding a mate, courting and expending its spermatophores, it may be beneficial for a salamander to return "home" to forage. Both males and females of a number of *Plethodon* species are known to forage during the courtship period (personal observations). Madison's (1975) curious observation that recognition of neighboring *P. jordani* breaks down during the breeding period may reflect a relaxation of territorial enforcement during courtship.

Premating isolating mechanisms seem to be strongly formed in salamanders. Although there are many parapatric borders between the distributions of sibling species of *Plethodon* in eastern North America, hybridization is relatively rare (Highton, 1972). Because olfactory communication appears to be greatly involved in courtship (Arnold, 1972, 1976), I suspect that species specific cues are transmitted between males and females either during tracking of a mate (possibly via skin or cloacal glands), or during the courtship ceremony itself (possibly via slapping with the metal hedonic gland), or both.

5. Interspecific Competition: A Model

Interspecific competition is thought, by many ecologists, to be a dominant force in organizing terrestrial vertebrate communities (see a number of papers in Cody & Diamond, 1975). One indication that such competition occurs is through the observation of parapatric distribution of similar species (e.g., Terborgh & Weske, 1975). Many sibling species of terrestrial *Plethodon* salamanders in eastern North America maintain such parapatry (Highton, 1972): abutting ranges with little geographic overlap. Jaeger (1971, 1972, 1974) examined the parapatric sibling species *P. cinereus* and *P. nettingi shenandoah* and concluded that competition occurs for a periodically limited food resource (see the discussion above on intraspecific competition). More recent research (Jaeger, MS; Jaeger & Gergits, MS) suggests that interspecific territories may exist, similar to those discussed above for intraspecific competition. Thurow (1976) suggested that interspecific aggression and territories occur among a number of species of *Plethodon*, and Jaeger & Gergits (MS) presented data showing that interspecific communication occurs between *P. cinereus* and *P. nettingi shenandoah*; members of the two species generally avoid substrates marked by congeners. The model, then, states that foraging territories are maintained between similar species of terrestrial salamanders through olfactory advertisement and aggressive encounters: the competitively superior species is thus able to displace the competitively inferior species from optimal habitats, as is seen between *P. cinereus* and *P. nettingi shenandoah* (Jaeger, 1970, 1971).

Species packing theory predicts, in very simple form, that the number of similar species that can coexist in an area is directly related to the predictability of the environment in that area. Thus, areas that have stochastic fluctuations in prey will contain a lower species diversity than areas that have more deterministic prey (May & MacArthur, 1972). Species packing increases in deterministic areas either by an increase in overlap along a single niche dimension, (such as prey size) (May & MacArthur, 1972), or by greater specialization by species along that dimension (Leigh, 1975). Jaeger (MS) proposed that salamander communities generally follow this theory: the greater the predictability of ambient moisture, the greater the diversity of species. The reasoning is that with less variability in ambient moisture, salamanders have a greater area over which to forage (free of the constraints of dessication) which in turn leads to less variability in prey availability. This leads to a more specialized foraging strategy, as predicted by Emlen's (1966) optimal choice model, which finally leads to compressed food and/or habitat niches. As this compression takes place, more species are able to exploit the food resources in the community, and species richness increases.

How does species packing theory relate to sensory ecology? By going one step farther, Schoener (1974) showed that for many communities, as species diversity increases species first partition the habitat niche dimension, then food, and finally time (such that some species are nocturnal, some diurnal, etc). In all of these cases, food can be effectively partitioned among species that forage in different habitats, feed on different sizes of prey, and forage at different times. Finer partitioning of these three dimensions occurs as more species pack into the community. If different species of salamanders in a community partition food on a temporal or spatial (habitat) basis, they will by necessity be foraging under different conditions of light intensity. Since the feeding response of salamanders is probably visually cued, species foraging under different intensity ranges of light should have visual systems that are optimally fit for those ranges. This is similar to the O.A.I. responses of anurans, discussed above. Some species should have their best visual perception under dim light (foraging at night or in dimly lit microhabitats) and others under brighter intensities. Indeed, Muntz (1963b) found two species of salamanders to have different phototactic responses: one preferring high intensities of white light and the other preferring low intensities. If increased specialization through species packing (Leigh, 1975) can be applied to visual physiology, then as species diversity in a community increases, one might expect to find an increase in visual specialization; i.e., species will forage under narrower ranges of illumination governed by their more specialized habitats or times of foraging. If, on the other hand, salamanders partition a food resource by size of prey, and if salamanders are found to track their optimal prey by olfactory cues, then one might expect an increase in species diversity to parallel a decrease (more specialization) in prey-type odors to which salamanders of a species will respond.

In short, the perception of environmental information in salamander species may depend to some extent on the number of species that share a community. This idea will be taken up again in discussing anurans.

6. Concluding Comments

The ideas presented in this section range from largely hypothetical to purely speculative. Most of them are presented in simplified form due to the constraints of space. My goal is merely to present ideas that may stimulate further research in population and community ecology of salamanders, research that will consider sensory modes as adaptive functions in intra- and interspecific interactions, such as competition, and in the foraging strategies of species.

ECOLOGICAL ENVIRONMENT OF FROGS

1. Constraints of the Physical Environment

Frogs use sound to transmit long distance information (attraction of a mate of the correct species) and short distance information (territorial calls). Vision is also used for long distance (celestial cues for homing) and short distance (prey and predator detection) cues. Since many species of frogs are more mobile than salamanders, they may be less constrained by the danger of desiccation, being able to move among patches of moisture. Thus, frogs tend to have more "niches" open to them, such as open fields, diurnal use of the water-shore interface, etc. Sound and photoreception provide excellent systems for gathering environmental information in relatively open habitats (relative to the secretive places inhabited by salamanders).

2. Intraspecific Competition

There is little evidence that temperate area frogs, such as *Rana* species, compete for a limited food resource. Many of these species feed on flying insects, which tend to be a rapidly renewable resource. Yet an increasing number of species are found or suspected to be territorial (e.g., Emlen, 1968; Wiewandt, 1969). In this case, interference competition for space, defended by aggression and vocal advertisement, appears to be a proximate factor in competition for mates (the ultimate factor). Presumably males can maximize their fitness by defending sites that will either attract the optimal number of females and/or provide optimal environments for their eggs and young.

3. Homing Behavior

As with aquatic-breeding salamanders, frogs that lay eggs in water return to the home pond each year or when displaced. This may be viewed as adaptive because the home pond is more predictable as a suitable habitat for young than are "unknown" ponds.

4. Interspecific Competition: A Model

I wish to examine phototactic behavior, as discussed earlier, in terms of species packing and niche dimensionality theories (as discussed under interspecific competition in salamanders). Each species of anuran has an O.A.I. value toward which it will move by phototaxis (Hailman & Jaeger, 1974; 1976). We hypothesized that the frog has its best perceptual abilities when at its O.A.I., and thus the optimal ability to detect prey and predators. Actually, the O.A.I. of a species is not a single value but is a range of values bounded by the light- and dark-adaptational abilities of the retina (Hailman & Jaeger, 1976). Consequently, some species seem to have an O.A.I. that moves over a small range of intensities while other species' O.A.I.'s range over several log units of illuminance, although data on this is still in the preliminary stage. I shall call these visual specialists and generalists, respectively. Now, by using species packing theory (May & MacArthur, 1972; Leigh, 1975) and niche dimensionality theory (Schoener, 1974), one can make several predictions about anuran communities. As species diversity increases, say from the north temperate area with only a few coexisting species to the neotropics with up to 81 sympatric species (Crump, 1974), one might expect to find an increase in the number of visual specialists due to niche compression. That is, in northern areas, species may forage under a broad range of illuminances (from quite dark to very bright intensities) perhaps sacrificing excellent perception at any given illuminance for reasonably good perception over a wide range of illuminances. These frogs optimize fitness by foraging under many diverse environmental conditions. However, in the tropics species may forage under only a narrow range of illuminances (dim, or bright, or somewhere inbetween). These specialists optimize fitness by foraging under a limited set of environmental conditions, which could be done if the tropics present a relatively deterministic environment, as is often proposed (May & MacArthur, 1972). With an increase in visual (i.e., O.A.I.) specialists in the tropics, more species could be packed into a community, by either more finely dividing foraging habitats or more finely partitioning time of day for foraging. Although there are few data yet concerning visual specialists and generalists (Hailman & Jaeger, 1976), their occurrence could be one reason why tropical anuran communities are so diverse in species.

Another prediction is that anurans in densely packed tropical communities will partition more niche dimensions than those in depauperate temperate communities. According to Schoener's (1974) ideas, species partition first habitats, then food (usually by size), then time of foraging as species packing increases. Taking the last of these (time) as the limiting case, one would expect to find a greater diversity of species' O.A.I.'s in the tropics than in temperate areas. There is evidence that this occurs. Jaeger & Hailman (1973) found that of the 66 tropical and subtropical species tested for O.A.I. values, 23% had low O.A.I.'s; of the 55 non-tropical species, only 1.8% (1 species) preferred low illuminance. Thus, increased species diversity in the tropics may be partially accomplished by the occurrence of a large number of species that are either nocturnal or inhabit very dimly lit microhabitats, such as tree cavities. Circadian rhythmicity and phototaxis, via a diversity of O.A.I. ranges, may account to some extent for the extensive partitioning of time and habitat niche dimensions by tropical species of frogs.

If food is also partitioned along a food size niche dimension, then this might be accomplished by species-specific adjustments in the "bug perceivers" (Lettvin et al., 1959), which are tuned to the size of the prey object in relation to the size of the visual receptive field. A small prey object may stimulate the feeding behavior of one anuran species but may not stimulate another species which feeds on a somewhat different size prey (Ewert, 1970).

There is further evidence that anurans do partition habitat and time niche dimensions in the tropics. Jaeger, Hailman & Jaeger (1976) and Jaeger & Hailman (MS) studied six species of sympatric ground-feeding anurans in Panama. *Bufo marinus* (Bufonidae) and *Leptodactylus pentadactylus* (Leptodactylidae) are similar in size, feed on large prey and are nocturnal; however, the former species forages in open areas under moonlight and the latter forages in the dimly lit forest. *Colostethus nubicola* and *Dendrobates auratus* (Dendrobatidae) both forage in the morning and evening under similar conditions of light, but they differ in body size (and thus presumably prey size ingested). *Bufo typhonius* (Bufonidae) and *Physalaemus pustulosus* (Leptodactylidae) both feed during the mid-day but differ in habitat. In sum, six species feed on a common resource base, yet effectively partition that resource in six ways along three niche dimensions (habitat, food size and time of foraging), as predicted by Schoener (1974).

5. Concluding Comments

Again, some of the ideas presented here are speculations and some are hypotheses based on an attempt by Hailman and myself to identify ecological adaptive functions for phototactic behavior in anurans. In either case, the ideas are amenable to rigorous empirical testing.

SENSORY ECOLOGY

My view of sensory ecology is that the sensory modes utilized by species are intimately adapted to the abiotic and biotic components of the environment under which the species exist. A greater understanding of sensory functions and environmental parameters should go hand in hand.

ACKNOWLEDGEMENTS

My own research discussed in this chapter was supported by N.S.F. grants GB-28824X and GB-38028, National Geographic Society grant 1127, American Philosophical Society grant 7113, State University of New York Faculty Research Fellowship and Grant-in-Aid 20-7331-A and S.U.N.Y. Faculty Research Grant 20-A050-A, and by funds from the Shenandoah Natural History Association.

REFERENCES

Adler, K. (1970). The role of extraoptic photoreceptors in amphibian rhythms and orientation: a review. J. Herpetol. 4: 99-112.
Adler, K. (1971). Pineal end organ: role in extraoptic entrainment of circadian locomotor rhythm in frogs. In M. Menaker [ed.] Biochronometry. Natl. Acad. Sci. U.S.A., Washington, D.C.; pp. 342-350.
Adler, K. (1976). Extraocular photoreception in amphibians. Photochem. Photobiol. 23: 275-298.
Adler, K. and D. H. Taylor (1973). Extraocular perception of polarized light by orienting salamanders. J. Comp. Physiol. 87: 203-212.
Arnold, S. J. (1972). The evolution of courtship behavior in salamanders. Ph.D. Dissertation, Univ. Michigan.
Arnold, S. J. (1976). Sexual behavior, sexual interference and sexual defense in the salamanders *Ambystoma maculatum*, *Ambystoma tigrinum* and *Plethodon jordani*. Z. Tierpsychol. 42: 247-300.
Blair, W. F. (1955). Mating call and stage of speciation in the *Microhyla olivacea - M. carolinensis* complex. Evolution 9: 469-489
Blair, W. F. (1958). Mating call in the speciation of anuran amphibians. Amer. Natur. 92: 27-51.
Blair, W. F. (1962). Non-morphological data in anuran classification. Syst. Zool. 11: 72-84.
Brown, C. W. (1968). Additional observations on the function of the nasolabial grooves of plethodontid salamanders. Copeia 1968: 728-731.
Burghardt, G. M. (1969). Comparative prey-attack studies in newborn snakes of the genus *Thamnophis*. Behaviour 33: 77-114.
Burghardt, G. M. (1970). Intraspecific geographical variation in chemical food cue preferences of newborn garter snakes *(Thamnophis sirtalis)*. Behaviour 36: 246-257.
Burghardt, G. M. (1973). Chemical release of prey attack: extension to naive newly hatched lizards, *Eumeces fasciatus*. Copeia 1973: 178-181.
Burghardt, G. M. and J. P. Abeshaheen (1971). Responses to chemical stimuli of prey in newly hatched snakes of the genus *Elaphe*. Anim. Behav. 19: 486-489.
Capranica, R. R. (1968). The vocal repertoire of the bullfrog *(Rana catesbeiana)*. Behaviour 31: 302-325.
Capranica, R. R. (1976). Morphology and physiology of the auditory system. In R. Llinás and W. Precht [eds.] Frog neurobiology. Springer-Verlag, New York; pp. 551-575.
Capranica, R. R., L. S. Frishkoph, and E. Nevo (1973). Encoding of geographic dialects in the auditory system of the cricket frog. Science 182: 1272-1275.
Cody, M. L. and J. M. Diamond [eds.] (1975). Ecology and evolution of communities. Harvard Univ. Pr., Cambridge, Mass.
Crump, M. L. (1972). Territoriality and mating behavior in *Dendro-*

bates granuliferus (Anura: Dendrobatidae). Herpetologica 28: 195-198.
Crump, M. L. (1974). Reproductive strategies in a tropical anuran community. Univ. Kansas Mus. Natur. Hist. Misc. Publ. 61: 1-68.
Dole, J. W. (1972). Evidence of celestial orientation in newly-metamorphosed *Rana pipiens*. Herpetologica 28: 272-276.
Dole, J. W. (1973). Celestial orientation in recently metamorphosed *Bufo americanus*. Herpetologica 29: 59-62.
Emlen, J. M. (1966). The role of time and energy in food preference. Amer. Natur. 100: 611-617.
Emlen, S. T. (1968). Territoriality in the bullfrog, *Rana catesbeiana*. Copeia 1968: 240-243.
Ewert, J.-P. (1968a). Der Einfluss von Zwischenhirndefekten auf die Visuomotorik im Beute- und Fluchtverhalten der Erdkröte *(Bufo bufo L.)*. Z. Vergl. Physiol. 61: 41-70.
Ewert, J.-P. (1968b). Verhaltensphysiologische Untersuchungen zum "stroboskopischen Sehen" der Erdkröte *(Bufo bufo L.)*. Pflügers Arch. 299: 158-166.
Ewert, J.-P. (1969). Quantitative Analyse von Reiz-Reaktionsbeziehungen bei visuellem Auslösen der Beutefang-Wendereaktion der Erdkröte *(Bufo bufo L.)*. Pflügers Arch. 308: 225-243.
Ewert, J.-P. (1970). Neural mechanisms of prey-catching and avoidance behavior in the toad *(Bufo bufo L.)*. Brain, Behav. Evol. 3: 36-56.
Ewert, J.-P. and H.-A. Härter (1969). Der hemmende Einfluss gleichzeitig bewegter Beuteattrappen auf das Beutefangverhalten der Erdkröte *(Bufo bufo L.)*. Z. Vergl. Physiol. 64: 135-153.
Ewert, J.-P. and D. Ingle (1971). Excitatory effects following habituation of prey-catching activity in frogs and toads. J. Comp. Physiol. Psychol. 77: 369-374.
Ferguson, D. E. (1967). Sun-compass orientation in anurans. In R. M. Storm [ed.] Animal orientation and navigation. Oregon State Univ. Pr., Corvallis; pp. 21-34.
Ferguson, D. E. (1971). The sensory basis of orientation in amphibians. Ann. New York Acad. Sci. 188: 30-36.
Ferguson, D. E., H. F. Landreth and J. P. McKeown (1967). Sun compass orientation of the northern cricket frog, *Acris crepitans*. Anim. Behav. 15: 45-53.
Ferguson, D. E., J. P. McKeown, O. S. Bosarge and H. F. Landreth (1968). Sun-compass orientation of bullfrogs. Copeia 1968: 230-235.
Fraser, D. F. (1976). Empirical evaluation of the hypothesis of food competition in salamanders of the genus *Plethodon*. Ecology 57: 459-471.
Gerhardt, H. C. (1975). Sound pressure levels and radiation patterns of the vocalizations of some North American frogs and toads. J. Comp. Physiol. 102: 1-12.
Gerhardt, H. C. (1976). Significance of two frequency bands in long distance vocal communication in the green treefrog. Nature 261: 692-694.

Gorman, R. R. and J. H. Ferguson (1970). Sun-compass orientation in the western toad, *Bufo boreas*. Herpetologica 26: 34-45.

Grant, D., O. Anderson and V. C. Twitty (1968). Homing orientation by olfaction in newts *(Taricha rivularis)*. Science 160: 1354-1356.

Grant, W. C. (1955). Territorialism in two species of salamanders. Science 121: 137-138.

Grubb, J. C. (1970). Orientation in post-reproductive Mexican toads, *Bufo valliceps*. Copeia 1970: 674-680.

Grubb, J. C. (1973a). Olfactory orientation in breeding Mexican toads, *Bufo valliceps*. Copeia 1973: 490-497.

Grubb, J. C. (1975) Olfactory orientation in southern leopard frogs, *Rana utricularia*. Herpetologica 31: 219-221.

Hailman, J. P. and R. G. Jaeger (1974). Phototactic responses to spectrally dominant stimuli and use of colour vision by adult anuran amphibians: a comparative survey. Anim. Behav. 22: 757-795.

Hailman, J. P. and R. G. Jaeger (1976). A model of phototaxis and its evaluation with anuran amphibians. Behaviour 56: 215-249.

Heusser, H. (1969). Die Lebenweise der Erdkröte, *Bufo bufo* (L.); das Orientierungsproblem. Rev. Suisse de Zool. 76: 443-518.

Highton, R. (1972). Distributional interactions among eastern North American salamanders of the genus *Plethodon*. In P. C. Holt [ed.] The distributional history of the biota of the southern Appalachians, Part III: vertebrates. Virginia Polytechnic Inst. Res. Div. Monogr. 4: 139-188.

Ingle, D. (1968). Visual releasers of prey-catching behavior in frogs and toads. Brain, Behav. Evol. 1: 500-518.

Ingle, D. (1970). Visuomotor functions of the frog optic tectum. Brain, Behav. Evol. 3: 57-71.

Ingle, D. (1971). A possible behavioral correlate of delayed retinal discharge in anurans. Vision Res. 11: 167-168.

Ingle, D. (1973). Two visual systems in the frog. Science 181: 1053-1055.

Ingle, D. (1975). Focal attention in the frog: behavioral and physiological correlates. Science 188: 1033-1035.

Jaeger, R. G. (1970). Potential extinction through competition between two species of terrestrial salamanders. Evolution 24: 632-642.

Jaeger, R. G. (1971). Competitive exclusion as a factor influencing the distributions of two species of terrestrial salamanders. Ecology 52: 632-637.

Jaeger, R. G. (1972). Food as a limited resource in competition between two species of terrestrial salamanders. Ecology 53: 535-546.

Jaeger, R. G. (1974). Competitive exclusion: comments on survival and extinction of species. BioScience 24: 33-39.

Jaeger, R. G. (1976). A possible prey-call window in anuran auditory perception. Copeia 1976: 833-834.

Jaeger, R. G. and J. P. Hailman (1971). Two types of phototactic

behaviour in anuran amphibians. Nature 230: 189-190.
Jaeger, R. G. and J. P. Hailman (1973). Effects of intensity on the phototactic responses of adult anuran amphibians: a comparative survey. Z. Tierpsychol. 33: 352-407.
Jaeger, R. G. and J. P. Hailman (1976a). Phototaxis in anurans: relation between intensity and spectral preferences. Copeia 1976: 92-98.
Jaeger, R. G. and J. P. Hailman (1976b). Ontogenetic shift of spectral phototactic preferences in anuran tadpoles. J. Comp. Physiol. Psychol. 90: 930-945.
Jaeger, R. G., J. P. Hailman and L. S. Jaeger (1976). Bimodal diel activity of a Panamanian dendrobatid frog, *Colostethus nubicola*, in relation to light. Herpetologica 32: 77-81.
Jordan, O. R., W. W. Byrd and D. E. Ferguson (1968). Sun-compass orientation in *Rana pipiens*. Herpetologica 24: 335-336.
Justis, C. S. and D. H. Taylor (1976). Extraocular photoreception and compass orientation in larval bullfrogs, *Rana catesbeiana*. Copeia 1976: 98-105.
Kauer, J. S. (1974). Response patterns of amphibian olfactory bulb neurones to odour stimulation. J. Physiol. 243: 695-715.
Kauer, J. S. and D. G. Moulton (1974). Responses of olfactory bulb neurones to odour stimulation of small nasal areas in the salamander. J. Physiol. 243: 717-737.
Kauer, J. S. and G. M. Shepherd (1975). Olfactory stimulation with controlled and monitored step pulses of odor. Brain Res. 85: 108-113.
Landreth, H. F. and D. E. Ferguson (1966). Evidence of sun-compass orientation of the chorus frog, *Pseudacris triseriata*. Herpetologica 22: 106-112.
Landreth, H. F. and D. E. Ferguson (1967). Movements and orientation of the tailed frog, *Ascaphus truei*. Herpetologica 23: 81-93.
Landreth, H. F. and D. E. Ferguson (1968). The sun compass of Fowler's toad *Bufo woodhousei fowleri*. Behaviour 30: 27-43.
Leigh, E. G., Jr. (1975). Population fluctuations, community stability, and environmental variability. In M. L. Cody and J. M. Diamond [eds.] Ecology and evolution of communities. Harvard Univ. Pr., Cambridge, Mass; pp. 51-73.
Lettvin, J. Y., H. R. Maturana, W. S. McCulloch and W. H. Pitts (1959). What the frog's eye tells the frog's brain. Proc. Inst. Radio Eng. 47: 1940-1951.
Littlejohn, M. J. (1959). Call differentiation in a complex of seven species of *Crinia*. Evolution 13: 452-468.
Littlejohn, M. J. (1965). Premating isolation in the *Hyla ewingi* complex (Anura, Hylidae). Evolution 19: 234-243.
Littlejohn, M. J. (1968). The systematic significance of isolating mechanisms. In Systematic biology, Nat. Acad. Sci. USA, Washington, D.C.; pp. 459-482.
Loeb, J. (1918). Forced movements, tropisms, and animal conduct. J. B. Lippincott Co., Philadelphia. Reprinted by Dover Publ. New York, 1973.

Lombard, R. E. (1977). Comparative morphology of the inner ear in salamanders (Caudata: Amphibia). Contrib. Vertebrate Evol. 2. Karger Publ., New York.

Lombard, R. E. and I. R. Straughan (1974). Functional aspects of anuran middle ear structures. J. Exp. Biol. 61: 71-93.

Madison, D. M. (1969). Homing behaviour of the red-cheeked salamander, *Plethodon jordani*. Anim. Behav. 17: 25-39.

Madison, D. M. (1972). Homing orientation in salamanders: a mechanism involving chemical cues. In S. R. Galler, K. Schmidt-Koenig, G. J. Jacobs and R. E. Belleville [eds.] Animal orientation and navigation. National Aeronautics and Space Administration, Washington, D.C.; pp. 485-498.

Madison, D. M. (1975). Intraspecific odor preferences between salamanders of the same sex: dependence on season and proximity of residence. Can. J. Zool. 53: 1356-1361.

Madison, D. M. (1977). Chemical communication in amphibians and reptiles. In D. Müller-Schwarze and M. M. Mozell [eds.] Chemical signals in vertebrates. Plenum Publ. Corp., New York; pp. 135-168.

Madison, D. M. and C. R. Shoop (1970). Homing behavior, orientation, and home range of salamanders tagged with Tantalum-182. Science 168: 1484-1487.

May, R. M. and R. H. MacArthur (1972). Niche overlap as a function of environmental variability. Proc. Nat. Acad. Sci. USA 69: 1109-1113.

McGavin, M. (1978). Recognition of conspecific odors by the salamander, *Plethdon cinereus*. Copeia, in press.

Miller, R. S. (1969). Competition and species diversity. Brookhaven Symp. Biol. 22: 63-70.

Muntz, W. R. A. (1962a). Effectiveness of different colors of light in releasing the positive phototactic behavior of frogs, and a possible function of the retinal projection of the diencephalon. J. Neurophysiol. 25: 712-720.

Muntz, W. R. A. (1962b). Microelectrode recordings from the diencephalon of the frog *(Rana pipiens)* and a blue-sensitive system. J. Neurophysiol. 25: 699-711.

Muntz, W. R. A. (1963a). The development of phototaxis in the frog *(Rana temporaria)*. J. Exp. Biol. 40: 371-379.

Muntz, W. R. A. (1963b). Phototaxis and green rods in urodeles. Nature 199: 620.

Muntz, W. R. A. (1964). The development of photopic and scotopic vision in the frog *(Rana temporaria)*. Vision Res. 4: 241-250.

Murray, M. J. and R. R. Capranica (1973). Spike generation in the lateral-line efferents of *Xenopus laevis*: evidence favoring multiple sites of initiation. J. Comp. Physiol. 87: 1-20.

Narins, P. M. and R. R. Capranica (1976). Sexual differences in the auditory system of the tree frog *Eleutherodactylus coqui*. Science 192: 378-380.

Organ, J. A. (1958). Courtship and spermatophore of *Plethodon jordani metcalfi*. Copeia 1958: 251-259.

Organ, J. A. (1960a). The courtship and spermatophore of the salamander, *Plethodon glutinosus*. Copeia 1960: 34-40.
Organ, J. A. (1960b). Studies on the life history of the salamander, *Plethodon welleri*. Copeia 1960: 287-297.
Porter, K. R. (1972). Herpetology. W. B. Saunders Co., Philadelphia.
Rabb, G. B. and M. S. Rabb (1963). Additional observations on breeding behavior of the Surinam toad, *Pipa pipa*. Copeia 1963: 636-642.
Savage, J. M. (1973). The geographic distribution of frogs: patterns and predictions. In J. L. Vial [ed.] Evolutionary biology of the anurans. Univ. Missouri Pr., Columbia, Missouri; pp. 351-445.
Schiøtz, A. (1973). Evolution of anuran mating calls: ecological aspects. In J. L. Vial [ed.] Evolutionary biology of the anurans. Univ. Missouri Pr., Columbia, Missouri; pp. 311-319.
Schoener, T. W. (1971). Theory of feeding strategies. Ann. Rev. Ecol. Syst. 2: 369-404.
Schoener, T. W. (1974). Resource partitioning in ecological communities. Science 185: 27-39.
Sever, D. M. (1975). Morphology and seasonal variation of the nasolabial glands in *Eurycea quadridigitata* (Holbrook). J. Herpetol. 9: 337-348.
Sever, D. M. (1976). Morphology of the mental hedonic gland clusters of plethodontid salamanders (Amphibia, Urodela, Plethondontidae). J. Herpetol. 10: 227-239.
Sever, D. M. Male cloacal glands of *Plethodon cinereus* and *Plethodon dorsalis* (Amphibia: Plethodontidae). Herpetologica, in press.
Straughan, I. R. (1973). Evolution of anuran mating calls: bioacoustical aspects. In J. L. Vial [ed.] Evolutionary biology of the anurans. Univ. Missouri Pr., Columbia, Missouri, pp. 321-327.
Straughan, I. R. (1975). An analysis of the mechanisms of mating call discrimination in the frogs *Hyla regilla* and *H. cadaverina*. Copeia 1975: 415-424.
Straughan, I. R. and W. R. Heyer (1976). A functional analysis of the mating calls of the neotropical frog genera of the *Leptodactylus* complex (Amphibia, Leptodactylidae). Papéis Avulsos de Zool. 29: 221-245.
Taylor, D. H. and K. Adler (1973). Spatial orientation by salamanders using plane-polarized light. Science 181: 285-287.
Taylor, D. H. and D. E. Ferguson (1969). Solar cues and shoreline learning in the southern cricket frog, *Acris gryllus*. Herpetologica 25: 147-149.
Taylor, D. H. and D. E. Ferguson (1970). Extraoptic celestial orientation in the southern cricket frog *Acris gryllus*. Science 168: 390-392.
Terborgh, J. and J. S. Weske (1975). The role of competition in the distribution of Andean birds. Ecology 56: 562-576.
Thurow, G. (1976). Aggression and competition in eastern *Plethodon* (Amphibia, Urodela, Plethodontidae). J. Herpetol. 10: 277-291.

Torelle, E. (1903). The response of the frog to light. Amer. J. Physiol. 9: 466-488.
Tristram, D. A. (1977). Intraspecific olfactory communication in the terrestrial salamander *Plethodon cinereus*. Copeia 1977: 597-600.
Twitty, V. C. (1955). Field experiments on the biology and genetic relationships of the California species of *Triturus*. J. Exp. Zool. 129: 129-147.
Twitty, V. C. (1959). Migration and speciation in newts. Science 130: 1735-1743.
Twitty, V. C. (1966). Of scientists and salamanders. W. H. Freeman & Co., San Francisco.
Wake, D. B. (1970). The abundance and diversity of tropical salamanders. Amer. Natur. 104: 211-213.
Wake, D. B. and J. F. Lynch (1976). The distribution, ecology, and evolutionary history of plethondontid salamanders in tropical America. Natur. Hist. Mus. Los Angeles Co. Sci. Bull. 25: 1-65.
Wiewandt, T. A. (1969). Vocalization, aggressive behavior, and territoriality in the bullfrog, *Rana catesbeiana*. Copeia 1969: 276-285.
Zug, G. R. (1972). Anuran locomotion: structure and function. I. preliminary observations on relation between jumping and osteometrics of appendicular and postaxial skeleton. Copeia 1972: 613-624.
Zweifel, R. G. (1970). Distribution and mating call of the treefrog, *Hyla chrysoscelis*, at the northeastern edge of its range. Chesapeake Sci. 11: 94-97.

REPTILE SENSORY SYSTEMS AND THE ELECTROMAGNETIC SPECTRUM

W.R.A. MUNTZ

Department of Biology, University of Stirling

Stirling, FK9 4LA, Scotland

Although some reptiles are secondarily aquatic or amphibious, the group as a whole are true land animals, basically independent of water as an environment. The present day reptiles are also characterised by being poikilothermic (though they may have a limited degree of temperature control by behavioural means). Many reptilian groups in the past were probably homeothermic, but these have either become extinct, as in the case of the dinosaurs, or they have developed further and become birds.

The independence of the reptiles from water has allowed them to radiate into a wide range of habitats previously unavailable to vertebrates. Many groups have furthermore changed their habitats several times in the course of their evolution, with each change leaving its traces as modifications of the sensory apparatus. Their poikilothermy has however resulted in their range being limited to comparatively warm habitats (Darlington 1966), and has also meant that, although a few species may glide, true flight is beyond their metabolic capabilities (the flying dinosaurs, such as *Herodactylus* were probably warm-blooded (Desmond 1975)).

The present chapter deals with those aspects of the responses of reptiles to the electromagnetic spectrum, including the infrared, that are both related to the environment and to these special reptilian characteristics. No attempt has been made to provide a comprehensive description of their sensory systems, although references to recent reviews are given where appropriate.

VISUAL ADAPTATIONS IN TERRESTIAL REPTILES

When the vertebrates emerged on to the land, in many ways their visual problems became much simpler. In the first place, the difference in the refractive indexes of air and water resulted in the cornea becoming the major focussing element of the eye, with the lens itself being in most cases relegated to the role of a fine focussing device. In man, for example, the total power of the eye is about 60 dioptres, of which over 40 dioptres are due to the cornea; and the relative contributions of the lens and cornea are probably similar for other terrestial vertebrates. This change results in the optical centre of the eye moving forward, with the consequence that an eye of a given size will have a bigger retinal image and be capable of resolving finer detail. In the second place, since water absorbs heavily in both the ultraviolet and the infrared, a terrestial environment results in a much broader band of electromagnetic radiation being available to the animal, with consequently a greater potential for the transmission of information, particularly in the realm of spectral analysis (colour vision). Finally water both absorbs and scatters light to a much greater degree than air, so that objects underwater have a much lower contrast, and will become invisible at much shorter ranges. The terrestial environment does of course have concommittant disadvantages. The eye becomes exposed to potentially damaged ultraviolet radiation, and the problems associated with drying and the lack of a cleansing effect from the water necessitate the development of eyelids and lachrymal glands.

The environmental adaptations of the reptilian visual system, especially of the retina, are particularly associated with the studies and theories of Walls (1942), and we may start by briefly summarising these, since to a large degree more recent work has both confirmed and extended his views (see Crescitelli 1972 for a review). Walls believed that the primitive reptiles were all diurnal, thus benefiting from the warmth and high light levels obtaining during the day. As such he supposed that they inherited the cones of their ancestors, and lacked rods. The cones had various characteristics that adapted them to high light levels, including, for example, small outer segments and the presence of coloured oil-droplets acting as filters. As, however, the competition on land increased, particularly from the evolving warm-blooded animals, many reptiles were forced to become either partially or completely nocturnal. This process was accompanied by an increase in the size of the outer segments of some of the cones, so that they gradually became rod-like in both function and appearance. The following quotation from Walls (1942) summarises this transmutation hypothesis: "the first rods in the world were produced by the transmutation of cones, and the process has been occasionally repeated, wherever needed, ever since the

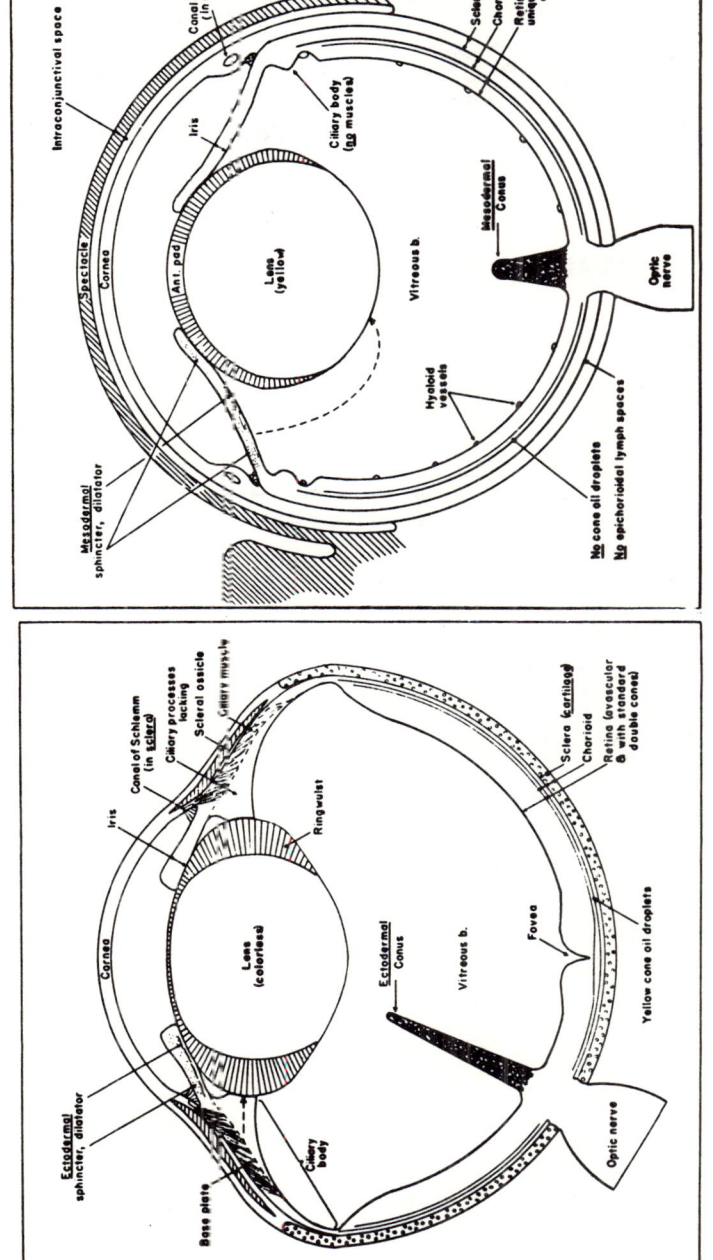

Fig. 1. Diagram of eyes of a lizard and a snake, to show contrasts and similarities resulting from the presumed loss and reacquisition by the snakes of features paralleling those of their ancestors. From Walls (1942).

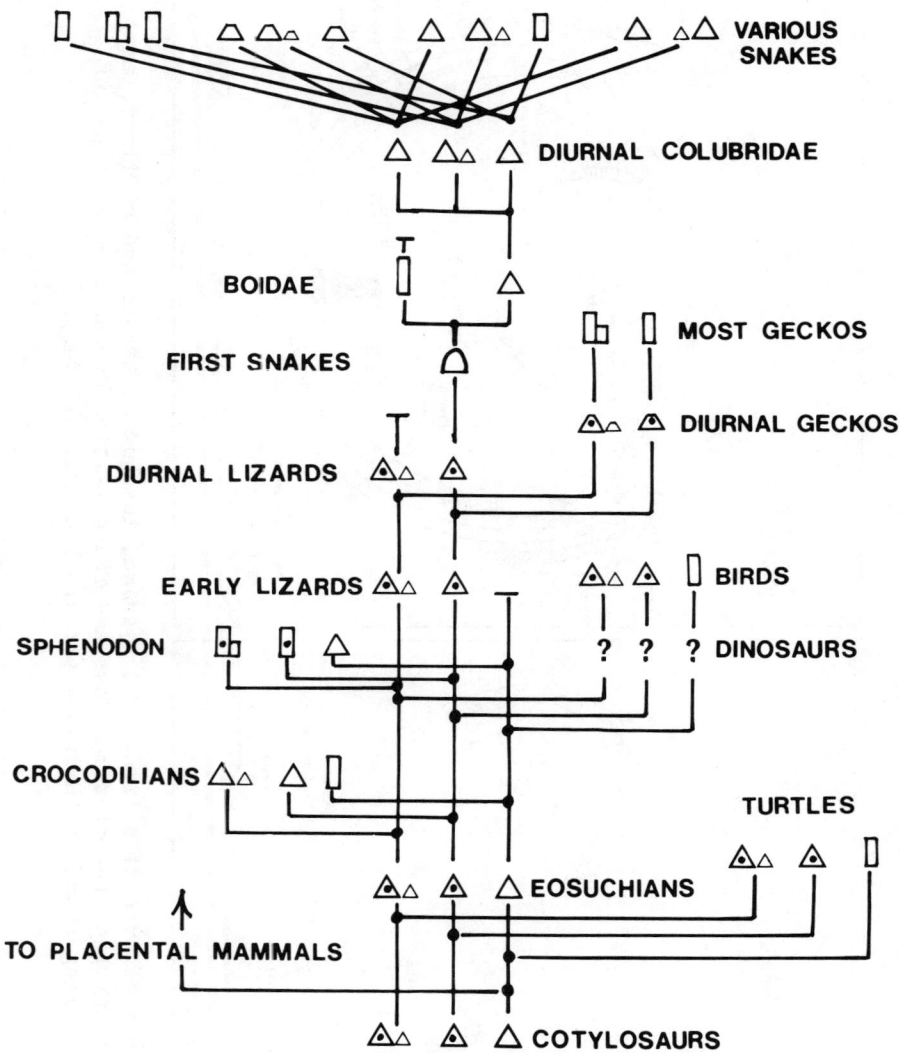

Fig. 2. Phylogenetic scheme for reptilian receptors, after Walls (1942). Rods and cones are shown by the rectangles and triangles respectively, and intermediate elements by truncated triangles. Dots indicate oil droplets.

vertebrates came on land". He thus considered the cones as the ancestral receptors of the vertebrate eye, and also that the visual apparatus shows great plasticity, altering rapidly to conform with the demands of the environment. Thus cones can be transmuted into rods if the animal takes up a nocturnal mode of life, but the reverse process can also occur: rods can be transmuted into cones if a previously nocturnal species become diurnal.

Different reptiles show various intermediate stages in these changes, providing "snap-shots" of the evolutionary process. Walls described such intermediate stages for lizards and geckoes, and more strikingly for the snakes. This last group probably evolved from an ancestral lizard that had become fossorial in habit. As a result of this change in habit the limbs became rudimentary and disappeared, as did the external ears and the parietal eye. The lateral eyes initially developed a protective spectacle, beneath which the eye also slowly atrophied. When later the snakes came back to the surface, they were faced with the necessity of reconstructing their eyes "from scratch", with the result that the well developed eyes of the present day diurnal snakes, while superficially similar, are in fact very different from those of other reptiles, giving us a remarkable example of convergent evolution (Fig. 1). As Walls puts it: "Zoologists have long been fond of citing the cephalopod molluscs, as showing how nearly an invertebrate group can imitate the vertebrate eye if it tries hard. They might at least give as much credit to the snakes; for in them we see a vertebrate group which has been under the necessity of duplicating the vertebrate eye, and has made a very good job of it". The various phylogenetic changes that Walls postulates occurred in the reptilian retina are summarised in Fig. 2, which is a simplified version of the scheme given in his book.

More recent work, which has been reviewed in detail by Crescitelli (1972), has in general extended and confirmed Walls' views. Among the snakes, Underwood (1951) agrees with Walls in assuming that the boid receptors represent the ancestral type, and although he disagrees about the detailed derivation of the types found in other species, he agrees with the general concept of transmutation. A considerable amount of more recent work has also been carried out on the geckos. A series of histological studies carried out on this group by Underwood (1951), Dunn (1965), Tansley (1964), and Crescitelli (1972) are summarised in Fig. 3. These results confirm Walls' findings of various intermediate types, ranging from cone-like to rod-like. They also extend his findings in two respects. Firstly, many species have "rods" containing oil-droplets (a typical cone feature), which we might expect on Walls' transmutation theory, since he considered gecko rods to be prime examples of this process. Secondly, it was found that gecko

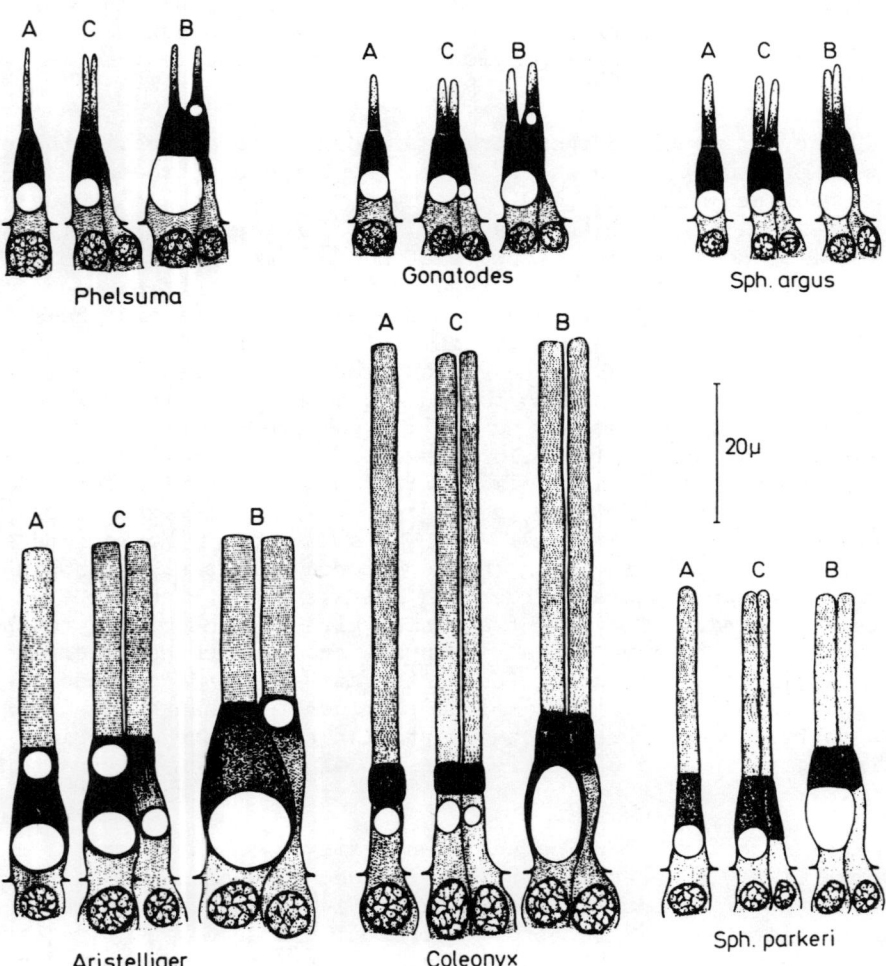

FIG. 3. Retinal receptors of various geckoes, illustrating the various intermediate forms that may be found in this group of animals. A, B and C are the singles, type B doubles and type C doubles, according to the nomenclature of Underwood (1951). From Crescitelli (1972).

retinas contain two types of double cone, not one, and in fact that triple and other multiple cells can also occur.

Other experiments on gecko vision support the general points that their visual systems are very variable, and often show characteristics intermediate between those of cones and rods. In particular, the visual pigment extracts of Crescitelli (1963, 1972) have revealed photopigments absorbing maximally over the wide spectral range from 490 nm to 530 nm, whereas typically terrestrial vertebrates have pigments absorbing maximally near 500 nm. Some species yielded two extractable visual pigments. Electrophysiological work has confirmed the presence of two receptor types which show some of the typical characteristics of rods and cones, even in "pure-rod" species such as *Hemidactylus turcicus* and *Tarentola mauritanica*, (Dodt and Jessen 1961).

In view of the very wide variations found in the visual systems of the lizards and snakes, it is unfortunate that almost no behavioural studies have been carried out on these groups. Tansley (1959), for example, pointed out that even in nocturnal geckos the ratio of visual cells to ganglion cells is low, and Dodt and Jessen (1961) have shown that high critical flicker fusion rates may be demonstrated electrophysiologically in such species. It would be of great interest to find out behaviourally what the absolute sensitivity and visual acuity of such animals is. The only behavioural experiment at present available in this area is due to Crozier and Wolf (1939) who showed that the flicker response curve of a gecko, *Sphaerodactylus inaquae*, was effectively identical to that of a turtle, *Pseudemys scripta*. They assumed that the former species was a pure-rod species, and the latter a pure-cone species. They failed however, to provide any histological data supporting this assumption, and Crescitelli (1972) points out that other species of *Sphaerodactylus* are not particularly nocturnal, and may have retinal receptors that resembles cones more than rods.

ADAPTATIONS TO A POIKILOTHERMIC EXISTENCE

(i) <u>Infrared detectors</u>: For most animals vision is possible from roughly 350 nm at the short wavelength end of the spectrum to about 750 nm at the long wavelength end. This is a very narrow range within the whole spectrum of electromagnetic radiations, and considerably narrower than even the spectrum of solar energy that penetrates the atmosphere and reaches the earth. The reasons for these limits have been discussed by many authors (e.g. Pirenne 1951, 1962). At the short-wavelength end chromatic aberration becomes serious, the lens and cornea start to absorb strongly, and the high quantal energy of the light may become damaging. At the long-

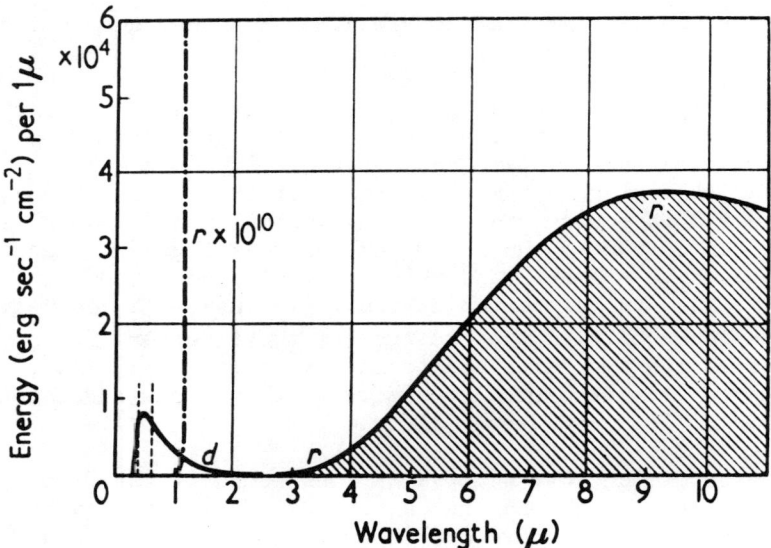

FIG. 4. Thermal radiation inside eye and retinal illumination from a white object in sunlight. From Pirenne (1951, 1962).

wavelength end the quantal energy of the light becomes low, so that photochemical reactions become insensitive. There is, however, a more interesting long wavelength limitation for homeothermic animals, namely the thermal radiation that fills the eye due to the body's own heat. This suggestion has been discussed by Pirenne (1951), and is illustrated in figure 4. The curve r represents the spectral energy distribution of a black body at 37°C (blood heat), and the curve d represents, for man, the spectral distribution of energy in the retinal image of a perfectly diffusing white object in full sunlight. At absolute threshold an object can be detected when it radiates roughly 10^{-10} times this energy. In order therefore to compare the energy distribution of a white stimulus at threshold with that of the body's heat, the curve d should be divided by 10^{10}, or the curve r multiplied by the same factor. The latter procedure is more convenient to depict, and is also shown in the figure. It is clear that there is a fundamental objection to homeothermic animals having eyes with sensitivities extending much beyond 1000 nm, since if this occurred the retinal image would become progressively degraded by the veiling thermal radiation of the eye itself.

If however, an animal is both poikilothermic and terrestial, the above limitation becomes less serious, and the useful detection of infrared radiation become possible. Thus, although infrared radiation is removed by water, it is readily transmitted by air, and if the animal is poikilothermic the thermal radiation from its own body is also reduced. Among the vertebrates the reptiles are thus uniquely placed to make use of infrared radiation, and two groups of snakes have indeed evolved specialised infrared receptors to make use of this possibility. The relationships shown in figure 4 are also relevant to the consideration of such receptors, but in the reverse direction. The curves r and r x 10^{10} now represent the energy distribution of the prey (mammals or birds) and so constitute the stimulus to which the snake responds. This energy distribution must be discriminated from the thermal radiation from the background, and from the snake itself. For example, if a mammal with a body temperature of $37^{o}C$ (as in the figure) is to be discriminated from its surroundings when these are at $27^{o}C$, the receptors must respond to the difference in the radiation from these two sources, which, assuming both the stimulus and the surroundings may be represented as black-bodies, will contain almost all its energy between about 3 u and 25 u (Hartline 1974). It is over this spectral range therefore that we would expect the infrared receptors to be sensitive. Furthermore, the curve d now represent light energy which comes from many sources, not only prey, and so represents an added source of noise that will confuse the detection of the stimulus. The properties of the infrared receptors should thus have complimentary properties to those of the eye: they should be sensitive to infrared but insensitive to wavelengths below about 1000 nm. Exactly this situation is found. Stimuli of less than 1000 nm are ineffective, while stimuli longer than about 1500 nm, even up to about 2.8 cm are effective (see Hartline 1974 and Loftus, this volume, for recent reviews).

Infrared receptors occur in two groups of snakes, the Boidae and the Crotalidae, and they are sufficiently different that it seems likely that they evolved separately in the two cases. In the Crotalidae (pit-vipers) they consist of two forward facing cavities, one on each side, with an inner and outer chamber separated by a thin membrane. The interior of each cavity is larger than its opening, so that the whole organ can act as a crude pin-hole camera. Optics such as those found in the eye cannot be used, since above about 1500 nm infrared is almost totally absorbed by water. The Boidae often also have pits along the upper and lower lips, but of a different form, and many species have no visible pits at all. The transduction process, by which the infrared energy is converted into nervous activity, is unknown, though the fact that the organs are responsive even to microwaves makes a photochemical reaction unlikely.

The snakes presumably use their infrared receptors for the detection of warm-blooded prey, and possibly also to seek out warm areas in the environment. It is therefore necessary that the organ should have sufficient sensitivity, and be capable of giving reasonably accurate localisation.

Estimates of sensitivity have been made on several occasions, and a summary is available in Hartline (1974). Early behavioural work (Noble and Schmidt 1937), using illuminated (warm) and cold light bulbs, covered to remove visual cues, showed that both boids, and crotalids could distinguish temperature differences of $0.2^\circ C$ or less, and localise the warm bulb well provided it was moving. Elimination of the various sense organs showed that the facial pits were responsible. Noble and Schmidt also showed that freshly killed warm rats could be distinguished from freshly killed chilled rats in this way. Electrophysiological recordings from peripheral nerves in crotalids suggests an absolute threshold of around 1.3×10^{-4} joules/cm^2 delivered in 0.1 sec, which is equivalent to 1.3×10^{-5} Watt/cm^2. For the less specialised boids such experiments gave an estimated threshold of 5.4 joules/cm^{-2} delivered over the same time (= 5.4×10^{-3} Watt/cm^2). It has also been shown that the peripheral nerves from the infrared receptors of boids and crotalids will respond to a human hand at 40-50 cm, which, with various assumptions, can be shown will deliver about 8×10^{-5} Watt/cm^2 to the receptors. It is likely that under natural conditions the snakes would prove more sensitive than this.

As we have seen, the structure of the receptors, at least in the crotalids, could permit them to act as crude pin-hole cameras, although to achieve the accuracy of localisation found experimentally some further neural processing would be required (Otto 1972). Hartline (1974) has reported that the pit-organ is capable of quite reasonable localisation, for the peripheral nerves from it show a concentric receptive field organisation, with a central "warm-on" zone, a rather larger central "cool-off" zone, and a peripheral "warm-off" "cool-on" zone. The size of the central "warm-on" zone is given at 15°, and the receptors are apparently specifically sensitive to movement. Finally, Terashima and Goris (1975) have shown that the infrared receptors of pit-vipers project to the optic tectum, where an orderly map of the outside world is formed. The eyes of course also project to the optic tectum to give an orderly map of the outside world, and the pit-organ map and the visual map are superimposed. The pit-organ map lies deeper, and the fineness of its localisation is less precise, but a given point on the tectal surface represents a given direction in space for both modalities. Since it is now considered that the optic tectum may be especially concerned with localisation in space, these results support the view that the infrared receptor system is capable of quite precise localisation.

(ii) <u>The parietal eye</u>: It is well known that many lizards have an eye-like structure on top of their heads, lying beneath a hole in the parietal bone and known as the parietal or third eye. This third eye has a similar structure in all those reptiles that possess it, consisting of a cup-shaped retina closed in above by a lens-like structure that joins the retina on either side. The retina contains receptors that resemble, at both the light and electronmicroscope level, the rods and cones of the lateral eyes, and is connected by a nerve to the diencephalon. The development of the parietal eye differs, however, in several ways from that of the lateral eyes: it is not cupped by invagination, and the lens is formed from the same outgrowth from the brain as the retina, instead of arising from the skin, as it does in the lateral eyes. One consequence of this is that the receptors lie nearest the cavity of the organ and point towards the entering light, unlike the situation in the lateral eyes (see Eakin 1973 for a general review).

Although simple experiments, in which the pineal eye is covered or strongly illuminated, have failed to show any obvious behavioural responses associated with it, evidence is now accumulating that it may function as a register of solar radiation, controlling and synchronising daily cycles of activity, seasonal changes in the endocrine glands, and the degree to which the animals expose themselves to light. Effects of this sort were first demonstrated by Stebbins and Eakin in 1958. They removed the parietal eye from a large number of fence lizards (*Sceloporus occidentalis*), and performed sham operations on a similar number of controls. The lizards were marked, and then released in the same area of countryside from which they had originally been captured. Subsequently observations made in the field at regular intervals, showed that the operated animals exposed themselves to light for longer periods and at higher intensities than the controls. The operated animals also showed less fright and escape reactions than the controls.

These results were confirmed in 1973 by Stebbins and Cohen, who also showed that parielectomy caused an increase in thyroid activity (as judged histologically), and an increased development of the ovaries in females: no effect could be detected on the male reproductive system. Changes in thyroid activity have been detected in other studies as well (see Eakin 1973 for references).

Along similar lines, Glaser (1958) demonstrated an increased locomotor activity in desert night lizards (*Xantusia vigilis*) when the third eye was shielded from the light by an aluminium foil cap. Specific effects on thermal preferences have been shown by Hutchinson and Kosh (1974) for *Anolis carolinensis*. In these experiments animals were kept in a 12 hr light 12 hr dark cycle, and a daily temperature cycle that varied from $15^\circ - 25^\circ C$. Their temperature preferences were tested behaviourally at regular intervals over the day in a temperature gradient. All animals selected higher temperatures during the light phase of the cycle, during which the ambient temperatures were also higher. Parielectomised animals selected higher temperatures than the controls, and their temperature preference cycle was also delayed in comparison to that of controls, with peak temperatures being selected later in the day.

Comparative studies are also compatible with this general picture of the function of the third eye. Thus Gundy et al. (1971) have shown that the third eye usually regresses in species that have adopted a fossorial habitat, when of course diurnal rhythms and thermoregulation through basking in the sun become unimportant. Gundy et al (1975) have further shown that lizards without parietal eyes tend to be restricted to low latitudes, whereas those having parietal eyes are successful at higher latitudes also. They suggest that the third eye facilitates survival at higher latitudes, through reproductive or thermoregulatory modulation.

These suggested functions, particularly those related to thermoregulation, are clearly more important to poikilotherms than to homeotherms. Although the third eye is probably best developed in the reptiles it, or associated diencephalic photosensitive structures, also occur among the fishes and amphibians. The third eye is however lacking among the birds and mammals, although in the former group the diencephalic part of the brain may retain some photosensitive function, apparently correlated with the control of the breeding cycle.

That the third eye is indubitably light sensitive has been demonstrated directly through electrophysiological recordings. The first such recordings were made by Miller and Wolbarscht (1962), who showed that both sustained potential changes and nerve impulses occurred on illumination of the third eye of *Anolis carolinensis*. More detailed studies have since been carried out by Dodt and Scherer (1968 a,b) on *Lacerta sicula campestris*, and by Hanasaki (1968, 1969 a,b) on *Iquana iquana*. The results were substantially the same for the two species: short wavelength light caused excitation (a positive sustained potential and an increase in the number of nerve impulses), whereas light of longer wavelengths caused inhibition (a negative sustained potential and a decrease in the number of nerve impulses). The excitatory component of the response was maximally sensitive at about 450 nm and the inhibitory component at about 520 nm. The spectral sensitivity of the lateral eyes differed between the two species, and in neither case was it the same as that of the parietal eye.

If the function of the parietal eye is to detect solar radiation, its spectral sensitivity is at first sight surprising. It is however possible that it is specialised for detecting changes in the ambient light at sunrise and sunset. At these times there is a brief increase in the ratio of blue to red light (Johnson et al. 1967, McFarland and Munz 1975), and a marked peak in the energy distribution develops at about 460 nm. This peak agrees well with the inhibitory component of the parietal eye response. The detection of the changes associated with twilight could be important in the control of diurnal rhythms.

THE RETURN TO WATER

Although the reptiles are the first vertebrate group to have become independent of the water as an environment, some forms (crocodiles, turtles, sea-snakes) have returned secondarily to an aquatic habitat. Such animals have, over evolution, passed through a terrestial stage, during which their sensory systems presumably became adapted to their environment, so it is of interest to see in what ways their secondary adaptation to the water resembles that of purely aquatic animals. Even the most aquatic reptiles, such as sea-turtles, are not however completely independent of the land, for they must come ashore to breed, and the hatchlings must pass through a brief terrestial period before they re-enter the sea. Other forms, such as many freshwater turtles and the crocodilians, while being basically aquatic may pass substantial periods of time ashore. For at least a short period of their lives therefore, all aquatic reptiles face the problems associated with an amphibious existence.

Most of the work allowing comparisons between the visual systems of freshwater, marine, and terrestrial reptiles has been carried out on the Chelonians. These animals have commercial importance, perform extensive migrations, and apparently rely on vision to reach the water during the critical phase of their life just after hatching (Carr and Ogren 1960, Mrosovsky 1972).

The focussing power of the cornea is lost underwater, so that image formation has to be carried out by the lens alone. However, if an amphibious animal is to focus adequately on land as well, where the cornea is effective, the lens also has to have an exceptional range of accommodation. In some cases this is achieved. Thus in some freshwater turtles the lens is unusually pliable, and can be squeezed by a powerful sphincter muscle in such a way that the front surface attains a very high curvature (Walls 1942). Ehrenfeld and Koch (1967) have shown retinoscopically that *Clemmys insculpta* can accommodate adequately in both air and water, and Dudziak (1955) has shown behaviourally that the visual acuity of *Emys orbicularis* is the same underwater as on land, being about 2.85 min in each case. In other species accommodation is only possible in one medium. The sea-turtle *Chelonia mydas* can apparently only accommodate accurately underwater (Ehrenfeld and Koch 1967), and must have very defective vision on land; and in the crocodilians the range of accommodation is small, and they can probably only see well in air (Walls 1942).

The spectral characteristics of the visual systems of freshwater animals also often differ from those of marine and terrestrial animals. The best known and most studied difference of this type is undoubtedly in the visual pigments, where it has been known, in outline, ever since the earliest work of Kühne (1878) and Köttgen and Abelsdorff (1896). The latter paper presented difference spectra for bile extracts of retinas from 16 different species: 4 mammals, 1 bird, 3 amphibians, and 8 fishes. Reptiles were not represented in their sample, and the fishes were all freshwater species. They found that the extracts from the 8 fishes were all similar to each other, absorbing maximally at about 540 nm; and differed markedly from the extracts from the other species, which were again very similar to each other but absorbed maximally at about 500 nm. The results give the impression of two, and only two, types of pigment, segregated in different animal groups.

It is clearly not possible to describe in this paper the wide variety of visual pigments that have since been described, and the difficulties that have been encountered in trying to relate them to the characteristics of the environment (see Dartnall 1975 for reviews). It is however true that there are only two basic classes of visual pigment, one based on vitamin A_1 and the other on vitamin A_2; and even though there is a considerable variation between the spectral absorbance properties of different pigments within each class, the A_2-based pigments tend to absorb at longer wavelengths than the A_1-based pigments. It is also true that, with rather few exceptions, A_2-based pigments are only found in freshwater animals (although such animals may have an A_1-based pigment as well). Extracted visual pigments usually come from rods, so we can say that there is a tendency for freshwater animals to be sensitive, when dark-adapted, to longer wavelengths than terrestial or marine animals. Recent work has further shown that the type of pigment found in the rods correlates with the type of pigment found in the cones (e.g. Liebman 1972, Loew and Dartnall 1976), so it is highly probable that freshwater animals will be sensitive at longer wavelengths than marine and terrestial animals under photopic conditions also.

The characteristics of the chelonian visual system fit in quite well with the general picture outlined above. Thus chemical tests (the Carr-Price reaction) have indicated that A_2-based pigments occur in two species of *Pseudemys*, which is a freshwater

genus (Wald et al 1953). Using microspectrophotometry, Liebman
and Granda (1971) have concluded that *Chelonia mydas* (marine) has
A_1-based pigments in both the rods and the cones, whereas *Pseudemys
scripta* (freshwater) has an A_2-based system of photopigments. The
Chelonia rod pigment was found to absorb maximally at 502 nm, and
three cone pigments were described, absorbing maximally 440 nm, 562 nm,
and 502 nm. For *Pseudemys* the rods absorbed at 518 nm, and the cones
at 450 nm, 518 nm and 630 nm. Electroretinogram studies have
yielded spectral sensitivity curves with maxima near 455 nm,
520 nm and 620 nm for *Chelonia mydas* (Granda and O'Shea 1972), and
at approximately 575 nm and 645 nm for *Pseudemys scripta* (Granda
1962). The latter species had a further ill-defined peak at
short wavelengths, that could be revealed using computer averaging
(Granda and Stirling 1966). Microelectrodes inserted into
individual cones or *Pseudemys scripta* have also revealed three
classes of cone, maximally sensitive at about 460 nm, 555 nm, and
640 nm. The short wavelength side of the sensitivity curves of
the 555 nm and 640 nm cones showed considerable variation, apparently
due to the coloured oil-droplets (Baylor and Hodgkin 1973). Finally,
behavioural studies have shown clear sensitivity maxima at about
640 nm in both *Pseudemys scripta* and *Chrysemys picta* (both freshwater forms), another maximum at shorter wavelengths (roughly 460 nm),
and possibly an inflection at about 550 nm (Sokol and Muntz 1966,
Muntz and Sokol 1967, Graf 1972). These various data are
summarised in Table 1.

The electrophysiological and behavioural spectral sensitivity
curves tend to have their maxima located at longer wavelengths than
the microspectrophotometrically determined maxima of the underlying
visual pigments. This is presumably a consequence of the coloured
oil-droplets of the cones, which are better developed in the
chelonians than they are in other reptiles. Interestingly enough,
it appears that marine and freshwater turtles also differ in their
oil-droplets. Thus while *Pseudemys scripta* (freshwater) possesses
red, orange-yellow, and colourless droplets, *Chelonia mydas* (marine)
possesses orange, yellow and colourless droplets: red droplets are
lacking (Granda and Haden 1970). Here again therefore there is a
tendency for the marine form to have a retina adapted for vision at
shorter wavelengths than the freshwater form. It is probable
that, as an inherent consequence of the water itself, long wavelength
stimuli ("reds") will always be more conspicuous than other stimuli
in turbid or coloured water, whereas stimuli of shorter wavelengths
("yellows") will be more visible in clear waters (Lythgoe 1975).
Freshwater is almost always of the former type, and oceanic water
of the later, and Lythgoe has suggested that this is one reason
why oceanic fishes often have yellow markings while freshwater
fishes have red markings. The comparison between marine and
freshwater turtles provides one example where the visual system also
appears to be adapted to this characteristic of the water in the
two environments.

Table 1. Summary of the spectral characteristics of the retinal receptors of *Chelonia mydas* and *Pseudemys scripta* obtained using various techniques. The figures show the wavelength of maximal sensitivity (in nm) of the different receptor types. For details and references, see text.

Experimental Technique	*Chelonia mydas*				*Pseudemys scripta*			
	rods	cones			rods	cones		
Microspectrophotometry	502	440	502	562	518	450	518	620
Electroretinogram	–	455	520	620	–	?	575	645
Microelectrodes	–	–	–	–	–	460	555	640
Behavioural	–	–	–	–	–	460	approx 550	640

SUMMARY

Reptilian sensory systems are highly plastic, and have changed markedly during evolution in response to environmental demands. Evidence of intermediate stages can often be seen, as in the retinas of geckos and snakes. Examples of convergent evolution are also common, as may be seen by comparing the eyes of snakes and other terrestially reptiles, or those of turtles and other aquatic animals.

The reptiles are unique among vertebrates in being both poikilothermic and terrestrial. It is possible that the development of the infrared receptors and the parietal eye are special adaptations to this way of life.

REFERENCES

Baylor, D.A., and Hodgkin, A.L. (1973). Detection and resolution of visual stimuli by turtle photoreceptors. J. Physiol. 234: 163-198.
Carr, A. and Ogren, L. (1960). The ecology and migrations of sea turtles. IV. The green turtle in the Carribean Sea. Amer. Mus. Nat. Hist. Bull. 121: 1-48
Crescitelli, F. (1963). The photosensitive visual pigment system of *Gekko gekko*. J. Gen. Physiol. 47: 33-52.
Crescitelli, F. (1972). The visual cells and visual pigments of the vertebrate eye. In Handbook of Sensory Physiology, vol. VII/1, Photochemistry of Vision, ed. H.J.A. Dartnall, Springer-Verlag: Berlin, Heidelberg, New York.
Crozier, W.J. and Wolf, E. (1939). The flicker response contour for the gecko (rod retina). J. Gen. Physiol. 22: 555-566.
Darlington, P.J. (1957). Zoogeography: the Geographical Distribution of Animals. Wiley and Sons Inc.: New York, London, Sydney.
Dartnall, H.J.A. (1975). Assessing the fitness of visual pigments for their photic environmenta. In Vision in Fishes, New Approaches in Research ed. M.A. Ali, Plenum Press: New York and London.
Desmond, A.J. (1975). The Hot-blooded Dinosaurs. Blond and Briggs: London.
Dodt, E. and Jessen, K.H. (1961). The duplex nature of the retina of the nocturnal gecko as reflected in the electroretinogram. J. Gen. Physiol. 44: 1143-1158.
Dodt, E. and Scherer, E. (1968a). Photic responses from the parietal eye of the lizard *Lacerta sicula campestris* (De Betta). Vision Res. 8: 61-72.
Dodt, E. and Scherer, E. (1968b). The electroretinogram of the third eye. Adv. Electrophysiol. Path. Visual System 6: 231-237.
Dudziak, J. (1955). Ostrosc widzenia u zolwia blotnego (*Emys orbicularis* L.) przypatrznym i wodnym. Folia Biol. 3: 205-228.

Dunn, R.F. (1965). Electron microscopy studies on the receptor cells of the gecko, *Coleonyx variegatus*. Ph.D. Thesis, Univ. of Calif., Los Angeles. (Quoted in F. Crescitelli, 1972).

Eakin, R.M. (1973). The Third Eye. Univ. Calif. Press: Berkeley Los Angeles, London.

Ehrenfeld, D.W. and Koch, A.L. (1967). Visual accomodation in the green turtle. Science 155: 827.

Glaser, R. (1958). Increase in locomotor activity following shielding of the parietal eye in night lizards. Science 128: 1577-1578.

Graf, V. (1972). Behavioural visual functions for *Chrysemys picta picta*. Preferences and frequency responses. Brain Behav. Evol. 5: 155-175.

Granda, A.M. (1962). Electrical responses of the light- and dark-adapted turtle eye. Vision Res. 2: 343-356.

Granda, A.M. and Stirling, C.E. (1966). The spectral sensitivity of the turtle's eye to very dim lights. Vision Res. 6: 143-152.

Granda, A.M. and Haden, K.A. (1970). Retinal oil globule counts and distributions in two species of turtles: *Pseudemys scripta elegans* (Wield) and *Chelonia mydas mydas* (Linneaus). Vision Res. 10: 79-84.

Granda, A.M. and O'Shea, P.J. (1972). Spectral sensitivity of the green turtle (*Chelonia mydas mydas*) determined by electrical responses to heterochromatic light. Brain Behav. Evol. 5: 143-154.

Gundy, G.C. and Ralph, C.L. (1971). A histological study of the third eye and related structures in scincid lizards. Herpetol. Rev. 3: 65.

Gundy, G.C., Ralph, C.L. and Wurst, G.Z. (1975). Parietal eyes in lizards: zoogeographic correlates. Science 190: 671-672.

Hamasaki, D.I. (1968). Properties of the parietal eyes of the green iguana. Vision Res. 8: 591-599.

Hamasaki, D.I. (1969a). Spectral sensitivity of the parietal eye of the green iguana. Vision Res. 9: 515-523.

Hamasaki, D.I. (1969b). Interaction of the slow responses of the parietal eye. Vision Res. 9: 1453-1459.

Hartline, P.H. (1974). Thermoreceptors in snakes. In Handbook of Sensory Physiology, vol. III/3, Electroreceptors and Other Specialised Receptors in Lower Vertebrates, ed. A. Fessard, Springer-Verlag: Berlin, Heidelberg, New York.

Hutchinson, V.H. and Kosh, R.J. (1974). Thermoregulatory function of the parietal eye in the lizard *Anolis carolinensis*. Oecologia (Berlin) 16: 173-177.

Johnson, T.B., Salisbury, F.B. and Connor, G.I. (1967). Ratio of blue to red light: a brief increase following sunset. Science 155: 1663-1665.

Köttgen, E. and Abelsdorf, G. (1896). Absorption und Zersetzung des Sehpurpurs bei den Wirbeltieren. Z. Psychol. Physiol. Sinnesorg. 12: 161-184.

Kühne, W. (1878). On the photochemistry of the retina and on visual purple. (Ed. with notes by Michael Foster). Macmillan & Co.: London.

Liebman, P.A. (1972). Microspectrophotometry of photoreceptors. In Handbook of sensory physiology, vol. VII/1, Photochemistry of Vision, ed. H.J.A. Dartnall, Springer-Verlag: Berlin, Heidelberg, New York.

Liebman, P.A. and Granda, A.M. (1970). Microspectrophotometric measurements of visual pigments in two species of turtle, *Pseudemys scripta* and *Chelonia mydas*. Vision Res. 11, 105-114.

Lowe, E.R. and Dartnall, H.J.A. (1976). Vitamin A_1/A_2-based visual pigment mixtures in the cones of the rudd Vision Res. 16, 891-896.

Lythgoe, J.N. (1975). Problems of seeing colours underwater. In Vision in Fishes, New Approaches in Research. Ed. M.A. Ali, Plenum Press: London, New York.

McFarland, W.N. and Munz, F.W. (1975). The visible spectrum during twilight and its implications to vision. In Light as an Ecological Factor II. ed. G.C. Evans, R. Bainbridge and O. Rackham. Blackwell Scientific Publ.: Oxford, London, Edinburgh, Melbourne.

Miller, W.H. and Wolbarsht, M.L. (1962). Neural activity in the parietal eye of a lizard. Science 135: 316-317.

Mrosovsky, N. (1972). The water-finding ability of sea turtle, behavioural studies and physiological speculations. Brain, Behav. Evol. 5: 202-225.

Muntz, W.R.A. and Sokol, S. (1967). Psychophysical thresholds to different wavelengths in light adapted turtles. Vision Res. 7: 729-741.

Noble, G.K. and Schmidt, A. (1937). Structure and function of the facial and labial pits of snakes. Proc. Amer. Phil. Soc. 77: 263-288.

Otto, I. (1972). Das Grubenorgan, ein biologisches System zur Abbildung von Infrarotstrahlern. Kybernetik 10: 103-106.

Pirenne, M.H. (1951). Limits of the visible spectrum. Research 4: 508-514.

Pirenne, M.H. (1962). Spectral luminous efficiency of radiation. In The Eye, vol. 2, ed. H. Davson, Academic Press: New York and London.

Sokol, S. and Muntz, W.R.A. (1966). The spectral sensitivity of the turtle *Chrysemys picta picta*. Vision Res. 6: 285-292.

Stebbins, R.C. and Eakin, R.M. (1958). The role of the "third eye" in reptilian behaviour. Amer. Mus. Nov., 1870: 1-40.

Stebbins, R.C. and Cohen, N.W. (1973). The effect of parielectomy on the thyroid and gonads in free-living western fence lizards (*Sceloporus occidentalis*). Copeia 1973: 662-668.

Tansley, K. (1959). The retina of two nocturnal geckos, *Hemidactylus turcicus* and *Tarentola mauritanica*. Pflüg. Arch. Ges. Physiol. 268: 213-220.

Tansley, K. (1964). The gecko retina. Vision Res. 4: 33-37.
Terashima, S.I. and Goris, R.C. (1975). Tectal organisation of pit viper infrared reception. Brain Res. 83: 490-494.
Underwood, G. (1951). Reptilian retinas. Nature (Lond.) 167: 183-185.
Wald, G., Brown, P.K. and Smith, P.H. (1953). Cyanopsin, a new pigment of cone vision. Science 119: 505.
Walls, G.L. (1942). The Vertebrate Eye and Its Adaptive Radiation. Cranbrook Inst. of Sci. Bull. 19: Bloomfield Hills, Mich.

SENSORY ECOLOGY OF BIRDS

R. A. SUTHERS

Physiology Section, Medical Sciences, Indiana University

Bloomington, Indiana 47401 U.S.A.

Birds inhabit a wide range of habitats; some are flightless, others annually migrate thousands of miles. Despite this variety, relatively few studies have specifically investigated their sensory ecology. Nevertheless, in the past few years important new information on avian sensory abilities has become available. Some of the most exciting developments have been associated with attempts to clarify the sensory basis for orientation during homing and migration. The majority of the following discussion is devoted to vision and audition since it is for these senses that the most data are available.

1. AUDITION

Vocal communication plays a very important role in avian behavior. Songs and call notes serve variously to attract mates, advertise territories, maintain contact between individuals, warn of predators, etc. Songs in particular may be very complex and suggest that birds must possess a well developed auditory apparatus.

1.1 Ecological Acoustics

The effectiveness of sound as a means of communication depends in part on the distance over which vocalizations or other biologically significant sounds can be heard. The broadcast range of a sound depends not only on the sensitivity of the ear, but also on the nature of the animal's environment. One might expect that open habitats such as plains would have the largest broadcast range while "cluttered" habitats such as forests would have a limited broadcast range. In natural environments, however, many factors influence the propagation of sound and intuition fails. Even field measurements of sound attenuation are difficult to repeat and to interpret in terms of their bioacoustical significance. For a more detailed discussion of this subject the reader is referred to Michelsen's chapter in this volume.

Deserts and grasslands often have high propagation losses, caused by thermal and wind gradients which produce atmospheric turbulence causing sound scattering, and by interference between the direct wave and waves reflected from the ground (Mokhtar and Marrous, 1955). In forest habitats scattering is an important cause of sound attenuation, especially at high frequencies whose wavelengths are short compared to the dimensions of branches, twigs and leaves. Foliage is especially important in this process (Konishi, 1970a; Aylor, 1972a, 1972b). Climatic conditions within forests are more stable and uniform than in open country. The propagation of sound in forests is thus affected less by gradients in temperature, humidity and wind speed or turbulence.

The vocalizations of some birds seem to have evolved to take maximum advantage of preferential sound channels that may exist in their habitat. Morton (1975) measured the amount various frequencies were attenuated by three tropical habitats: monsoon forest, edge and grasslands. He calculated the amount of attenuation in excess to that predicted assuming spherical spreading losses according to the inverse square law. Propagation losses greater than those predicted by the inverse square law were termed "excess attenuation." Transmission losses in grassland and edge habitat were not significantly different and showed no band of markedly reduced attenuation. When the sound source was within a few feet of the ground in monsoon forest, however, there was a marked reduction in excess attenuation between 1585 and 2500 Hz (see Fig. 11 in Michelsen's chapter). This frequency dependent reduction was not noticeable more than 45 feet (13.7 m) above the ground.

Birds living near the forest floor apparently take advantage of this sound channel. Analysis of the vocalizations of 177 neotropical birds showed that the frequencies emphasized by the forest species, especially those of the low forest, tended to lie within the band of minimum attenuation where their broadcast range is greatest (see Fig. 12 in Michelsen's chapter). Selection of these frequencies could substantially affect the broadcast range. At ground level, frequencies between 1585 and 2500 Hz may travel about twice as far as either a 500 Hz or a 3000 Hz tone before they reach the ambient noise level. The songs of forest birds are predominantly pure tone-like. Morton suggests that the relatively stable climatic conditions in this habitat permit these birds to concentrate acoustic energy into a narrow band which could improve signal-to-noise ratio. This is in contrast to the vocalizations of birds living in edge habitat whose songs vary greatly in the frequencies emphasized and which contain both frequency modulated and pure tone components.

Another aspect of sound propagation which may be important to birds is the audibility to migrating birds of sound produced on the earth's surface. This question has been investigated by Griffin and his colleagues (Griffin and Hopkins, 1974; Griffin, 1976) using a radio microphone suspended aloft by a helium-filled balloon. One of their interesting findings is that the intensity of sound propagated upward does not decrease as rapidly as the inverse square law predicts and is more predictable than propagation along the surface. Frog choruses, for example, are audible to migrants at altitudes of at least 1 km. Thus, migrating birds may hear a variety of natural sounds - such as frogs, waves, insects and the wind in vegetation - which could provide useful information to them, allowing them to compensate for wind drift or otherwise assisting in their orientation.

1.2 The Avian Ear

The anatomy of the bird's ear has been discussed elsewhere (*e.g.*, Takasaka and Smith, 1971; Schwartzkopff, 1973) and so will not be treated here except to mention a few of the more obvious differences from mammalian auditory organs.

There is only a single ossicle, the columella, in the middle ear of birds. The columella is less efficient than the three middle ear ossicles of mammals in transmitting high frequencies (Saunders and Johnstone, 1972). In some penguins it contacts the wall of the tympanic cavity along its entire length, perhaps thus permitting bone conduction of sound which could be useful underwater (Anisimov, 1976). Diving birds show various adaptations for protection of the ear from damage by high pressures. In some species which dive deeply for fish the outer ear is arranged so that the external opening of the meatus is closed by a valve-like flap when exposed to water pressure (Kartashev and Ilichev, 1964).

The avian basilar papilla is small and contains a very short basilar membrane (3 mm long in the pigeon, *Columba livia*, vs. 35 mm long in man) (Schwartzkopff, 1955; Pumphrey, 1961). This short basilar papilla, however, contains many more rows of hair cells (14 rows proximally increasing to 54 near the distal end in the pigeon) than are present in mammals (Takasaka and Smith, 1971).

1.3 Auditory Performance

Auditory threshold. The high frequency cut off for most avian ears appears to be considerably lower than that of most mammals. Behaviorally determined auditory thresholds of the starling (*Sturnus vulgaris*) (Trainer, 1946), bullfinch (*Pyrrhula pyrrhula*),

(Schwartzkopff, 1949), pigeon (Heise, 1953; Harrison and Furumoto, 1971), great horned owl (*Bubo virginianus*) (Trainer, 1946), canary (*Serinus canarius*) (Dooling, Mulligan and Miller, 1971), and parakeet (*Melopsittacus undulatus*) (Dooling, 1973) all have high frequency cutoffs near 8 to 10 kHz (Fig. 1). The cochlear microphonic of black-footed penguins (*Spheniscus demersus*) suggests this species may be able to hear up to 15 kHz (Wever, Herman, Simmons and Hertzler, 1969) but a behavioral test is needed to confirm this. The tuning curves of single units in various song birds are also consistent with this conclusion (Konishi, 1970b). Dooling, *et al*. (1971) hypothesize that the lowest thresholds have evolved to match the frequencies in the songs rather than the call notes. They argue that the former are used for long distance communication whereas the latter are used at short range. The high threshold and extended low frequency response of the pigeon are compatible with the observation that it uses vocalizations of low frequency and only over short ranges. Konishi (1970b) found that the auditory tuning as judged by single units is not narrowly matched to the frequency of the song.

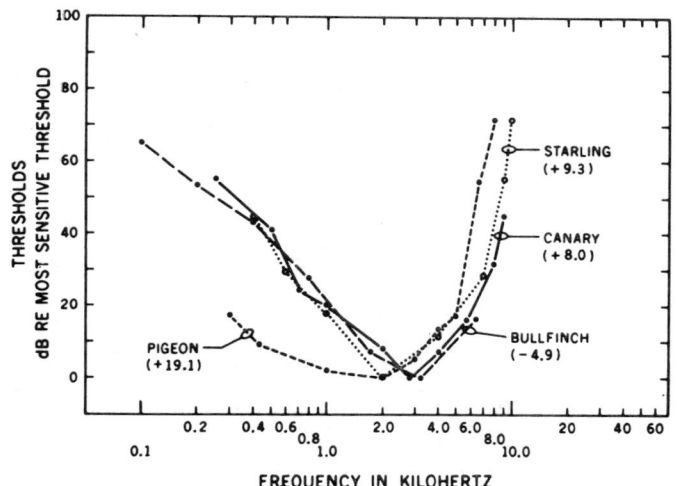

FIG. 1. Auditory sensitivity of three song birds and the pigeon. Curves are plotted so that maximum sensitivity is defined as zero dB. Each curve can be converted to SPL re 0.0002 μbar by adding the constant under each bird's name to the appropriate ordinates. Data for Starling are from Trainer (1946); for bullfinch are from Schwartzkopff (1949); for canary are from Dooling, Mulligan and Miller (1971) and for pigeon are averages from Trainer (1946) and Heise (1953). Figure is from Dooling, Mulligan and Miller (1971).

Recent experiments have suggested that birds may be able to hear very low frequencies. Yodlowski, Kreithen and Keeton (1977) have demonstrated conditioned cardiac responses by homing pigeons to frequencies below 10 Hz. They report that a 50% response level was obtained with SPL's of 50 dB at 10 Hz, 60 dB at 5 Hz and 80 dB at 1.5 Hz. Birds with their cochleas surgically removed or with their columellas bisected could not be conditioned to infrasound. The biological significance of infrasound detection by pigeons is not yet known. There are many natural sources of infrasound in the environment (wind, weather fronts, ocean waves, etc.) which a homing pigeon might detect. Since infrasound can be propagated over hundreds of kilometers with relatively little attenuation it could provide a bird with useful information about approaching storms and perhaps assist in long distance orientation (Griffin, 1969). The conditioning experiments of Yodlowski, *et al.* (1977) are performed in a sealed chamber which insulates the pigeon from the air currents and small pressure fluctuations of the natural environment. It is uncertain if a pigeon can detect infrasonic signals from distant sources in the presence of low frequency noise created, for example, by air movement around its head. Furthermore, there are formidable physiological problems in localizing the source of such long wavelength signals. Future research will hopefully provide answers to these questions.

There is some evidence suggesting that auditory sensitivity may undergo seasonal changes in certain birds. Zablotskaya (1974) used a chronically implanted electrode to monitor the summed responses of the auditory nerve in black-headed and grey-headed goldfinches (*Carduelis carduelis* and *Carduelis caniceps*). For each species he reported a low frequency peak, 2 and 3 kHz, respectively, in the N_1 and N_2 components that was stable at all seasons, but a higher frequency peak, 5 and 4.5 to 5 kHz, respectively, that began to increase in May, reached a maximum in late August, and then decreased during the winter months. He suggested that this aspect of the goldfinches' auditory threshold may be controlled by the gonadal cycle. The possibility exists that the endocrine system may tune the auditory system to the most important vocalizations during the breeding season.

Frequency discrimination. Von Békésy (1944) observed that the basilar membrane of a chicken (*Gallus domesticus*) increased in stiffness from apex to base and that as stimulus frequency increased the point of maximum displacement moved from apex to base. Nevertheless, the shortness and bulkiness of the basilar membrane has caused some to question whether or not frequency discrimination may have a different mechanism in birds than in mammals (Schwartzkopff, 1955; Jahnke, Lundquist and Wersall, 1969). Noise induced threshold shifts in the parakeet differ from those in mammals, suggesting important differences in peripheral auditory processing in birds and mammals. When mammals are exposed to a band of noise, the maximum threshold shift occurs one half to one and one half octaves above the center frequency of the noise band and spreads to frequencies above the noise band more than to those below it. When parakeets were exposed to a one-third octave noise band centered on 2 kHz, however, the maximum threshold shift was also at 2 kHz with very little threshold shift at higher or lower frequencies (Saunders and Dooling, 1974). Saunders and Dooling suggest that this difference may be due to a very "sharp" traveling wave along the basilar membrane (Greenwood, 1961, 1962) or to a more restricted innervation of hair cells in the longitudinal direction along the basilar membrane by single nerve fibers (Takasaka and Smith, 1971).

Psychophysical data suggest that songbirds can detect a smaller change in frequency (ΔF) than can birds with less elaborate vocalizations. Thus the laboratory pigeon can detect a ΔF of about 2.4% which is about six times the ΔF of certain songbirds. Dooling and Saunders (1975a) found that parakeets have good frequency discrimination between 1 and 4 kHz (Fig. 2). Within this range they were able to detect a frequency difference of less than 1% which is better than the frequency resolution of the pigeon or cat and nearly as good as that of man.

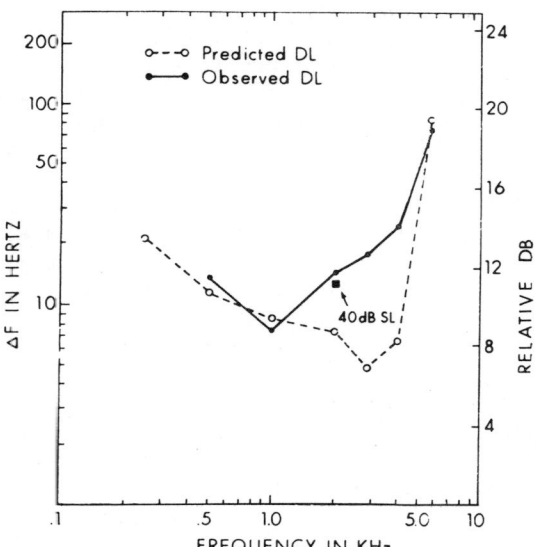

FIG. 2. Frequency discrimination by the parakeet. Observed difference limens (DL) are compared with those predicted from the critical ratio. The threshold value of ΔF at 2.86 kHz and a 40 dB sensation level is also indicated (from Dooling and Saunders, 1975a).

Temporal resolution. The temporal pattern of elements within a bird's song is often very precise and constant. Analysis of many songs show they contain a wealth of details and in some species, which song antiphonally, temporal coordination is accurate to within a few milliseconds. For these and other reasons, the temporal resolution of the avian ear has been presumed to be very good. There are few data on which to assess this claim except for those of Wilkinson and Howse (1975) who conditioned bullfinches, greenfinches (*Carduelis chloris*), and pigeons to discriminate a single click from a double click. Discrimination broke down if the interval between the pair of clicks constituting the double click was less than 6 to 10 msec for the finches or 2 to 10 msec for the pigeon. Human subjects tested under similar conditions reported they could not discern the clicks as separate events if the interval between them was less than 12 to 50 msec, but two subjects said the double click still sounded different than the single click when interclick intervals were greater than 2 to 5 msec.

Intensity discrimination. The ability of birds to discriminate differences in intensity does not appear to exceed that of mammals. Dooling and Saunders (1975b) measured the intensity difference limen in parakeets for successively presented tone bursts and found it to be 2.7 dB. This is about 1 to 1.5 dB larger than the threshold ΔI obtained by other investigators for certain mammals - including man (Fig. 3).

1.4 Passive and Active Acoustic Orientation

Some nocturnal and cave-nesting birds rely on acoustic information to locate prey or guide their flight in darkness. In the case of owls, passive listening is an important means of locating prey. South American oilbirds and Asian swiftlets on the other hand have evolved an active sonar system for navigating through their dark nesting caves.

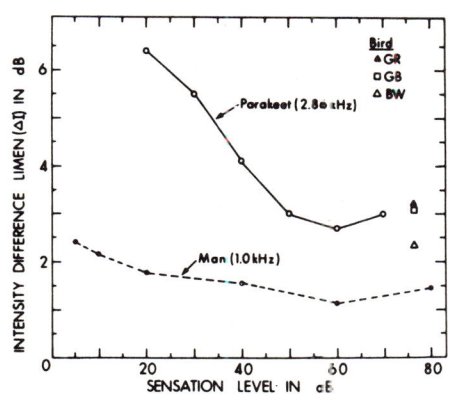

FIG. 3. Intensity discrimination by the parakeet and by man. Symbols at right indicate the mean performance of individual birds pooled over 50, 60 and 70 dB sensation levels. Human thresholds were obtained by the loudness-memory procedure (from Dooling and Saunders, 1975b).

Acoustic prey detection by owls. Owls are specialized in their ability to use sound to detect prey on dark nights. Payne (1971) has shown that barn owls (*Tyto alba*) can locate a deer mouse in total darkness by the rustling sound the mouse makes in dry leaves. The accuracy of the strike under these conditions was within less than 1° in both the vertical and horizontal planes. In some experiments the mouse was replaced by a loudspeaker which played recordings of rustling leaves which could be passed through a high or low pass filter. These tests showed that the barn owl depends on high frequencies in order to localize the sound source. When frequencies above 8500 Hz were filtered out the owl missed by 6 or 7° in ten out of ten strikes. When frequencies above 5000 Hz were removed the owl refused to attempt a strike.

Owls appear to be aided in sound localization by asymmetries in their external ears. In the barn owl, one ear opening is higher than the other and in front of these are flaps of skin, the operculi, which are positioned differently and oriented at different angles on each side of the head. Furthermore, the front of the ear opening is covered by thin, acoustically transparent, auricular feathers. Behind these is a heart-shaped parabola of very dense feathers which acts to reflect sound into the ear. By placing a microphone probe at the position of the ear drum in a dead owl, Payne was able to measure the way in which the direction of a sound source affected the intensity patterns at the two ears. He found that at frequencies above 8.5 kHz the ear was highly directional and that small movements of the sound source could cause large changes, several dB per degree, in the intensity at the ear drum. There was always an area around the owl's line of sight where intensity at all frequencies was greatest, but this area became smaller and smaller with increasing frequency. Thus an owl would directly face the source of a sound if it first adjusted the position of its head for maximum intensity while listening to all frequencies and then refined this initial position by further adjusting its head for maximum intensity at the high frequencies. For each ear there are multiple regions of low sensitivity, especially at high frequencies, but these are not aimed at the same points in auditory space. The asymmetry of the ears thus makes it possible to improve localization by having steep intensity gradients for small head movements with intensity increasing in one ear while it is decreasing in the other.

Payne believes barn owls rely on interaural intensity differences to localize a sound source. The possibility that time of arrival differences between ears may be used instead has also been investigated. Norberg (1968), who studied the effects of asymmetrical ears using a model of the head of Tengmalm's owl (*Aegolius funereus*), concluded that time differences might be used and that the regions of maximum sensitivity were too large

to provide adequate localization. Payne (1971) points out that
Norberg, since he measured only in the horizontal and vertical
planes, would have missed intervening low sensitivity areas that
would have improved directional resolution.

Evidence that barn owls do not depend on the arrival time or
on phase differences between ears for localizing sound was obtained
by Konishi (1973) who conditioned owls to strike at tones played
back from hidden speakers. His barn owls were not able to localize
tone bursts better than continuous tones. Pure tones of 7 or 8 kHz
were located more accurately than higher or lower frequencies and
wide band noises were located more accurately than narrow band or
pure tone signals. The improved ability to locate the sound with
increasing frequency and with increasing bandwidth are expected if
intensity differences are being relied on. It is not clear why
strike accuracy decreases at frequencies above 8 kHz, however.
Konishi verified that the high frequencies were audible to the barn
owls. He obtained behavioral auditory thresholds of -14 dB SPL
at 3 kHz, -20 dB at 7 kHz and 0 dB at 10 kHz. Payne (1971) elicited
a cochlear microphonic up to 20 kHz but these cochlear potentials
do not always accurately indicate the range of auditory perception.

The acute hearing of owls and its importance to prey capture
is further emphasized by the observation of Oeming (quoted by Payne,
1971) of great grey owls (*Strix nebulosa*) diving into deep snow and
catching lemmings beneath the surface. Payne speculates that the
owls locate the lemmings by hearing the sound made by them as they
chew on seeds.

Echolocating birds. There are about 230 species of nocturnal
birds (Fenton and Flemming, 1976) but only two genera are known to
utilize an active system of echolocation. These are the oilbird
Steatornis caripensis (Steatornithidae) and swifts of the genus
Collocalia (Apodidae). Neither of these birds are thought to rely
on echolocation for feeding but rather to use it primarily as a
means of navigating within caves where they nest.

Steatornis is a nocturnal bird that lives in South America
and feeds on fruit (Snow, 1962) during the night. It has large,
well developed eyes that are adequate for most nocturnal vision
but when flying in a dark cave it emits trains of clicks. The
intervals between clicks within these bursts varies from 1.7 to
4.4 msec and the duration of each click is about 1 msec. Clicks
contain frequencies between about 6 and 10 kHz with most of the
energy at 7 or 8 kHz (Griffin, 1953). Observations in caves when
a single bird was flying indicate that the click repetition rate
increases to 11 - 12 bursts/sec as the bird approaches a ledge on
which to land and is lowest (4 - 7 bursts/sec) when the bird is
flying in the middle of the chamber away from its walls (Sales
and Pye, 1974; Suthers, unpublished data). Griffin found that

when their ears are plugged, captive oilbirds become disoriented and fly into walls. Usually no clicks are heard at trees where the oilbirds are feeding (Snow, 1961) but there is at least one report of these birds clicking at their feeding trees (J. Dunston, personal communication). It would be interesting to know more about the sensitivity of echolocation by *Steatornis* and the extent to which it is used outside the nesting cave. Unfortunately, the difficulty of maintaining these birds in captivity has limited experiments on their acoustic orientation.

Cave swiftlets in the genus *Collocalia* also produce clicks when flying inside their nesting caves. Unlike *Steatornis*, *Collocalia* are diurnal insectivores which catch insects on the wing, presumably without echolocation. The ability to echolocate, however, may extend the foraging time available to these swifts by allowing them to return to their roost later in the evening or even to feed at night if conditions permit visual detection of insect prey (Medway, 1967). Fenton (1975) points out that the distribution of *Collocalia* corresponds with a region of relatively depauperate bat fauna, particularly in the Molossidae which are most similar to swifts in their foraging behavior. *Collocalia* do, however, click in daylight when they are flying in dense traffic at the mouth of the cave, sweeping low over water or during courtship and "play" flights (Harrisson, 1966). It is thus likely that clicks also have a social function. Various species of *Collocalia* differ in the extent to which they use caves and also in the degree to which they use echolocation. *C. esculenta*, for example, which nest in partially lighted sites, are unable to echolocate (Medway, 1967).

The clicks of *C. vanikorensis* have most of their sound energy between 4.5 and 7.5 kHz. Each click is double having an initial portion a few milliseconds long followed after several msec by a second higher amplitude portion lasting 4 to 8 msec (Fig. 4). Captive birds flying through an obstacle course vary their click repetition rate from 3 to 20 clicks per second. Obstacle avoidance scores indicate *C. vanikorensis* is able to echolocate rods having a diameter as small as 6 mm (Griffin and Suthers, 1970). The frequencies used by different swiftlets vary somewhat. Clicks of *C. maxima* have most of their energy between 2 and 4.5 kHz (Medway, 1959), those of *C. brevirostus* are between 4 and 5 kHz (Novick, 1959). The mechanism responsible for click production is not known.

2. VISION

Vision is a major sensory modality of almost all birds. Even the swiftlets and oilbirds, which under some conditions use

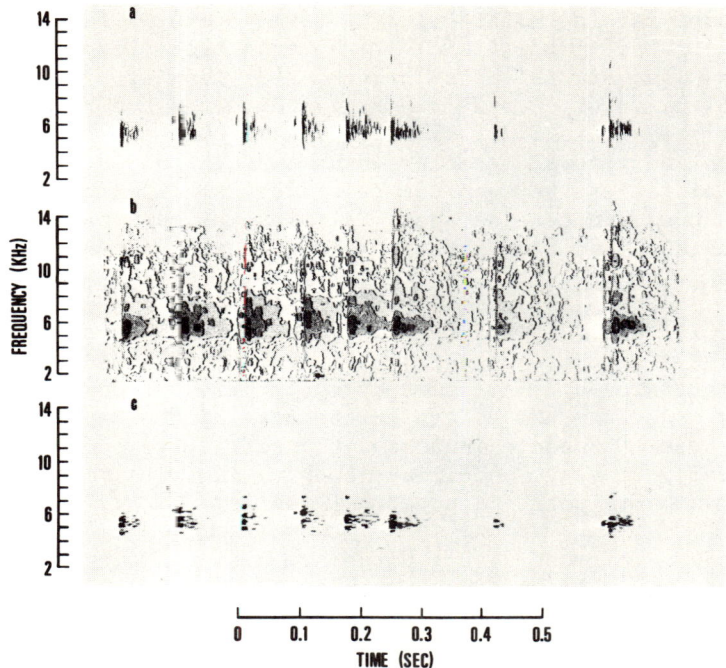

FIG. 4. Sound spectrograph of eight successive orientation clicks of a single *Collocalia vanikorensis* flying in the dark; a) wide bandwidth, b) intensity contour, c) narrow bandwidth displays (from Griffin and Suthers, 1970).

echolocation, have highly developed eyes. The widespread occurrence of cryptically or aposematically colored prey and of visual mimics indicates the importance of vision in prey selection just as the extensive use of visual displays confirms its major role in avian communication.

2.1 Ocular Specializations

Reviews of the avian visual system are available elsewhere (*e.g.*, Duke-Elder, 1958; Sillman, 1973). Birds' eyes are relatively very large. They occupy a major portion of the skull and the adjacent ocular globes may even touch in the midline. Depending on the design of the dioptric system and of the retina, an eye whose absolute size is large can provide better vision for both a diurnal and a nocturnal animal. Walls (1942) classified avian eyes into three categories according to their shape: 1) Flat eyes with a short antero-posterior axis are present in ground dwelling birds such as Columbidae and Galliformes which have narrow heads.

Their shape may facilitate fitting them laterally into such a head to maximize the visual field. 2) Globose eyes are present in many passerines and diurnal birds of prey. 3) Tubular eyes in which the axial diameter exceeds the radial diameter occur in nocturnal birds of prey (Duke-Elder, 1958). The aspherical shape is maintained by the scleral ossicles which support the intermediate portion of the eye between the relatively small cornea and the large fundus. Increasing the axial diameter permits a longer focal distance and larger retinal image - thus potentially improving visual acuity.

Special mechanisms for rapid accommodation involving both the cornea and lens are especially well developed in diurnal birds. Corneal curvature is increased by contraction of Crampton's muscle which extends from the sclera to the edge of the cornea. Contraction of Brucke's muscle increases the refractive power of the lens by actively pressing the ciliary body against its margin. These muscles are very small in nocturnal species.

The eyes of most diving birds are designed to provide good aerial vision. Plunge divers such as the brown pelican (*Pelicanus occidentalis*) which locate their prey before entering the water do not need good underwater vision and lack adaptations to compensate for the loss of corneal refraction in water (Sivak, Lincer and Bobier, 1977). Species which pursue their prey underwater, however, have adaptations to improve underwater vision. Thus cormorants (*Phalacrocorax carbo* and *P. auritus*) have an increased accommodative ability created by a hypertrophied ciliary muscle and by squeezing the anterior surface of the lens with the iris sphincter muscle (Hess, 1912; Pumphery, 1961; Sivak, *et al.*, 1977). A similar mechanism exists in the dipper (*Cinclus mexicanus*) (Goodge, 1960). Loons and diving ducks may combine this technique for increasing the refractive power of the lens with a window in their nictitating membrane having a high refractive index (Ishreyt, 1913). Penguins have gone about solving the problem of amphibious vision by evolving a relatively flat cornea which minimizes the refractive change caused by going from air to water. They are emmetropic in air and are probably able to compensate for moderate underwater hyperopia by accommodation (Sivak, 1976; Sivak and Millodot, 1977).

The retina is avascular, being nourished by the choroidal circulation and the highly vascular pecten oculi which protrude into the fundus from the optic disc. The choroid is especially thick in birds and in woodpeckers it contains a mucus-like substance which Walls (1942) suggested may enable the retina to withstand the stresses imposed on it during pecking and help to prevent retinal detachment.

<u>Areas and foveas</u>. Specialized regions - areas and foveas - also exist in the avian retina. Both are associated with an

increased density of receptor cells. Areas are further characterized by an increased number of ganglion cells and other neurons, making the retina thicker there than elsewhere. Foveas, when present are located within an area and are regions in which the retinal elements proximal to the receptors are centrifugally displaced to form a pit in the vitreal surface of the retina. Most birds have a fovea located within the area. Areas and foveas in birds have been investigated by Wood (1917), Walls (1942), and Duke-Elder (1958) among others. Some more recent findings are summarized by Sillman (1973) and Fite and Rosenfield-Wessels (1975).

The presence, position, and shape of these specialized regions varies according to the bird's sensory ecology. Almost all birds seem to have an area of some kind but some lack a fovea. These include some domesticated species such as the domestic fowl and some ground feeding species including certain quail, the turkey, and a guinea hen. According to Nye (1968) some varieties of pigeons have foveas whereas others do not. A second category of afoveate birds includes seabirds such as the northern fulmar (*Fulmarus glacialis*), manx shearwater (*Puffinus puffinus*) and puffin. These, plus various water birds and some inhabiting open plains, have an elongate horizontal area across the middle of the retina (Lockie, 1952). Such an area is presumed to normally receive the image of the bird's horizon which is the most important part of its visual field and in seabirds these retinal areas can be very highly developed.

Most birds have a single fovea located in the central retina and surrounded by an area. Owls, which have a large binocular field and less than 0.5 degree of eye movement (Steinbach and Money, 1973), are also monofoveate but it is located temporally instead of centrally (Duke-Elder, 1958). Nocturnal owls have a rather shallow, rounded fovea which contrasts with that of the diurnal owl *Asio flammeus* (Oehme, 1961). Finally there is a group of birds that are unique among vertebrates in having both central and temporal foveas. These bifoveate species include diurnal birds of prey (hawks and eagles) and other wing-feeding species such as swallows, kingfishers, and terns which dive for food, as well as bitterns and hummingbirds (Walls, 1942) (Fig. 5). The central fovea is better developed and is thought to be used in monocular fixation whereas the temporal foveas are presumably used for binocular fixation of near objects.

Both diurnal and nocturnal birds have a duplex retina but the proportion of cones is low in nocturnal species. Receptor cell densities have been reported for various species (see Franz, 1934; Oehme, 1961, 1962; Galifret, 1968; Binggeli and Paule, 1969; Fite, 1973; Fite and Rosenfield-Wessels, 1975). Problems of estimating the foveal receptor density are discussed by Fite and Rosenfield-Wessels (1975) and one must be somewhat cautious in comparing

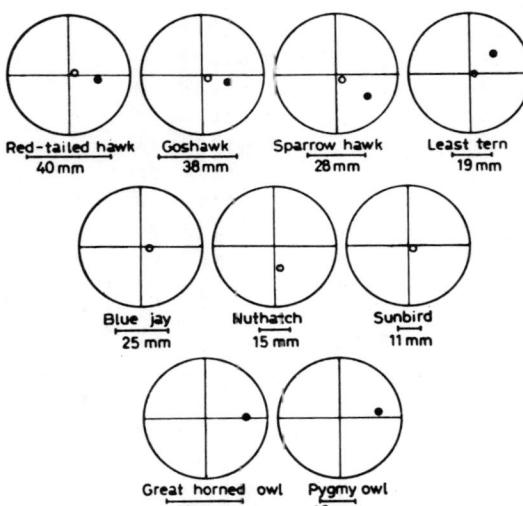

FIG. 5. Foveal location in several species of birds with regard to the horizontal and vertical retinal meridians of left eye. Eye size is indicated by the length of retina measured from a horizontal section cut through the center of the eye. Open Circle = central fovea, solid circle = temporal fovea (from Fite and Rosenfield-Wessels, 1975).

estimates obtained by different authors. There are however significant differences between species. In bifoveate species the receptor density is higher in the central fovea than in the temporal. Estimates range from 1,000,000 receptors/mm^2 in large diurnal raptors such as the golden eagle (*Aquila chrysaetos*) (Polyak, 1957) to less than 93,000/mm^2 in the temporal fovea of the great-horned owl (Fite and Rosenfield-Wessels, 1975). Foveal receptor densities for most birds studied lie between 200,000 and 400,000/mm^2. Fite and Rosenfield-Wessels found a four-fold variation in foveal cell counts between the species they studied. Rods as well as cones are present in the fovea of the great-horned owl. Birds with large eyes average more receptors/visual degree^2 as well as a higher coincidence ratio of receptors to ganglion cells suggesting that retinal convergence in small eyes may be reduced to maintain a high acuity.

   Function of the convexiclivate fovea. Various ideas have been put forth regarding the function of the deep foveal pit – the convexiclivate fovea – present in many birds. It is sometimes assumed that the removal *per se* of the inner retinal tissue from the light path to the receptors improves foveal vision. In the living eye, however, this retinal tissue is optically homogeneous and its transparency is similar to that of the vitreous. Its absence, therefore, can hardly have a significant effect on the amount of light reaching the outer segments (Walls, 1937). Walls (1940, 1942) proposed another function for the very deep convexiclivate central fovea of falciform birds. He noted that since the refractive index of the retina was slightly greater than that of the vitreous (ratio 1.006) light rays would be refracted at the retinal surface. This refraction would magnify the image on the receptors beneath the fovea and should thus improve foveal resolution (Fig. 6). He argued that the high density of visual cells combined with this magnification effect could increase the foveal acuity of some hawks and eagles to at least eight times that of man (Walls, 1942).

Walls' hypothesis was rebutted by Pumphery (1948a, 1948b) who pointed out that Walls had ignored optical aberations caused by refraction at a convexiclivate fovea. Pumphery calculated that as a result of this distortion the resolution at the convexiclivate fovea should be reduced rather than enhanced (Fig. 7). He argued that, though it is poorly designed for improving acuity, this kind of fovea is ideal for maintaining fixation and detecting angular movement of retinal images since the contours of an image moving across it would fluctuate irregularly in a way that should attract the bird's attention. In the golden eagle, on which Pumphrey based his calculation, this region of distortion is confined to a very small central portion of the fovea containing only about 1000 of the approximately 500,000 cones in the entire fovea. High acuity due to dense receptor packing can therefore still be realized over more than 95% of the central fovea. Birds such as the flamingo, shearwater, herring gull (*Larus argentatus*), curlew, snow goose (*Anser caerulescens*) and cormorant in which the fovea is elongated horizontally might use it to fixate the horizon and to detect objects moving vertically relative to the horizon. Distortion of this nature should not be a problem in the shallow temporal foveas of bifoveate birds or the fovea of primates.

Pumphrey proposed his hypothesis before it was appreciated how complex and sophisticated is the neural processing of visual information. A great deal of this processing occurs within the

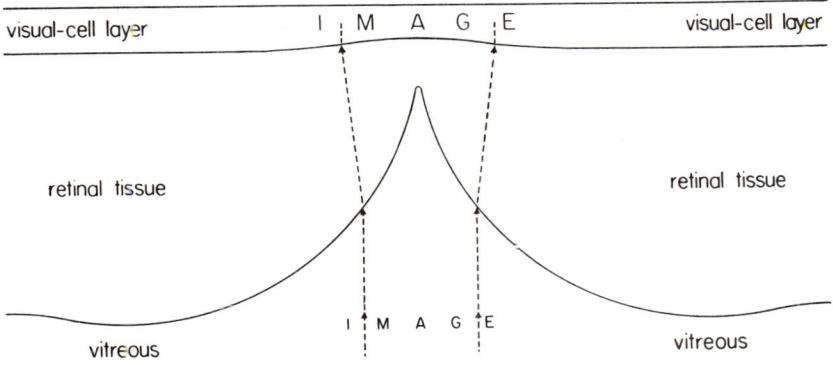

FIG. 6. Magnifying action of the convexiclivate central fovea of the hawk, *Buteo borealis* (from Walls, 1942).

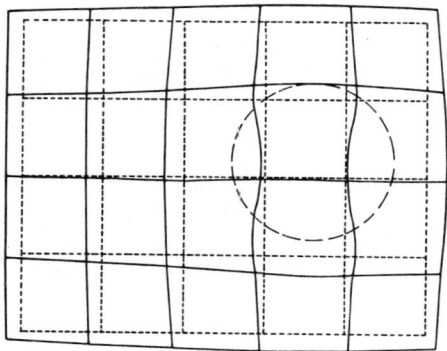

FIG. 7. Distortion of retinal image (solid lines) of a grid (dashed lines) imaged on the convexiclivate forea of a golden eagle. Circle indicating center of fovea has a diameter of 20 μ (from Pumphrey, 1948).

retina itself. In light of these further insights into visual integration it is premature to rule out the possibility that birds with convexiclivate foveas may have neural mechanisms to compensate for the distortion such foveas impose on the retinal image.

2.2 Visual Performance

Visual acuity and absolute threshold. Behavioral aspects of vertebrate vision have been summarized by Muntz (1974). Behavioral measures of visual acuity in birds, with a few exceptions, do not support claims that their visual resolution is superior to that of man. Minimum separable angles - determined using gratings - are reported to be between 1.16 and 4.0 minutes for arc for the pigeon (Blough, 1971); 4.23 for the hen (Johnson, 1914) and from 1.3 to 3.8 for various small diurnal passerines (Donner, 1951). These compare to a minimum separable angle of about 0.5 min in the case of man. Similar acuity seems to be achieved by nocturnal owls. Thus Fite (1973) reported the minimum separable angle of the great-horned owl reached an asymptote of 4 to 5 min between 0 and 1.0 log mL and Martin and Gordon (1974) determined the acuity of two tawny owls (*Strix aluco*) to be 2.7 and 3.7 min of arc, respectively at 1.7 log mL. One of these birds resolved 3.4 min at 0.2 mL. When tested in the same apparatus, human subjects had minimum separable angles of 0.82 and 1.80 min at both of these luminances.

Among the species examined to date, diurnal birds of prey have the best visual acuity and do in fact have better visual resolution than man. Fox, Lehmkuhle and Westendorf (1976) have shown that the minimum separable angle of the American kestrel (*Falco sparvorius*) is 0.19 min of arc in bright illumination (average luminance = 350 cd/m^2). This acuity is 2.6 times better than that of man, but declines more rapidly than does human acuity when light intensity is decreased. Fox, *et al.* report that their kestrel appeared to examine the gratings with its central fovea, turning its head to one side, before making a choice. This is consistent with the hypothesis that the central convexiclivate fovea is the retinal region of highest resolution.

The eyes of large diurnal raptors may have resolving powers even higher than that of the kestrel. Shlaer (1972), after a careful ophthalmoscopic examination of the eye of a living African serpent eagle (*Dryotriorchis spectabilis*), predicted it should have an acuity 2.0 to 2.4 times that of man and that the larger martial eagle (*Spizaetus bellicosus*), with an eye 36 mm long, may have a resolving power 3.0 to 3.6 times that of man.

Dice (1945) attempted to measure the visual threshold of owls by testing their ability to see dead prey at various low levels of illumination. He concluded that the eyes of nocturnal owls are 10 to 100 times more sensitive than those of man. He was not able to accurately calibrate his measurements, however, and his conclusions are not supported by more recent experiments. Martin (1977) determined the absolute visual threshold of two dark adapted tawny owls to be 0.84×10^{-6} and 0.28×10^{-6} cd/m^2. This was 1.4 and 4.4 times, respectively, the mean human threshold measured in the same apparatus and 100 fold more sensitive than a pigeon. Martin argues that the owl's greater visual sensitivity relative to man is due primarily to an increase in its light gathering power rather than to either greater density of visual pigment or to different neural mechanisms within the retina. When its pupil is dilated the tawny owl's eye has an f-number of 1.30 while that of man is 2.10. The retinal illumination of the owl is thus 2.6 times that of man, which is enough to account for the mean difference in visual sensitivity.

There is evidence that the tawny owl possesses color vision. A behavioral measure of its spectral sensitivity shows a peak at 580 nm and is very similar to that of the pigeon. Electroretinographic responses to flickering stimuli suggest the owl's photopic system peaks at about 600 nm and that its scotopic system has a broad peak between 500 and 525 nm (Martin and Gordon, 1974b; Martin, Gordon and Cadle, 1975).

Oil droplets. Many cones in the avian retina contain an oil droplet at the distal end of their inner segment. Similar droplets are present in the retinas of assorted fish, amphibians, reptiles, monotremes and marsupials, but only in amphibians, reptiles, and birds are they colored. Four types of droplets have been reported in most diurnal birds: red, orange, golden or yellowish green and colorless. They are composed of lipid containing a carotenoid pigment. Japanese quail (*Coturnix*) raised on a carotenoid-free diet supplemented with vitamin A develop normally but have only colorless droplets (Meyer, 1971). Measurement of the absorption spectrum of colored droplets by microspectrophotometry shows that they act as cut off filters screening out the short wavelengths complementary to their color (Mayr, 1972). Maximum absorption of different colored droplets is similar in the pigeon and amounts to between 70 and 94% (Strother, 1963; King-Smith, 1969). There is some evidence that each color droplet occurs in a particular type of cone (Morris and Shorey, 1967; Mayr, 1972). Droplets occur in single cones and the chief member of double cones. There is disagreement as to whether they are also present in accessory cones (see Meyer and Cooper, 1966; Sillman, 1973).

Bowmaker and Knowles (1977) found six types of oil droplets in their study of the chicken retina using microspectrophotometry. Three of the five colored droplets occur in single cones where they act as cut off filters and have a T_{50} (wavelength at which 50% transmission occurs, shorter wavelengths being eliminated) of 454, 520 and 585 nm, respectively. One of the remaining two types of colored droplet is associated with the chief member of double cones. It likewise acts as a cut off filter and transmits wavelengths longer than about 497 nm. The fifth droplet occurs in the accessory member of double cones. Its maximum absorbance varies from about 0.2 to 0.6 and occurs at about 451 nm (Fig. 8). No droplets were found in rods.

In many birds the proportion of droplets of a particular color varies in different regions of the retina. The dorsal posterior part of the pigeon retina contains a high proportion of red droplets whereas the ventral and anterior retina contains many yellowish droplets (Waelchi, 1883; Peiponen, 1964; King-Smith, 1969). These parts of the pigeon retina are often referred to as red or yellow field, respectively. The retinotopic distribution of particular colors of droplets is different in different species. No regional differences are apparent in the chicken (Bowmaker and Knowles, 1977). Ecological factors seem to have played a role in the evolution of these droplet distributions and in the relative abundance of different colors (Walls and Judd, 1933; Peiponen, 1964), though the causative relationship is not fully understood. Muntz (1972) has summarized the ecological correlates of oil droplet distribution as follows:

a. Nocturnal species have few colored droplets. The goatsucker (*Caprimulgus europeaus*) has only 10% red or yellow droplets and the owls *Strix aluco* and *Tyto alba* have none.

b. Birds which specialize in catching insects on the wing have few red and orange droplets. In swifts and swallows only 2 to 15% of droplets are these colors.

c. Diving birds that need to see into water in search of fish have a high proportion of red droplets. In gulls and terns

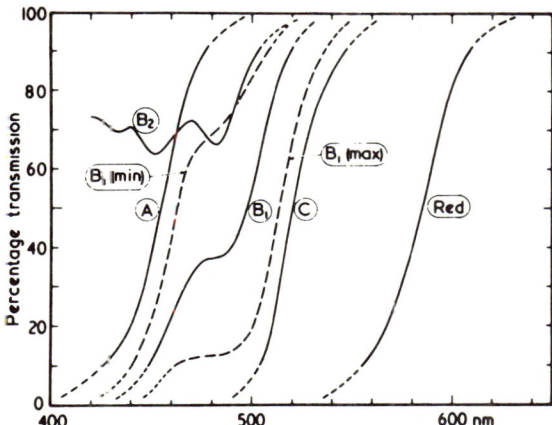

FIG. 8. Transmission spectra of oil droplets in chicken retina. Droplet types A, C and red occur in single cones; type B_1 occurs in the chief member of double cones and type B_2 occurs in the accessory member (from Bowmaker and Knowles, 1977).

75 to 80% of the droplets are red or yellow; in the kingfisher 60% are red.

It is interesting that water birds such as the duck (*Nycera fuligula*) and European water ousel that do not need to see into the water have only 20 and 24% red droplets, respectively. The gannet (*Sula bassana*) which dives for fish has 50 to 57% red or orange-yellow droplets whereas the closely related cormorant (*Phalacrocorax aristotelis*) which swims underwater to feed has only 19% red or orange droplets (Cullen quoted by Muntz, 1972). The heron (*Ardea cinerea*) is an exception, for though it must see into the water, only 20% of its droplets are red. It has been suggested that red droplets serve to reduce the glare from the water surface.

What is the visual function of oil droplets? Various possibilities have been suggested. Oil droplets in the pigment epithelium of the frog retina have been found to concentrate retinol and may serve as a storage reservoir of retinol adjacent to the photoreceptor lamellae (Young and Bok, 1970). Colored droplets act as spectral filters of light stimulating their respective cones since essentially all light reaching the outer segment must pass through them. Since birds lack the preretinal filters present in many mammals, Walls and Judd (1933) suggested yellow oil droplets reduce chromatic aberation and also, by making the sky appear dark, increase the contrast of objects viewed against it. Muntz (1972) has pointed out, however, that birds must usually see such objects as silhouettes in which case darkening the sky would decrease instead of increase their contrast. Walls and Judd (1933) proposed that red droplets serve to decrease glare from water surfaces but it is not clear why red has any advantage over other colors or a neutral density filter for this purpose.

Color vision. Oil droplets must play an important role in color vision since the spectral sensitivity of each cone depends not only on the absorption spectrum of its visual pigment but also on that of the oil droplet through which light must pass before reaching the photosensitive molecules. A bird's ability to discriminate wavelengths might be significantly enhanced if cones having the same visual pigment are divided into sub-groups, each filtering light with a differently colored oil droplet.

Cone visual pigments have been studied by several investigators. Wald (1958) and Liebman (1972) identified a cone pigment with maximum absorption (λ_{max}) at 562 nm in the pigeon. A second cone pigment with a λ_{max} about 400 nm has been suggested on the basis of electrophysiological evidence by Graf and Norren (1974) who recorded the pigeon electroretinogram, and from operant conditioning experiments by Romeskie and Yager (1976) in which they measured the photopic spectral sensitivity in the red field. Norren (1975) obtained evidence of a similar blue sensitive system in the chicken and the daw (*Corvus monedula*).

Govardovskii and Zueva (1977) monitored the early receptor potential of isolated chicken and pigeon retina after selective bleaching. They postulated the presence of four visual pigments with absorption maxima at 413, 467, 507, and 562 nm. They believe the 562 nm absorption peak is due to iodopsin and the 507 nm peak is caused by rhodopsin, some of which they think is contained in cones. They were unable to determine how these pigments are combined with oil droplets of various colors, but both Romeski and Yager (1976) and Govardovskii and Zueva (1977) point out that the short-wave blue sensitive pigment must reside in receptors with colorless droplets since colored droplets absorb nearly all the light to which this pigment is sensitive.

Bowmaker and Knowles (1977) measured the absorption spectra of single cones in the chicken retina using a microspectrophotometer. They report that cone outer segments contained either a green pigment (λ_{max} 497 nm) or a yellow pigment (λ_{max} 569 nm) (Fig. 9). The photopigment in rods had a λ_{max} of 506 nm. No blue sensitive pigment was present. They hypothesize that the blue sensitive system reported by authors using other techniques

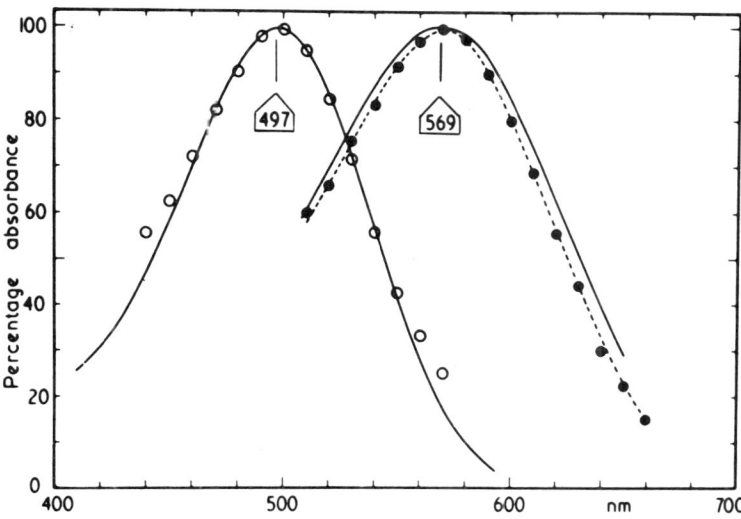

FIG. 9. Absorbance spectra of the two cone visual pigments found in chicken retina. Solid lines are the Dartnall nomogram curves with λ_{max} = 569 and 497 nm (from Bowmaker and Knowles, 1977).

might result from the interaction of cones containing different pigments and oil droplets. The two cone visual pigments combine with different oil droplets in such a way as to result in five different types of cones (Fig. 10). Nearly half of the cones are double cones which thus dominate the retina. These have a broad spectral bandwidth with a maximum sensitivity at 569 nm. They are little affected by their oil droplets which only screen out light below about 55 nm.

From the viewpoint of sensory ecology it is particularly interesting to consider the relationship between a bird's habitat and its visual sensitivity. Bowmaker and Knowles point out that hue discrimination within the broad spectral band dominated by double cones would be very poor were it not for the presence of two types of single cones - one combining a red oil droplet with yellow visual pigment and the other combining a yellowish droplet with green pigment. This results in two types of receptors having narrow band spectral sensitivity peaking at 606 and 533 nm, respectively, and with only a little overlap (Fig. 10). These should greatly enhance wavelength discrimination in that part of the spectrum where other visual pigments overlap extensively.

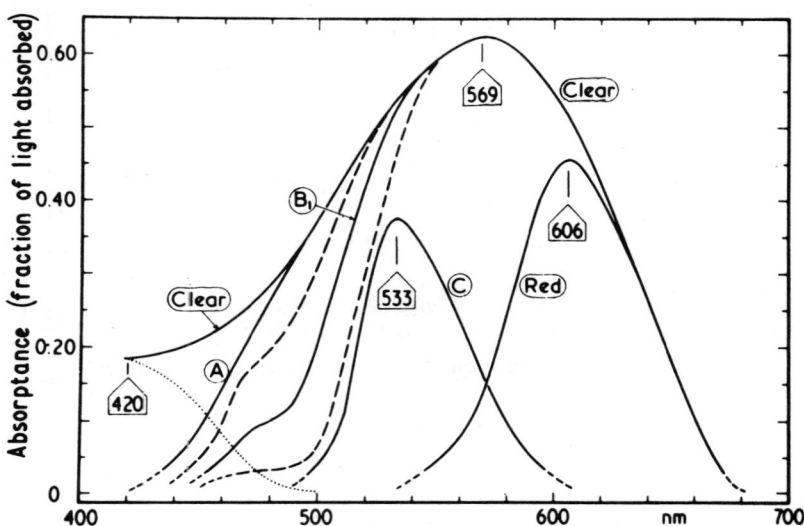

FIG. 10. Calculated absorptance for the pigment-oil droplet combinations found in the five types of cones recognized in the chicken retina. Clear: absorptance of cone with clear oil droplet and P569 in outer segment; A: type A droplet plus P569; B: type B_1 droplet plus P569 (dashed lines indicate effect of variation in B_1 droplet absorbance); C: type C plus P497; Red: red droplet plus P569. The dotted line indicates the possible blue sensitive response with a maximum at 420 nm which could result from interaction between the "clear" and "type A" cones. All calculations assumed an end-on absorbance of 0.42 in the outer segment (from Bowmaker and Knowles, 1977).

The photic environment of jungle fowls, from which the domestic chicken is derived, is influenced by the green canopy of vegetation. McFarland and Munz (1975) showed that the light in forest shade has a maximum irradiance, determined by the transmission of chlorophyll, at about 550 nm. Irradiance changes rapidly on either side of this peak and it is in these regions that the greatest spectral differences occur in light reflected from various kinds of vegetation. Bowmaker and Knowles point out that the narrow band 606 and 533 nm receptors of the chicken retina occur within these regions and may significantly improve hue discrimination at wavelengths which are particularly important in the bird's ancestoral environment.

Detection of ultraviolet and polarized light. A number of invertebrates are known to be able to detect ultraviolet (UV) light, but only recently has it been realized that birds also possess this ability. Evidence for a blue sensitive mechanism in the avian retina has already been discussed. Govardovskii and Zueva (1977) determined that the cornea, lens, and vitreous body of chickens and pigeons are transparent down to 340 nm and that colorless oil droplets transmit down to 350 nm. This is in contrast to man and many other vertebrates which have a yellowish lens pigment that absorbs UV light.

A better understanding of the basis of UV sensitivity in birds is needed before its behavioral significance is known. Huth and Burkhardt (1972) demonstrated, using behavioral conditioning, that the hummingbird (*Colibri serrirostris*) can see light at wavelengths shorter than those visible to man. If detection is the result of fluorscence of the lens no image would be formed and the perception of objects or pattern vision using UV would be impossible. The human lens fluoresces when exposed to high intensity UV, producing an effect known as "blue haze." It is quite possible however that birds perceive UV light in a manner similar to longer wavelengths. If so, UV sensitivity could be useful to the animal in a variety of ways. Many flowers reflect UV which is thought to be important in attracting pollinating insects (Eisner, 1969) and may conceivably play a similar role for hummingbirds. Insectivorous birds might also profit from short wavelength sensitivity. Some otherwise cryptically colored butterflies have UV spots on their wings (Gogala, personal communication). Since these spots are visible to other butterflies, they are thought to provide a secret channel of communication not detectable by predators. If birds have UV pattern vision, it may assist them in finding such insects.

UV sensitivity could also be important in bird navigation. Insects such as bees (von Frisch, 1967) use the pattern of polarized light in the sky to orient. Homing pigeons are known to be able to detect polarized light (Kreithen and Keeton, 1974; Delius, Perchard and Emmerton, 1976). UV light penetrates haze better than longer wavelengths and sky light away from the sun peaks in the UV. The polarization pattern of the sky is thus most pronounced at these wavelengths. If a bird can perceive the plane of polarization of light at very short wavelengths it will have the best opportunity to use this information as an aid in direction finding.

3. OTHER SENSES

3.1 Olfaction

Comparative studies of nasal anatomy and of the relative size of olfactory lobes strongly suggest that substantial interspecific differences exist in the olfactory ability of birds (*e.g.*, Bang and Cobb, 1968; Bang, 1971; see Wenzel, 1973 for a review of avian chemoreception). There is a twelve-fold variation in the olfactory ratio (longest diameter of the olfactory lobe to the longest diameter of the ipsilateral cerebral hemisphere) for species studied. About 80% of the species with olfactory ratios greater than 25% are ground nesting birds, water associated birds, colonial species and those having carnivorous or piscivorous diets. Similarly, about 80% of the birds having olfactory ratios less than 15% are tree nesting, land living, omnivorous or granivorous and solitary breeders. The only species with habits corresponding to the first group, but having olfactory ratios less than 15%, are the Pelicaniformes in which the external nares are closed. Other species such as the kiwi (*Apteryx*), turkey vulture (*Cathartes aura*), emu and whip-poor-will (*Caprimulgus vociferus*) have high olfactory ratios but are not colonial nor are they associated with water (Bang, 1971).

Evidence for the behavioral function of olfaction is available for only a few birds. Several species utilize chemoreception in finding food. The flightless kiwi feeds at night on worms and insects obtained by probing the soil with its bill. Its eyes are small and poorly developed. It has relatively large olfactory lobes and experiments indicate that it utilizes its sense of smell to locate food (Wenzel, 1968, 1972). New world vultures have also often been assumed to locate carrion by smell. Field experiments suggest this is true of turkey vultures and perhaps the king vulture (*Sarcoramphus papa*), but not in the case of the black vulture (*Coragyps atratus*) or of condors (Stager, 1964). It would be interesting to know more about interspecific differences in the sensory ecology of feeding by these birds. Procellariiform seabirds such as albatrosses, petrels, and shearwaters have well developed olfactory lobes. They can be attracted to ships at sea by pouring warm animal fat on the water (Stager, 1967). Grubb (1972) obtained evidence that shearwaters and petrels are attracted to a sponge soaked in cod liver oil. Control tests used a sponge soaked in sea water. Finally, African honeyguides (family Indicatoridae) which feed on beeswax seem to be attracted by the odor of burning beeswax candles which they then proceed to feed upon (Stager, 1967).

Birds having a well developed olfactory sense probably utilize it in other ways in addition to finding food. Likely possibilities include the recognition of individual nest sites, especially in cave nesting species such as oilbirds and swifts, or the location of the breeding colony in the case of seabirds. A role in social interaction is also possible.

Recent experiments with homing pigeons by Papi and his colleagues (summarized by Papi, 1976) have provided strong evidence that olfactory cues can provide one of several mechanisms for direction finding. On the basis of a series of experiments, Papi concludes that young homing pigeons learn the odor of their loft and associate foreign odors with the direction from which they come on prevailing winds. Papi believes this provides the bird with the map component of its orientation system so that when it is transported to an unfamiliar area it can, by recognizing the local odor, determine the direction of that area from the home loft. Having thus determined the direction of its displacement with its olfactory map, the pigeon could use solar or magnetic cues to locate the home direction. Pigeons which are prevented from breathing through their nostrils at their home loft fail to show a correct initial orientation when released in unfamiliar territory. Young birds kept in cages with wind deflectors that changed the direction of the wind entering the cage showed a corresponding deflection in their initial orientation upon release. These and other experiments provide convincing evidence for olfactory orientation by Papi's birds. Papi argues that this is the only true navigational system demonstrated in animals, in that it provides the bird with a map rather than simply a compass direction, and that it is probably fundamental to all homing pigeons. Attempts by others to repeat these experiments have had limited success, however, suggesting that different strains of homing pigeons or birds in different geographical regions may differ in the extent to which they use olfactory cues (*e.g.* Keeton and Brown, 1976; Keeton, Kreithen and Hermayer, 1977).

3.2 Detection of Geomagnetism

The question of whether or not birds are able to use the earth's magnetic field for navigation has had a long and controversial history. Only in recent years has convincing evidence for magnetic field detection been obtained (see reviews by Keeton, 1974 and Emlen, 1975). Young homing pigeons displaced from their loft for their first homing flight are disoriented if bar magnets are attached to their backs whereas control birds wearing brass bars home successfully. This is true whether the sky is sunny or overcast. The homing of experienced pigeons, however, is not affected by attaching magnets to them if the sun is visible. These birds become disoriented, however, under overcast sky. These and other experiments lead Keeton to conclude that homing pigeons need both sun and magnetic cues in order to learn to home, but after some experience can home successfully when one or the other of these cues is unavailable.

Walcott and Green (1974) released pigeons with miniature Helmholtz coils around their heads. When these were connected to a small battery carried by the bird, they generated a magnetic field of about 0.6 gauss. In one group of birds magnetic north was toward the top of the head, in a second group it was toward the neck. The behavior of these pigeons was not affected when they were released in unfamiliar territory under sunny skies, but under overcast conditions one of the groups (north toward neck) oriented homeward whereas the other group (north toward top of head) oriented in almost the opposite direction. Magnetic disturbances cause some normal fluctuation to occur in the earth's magnetic field. Larkin and Keeton (1976) believe that these fluctuations are responsible for some of the variation in the headings taken by homing pigeons. They have shown that the headings of pigeons wearing magnets do not vary in response to magnetic disturbances as much as those of control birds wearing brass bars. They conclude that the bar magnets mask these natural magnetic fluctuations.

Moore (1977) has found that the variation in the flight direction of migrating passerines increases on nights when there are geomagnetic disturbances. Larkin and Sutherland (1977) also report evidence that the headings of migrating birds are affected by the low intensity alternating-current electromagnetic field associated with the Project Seafarer antenna of the U.S. Navy.

Geomagnetism may also play a role in the orientation of other birds. Southern (1972a, 1972b, 1975) reports that small disturbances of the magnetic field affect the orientation of displaced ring-billed gull chicks (*Larus delawarensis*) released in a circular enclosure that allows them to see only the sky. Merkel, the Wiltschko's and Emlen have also reported tentative evidence for a geomagnetic mechanism in the migratory orientation of European robins (*Erithacus rubecula*) and indigo buntings (*Passerina cyanea*). When visual cues are eliminated, changes in the magnetic field have a weak effect on the orientation of the migratory restlessness of these species as recorded in circular activity cages (see Emlen, 1975).

In subsequent studies, Wiltschko and Wiltschko (1975a, 1975b) report that the normal orientation of migratory restlessness of both European robins and warblers, tested outdoors in activity cages under a starry sky, is altered in a predictable way if the local magnetic field is artificially changed. These birds oriented according to the magnetic field even though they continued to view the natural starry sky. The Wiltschko's believe that star orientation by these birds is a secondary orientation mechanism which is learned by associating star patterns with the earth's magnetic field. In evaluating this hypothesis, they presented wild-caught European robins with an arbitrary pattern of 16 star-like points of light in an activity cage without other visual cues for orientation. When the local magnetic field was of a reduced intensity, the robin's activity was random indicating the "star" pattern contained no directional information. When Helmholtz coils were used to induce a magnetic field of approximately normal strength, the robins oriented in the appropriate magnetic direction for spring migration. If after a night of oriented activity the local magnetic field was reduced to its previous low level, the robin continued to orient correctly using the artificial star pattern (Wiltschko and Wiltschko, 1976).

Until recently, attempts to elicit responses from restrained homing pigeons to changes in magnetic field in the laboratory have been unsuccessful. Reille (1968) reported obtaining a conditioned heart rate response but attempts to repeat his experiment have failed (Kreithen and Keeton, 1974; Beaugrand, 1976). Bookman (1977) has made an important advance in successfully conditioning homing pigeons to select the correct one of two feeding boxes at the end of a tunnel depending on whether the magnetic field strength in the tunnel was either 0.02 or 0.5 G. Only pigeons which fluttered while in the tunnel were able to reliably choose the correct food box. Although the significance of the fluttering is not known, these experiments may provide a way of investigating the sensory mechanism for detecting magnetic fields. One such theoretical mechanism was recently proposed by Leask (1977).

3.3 Detection of Changes in Atmospheric Pressure

An ability to detect small changes in atmospheric pressure could be especially advantageous for a bird. Kreithen and Keeton (1974a) have obtained conditioned cardiac responses from homing pigeons to pressure changes as small as about 10 mm H_2O, or lower, presented over an interval of 5 sec. This corresponds to a change of altitude of about 10 m. Both positive and negative pressure changes are effective but it is not clear if the pigeon can discriminate one from the other. Kreithen and Keeton point out that sensitivity to barometric pressure changes could help explain the ability of birds to synchronize migratory flights with the occurrence of favorable meterological conditions and warn them of approaching weather fronts. It could also enable migrants to maintain a constant altitude when other cues are not available and might permit birds to locate thermal updrafts, detect air turbulence or, if they can maintain a reference pressure, even to navigate by pressure-pattern flying.

The nature of the barometric pressure receptors is not known. Kreithen and Keeton found the performance of their pigeons improved in the presence of background noise and suggest that one possible mechanism might be via a modulating effect of air pressure on auditory sensitivity.

4. REFERENCES

Anisimov, V.D. (1976). The morphology of the middle ear in penguins. Vestn. Mosk. Univ. Ser. VI Biol. Pochvoved. 31:16-19 (in Russian).

Aylor, D. (1972a). Noise reduction by vegetation and ground. J. Acoust. Soc. Am. 51:197-205.

Aylor, D. (1972b). Sound transmission through vegetation in relation to leaf area density, leaf width, and breadth of canopy. J. Acoust. Soc. Am. 51:411-414.

Bang, B.G. (1971). Functional anatomy of the olfactory system in 23 orders of birds. Acta Anat. 79 (Suppl. 58):1-76.

Bang, B.G. and S. Cobb (1968). The size of the olfactory bulb in 108 species of birds. Auk. 85:55-61.

Beaugrand, J.P. (1976). An attempt to confirm magnetic sensitivity in the pigeon, *Columba livia*. J. Comp. Physiol. 110:343-355.

Binggeli, R.L. and W.J. Paule (1969). The pigeon retina: Quantitative aspects of the optic nerve and ganglion cell layer. J. Comp. Neurol. 137:1-18.

Blough, P.M. (1971). The visual acuity of the pigeon for distant targets. J. Exp. Anal. Behav. 15:57-67.

Bookman, M.A. (1977). Sensitivity of the homing pigeon to an earth-strength magnetic field. Nature (Lond.) 267:340-342.

Bowmaker, J.K. and A. Knowles (1977). The visual pigments and oil droplets of the chicken retina. Vision Res. 17:755-764.

Delius, J., R. Perchard and J. Emmerton (1976). Polarized light discrimination by pigeons and an electroretinographic correlate. J. Comp. Physiol. Psychol. 90:560-571.

Dice, D.L. (1945). Minimum intensities of illumination under which owls can find dead prey by sight. Am. Nat. 79:384-416.

Donner, K.O. (1951). The visual acuity of some passerine birds. Acta. Zool. Fenn. 66:1-40.

Dooling, R.J. (1973). Behavioral audiometry with the parakeet *Melopsittacus undulatus*. J. Acoust. Soc. Am. 53:1757-1758.

Dooling, R.J. and J.C. Saunders (1975a). Hearing in the parakeet (*Melopsittacus undulatus*): Absolute threshold, critical ratios,

frequency difference limens and vocalizations. J. Comp. Physiol. Psychol. 88:1-20.
Dooling, R.J. and J.C. Saunders (1975b). Auditory intensity discimination in the parakeet (*Melopsittacus undulatus*). J. Acoust. Soc. Am. 58:1308-1310.
Dooling, R.J., J.A. Mulligan and J.D. Miller (1971). Auditory sensitivity and song spectrum of the common canary (*Serinus canarius*). J. Acoust. Soc. Am. 50:700-709.
Duke-Elder, S. (1958). The eyes of birds. In: System of Ophthalmology: The Eye in Evolution, Vol. 1, pp. 397-427. Kimpton, London.
Eisner, T. (1969). Ultraviolet video-viewing: The television camera as an insect eye. Science 166:1172-1174.
Emlen, S.T. (1975). Migration: Orientation and navigation. In: Avian Biology. D.S. Farner and J.R. King, editors. Vol. V, pp. 129-219. Academic Press, New York.
Fenton, M.B. (1975). Acuity of echolocation in *Collocalia hirundinacea* (Aves: Apodidae) with comments on the distributions of echolocating swiftlets and molossid bats. Biotropica 7:1-7.
Fenton, M.B. and T.H. Fleming (1976). Ecological interactions between bats and nocturnal birds. Biotropica 8:104-110.
Fite, K.V. (1973). Anatomical and behavioral correlates of visual acuity in the great horned owl. Vision Res. 13:219-230.
Fite, K.V. and S. Rosenfield-Wessels (1975). A comparative study of deep avian foveas. Brain Behav. Evol. 12:97-115.
Fox, R., S. W. Lehmkuhle and D.H. Westendorf (1976). Falcon visual acuity. Science 192:263-265.
Franz, V. (1934). III Höhere Sinnesorgane. 1. Vergleichende Anatomie des Wirbeltierauges. In: Handbuch der vergleichenden Anatomie der Wirbeltiere. L. Bolk, E. Göppert, E. Kallius and W. Lubosch, editors. Vol. 2, part 2, pp. 989-1292. Urban and Schwarzenberg, Berlin.
Galifret, Y. (1968). Les diverses aires fonctionelles de la retine du Pigeon. Z. Zellforsch. 86:535-545.
Goodge, W.R. (1960). Adaptations for amphibious vision in the dipper (*Cinclus mexicanus*). J. Morphol. 107:79-91.
Govardovskii, V.I. and L.V. Zueva (1977). Visual pigments of the chicken and pigeon. Vision Res. 17:537-543.
Graf, V. and D.V. Norren (1974). A blue sensitive mechanism in the pigeon retina: λ_{max} 400 nm. Vision Res. 14:1203-1209.
Greenwood, D. (1961). Critical bandwidth and the frequency coordinates of the basilar membrane. J. Acoust. Soc. Am. 33: 1344-1356.
Greenwood, D. (1962). Approximate calculation of the dimensions of traveling-wave envelopes in four species. J. Acoust. Soc. Am. 34:1364-1369.
Griffin, D.R. (1953). Acoustic orientation in the oil bird, *Steatornis*. Proc. Nat. Acad. Sci. 39:884-893.

Griffin, D.R. (1969). The physiology and geophysics of bird navigation. Q. Rev. Biol. 44:255-276.
Griffin, D.R. (1976). The audibility of frog choruses to migrating birds. Anim. Behav. 24:421-427.
Griffin, D.R. and C. Hopkins (1974). Sounds audible to migrating birds. Anim. Behav. 22:672-678.
Griffin, D.R. and R. Suthers (1970). Sensitivity of echolocation in cave swiftlets. Biol. Bull. (Woods Hole) 139:495-501.
Grubb, T.C., Jr. (1972). Smell and foraging in shear waters and petrels. Nature (Lond.) 237:404-405.
Harrison, J.B. and L. Furumoto (1971). Pigeon audiograms: Comparison of evoked potential and behavioral thresholds in individual birds. J. Aud. Res. 11:33-42.
Harrisson, T. (1966). Onset of echo-location clicking in *Collocalia* swiftlets. Nature (Lond.) 212:530.
Heise, G.E. (1953). Auditory thresholds in the pigeon. Am. J. Psychol. 66:1-19.
Hess, C. (1912). Vergleichende Physiologie des Gesichtsinnes. Gustav Fischer, Jena.
Huth, H. and D. Burkhardt (1972). Der spektrale Sehbereich eines Violettohr-Kolibris. Naturwissenschaften 59:650.
Ishreyt, G. (1913). Zur vergleichenden Morphologie des Entenauges. Arch. Vergl. Ophthalmol. 3:39-76.
Jahnke, V., P.G. Lundquist and J. Wersall (1969). Some morphological aspects of sound perception in birds. Acta. Oto-Laryngol. 67:583-601.
Johnson, H.M. (1914). Visual pattern discrimination in the vertebrates. II. Comparative visual acuity of the dog, the monkey and the chick. J. Animal Behav. 6:169-188.
Kartashev, N.N. and V.D. Ilichev (1964). Über das Gehörorgan der Alkenvögel. J. Ornithol. 105:113-136.
Keeton, W.T. (1974). The orientational and navigational basis of homing in birds. In: Advances in the Study of Behavior. D.S. Lehrman, J.S. Rosenblatt, R.A. Hinde and E. Shaw, editors. Vol. 5, pp. 47-132. Academic Press, New York.
Keeton, W.T. and A.I. Brown (1976). Homing behavior of pigeons not disturbed by application of an olfactory stimulus. J. Comp. Physiol. 105:259-266.
Keeton, W.T., M. Kreithen and K. Hermayer (1977). Orientation by pigeons deprived of olfaction by nasal tubes. J. Comp. Physiol. 114:289-299.
King-Smith, P.E. (1969). Absorption spectra and function of the coloured oil drops in the pigeon retina. Vision Res. 9:1391-1399.
Konishi, M. (1970a). Evolution of design features in the coding of species-specificity. Am. Zool. 10:67-72.
Konishi, M. (1970b). Comparative neurophysiological studies of hearing and vocalization in songbirds. Z. Vgl. Physiol. 66:257-272.
Konishi, M. (1973). Locatable and non-locatable acoustic signals for barn owls. Am. Nat. 107:775-785.

Kreithen, M.L. and W.T. Keeton (1974a). Detection of changes in atmospheric pressure by the homing pigeon, *Columba livia*. J. Comp. Physiol. 89:73-82.

Kreithen, M.L. and W.T. Keeton (1974b). Detection of polarized light by the homing pigeon, *Columba livia*. J. Comp. Physiol. 89:83-92.

Kreithen, M.L. and W.T. Keeton (1974c). Attempts to condition homing pigeons to magnetic stimuli. J. Comp. Physiol. 91:355-362.

Larkin, R. and P. Sutherland (1977). Migrating birds respond to project seafarer's electromagnetic field. Science 195:777-773.

Larkin, T.S. and W.T. Keeton (1976). Bar magnets mask the effect of normal magnetic disturbances on pigeon orientation. J. Comp. Physiol. 110:227-231.

Leask, M.J.M. (1977). A physicochemical mechanism for magnetic field detection by migratory birds and homing pigeons. Nature (Lond.) 267:144-145.

Liebman, P.A. (1972). Microspectrophotometry of photoreceptors. In: Handbook of Sensory Physiology. H.J.A. Dartnall, editor. Vol. VII, part 1, pp. 481-528. Springer-Verlag, Berlin.

Lockie, J.D. (1952). A comparison of some aspects of the retinae of the Manx Shearwater, Fulmar Petrel and House Sparrow. Q. J. Microscop. Sci. 93:347-356.

McFarland, W.N. and F.W. Munz (1975). The visible spectrum during twilight and its implications to vision. In: Light as an Ecological Factor: II. G.C. Evans, R. Bainbridge and O. Rackham, editors. pp. 249-270. Blackwell Scientific, Oxford.

Martin, G. (1977). Absolute visual threshold and scotopic spectral sensitivity in the tawny owl *Strix aluco*. Nature (Lond.) 268:636-638.

Martin, G. and I.E. Gordon (1974a). Visual acuity in the tawny owl (*Strix aluco*). Vision Res. 14:1393-1397.

Martin, G. and I.E. Gordon (1974b). Increment-threshold spectral sensitivity in the tawny owl (*Strix aluco*). Vision Res. 14:615-621.

Martin, G., I.E. Gordon and D.R. Cadle (1975). Electroretinographically determined spectral sensitivity in the tawny owl (*Strix aluco*). J. Comp. Physiol. Psychol. 89:72-78.

Mayr, I. (1972). Verteilung, Lokalisation und Absorption der Zapfenoelkugeln bei Voegeln (Ploceidae). Vision Res. 12:1477-1484.

Medway, Lord (1959). Echolocation among *Collocalia*. Nature (Lond.) 184:1352-1353.

Medway, Lord (1967). The function of echonavigation among swiftlets. Anim. Behav. 15:416-420.

Meyer, D.B. (1971). The effect of dietary carotenoid deprivation on avian retinal oil droplets. Ophthalmic Res. 2:104-109.

Meyer, D.B. and T.G. Cooper (1966). The visual cells of the chicken as revealed by phase contrast microscopy. Am. J. Anat. 118:723-734.

Mokhtar, M. and M.A. Marrous (1955). The attenuation of sound in a turbulent atmosphere over a desert terrain. Acoustica 5:179-181.

Moore, F.R. (1977). Geomagnetic disturbance and the orientation of nocturnally migrating birds. Science 196:682-683.

Morris, V.B. and C.D. Shorey (1967). An electron microscope study of types of receptor in the chicken retina. J. Comp. Neurol. 129:313-340.

Morton, E.S. (1975). Ecological sources of selection on avian sounds. Am. Nat. 109:17-34.

Muntz, W.R.A. (1972). Inert absorbing and reflecting pigments. In: Handbook of Sensory Physiology, Photochemistry of Vision. H.J.A. Dartnall, editor. Vol. VII, part 1, pp. 529-565. Springer-Verlag, New York.

Muntz, W.R.A. (1974). Comparative aspects in the behavioral study of vertebrate vision. In: The Eye. H. Davson and L.T. Graham, Jr., editors. Vol. 6, pp. 155-226. Academic Press, New York.

Norberg, A. (1968). Physical factors in directional hearing in *Aegolius funereus* (Linne) (Strigiformes), with special reference to the significance of the asymmetry of the external ears. Ark. Zool. 20(3/4):181-204.

Norren, D.V. (1975). Two short wavelength sensitive cone systems in pigeon, chicken and daw. Vision Res. 15:1164-1166.

Novick, A. (1959). Acoustic orientation in the cave swiftlet. Biol. Bull. (Woods Hole) 117:497-503.

Nye, P.W. (1968). The binocular acuity of the pigeon measured in terms of the modulation transfer function. Vision Res. 8: 1041-1053.

Oehme, H. (1961). Vergleichend-histologische Untersuchungen an der Retina von Eulen. Zool. Jahrb. Abt. Anat. Ontog. Tiere 79: 439-478.

Oehme, H. (1962). Eas Auge von Mauersegler, Star und Amsel. J. Ornithol. 103:189-212.

Papi, F. (1976). The olfactory navigation system of the homing pigeon. Verh. Dtsch. Zool. Ges. 69:184-205.

Payne, R.S. (1971). Acoustic location of prey by barn owls, (*Tyto alba*). J. Exp. Biol. 54:535-573.

Peiponen, V.A. (1964). Zur Bedeutung der Olkugeln im Farbensehen der Sauropsiden. Ann. Zool. Fenn. 1:281-302.

Polyak, S.L. (1957). The Vertebrate Visual System. University of Chicago Press.

Pumphrey, R.J. (1948a). The sense organs of birds. Ibis. 90:171-199.

Pumphrey, R.J. (1948b). The theory of the fovea. J. Exp. Biol. 25:299-312.

Pumphrey, R.J. (1961). Part I. Sensory organs: Vision. In: Biology and Comparative Physiology of Birds. A.J. Marshall, editor. Vol. II, pp. 5568. Academic Press, New York.

Pumphrey, R.J. (1961). Part II. Sense organs: Hearing. In: Biology and Comparative Physiology of Birds. A.J. Marshall, editor, Vol. II, pp. 69-86. Academic Press, New York.

Reille, A. (1968). Essai de mise en évidence d'une sensibilité du pigeon au champ magnétique à l'aide d'un conditionnement nociceptif. J. Physiol. (Paris) 60:85-92.

Romeskie, M. and D. Yager (1976). Psychophysical studies of pigeon color vision. I. Photopic spectral sensitivity. Vision. Res. 16:501-505.

Sales, G. and D. Pye (1974). Ultrasonic Communication by Animals. Chapman and Hall, London.

Saunders, J. and R. Docling (1974). Noise induced threshold shift in the parakeet (*Melopsittacus undulatus*). Proc. Natl. Acad. Sci. 71:1962-1965.

Saunders, J. and B. Johnstone (1972). A comparative analysis of middle-ear function in non-mammalian vertebrates. Acta. Oto-Laryngol. 73:353-361.

Schwartzkopff, J. (1949). Über Sitz und Leistung von Gehör und Vibrationssinn bei Vögeln. Z. Vgl. Physiol. 31:527-608.

Schwartzkopff, J. (1955). Schallsinnesorgane, ihre Funktion und biologische Bedeutung bei Vögeln. Acta 11th Congr. Intern. Ornithol. 1954. pp. 189-208.

Schwartzkopff, J. (1973). Mechanoreception. In: Avian Biology. D.S. Farner and J.R. King, editors. Vol. III, pp. 417-477. Academic Press, New York.

Shlaer, R. (1972). An eagle's eye: Quality of the reinal image. Science 176:920-922.

Sillman, A.J. (1973). Avian vision. In: Avian Biology. D.S. Farner and J.R. King, editors. Vol. III, pp. 349-387. Academic Press, New York.

Sivak, J.G. (1976). The role of the flat cornea in the amphibious behavior of the black-footed penguin, *Spheniscus demersus*. Can. J. Zool. 54:1341-1345.

Sivak, J.G. and M. Millodot (1977). Optical performance of the penguin eye in air and water. J. Comp. Physiol. 119:241-247.

Sivak, J.G., J.L. Lincer and W. Bobier (1977). Amphibious visual optics of the eyes of the double-crested cormorant (*Phalacrocorax auritus*) and the brown pelican (*Pelecanus occidentalis*). Can. J. Zool. 55:782-788.

Snow, D.W. (1961). The natural history of the oilbird, *Steatornis caripensis*, in Trinidad, W.I. Part 1. General behavior and breeding habits. Zoologica 46:27-48.

Snow, D.W. (1962). The natural history of the oilbird, *Steatornis caripensis*, in Trinidad, W.I. Part 2. Population, breeding ecology and food. Zoologica 47:199-221.

Southern, W.E. (1972a). Influence of disturbances in the earth's magnetic field on Ring-billed Gull orientation. Condor. 74:102-105.

Southern, W.E. (1972b). Magnets disrupt the orientation of juvenile Ring-billed Gulls. Bioscience 22:476-479.

Stager, K.E. (1964). The role of olfaction in food location by the Turkey Vulture (*Cathartes aura*). Los Angeles County Mus. Contrib. Sci. No. 81. pp. 3-63.

Stager, K.E. (1967). Avian olfaction. In: Symposium on Vertebrate Olfaction, Washington, D.C., December 30, 1966. Am. J. Zool. 7:415-419.

Steinbach, M.J. and K.E. Money (1973). Eye movements of the owl. Vision Res. 13:889-891.

Strother, G.K. (1963). Absorption spectra of retinal oil globules in turkey, turtle and pigeon. Exp. Cell Res. 29:349-355.

Takasaka, T. and C.A. Smith (1971). The structure and innervation of the pigeon's basilar papilla. J. Ultrastruct. Res. 35: 20-65.

Trainer, J.E. (1946). The auditory acuity of certain birds. Ph.D. Thesis, Cornell University, Ithaca, New York.

von Békésy, G. (1944). Über die mechanische Frequenz-analyse in der Schnecke verschiedner Tiere. Akust. Z. 9:3-11.

von Frisch, K. (1967). The Dance Language and Orientation of Bees. Harvard University Press, Cambridge.

Waelchi, G. (1883). Zur Topographie der gefärbetn Kugeln der Vogelnetzhaut. Albrect von Graefes Arch. Ophthalmol. 29:205-224.

Walcott, C. and R.P. Green (1974). Orientation of homing pigeons altered by a change in the direction of an applied magnetic field. Science 184:180-182.

Wald, G. (1958). Retinal chemistry and the physiology of vision. in: Visual Problems of Colour. Symp. No. 8. Nat. Physics Lab. (U.K.) H.M.S.O., London. pp. 7-61.

Walls, G. (1937). Significance of the foveal depression. Arch. Ophthalmol. 18:912-919.

Walls, G. (1940). Postscript on image expansion by the foveal clivus. Arch. Ophthalmol. 23:831-832.

Walls, G. (1942). The Vertebrate Eye and its Adapative Radiation. Cranbrook Inst. of Science, Michigan.

Walls, G. and H. Judd (1933). The intra-ocular color filters of vertebrates. Br. J. Ophthalmol. 17:641-675 and 705-725.

Wenzel, B.M. (1968). The olfactory prowess of the Kiwi. Nature (Lond.) 220:1133-1134.

Wenzel, B.M. (1972). Olfactory sensation in the Kiwi and other birds. Ann. N.Y. Acad. Sci. 188:183-193.

Wenzel, B.M. (1973). Chemoreception. In: Avian Biology. D.S. Farner and J.R. King, editors. Vol. III, pp. 389-415. Academic Press, New York.

Wever, E. G., P. Herman, J.A. Simmons and D.R. Hertzler (1969). Hearing in the blackfooted penguin, *Spheniscus demersus*, as represented by the cochlear potentials. Proc. Nat. Acad. Sci. U.S.A. 63:676-680.

Wilkinson, R. and P.E. Howse (1975). Time resolution of acoustic signals by birds. Nature (Lond.) 258:320-321.

Wiltschko, W. and R. Wiltschko (1975a). The interaction of stars and magnetic field in the orientation system of night migrating birds. I. Autumn experiments with European warblers. (Gen. *Sylvia*). Z. Tierpsychol. 37:337-355.

Wiltschko, W. and R. Wiltschko (1975b). The interaction of stars and magnetic field in the orientation system of night migrating birds. II. Spring experiments with European robins (*Erithacus rubecula*). Z. Tierpsychol. 39:265-282.

Wiltschko, W. and R. Wiltschko (1976). Interrelation of magnetic compass and star orientation in night-migrating birds. J. Comp. Physiol. 109:91-99.

Wood, C.A. (1917). The Fundus Oculi of Birds. Lakeside Press, Chicago.

Yodlowski, M., M. Kreithen and W. Keeton (1977). Detection of atmospheric infrasound by homing pigeons. Nature (Lond.) 265:725-726.

Young, R.W. and D. Bok (1970). Autoradiographic studies on the metabolism of the retinal pigment epithelium. Invest. Ophthalmol. 9:524-536.

Zablotskaya, M.M. (1974). Seasonal responses in summation responses of the auditory nerve in *Carduelis carduelis* L. and *Carduelis caniceps* Vig. Vestn. Mosk. Univ. Ser. 6. Biol. Pochvoved 29:22-26 (in Russian).

ACKNOWLEDGEMENT

The author thanks the following for permission to reproduce figures. Fig. 2: Copyright 1975 by the American Psychological Association, Reprinted by permission; Fig. 4: The Biological Bulletin, reprinted by permission; Fig. 5: Brain, Behavior and Evolution, reprinted by permission S. Karger, A.G. Basel; Fig. 6: The Cranbrook Institute of Science, reprinted by permission.

SENSORY ECOLOGY OF MAMMALS

R. A. Suthers

Physiology Section, Medical Sciences, Indiana University

Bloomington, Indiana 47401 U.S.A.

Mammals inhabit a wide range of environments and exhibit a variety of sensory adaptations for specialized ecological niches. The discussion in the following pages will be restricted to vision and audition. Since even these senses cannot be fully treated in the space available, attention will be focused on the sensory ecology of selected mammalian groups which illustrate sensory adaptations to certain diverse habitats.

1. SENSORY ADAPTATIONS OF TERRESTRIAL MAMMALS

1.1 <u>Vision</u> <u>in</u> <u>Dim</u> <u>and</u> <u>Bright</u> <u>Environments</u>

It is well known that the eyes of nocturnal mammals exhibit various anatomical and physiological adaptations which improve their performance in dim light. Specializations for nocturnal and diurnal vision have been discussed in detail by previous authors (see, for example, Walls 1942) and need only to be briefly mentioned here. Whereas diurnal species can sacrifice visual sensitivity for improved acuity, nocturnal mammals are limited in their visual acuity by the necessity of achieving maximum sensitivity. One of the classical adaptations for nocturnal vision is a large pupil to admit as much light as possible. Nocturnal eyes are typically both relatively and absolutely large to increase their light gathering ability, but the lens is almost spherical, giving them a short focal distance. The retinal image is thus small compared to a diurnal eye of similar size, so that the available light is concentrated onto the receptors. The retina itself contains a high density of rods. Cones may be scarce or absent. The coincidence ratio of receptors to ganglion cells is high, suggesting a great deal of convergence between receptors and the cells forming the optic nerve.

Many mammals with a need to see in dim light - including ungulates, elephants, carnivores, whales, prosimians, and certain nocturnal monkeys - have a tapetum lucidum in the choroid which reflects light back through the retina. Fruit bats (Megachiroptera) and the common opossum (*Didelphis virginiana*) have reflecting granules in their pigment epithelium (Walls, 1942; Pedler and Tilley, 1969; Muntz, 1972). In many cases these tapeta may substantially increase the sensitivity of the eye. The absolute threshold of the cat has been determined through behavioral experiments to be 5 to 10 times lower than that of man (Bridgeman and Smith, 1942; Gunter, 1951). The reflectance of the cat's tapetum was measured by Weale (1953). Muntz (1972) points out that tapetal reflection, combined with a brighter retinal image resulting from the higher optical power of the cat's eye compared to that of man (Vakkur and Bishop, 1963), could easily account for differences between human and feline absolute thresholds without invoking any special retinal mechanisms. Comparison of the electroretinogram (ERG) obtained from areas backed by a tapetum, with the ERG from non-tapetal retina suggests that tapetal retina is about 0.5 log unit more sensitive than non-tapetal retina (Dodt and Walther, 1958).

Diurnal mammals which depend heavily on vision may also have large eyes, but the lens is flattened and placed well forward. This, combined with a large fundus, gives the diurnal eye a long focal length and larger retinal image, thus improving visual resolution. The retina contains cones which may be especially densely packed in a specialized region known as the area centralis where visual resolution is best. In some species (*e.g.*, the cat and primates) this thickened portion of the retina is of a circular or oval shape. In others it forms an elongate region and is called the visual streak. A visual streak is present in ungulates (pig, sheep, ox and horse) (Hebel, 1976), in herbivores such as the rabbit (Whitteridge, 1965) and in carnivores such as the leopard (Rodieck, 1973) (Fig. 1). Even in the cat there is a horizontal elongation of the area centralis (Stone, 1965) while dogs have a horizontal visual streak which, near its temporal end, contains a circular region with a higher density of retinal cells (Rodieck, 1973; Hebel, 1976).

In those species possessing a visual streak, its orientation is such that it must normally receive an image of the most important part of the environment. In large mammals it forms a horizontal band dorsal to the optic disc in retina that corresponds to the animal's visual horizon. In the rabbit, on the other hand, the

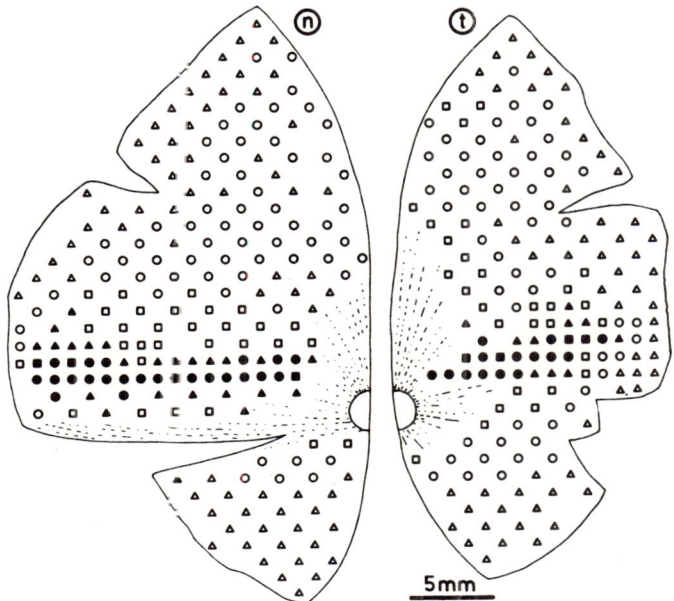

FIG. 1. Distribution of ganglion cells in the retina of the left eye of the sheep. Cell density per mm^2 is indicated by symbols as follows: open triangle = 500–1000; open circle = 1,000–2,000; open square = 2,000–3,000; solid triangle = 3,000–4,000; solid circle = 4,000–5,000; solid square >5,000. Dotted lines show course of nerve fibers obscuring ganglion cells as they radiate from the optic disc. t = temporal, r = nasal retina (From Hebel, 1976).

visual streak is ventral to the optic disc, presumably because in this small mammal the horizon lies higher in the visual field.

The area centralis of many higher primates contains a pit-like depression, the fovea. The primate fovea is much shallower than the convexiclivate fovea of birds and its function in vision is a matter of some debate (Weale, 1966). The absence of a fovea does not necessarily indicate poor visual acuity. DeValois and Jacobs (1971) point out that this is well demonstrated by the work of Ordy and Samorajski (1968) who related visual acuity to retinal structure in six species of primates. The minimum separable visual angle of squirrel monkeys (*Saimiri sciureus*) and marmosets (*Callithrix jacchus*) was found to be 0.5 to 1.5 min of arc. Both species are diurnally active and have a well defined fovea in a duplex retina. Nocturnal owl monkeys (*Aotes trivirgatus*) and the bush baby (*Galago crassicaudatus*), on the other hand, have almost all-rod retinas containing an area centralis but no fovea. As might be expected their visual acuity, 3.5 to 8.0 min of arc, was poorer than that of the previous species. Common tree shrews (*Tupaia glis*) and ring-tailed lemurs (*Lemur catta*), however, do not conform to this presumed relationship between high acuity and the presence of a fovea. Both the tree shrew's cone retina and the lemur's rod retina possess an area but no fovea, yet the acuity of these species was very good (0.5 to 1.5 min of arc). Ordy and Samorajski suggest that the morphological feature most consistently associated with high acuity is the extent to which there is convergence of receptors onto ganglion cells rather than the presence of a fovea.

Just as the position and shape of the area centralis or the visual streak may be adapted to the animal's ecology, so likewise is that of the optic disc sometimes modified according to a species' habitat. This is most strikingly demonstrated in squirrels whose retinas exhibit adaptations to minimize the visual significance of the scotoma associated with the optic disc. Such scotomas must present a particularly serious liability to animals whose visual fields are largely monocular. In diurnal squirrels the optic disc, instead of being near the center of the fundus, is moved well up into the dorsal part of the retina where, instead of forming a disc, it forms a long narrow horizontal streak. The length of this streak varies with the number of optic nerve fibers, being longest in ground squirrels and prairie dogs, of intermediate length in the woodchuck and shorter still in the forest dwelling gray squirrel (Fig. 2). In each case, however, the optic "disc" is very thin, minimizing the possibility that the image of an object will fall entirely within its scotoma and thus be undetected. Nocturnal flying squirrels have conventional circular optic discs located near the center of the retina (Walls, 1942).

A number of diurnal mammals have pigmented structures interposed in the light path between the cornea and the retina (see review by Muntz, 1972). In primates the macula lutea is formed by a yellow pigment - the xanthophyll, lutein - present in the retina over and around the fovea. Primates, tree shrews, and diurnal squirrels have a pigment in their lens which absorbs in the blue-violet thus making the lens appear yellow (Walls and Judd, 1933, Walls, 1942; Cooper and Robson, 1969a and 1969b, Yolton, Yolton, Renz and Jacobs, 1974). Walls (1942) believes that one function of this pigment is to reduce chromatic aberation by filtering out short wavelengths. He argues that this is a luxury only diurnal animals can afford and points to variation in the amount of lens pigment among the squirrels, which again present a particularly interesting case. Subsequent spectrophotometric measurements of preretinal absorbance by Yolten *et al.* (1974) confirm Walls' finding that the prairie dog's lens has more yellow pigment than that of the ground squirrel which in turn has

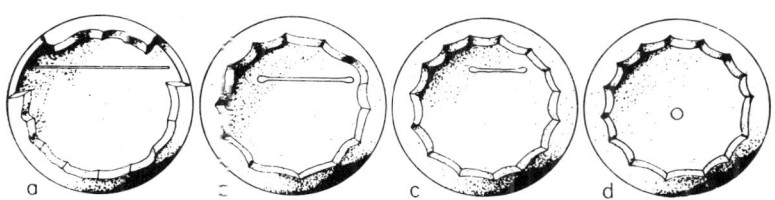

FIG. 2. Schematic view of fundus of left eye to show optic disc of certain members of the squirrel family. a) prairie dog (*Cynomys ludovicianus*) which inhabits very bright open spaces; b) woodchuck (*Marmota monax*) inhabitant of less bright places ; c) gray squirrel (*Sciurus carolinensis*) lives in dense woods and d) flying squirrel (*Glaucomys volans*) a nocturnal species. Eyes are not drawn to same scale (From Walls, 1942).

more than the arboreal gray squirrel. The lens and cornea of the nocturnal flying squirrel is essentially colorless. The filtering out of short wavelengths thus seems to be proportional to the brightness of the habitat in the case of these animals. Unfortunately, not all mammals fit this chromatic aberation hypothesis as well as do the squirrels. For example, the lens of the cat becomes yellower with age and pigmentation of the human lens varies with age (Cooper and Robson, 1969b). A metabolic cause for the pigmentation cannot be ruled out (Weale, 1974). Visual pigment photoproducts have an absorbance peak that matches that of the lens pigment, suggesting that such lenses may be important in filtering out wavelengths that might otherwise interfere with visual pigment regeneration (Cooper and Robson, 1969a).

A variety of mammals have at least some ability to discriminate wavelengths of light. Data on the distribution of color vision in mammals are incomplete, however, making it difficult to assess the effect, if any, of subtle ecological or behavioral factors on the evolution of color vision or on the particular kind of color vision (*e.g.*, dichromatic vs. trichromatic) present in a given group of mammals. Autrum and Thomas (1973) provide a summary of mammals for which color vision has been demonstrated and point cut some of the difficulties in obtaining convincing evidence of wavelength discrimination. More data are also needed on the photic qualities of various terrestrial environments, (see Lythgoe, 1972; McFarland and Munz, 1975) in order to better evaluate ecological influences on the evolution of wavelength discrimination.

1.2 Auditory Specializations of Desert Rodents

The inhospitable daytime temperatures and aridity of deserts force most small mammals to take refuge in burrows during the day and to be active at night. Under these conditions, especially when cover is sparse, they are particularly susceptible to nocturnal predators. Desert rodents of the family Heteromyidae in the New World and gerbelline rodents of the family Cricetidae in the Great Palaearctic Desert of the Old World exhibit a variety of auditory specializations which are believed to increase their sensitivity to low frequency sounds, thus improving their chances of detecting nocturnal predators.

The extent of these specializations varies among species. One of the more specialized heteromyids, and one that has been studied extensively by Webster and Webster, is the kangaroo rat, *Dipodomys*, which inhabits the arid and semi-arid regions of Mexico and western North America.

Middle ear specializations. The auditory specializations of *Dipodomys* involve both the middle and inner ear. In the former they have the effect of reducing the stiffness and mass of the sound transmitting apparatus - resulting in improved low frequency sensitivity. The mastoid portion of the temporal bone is hypertrophied and composed of thin bone containing a greatly enlarged middle ear cavity (Webster, 1961, Webster and Webster, 1975). This enlarged air space does not seem to serve as a resonance chamber, for reducing its volume or perforating it with a hole does not alter the frequencies at which sensitivity peaks of the cochlear microphonic occur. Rather it appears that its primary function may be to decrease the stiffness or resistance of the middle ear by reducing the damping effect of the middle ear air cushion on the tympanic membrane. When the middle ear cavity of *Dipodomys* is partially filled with dental cement or plasticene so that its volume is reduced about 75%, the amplitude of the cochlear microphonic (CM) is also reduced, particularly between 1 and 3 kHz (Webster, 1962).

Auditory thresholds determined by shock avoidance conditioning confirm that kangaroo rats have a high sensitivity to low frequency sounds when compared to other mammals (Fig. 3). When middle ear volume was reduced these behavioral thresholds increased more than 10 dB between 125 and 1,000 Hz (Webster and Webster, 1972).

The adaptive value of low frequency hearing is suggested by behavioral observations before and after middle ear reduction. Kangaroo rats often avoid being caught by owls or snakes by executing an almost vertical leap into the air just as the predator is about to strike. Normal *Dipodomys* placed in an enclosure with a

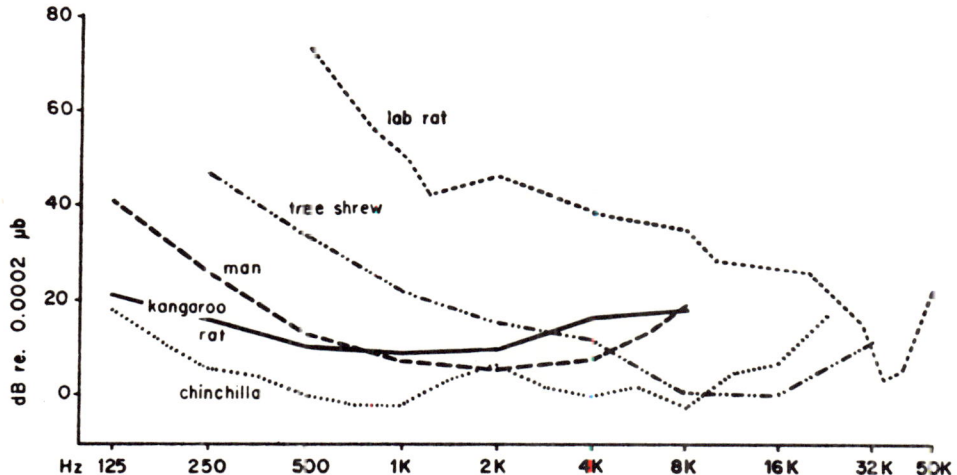

FIG. 3. Behavioral auditory threshold of a normal kangaroo rat (Webster and Webster, 1972); the chinchilla (Miller, 1970) which is larger and also has inflated auditory bullae; man (Sivian and White, 1933, as cited by Price 1963); tree shrew (Heffner et al. 1969) and laboratory white rat (Gourevitch and Hack 1966; Gourevitch, 1965, as cited) (From Webster and Webster 1972).

rattlesnake or owl were largely successful in avoiding their predators through the use of this escape behavior pattern. However, those with reduced middle ear volumes were usually caught and seldom initiated escape behavior until after they were struck by the predator. Both owls (see Gruschka, Borchers and Coble, 1971) and snakes produce faint sounds containing frequencies up to 1200 to 2000 Hz just before they strike their prey. Apparently decreasing the size of the middle ear air cushion raises the threshold enough to prevent *Dipodomys* from hearing its predator before it strikes (Webster, 1962).

Lay (1974) compared the susceptibility of captive gerbils and deer mice to predation by little owls (*Athene brahma*). Gerbils, *Meriones libycus*, were placed in cages with a little owl. Despite many attacks by the owl, neither blind nor normal *Meriones* were captured. The gerbils evaded the owl by running away or jumping as the owl attempted to strike. In control experiments, both blind and normal *Peromyscus leucopus* were easily captured by the owls. The hypothesis that the enlarged auditory bullae of *Meriones* are important in enabling it to hear the owl's approach receives some support from the fact that three young *Meriones* in which the bulla was not yet enlarged were caught on the first attack. Most normal adult gerbils were able to survive when left in the owl's cage from 3 to 17 days.

Kangaroo rats also use vision to detect predators - even in dim light. Further experiments suggest that rattlesnakes can be avoided by *Dipodomys* which either have been surgically blinded or have had their middle ear volumes reduced, but not by those with both of these treatments. Reduction of the middle ear volume alone, of *Dipodomys* released into their natural habitat and then retrapped, resulted in high predation during the dark phase of the moon (Webster and Webster, 1971), indicating vision is of little help on dark nights when the animal must rely heavily on audition. The sample size in these experiments was small, however, making the conclusions somewhat tentative.

Other middle ear specializations which reduce stiffness and mass include light weight ossicles held in place by only two ligaments instead of the usual four, a very fragile annular ligament attaching the unusually thin stapes footplate to the oval window and small middle ear muscles with the stapedius being absent in some genera.

Webster (1961) has measured the effective transformer ratio of the middle ear in *Dipodomys merriami* and found it to be a high 97.2:1 compared to 18.3 in man and 60.7 in the cat (Wever and Lawrence, 1954). In other words, ignoring friction, a force on the tympanic membrane of *D. merriami* would be multiplied 97.2 times at the oval window. More importantly, the middle ear provides an excellent impedance match between the air and fluid of the inner ear. Calculation of the impedance transform ratio indicates that the middle ear of *Dipodomys, Microdipodops* and *Perognathus* can theoretically transmit 94 to 100% of the incident acoustic energy at its resonant frequency and that of *Liomys*, a less specialized genus, can theoretically transmit 78 to 80% (Webster and Webster, 1975). This may be compared with estimates of 94% in the cat and 60% in man and indicates an extremely efficient transfer of energy by desert rodents. The rodents' 40% greater efficiency compared to man may represent a 3 to 4 dB increase in auditory sensitivity.

<u>Inner ear specializations</u>. Desert rodents also have some unusual specializations of the inner ear. The functional significance of these is less clear than is the case with middle ear specializations. They include increased cross sectional area of the inner ear ducts which may reduce friction during displacement of the cochlear fluid, an elevated reticular lamina suggested to increase cochlear sensitivity (Lay, 1972; Webster

and Webster, 1977) and the presence of a localized thickening – a hyaline mass – in the zona pectinata of the basilar membrane. This mass is thickest midway along the length of the basilar membrane and is thin or absent at the base and apex.

Its function is not clear but it is interesting that a similar thickening has evolved independently in certain other mammalian families including echolocating bats (Brown and Pye, 1975). In bats, which use high frequencies for echolocation, the hyaline mass is confined to the basal turn. Bats emitting constant frequency echolocative pulses have a much thicker mass in this region than do bats using frequency modulated pulses covering an octave or more.

The hyaline mass in the basilar membrane may help protect the very sensitive Organ of Corti from damage by intense sound through absorbing energy and damping the basilar membrane so it does not vibrate too much. According to this hypothesis, the ear's sensitivity might be maintained if the zona tecta, being more lightly damped, vibrated independently of the zona pectinata and stimulated the hair cells (Webster, 1961). Legouix and Wisner (1955) proposed a similar protective function for the mass added by hypertrophied Hensen's cells.

Lay (1972) – arguing by analogy to the behavior of circular diaphragms whose sensitivity can be increased in water by appropriately loading their centers – further suggested that the hyaline mass may in some similar fashion increase the sensitivity of the Organ of Corti and/or tune it to certain frequencies. He reported that in gerbilline rodents, increased amplitude of the cochlear microphonic was related to increasing anatomical specialization in terms of middle ear volume, presence and size of hyaline mass, height of Cells of Hensen, etc.

It is possible that the highly specialized cochleas of these desert rodents may operate in a way fundamentally different from most other cochleas. Webster and Webster (1977) have suggested that the small helicotrema of specialized desert rodents may act as an impedance discontinuity which, due to the abrupt change in impedance from the basilar membrane, may cause the acoustic pressure wave to be reflected back toward the stapes. If this happens it would cause a standing wave to develop along the basilar membrane. Such a possibility is compatible with the fact that the width of the basilar membrane increases very little after the first half turn in *Dipodomys* and *Microdipodops*. This fact, as well as the morphology and demonstrated sensitivity to low frequencies, suggests more of the basilar membrane responds to low frequencies than to high. Although there is no direct evidence for a standing wave in the kangaroo rat cochlea, the Websters point out that such a phenomenon could help explain frequency dependent changes in the symmetry of auditory neuron tuning curves as reported by Moushegian and Rupert (1970) and Caspary (1972). Further study of the cochlear mechanics of desert rodents is needed.

Ecological correlates of auditory specialization. To what extent is the degree of auditory specialization present in various desert rodents correlated with a species' ecological niche? In the case of the Gerbillinae, Lay (1972) concluded that "a very high correlation exists for degree of specialization and aridity of habitat." All the species he studied are nocturnal granivores except for three herbivores (*Rhombomys opimus*, *Psammomys obesus* and *Meriones hurrianae*) which are diurnal. Although these three diurnal species belong to the gerbilline group that is most specialized, and thus seem to contradict Lay's statement, he believes that their diurnal habits and herbivorous diet evolved after they had developed their auditory specializations.

Webster and Webster (1975), however, failed to find a strong correlation between aridity of habitat and middle ear inflation in 26 species of heteromyids. Some species are found in a variety of habitats ranging from sparse to dense vegetation and some are sympatric. For example *D. merriami* with a relative middle ear volume (cube root of middle ear volume divided by nasooccipital length of skull) of 0.29 is found in some of the same habitats with a variety of other species including *Perognathus penicillatus* which has a relative middle ear volume of only 0.15. It may be that this correlation with habitat would improve if more were known about details of each species' ecological niche, particularly its foraging behavior and predators. It is also well to remember that middle ear inflation is only one of a number of auditory specializations in these rodents. Some heteromyids with small middle ears have other specializations that are absent in species with inflated bullae. For example, *Liomys* and most *Perognathus* have relatively small middle ear volumes compared to *Microdipodops* and *Dipodomys*, yet the middle ear musculature of the former species is more specialized for they have lost the stapedius muscle altogether, whereas the latter species have not. Whether or not some formula including a variety of middle and inner ear specializations would give further insights into the effect of habitat on natural selection must remain a matter of conjecture for the present.

1.3 Sensory Ecology of Echolocating Bats

Ecology and echolocation. Among terrestrial mammals, active use of sound for orientation in darkness is most highly developed in the echolocating bats of the suborder Microchiroptera. Their orientation sounds, or pulses, consist of brief ultrasonic

tone bursts generated in the larynx. Space does not permit a
review of all the important aspects of microchiropetran echolocation
here (see Griffin 1958; Sales and Pye, 1974). The following
discussion will therefore focus on some of the ecological factors
that may influence the design of bat sonar systems.

It is important to emphasize, at the outset, however, that
in only a few species are there adequate data on the kinds of
sonar pulses they produce under natural conditions, as opposed
to those emitted in the laboratory. Moreover, much remains to be
learned about the feeding behavior and precise ecological niches
of most bats. Despite these gaps in our knowledge, several aspects
of the interrelationship between ecology, behavior and sonar pulse
design are already discernable.

Pulse intensity is one factor which determines the effective
range of echolocation. Fast flying insectivorous species that
pursue their prey on the wing have high intensity pulses compared
to those of slower flying species that feed on fruit or nectar,
the blood of large birds or mammals, or pick insects off the surface
of vegetation (Griffin, 1958). Fruit and nectar feeding bats have
pulse intensities between about 70 and 90 dB SPL measured several
centimeters in front of the mouth, whereas the peak pulse
intensities of insectivorous species usually lie between about
90 and 140 dB SPL (Diercks, 1972). Griffin (1958) points out that
the energy in an insectivorous bat's orientation pulse may often
be as much as a thousand times that in the pulse of a frugivorous
species. Foraging behavior is also important. *Plecotus* is a
member of the Vespertilionidae which are insectivorous and typically
emit high intensity pulses, but *Plecotus* feeds by hovering around
vegetation and picking insects off the surfaces of plants. Its
sonar is thus designed to operate over short distances and to
minimize interference by large echoes from the substrate on which
the insect is sitting. It is not surprising that the pulses of
Plecotus townsendii are 40 to 50 dB less intense than those of
Myotis lucifugus which pursues flying insects (Grinnell, 1963).

The design of a sonar signal affects in various ways the
information it can carry (Simmons, 1973; Simmons, Howell and
Suga, 1975). Simmons has pointed out that, with very few
exceptions, bat sonar systems can be categorized into one of
three groups, depending on the nature of their constant frequency
(CF) and frequency modulated (FM) components: 1) FM/short-CF
bats characteristically emit relatively short duration downward
sweeping FM pulses followed by a short low frequency CF portion;
2) Short-CF/FM bats in which the short CF is at a high frequency
and preceeds the downward FM sweep and 3) long CF/FM bats which
emit long CF pulses having a duration between about 6 and 70 msec
and terminated by a brief downward FM. Wide band (FM) signals

can carry more information in their echoes than can CF signals but the latter are advantageous for obtaining velocity information using Doppler shifts. Since echoes are reflected most strongly from objects that are large relative to the wavelength of the sound, the use of ultrasonic frequencies substantially improves the resolving power of a sonar system. High frequencies are attenuated more rapidly in air than are low frequencies, however, thus setting an upper limit to the range of frequencies useful for echolocation. The atmospheric attenuation coefficient at 10 kHz is 0.20 dB/m but at 100 kHz it is 3.6 dB/m (Griffin, 1958). Objects in the environment of an echolocating animal act as high pass filters whose low frequency cutoff is determined by the size and shape of the target. There is evidence that a bat's ability to discriminate between various targets depends on the analysis of subtle differences between the frequency structure of the echo and the emitted pulse (Griffin, Friend and Webster, 1965; Bradbury, 1970; Simmons et al., 1974).

Atmospheric conditions characteristic of particular environments play a role in determining the optimal design of orientation sounds. Many tropical habitats are characterized by high humidity which affects the transmission of sound through air. The relationship between humidity and atmospheric attenuation is complex and varies with frequency (Griffin, 1971). At the ultrasonic frequencies used by echolocating bats, attenuation is greatest when the humidity is high. In hot, humid, tropical rain forests atmospheric attenuation amounting to several dB/meter must not be unusual. Under extreme conditions atmospheric attenuation at 150 kHz could reach 15 dB/m. Griffin (1971) points out that in such environments the additional resolution provided by high frequencies may be offset by their limited range, which even for a perfect reflector could be as short as 2 or 3 m in the case of a bat emitting faint (74 dB) pulses. One might expect the orientation pulses of these tropical bats to contain very low frequency components which would suffer less attenuation. Indeed one unidentified South American bat has been recorded emitting FM pulses sweeping from 8 to 4 kHz while flying over a tributary of the Amazon (P. Hartline reported in Griffin, 1971). Atmospheric attenuation does have a potential advantage for echolocation, however, in that it limits the echoes returning from more distant objects that could otherwise interfere with the detection of nearby objects of more immediate importance. Rough estimates of the approximate distance at which a large flat perfectly reflecting target might be echolocated suggest the maximum range of detection for various bats probably lies between a few meters and a few hundred meters (Suthers, 1970). The reader is referred to Michelsen's chapter in this volume for a further discussion of the factors affecting sound propagation.

Simmons, *et al.* (1975) point out that most short-CF/FM bats live in the tropics. These authors speculate that the CF component, which is at the highest frequencies in the pulse, may enable tropical bats to partially compensate for the high atmospheric attenuation by increasing the acoustic power at the highest frequencies and so increase the range at which small objects can be detected. For this advantage the animal pays the price of degrading somewhat the bandwidth of its pulse, but this reduction is not enough to severely affect the performance of its sonar.

<u>Ecology and the role of vision</u>. Auditory stimuli control many kinds of behavior in echolocating bats that in other animals are controlled by visual stimuli. The sophistication of their sonar led many to assure vision played little, if any, role in microchiropteran behavior. It is now clear that vision is important to many echolocating bats which probably use it as a supplement to, and perhaps at times as a substitute for, acoustic orientation. While echolocation has the great advantage of allowing the animal to fly and feed in darkness and is better than vision for detecting nearby objects that contrast little with their background. Its useful range is severely limited by the spreading and attenuation losses of the transmitted pulse and the returning echo. An echolocating bat must also continuously emit vocalizations if it is to obtain acoustic information. Vision is thus probably energetically less expensive than echolocation and does not advertise the animal's presence to potential predators. It is also diffcult to see how echolocation alone could provide an adequate basis for orientation during homing, migration, or even long nightly foraging flights which in some species may be at altitudes of up to 3000 m or more (Williams and Williams, 1970; Williams, Ireland and Williams, 1973). It is probable that many bats rely on echolocation for detection of nearby objects but depend on vision to detect large distant objects.

The vision of Chiroptera has been reviewed elsewhere (Suthers, 1970). The microchiropteran eye has many attributes of a good nocturnal eye. The cornea is large relative to the size of the globe. The ciliary muscle is very poorly developed but the curved cornea, large spherical lens and the short focal length give an eye a high refractive power and great depth of focus so that accommodation is not necessary (Suthers and Wallis, 1970; Suthers, 1970). The avascular retina is densely packed with rods and shows a high degree of convergence.

In considering the sensory ecology of Microchiroptera it is essential to include vision as well as audition. Olfaction is also important to these animals, but little is known about it. microchiropteran visual abilities depend on the bats' ecology. Nocturnally active, insectivorous species have smaller eyes relative to their body size than do frugivorous, nectivorous, sanguinivorous or diurnally active insectivorous species (Suthers 1970; Chase 1972). Thus nectar feeding *Anoura geoffroyi* have eyes nearly twenty times larger than those of the similar sized but insectivorous *Pteronotus parnelli*. Relative to body weight, the eyes of the frugivorous *Chiroderma villosum* are forty times as large as those of *Pteronotus* (Table I). Chase (1972) compared the ocular development of 30 species of Microchiroptera in several New World families. Her data show that nocturnal insectivorous bats (families Vespertilionidae, Molossidae, Natalidae and Mormoopidae) have the smallest eyes with eye weight: body weight ratios between 0.5 and 4.4 x 10^{-4} whereas the insectivorous Emballonuridae which roost in well-lighted places such as on the trunks of trees have eye weight: body weight ratios of 14 to 18.5 x 10^{-4}.

TABLE I. The relationship between ecology and eye size in various echolocating bats.*

Family**	Mean Eye Wt. (mg)	Mean Body Wt. (g)	$\frac{Eye\ Wt.}{Body\ Wt.} \times 10^4$
Emballonuridae (3) (diurnal, insectivorous)	7.5	4.7	16.1
Mormoopidae (3) (nocturnal, insectivorous)	1.3	13.7	1.2
Phyllostomidae (nocturnal)			
Nectivorous (2)	10.1	12.0	8.1
Frugivorous (8)	26.7	33.5	8.6
Desmodontidae (2) (nocturnal, sanguinivorous)	12.5	34.5	3.7
Vespertilionidae (3) (nocturnal, insectivorous)	3.5	10.7	3.9

*Based on data of Chase, 1972.
**Number of species measured and habits in parentheses.

The retinal anatomy further reflects these differences in ecology and in the dependence on acoustic orientation. Chase (1972) found that diurnal emballonurids, which flit among the forest catching insects, have receptor cell densities that are about one-half those of the nocturnal fruit and nectar eating phyllostomids (Table II). The degree of retinal convergence in these families is also very different. The receptor to ganglion cell ratio of emballonurids is about 30:1 compared to ratios ranging from about 77:1 to 209:1 in phyllostomids. The phyllostomid with the lowest receptor to ganglion cell ratio, *Micronycteris megalotis*, roosts only in well-lit parts of caves, under bridges or near the openings in hollow trees (Goodwin and Greenhall, 1961). The nocturnal insectivorous vespertilionids have receptor cell densities comparable to many phyllostomids, but their eyes are both relatively and absolutely smaller. Chase further points out that the cells of the inner nuclear layer are twice as numerous in partially diurnal bats as they are in nocturnal species. In the former group the inner nuclear layer cells may nearly equal the number of receptor cells, a condition suggesting specialization for acuity.

TABLE II. The relationship between ecology and cell densities in the retinas of various echolocating bats.*

Family**	Nuclei per mm^2 (all numbers x 10^3)			Receptors Ganglion Cells
	R	I	G	R:G
Emballonuridae (2) (diurnal, insectivorous)	167	118	7	29:1
Mormoopidae (1) (nocturnal, insectivorous)	339	98	5	118:1
Phyllostomidae (nocturnal)				
Nectivorous (2)	322	60	3	146:1
Frugivorous (7)	408	76	4	147:1
Desmodontidae (1) (nocturnal, sanguinivorous)	465	75	5	124:1
Vespertilionidae (4) (nocturnal, insectivorous)	313	79	4	111:1

*Based on data of Chase, 1972. R = receptor cell layer; I = intermediate cell layer; G = ganglion cell layer.
**Number of species studied and habits in parentheses.

Behavioral data comparing the visual performance of various Microchiroptera are very limited. Suthers, (1966) and Chase (1972) showed that optomotor responses could be elicited from various phyllostomids and emballonurids with stripes subtending visual angles between 0.7 and 3.0 degrees and from *Myotis lucifugus* when stripes subtended 6.0 degrees or more. Form vision has been demonstrated in two phyllostomids, *Anoura* and *Carollia*, (Suthers, Chase and Braford, 1969). It has also been found that objects can be visually detected and avoided without the aid of auditory cues during flight. *Myotis lucifugus* can visually detect 2 mm diameter strings strung across its flight path (Bradbury and Nottebohm, 1969) while *Carollia perspicillata* and *Phyllostomus hastatus* can avoid 30 cm wide strips of white cloth (Chase and Suthers, 1969). The brightness discrimination threshold of *Eptesicus fuscus* is comparable to that of rats and mice (Ellins and Masterson 1974). The vampire bat's (*Desmodus rotundus*) minimum separable angle determined using test gratings is 48 min of arc at 310 lx and 2 degree 31 min of arc at 0.04 lx (Manske and Schmidt, 1976).

2. SENSORY ADAPTATIONS OF AMPHIBIOUS MAMMALS

2.1 Visual Adaptations

Amphibious mammals, such as pinnipeds and otters, need to have relatively good vision both in air and water. Many are diurnal predators which depend on underwater vision for catching fast moving prey under conditions of dim illumination, but also breed on brightly lit coasts. Their eyes must thus be able to function over an extremely wide range of illumination. Furthermore, the cornea, which is the most important refractive structure in the eye of terrestrial mammals, loses most of its refractive power in water. Special adaptations are therefore necessary to provide a sharp retinal image in both aerial and aquatic environments. The following discussion will be primarily concerned with pinnipeds.

The vision of pinnipeds, which include the seals, sea lions and walruses, is better studied than that of most other amphibious mammals (see Pütter, 1902; Walls, 1942; Jamieson and Fisher, 1972). Pinniped eyes are large and are positioned so that the visual axis is directed upward above the horizontal.

The cornea, which is ineffective for refraction underwater, is only slightly curved and has a heavily cornified epithelium. The eye's refractive power and accommodation in water depends on the lens which is thus almost spherical, though not as large as is characteristic of nocturnal mammals. The muscles of the iris are extremely well developed. In the harbour seal (*Phoca vitulina*), for example, the sphincter muscle is especially large. The iris is attached to the cornea by the pectinate ligament. In dim light the dilated pupil of the harbour seal is an oval shape with its long axis vertical. In bright light it constricts to a vertical slit with an enlarged aperature at the top. In very bright light the slit closes leaving only a pinhole at its upper end. The slit pupil is orientated diagonally in the bearded seal (*Phoca barbata*) and is horizontal in the walrus. The ciliary body contains circular as well as radial muscle fibers. These do not pull on the choroid during accommodation, but connect directly to the sclera.

Most retinoscopic measures of refraction in seal eyes indicate severe astigmatism in air. Johnson (1893) reported 13 D of myopia in the horizontal axis compared to 4 D of myopia in the vertical axis of the harbour seal's eye. Walls (1942) suggested that this corneal astigmatism and myopia could be minimized in air by the vertical slit pupil which can be constricted to approach a pinhole and thus give great depth of focus. In water the corneal astigmatism should disappear and the almost spherical lens would make the eye emmetropic. Accommodation in water is presumably effected by the ciliary body, whereas in air the need for accommodation is minimized by the stenopic pupil. This hypothesis is supported by Piggins (1970) who verified that the 5 to 10 D of astigmatism present in the harp seal's (*Pagophilus groenlandicus*) eye in air, disappears in water. Whether these measurements accurately indicate the degree of astigmatism in unrestrained, living seals - and hence support Wall's hypothesis - is debatable, however. Jamieson and Fisher (1972) suggest astigmatism may be simply a by-product of streamlining the eye to reduce the pressure exerted on the cornea by water flowing across it during swimming. They point out that the cornea is oblong with its meridian of least curvature (vertical in seals and horizontal in cetacea) in line with the axis along which water flows.

The underwater visual acuity of the California sea lion (*Zalophus californianus*), stellar sea lion (*Eumetopias jubatus*) and harbour seal has been measured behaviorally by Schusterman (1972) who found the threshold acuity for these species to lie between 5 and 9 min of arc over a range of luminance from about 3 to 130 mL. Although the resolving power of these pinnipeds is inferior to that of many primates, it is comparable to the aerial acuity of many highly visual terrestrial mammals including some carnivores whose ancestors gave rise to the pinnipeds.

Further experiments indicate that, under appropriate conditions of illumination, at least some pinnipeds have a similar acuity in both aerial and aquatic habitats. The ability of harbour seals (Jamieson and Fisher, 1970) to detect a gap between two well illuminated parallel lines is similar in both air and water. Schusterman and Balliet (1970, 1971), who used gratings to measure the visual acuity of the California sea lion, found its minimum separable angle to be 5.5 min of arc in both air and water. The aerial acuity of the sea lion, however, deteriorates much more rapidly when luminance is descreased than does submarine acuity. Decreasing luminance from 3 mL to 10^{-4} mL had little effect on acuity in water but greatly decreased aerial acuity. These behavioral data are thus consistent with the stenopaic theory of aerial vision proposed by Johnson (1901) and Walls (1942).

Pinniped eyes are well adapted to function in the dimly lit submarine environment where these diurnal animals pursue their prey. The retina contains mainly rods characterized by very long outer segments. The ratio of receptor nuclei to those in the outer and inner nuclear layers is 100:10:1 in the harbour seal, suggesting ample opportunity for lateral interaction and convergence. The choroid contains a tapetum cellulosum which is among the most extensive tapeta found in mammals (Walls, 1942; Jamieson and Fisher, 1972). Lythgoe and Dartnall (1970) have found different absorption peaks in the rhodopsin of weddell seals (*Leptonychotes weddelli*) and elephant seals (*Mirounga leonina*). The rhodopsin of weddell seals (λ_{max} = 495-496 nm) is similar to that of coastal fishes, while that of the elephant seal (λ_{max} = 485-486 nm) is similar to deep sea fish. Although the elephant seal seems more highly adapted to deep diving, both species dive to depths where the fish have deep sea rhodopsins. From this the authors infer that the deep sea rhodopsin in the elephant seal may be an adaptation to the light emitted by bioluminescent squids upon which it, but not the weddell seals, feeds.

Not all amphibious mammals have undergone the same visual adaptations as the pinnipeds. The sea otter (*Enhydra lutris*), for example, is emmetropic in air and is thought to increase the refractive power of its eye underwater by using its very well developed ciliary and iris sphincter muscles to distort the anterior surface of its lens (Walls, 1942; Gentry and Peterson, 1967). The acuity of the Asian "clawless" otter (*Amblonyx cineria*) (13 to 16 min of arc) is approximately similar in both air and water. This acuity is poor compared to pinnipeds but is comparable to that of other mustelids. In contrast to the sea lions, the acuity of *Amblonyx* deteriorates more rapidly in water than in air when luminance is decreased. Perhaps the dilated pupil does not squeeze the lens hard enough to produce the necessary accommodation. Alternatively, water may elicit a voluntary or reflexive pupillary constriction, thus limiting the aperature (Balliet and Schusterman, 1971; Schusterman and Barrett, 1973).

2.2 Auditory Adaptations

Vocalizations, including barks, humming and click-like sounds form an important part of the social behavior of pinnipeds both in air and underwater (*e.g.*, Bartholomew and Collias, 1962; Schevill, Watkins and Ray, 1963, 1966; Evans, 1967; Schusterman and Balliet, 1969; Schusterman, Gentry and Schmook, 1967; Schusterman, Balliet and St. John, 1970). Most of these vocalizations are of relatively low frequency with their energy lying below about 12 kHz, although ultrasonic frequencies are present in some cases. Because of the dim or murky environment in which seals often feed, it is natural to wonder if they can echolocate their prey. Poulter and his colleagues (Poulter, 1963; 1966, 1967, 1969, Poulter and Jennings, 1969; Shaver and Poulter, 1967, 1968) proposed the clicks of sea lions and fur seals (*Callorhinus ursinus*) are used for echolocation. Attempts by other investigators to substantiate this claim have not been successful, however (*e.g.*, Evans and Haugen, 1963; Schevill, Watkins and Ray, 1963; Schusterman, 1967). The underwater clicks of sea lions may have important social functions. Although the present evidence for echolocation is not convincing, this possibility should not be completely ruled out. Evans and Haugen (1963) point out that some pinnipeds may conceivably learn to use sonar under natural conditions.

The need to hear both aerial and underwater sound imposes special demands on the pinniped ear. Otological adaptations of these mammals for an amphibious existence are discussed in some detail by Repenning (1972) and by Ramprashad, Corey and Ronald (1972). The following discussion relies heavily on their reviews. The ancestors of pinnipeds, being terrestrial, had ears designed to function in air. If such an ear is to also operate underwater, it is necessary that it: a) Evolve a means of matching its impedance to that of an aqueous environment; b) evolve a mechanism to protect the cochlea from high hydrostatic pressures (some pinnipeds dive to 60 atmospheres whereas the human ear drum may rupture at one-third atm); and, c) retain preferential pathways by which sound energy reaches each cochlea in order to facilitate localization of the source. Many specializations of the pinniped ear can be interpreted as adaptations to meet these requirements.

The pinna of seals is greatly reduced or absent. The auricular muscles are able to close the auditory meatus. Sound in water may be conducted to the tympanic membrane by bone conduction along the walls of the meatus. It has been suggested that blood sinuses associated with the meatus wall may assist in conducting sound. These distensible sinuses presumably fill with blood during dives and bulge into the outer and middle ear cavities thus preventing what would otherwise be a relative partial vaccum in these spaces which could severely damage the ear. The middle ear ossicles are more massive than in terrestrial mammals. The inertia of such large ossicles must increase the stimulation of the inner ear when the head resonates to underwater sound. The basal turn of the cochlea is enlarged and in at least some seals contains a large spiral osseous lamina and thick spiral ligament similar to that reported in bats and cetacans. The round window is especially large and is recessed in a fossula.

Special provisions have also evolved which limit the pathways of bone conduction through the head to the cochlea. This is important in maintaining the capacity to accurately localize sound sources underwater. Certain bones of the pinniped skull have enlarged surfaces and angular contacts that seem well designed to selectively reflect sound, thus increasing interaural intensity differences. Furthermore, the petrosum forms a solid contact only with these bones and not with other bones of the skull. The petrosum is also designed in such a way that it will be maximally effective in distorting the cochlear capsule in response to sound.

The auditory performance of pinnipeds in air and water has been investigated in only a few species. Møhl (1964) trained a harbour seal to press a lever in order to hear a 0.58 sec duration tone from one of two transducers. The seal was rewarded with food if it swam toward the correct transducer. By varying the angle between the transducers and conducting experiments both in air and water, Møhl determined that the minimum audible angle is not significantly different in these two media - being 3.1° at 2 kHz in water and 4.8° at 500 Hz in air. Allowing for the higher velocity of sound in water compared to air and assuming the minimum detectable interaural time difference in the seal is similar to that of other mammals, this angular resolution suggests that even in water the effective distance between the ears is that between the external ear openings rather than that between the tympanic membranes.

In other experiments, Møhl (1968) measured the auditory threshold of harbour seals in water and in air (Fig. 4). In air the seal was most sensitive at about 12 kHz where its threshold was 11 dB SPL. Except for a region around 2 kHz, the seal's threshold in water was about 15 dB more sensitive than

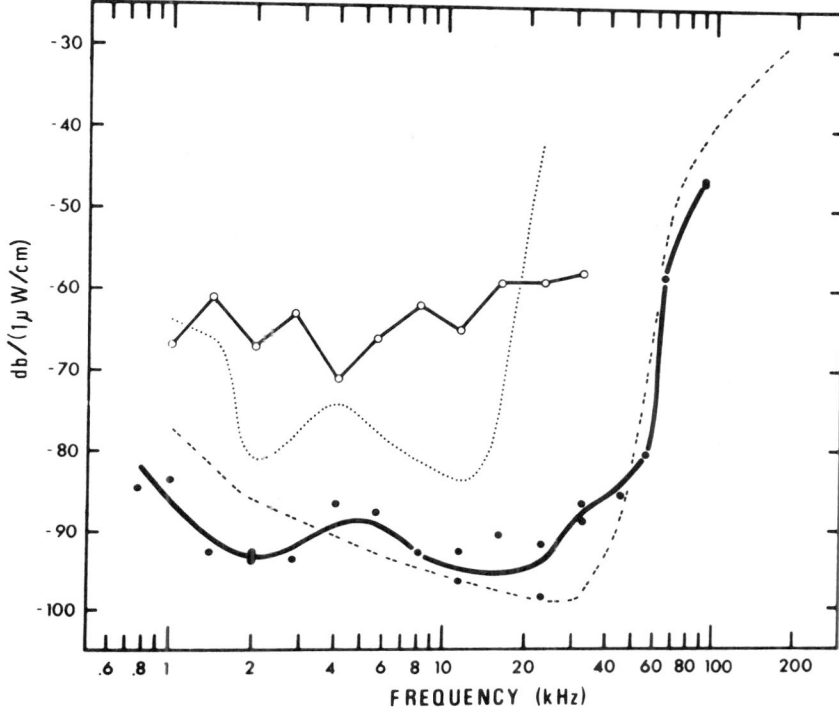

FIG. 4. Behavioral auditory threshold of the harp seal underwater (= solid dots) and in air (= open circles) compared with that of the harbour seal underwater (= dashed line) and in air (= dotted line). (Harbour seal data from Møhl, 1968; harp seal data and figure from Terhune and Ronald 1972).

that in air. The lowest threshold in water is at about 32 kHz and is within about a decibel of man's lowest threshold in air. Møhl points out that, based on these data, the seal's ear seems primarily designed for use underwater, but has some ability to accommodate to function in air. Frequency discrimination by seals in water is similar to that of other carnivores such as the cat, but is poor compared to human discrimination in air. The seal's Weber fraction, $\Delta F/F$, is about 13×10^{-3} from 1 to 57 kHz. Above 60 kHz frequency discrimination deteriorates. Sound detection above this frequency may be similar to hearing via bone conduction in submerged humans who are known to be able to detect underwater sound between 20 and 50 kHz but are unable to discriminate between frequencies within this range (Møhl, 1967).

Auditory thresholds of ringed seals (*Pusa hispida*) and harp seals have also been studied. Terhune and Ronald (1975a and b) report that ringed seals have a relatively uniform underwater auditory sensitivity between 1 and 45 kHz with the lowest threshold of -32 dB re 1 µbar at 16 kHz. Above 45 kHz the threshold rises at a rate of 45 dB/octave to 90 kHz, the highest frequency at which it was measured. The harp seal's threshold in air is relatively flat between 1 and 32 kHz, ranging from 29 dB SPL at 4 kHz to 42 dB SPL at 32 kHz. Its hearing range appears to extend up to as least 100 kHz. The poor auditory sensitivity in air may be due to blocking of the external auditory meatus by wax. The harp seal's ear is more sensitive underwater where its audiogram closely parallels that of the harbour seal, but with a maximum sensitivy at 15 kHz and a steeply rising threshold above 64 kHz (Terhune and Ronald 1971, 1972).

The California sea lion is maximally sensitive to underwater sounds between about 1 and 28 kHz with a sensitivity loss of 60 dB/octave between 28 and 36 kHz. At its most sensitive frequency (16 kHz) the sea lion is about 15 dB less sensitive than the harbour seal is at its best frequency of 32 kHz. It is not clear whether this difference is due to differing experimental procedures, individual variation or species differences (Schusterman, Balliet and Nixon, 1972). The sea lion's hearing is also much less sensitive in air than in water. Schusterman (1974) reports a hearing loss of from 9 to 22 dB between 4 and 32 kHz in air compared to water.

3. SENSORY ADAPTATIONS OF MARINE MAMMALS

3.1 Visual Adaptations

Preeminent among marine mammals is the order cetacea which includes the Odontoceti, or toothed whales, and the Mysticeti, or baleen whales. Little is known about the sensory abilities of sea cows - the dugong and manatee - in the order Sirenia. Even among the Cetacea, information regarding baleen whales is very meager. Vision may be better developed in the toothed whales, such as the diurnal dolphins and porpoises which pursue fish in relatively shallow water, than in the deep diving baleen whales. An exception is the Ganges river dolphin (*Platanista gangetica*) which lives in muddy water and has a very small rudimentary eye lacking even a lens (Dral and Beumer, 1974).

Adaptations of the cetacean eye to a marine existence include the loss of the lachrymal glands, the presence of a cornified protective layer on the cornea and conjunctiva and a very thick sclera capable of withstanding high pressures. Vision in dim light is facilitated by a retina composed primarily of large rods and by a tapetum lucidum in the choroid (Walls, 1942;

Slijper, 1962). The absorption maxima of visual pigments
in marine cetaceans appear to be roughly correlated with the
hues which predominate in the habitat of each species
(McFarland, 1971). Behavioral studies support the assumption
of reasonably good vision in those cetaceans which must pursue
rapidly moving prey. Killer whales (*Orcinus orca*), for example,
can detect the gap between two dark bars when it subtends as
little as 5.5 min of arc (stimulus intensity 10-20 ft-c).
(White, Cameron, Spong and Bradford 1971). The bottle-nosed
dolphin (*Tursiops truncatus*), while proficient at echolocation
also has well developed vision (*e. g.*, Kellogg and Rice 1964,
Dral 1972, Pepper and Simmons 1973). The dolphin's eye is
adapted to function well either above or below the water
surface. Their visual adaptations are in this sense more
nearly amphibious than marine in nature. The visual acuity
of the bottle-nosed dolphin, measured using gratings, is
comparable in water (8-14 min of arc) and air (12-19 min of
arc) (Herman, Peacock, Yunker and Madsen, 1975). This is
surprising since the dolphin, lacking a ciliary muscle, has
very little accommodative ability. Rivamonte (1976) has
proposed a solution to this apparent paradox based on the use
of different optic pathways for aerial and underwater viewing
and the fact that the core and periphery of the lens have
different refractive powers. In dim submarine light the pupil
is dilated, allowing light rays to pass along the optic axis
through the center of the cornea and lens. Ophthalmoscopic
examination indicates the eye is approximately emmetropic
along this axis in water, but quite myopic in air. When the
eye is exposed to bright light, typical of aerial vision, the
pupil constricts in such a way as to become crescent-shaped
with part of the iris (the operculum) blocking rays entering along
the central axis of the dioptric system (Fig. 5). Light rays
are thus prevented from passing through the lens core, but
must instead pass through the margin of the lens which has a
significantly lower refractive power than the core, thus making
the eye emmetropic in air without requiring any active accommodation (Fig. 6).

3.2. Auditory adaptations

The auditory sense plays an important role in cetacean
behavior. Odontocete whales produce a variety of sounds including
pure tones, broad-band clicks and rapid trains of clicks or
burst-pulsed signals in which individual clicks are emitted at
so high a repetition rate that the resulting vocalization sounds
like a chirp, squawk or bark. The pure tone whistles and other
signals having a complex frequency structure are thought to be
used primarily for communication, while the clicks and burst-
pulsed signals have an echolocative function. Acoustic orienta-

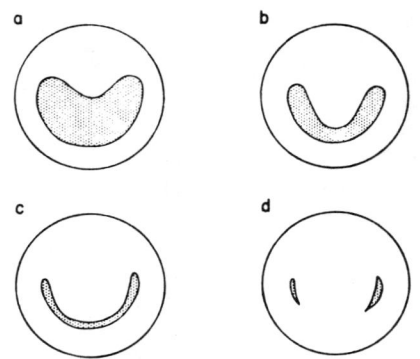

FIG. 5. Shape of the dolphin pupil at various stages of contraction shown superimposed on the outline of the spherical lens. Note the operculum which shields the lens core in bright light but not in dim light (From Rivamonte, 1976).

tion is of obvious advantage to marine mammals such as these toothed whales which must often pursue rapidly moving food in deep or turbid water where little light can penetrate.

Echolocation is accomplished by the use of high frequency clicks having a duration of about 1 msec and produced at repetition rates from a few per second up to 500/sec or more when catching food. The clicks made by *Tursiops* contain sound energy up to at least 170 kHz, although the peak energy is at about 35 kHz (Norris, 1969; Dierks, Trochta, Greenlaw and Evans, 1971; Sales and Pye, 1974). The use of high frequencies underwater is advan-

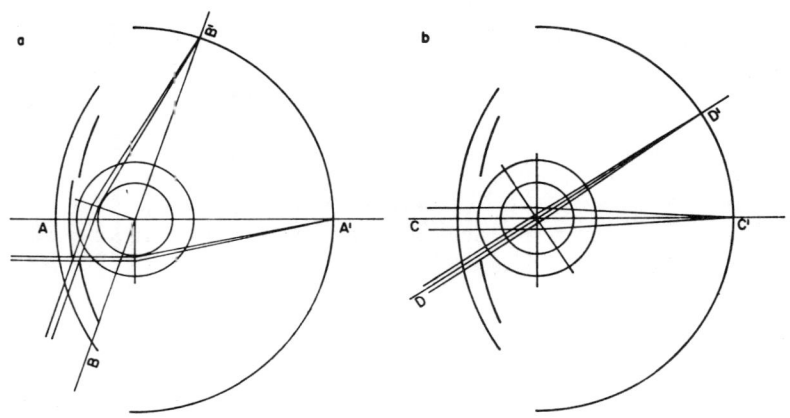

FIG. 6. Highly schematic horizontal section of dolphin eye. Representative ray traces: a. illustrates axial (AA') and oblique (BB') emmetropic viewing directions in air and high illumination; b. illustrates axial (CC') and typical (DD') emmetropic viewing directions in water and low illumination. (From Rivamonte, 1976).

tageous in improving resolution and reducing noise. Since the speed of sound is four and one half times that in air, the wavelength at a given frequency is proportionally longer in water than in air. High frequency pulses are thus advantageous in detecting small objects. The ambient sea noise is mostly below 10 kHz and so will mask high frequencies less than low frequencies (Evans, 1967). A variety of experiments have demonstrated the ability of odontocetes such as *Tursiops* to locate fish, avoid obstacles and discriminate between objects on the basis of echolocation (*e.g.*, Kellogg, 1961; Norris, Evans and Turner, 1967).

The means by which the echolocative clicks are produced is unclear. Both the larynx (Lawrence and Schevill, 1956; Purves, 1967) and the blowhole (Norris, 1969) have been suggested as probable sites. The larynx of *Tursiops* lacks vocal folds but Blevins and Parkins (1973) believe that it may still be involved in the production of vocalizations such as clicks and whistles. The oral cavity is not involved in phonation since it is isolated from the respiratory tract by a nasopharyngeal sphincter muscle. The nasal passages and tissues associated with the blowhole probably have important functions in phonation. A special problem for echolocating marine mammals arises from the fact that the body tissues of a submerged animal must be relatively transparent to sound. Porpoises must thus solve the problem of directing the energy of their echolocative clicks forward. There is now strong evidence that this is accomplished by a pad of fatty tissue--the melon--on the forehead. The melon is composed of lipids which slow the velocity of sound and thus act as an acoustic lens. This was confirmed by Norris and Harvey (1974) who used small hydrophones implanted in *Tursiops*' tissue to measure sound transmission and velocity. They found that the melon served to focus emitted sound energy forward and to provide a good impedance match with the seawater.

Although echolocation has not been demonstrated in baleen whales, sound seems to be an important means of communication. The sounds they produce are typically of low frequency and long duration. The humpback whale (*Megaptera novaeangliae*) is among the better studied mysticeti in this regard. During breeding season it emits a variety of complex, frequency modulated sounds including stereotyped songs which may last as long as 20 min - the same song being sung by all the whales in a given area (Payne and McVay, 1971) - as well as very low frequency sounds of about 20 kHz which may be used for long range communication (Payne and Webb, 1971). Sperm whales (*Physeter catodon*) also produce stereotyped series of clicks, or "codas". Each individual apparently has its own unique "coda," distinguished by the temporal pattern of its clicks, which may thus permit individuals to recognize one another (Watkins and Schevill, 1977). Fin whales (*Balaenoptera physalus*) emit very regularly spaced 20 Hz pulses over long periods of time, but the function of these sounds is unknown (Schevill, Watkins and Backus, 1964).

The cetacean ear exhibits a variety of adaptations for underwater hearing (see Reysenbach de Haan, 1957; Fraser and Purves, 1960; Dudok van Heel, 1962; Purves, 1966). The external auditory canal is a very narrow, S-shaped structure in Odondoceti and is filled with mucus or sea water. In Mysticeti it is reduced to a string-like structure containing a plug and is closed throughout most of its length. The middle ear is adapted to transmit high frequencies by having a very rigid and massive ossicular chain compared to that of terrestrial mammals. The petromastoid bone, containing the inner ear, is suspended from the cranial bones by a ligament, being otherwise acoustically isolated from the rest of the skull by an air and foam-filled cavity. The cochlea is well developed and also adapted for detection of high frequencies. The basilar membrane of the bottle-nosed dolphin and Pacific white-sided dolphin (*Lagenorhynchus obliquidens*) varies greatly in width, being extremely narrow and stiff at its basal end. Together with other features, such as a large population of hair cells, the anatomy of the dolphin cochlea suggests an excellent auditory capacity including good pitch discrimination (Wever, McCormick, Palin and Ridgway, 1971a, b, c, and 1972).

Behavioral measures of auditory performance in odontocete whales likewise indicate an acute sense of hearing. Only a few of these need be mentioned here. The frequency range of the bottle-nosed dolphin extends up to 126 or perhaps even above 150 kHz (Schevill and Lawrence 1953; Johnson 1967). The Amazon dolphin, *Inia geoffrensis*, can respond to sounds from 1 to 105 kHz, being most sensitive between 75 and 90 kHz (Jacobs and Hall, 1972). The audiogram of a single killer whale extended from 500 Hz to 31 kHz with a maximum sensitivity at 15 kHz, suggesting a lower high frequency cutoff in this species. Bullock *et al.* (1968) obtained neurophysiological evidence of hearing up to 140 kHz in *Tursiops gilli*. Jacobs (1972) found excellent frequency discrimination in a bottle-nosed dolphin. The smallest Weber ratios ($\Delta F/F$) were, surprisingly, between 2 and 20 kHz - in the frequencies of communication sounds rather than at the higher frequencies used in echolocation and where the ear is most sensitive. The difference limen of this species for auditory temporal discrimination is also about one half that of man when tested under comparable conditions.

The means by which sound energy reaches the inner ear was for some time a subject of considerable controversy. Fraser and Purves (1954, 1960) proposed that the structures of the external auditory canal, including its plug and walls, act as a wave guide channeling sound to the middle ear. Reysenbach de Haan (1957, 1966) and Dudok van Heel (1962), on the other hand, argued that sound travels to the bulla through the blubber and tissue of the head. This theory receives support from Norris and Harvey (1974) who obtained evidence that the subcutaneous blubber coat acts as a sound channel over the surface of the body - most sound energy being reflected at the blubber-muscle interface. Air spaces around the bulla acoustically isolate the inner ear from sound transmitted through the body. Sound reaching the ear seems to be channeled through thick blubber on the throat and lower jaw. The existence of these well defined acoustic channels to the inner ear are doubtless important for directional hearing.

Experiments of McCormick, Wever, Palin and Ridgway (1970) on the Pacific white-sided dolphin likewise indicate that neither the external auditory canal, the tympanic membrane, nor even the malleus play an important role in sound conduction. Sound energy is conducted through soft tissues on the side of the head and lower jaw--setting up vibrations in the incus (a process of which contacts the bony wall of the middle ear), the stapes, and in the fluids of the cochlea. McCormick et al. hypothesize that vibrations thus induced in the auditory ossicles and the cochlea, respectively, differ in amplitude and phase, creating a relative motion between the stapes footplate and the cochlear capsule. This motion displaces the cochlear fluid, resulting in stimulation of the hair cells. Cochlear fluid displacement is facilitated by the presence of a compressible gas pocket near the round window.

4. REFERENCES

Autrum, H. and I. Thomas (1973). Comparative physiology of colour vision in animals. In: Handbook of Sensory Physiology. R. Jung, editor, Vol. VII/3, Part A, pp. 661-692. Springer-Verlag, N.Y.
Balliet, R.F. and R.J. Schusterman (1971). Underwater and aerial visual acuity in the Asian "Clawless" Otter (*Amblonyx cineria cineria*). Nature (Lond.) 234: 305-306.
Bartholomew, G.A. and N.E. Collias (1962). The role of vocalization in the social behavior of the Northern elephant seal. Anim. Behav. 10:7-14.
Blevins, C.E. and B.J. Parkins (1973). Functional anatomy of the porpoise larynx. Am. J. Anat. 138: 151-164.
Bradbury, J.W. (1970). Target discrimination by the echolocating bat *Vampyrum specturm*. J. Exp. Zool. 173: 23-46.
Bradbury, J.W. and F. Nottebohm (1969). The use of vision by the little brown bat, *Myotis lucifugus*, under controlled conditions. Anim. Behav. 17: 480-485.
Bridgeman, C.S. and K.U. Smith (1942). The absolute threshold of vision in the cat and man with observations on its relation to the optic cortex. Am. J. Physiol. 136:463-466.

Brown, A.M. and J.D. Pye (1975). Auditory sensitivity at high frequencies in mammals. In: Advances in Comparative Physiology and Biochemistry. Vol. 6, pp. 1-73. O. Lowenstein, editor, Academic Press, N.Y.

Bullock, T.H., A.D. Grinnell, E. Ikezono, K. Kameda, Y. Katsuki, M. Nomoto, O. Sato, N. Suga and K. Yanagisawa (1968). Electrophysiological studies of central auditory mechanisms in cetaceans. Z. vgl. Physiol. 59: 117-156.

Caspary, D. (1972). Classification of subpopulations of neurons in the cochlear nuclei of the kangaroo rat. Exp. Neurol. 37: 131-151.

Chase, J. (1972). The role of vision in echolocating bats. Ph.D. Thesis. Indiana University, Bloomington, Indiana.

Chase, J. and R.A. Suthers (1969). Visual obstacle avoidance by echolocating bats. Anim. Behav. 17: 201-207.

Cooper, G.F. and J.G. Robson (1969a). The yellow colour of the lens of the grey squirrel (*Sciurus carolinensis leucotis*) J. Physiol. (Lond.) 203: 403-410.

Cooper, G.F. and J.G. Robson (1969b). The yellow colour of the lens of man and other primates. J. Physiol. (Lond.) 203: 411-417.

DeValois, R.L. and G.H. Jacobs (1971). Vision. In: A. Schrier and F. Stollnitz, editors. Behavior of Non Human Primates. Vol. 3, pp. 107-157. Academic Press, N.Y.

Diercks, K.J. (1972). Biological sonar systems: A bionics survey. Publication ARL-TR-72-34. Applied Research Laboratories, Univ. of Texas, Austin, Texas.

Diercks, K.J., R.T. Trochta, C.F. Greenlaw and W.E. Evans (1971). Recording and analysis of dolphin echolocation signals. J. Acoust. Soc. Am. 49: 1729-1732.

Dodt, E. and J. Walther (1958). Spektrale Sensitivität und Blutreflexion. Pflügers Arch. Ges. Physiol. 266: 187-192.

Dral, A.D.G. (1972). Aquatic and aerial vision in the bottlenosed dolphin. Neth. J. Sea Res. 5: 510-513.

Dral, A.D.G. and L. Beumer (1974). The anatomy of the eye of the Ganges River Dolphin, *Platanista gangetica* (Roxburgh, 1801). Z. Säugetierkd. 39: 143-167.

Dudok van Heel, W.H. (1962). Sound and cetacea. Neth. J. Sea. Res. 1: 407-507.

Ellins, S.R. and F.A. Masterson (1974). Brightness discrimination thresholds in the bat, *Eptesicus fuscus*. Brain Behav. Evol. 9: 248-263.

Evans, W.E. (1967). Vocalization among marine mammals. In: Marine Bio-Acoustics. W.N. Tavolga, editor. Vol. 2 pp. 159-186. Pergamon Press.

Evans, W.E. and R.M. Haugen (1963). An experimental study of the echolocation ability of a California sea lion *Zalophus californianus* (Lesson). Bull. South. Calif. Acad. Sci. 62: 165-175.

Fraser, F.C. and P.E. Purves (1954). Hearing in Cetaceans. Bull. Brit. Mus. (Nat. His.), Zool. 2: 103-116.

Fraser, F.C. and P.E. Purves (1960). Hearing in Cetaceans. Bull. Brit. Mus. (Nat. His.), Zool. 7: 1-140.

Gentry, R.L. and R.S. Peterson (1967). Underwater vision of the sea otter. Nature (Lond.) 216: 435-436.

Goodwin, G.G. and A.M. Greenhall (1961). A review of the bats of Trinidad and Tobago. Bull. Am. Mus. Nat. Hist. 122 (Article 3): 191-301.

Gourevitch, G. and M. Hack (1966). Audibility in the rat. J. Comp. Physiol. Psychol. 62: 289-291.

Griffin, D.R. (1958). Listening in the Dark. The Acoustic Orientation of Bats and Men. Yale University Press. New Haven.

Griffin, D.R. (1971). The importance of atmospheric attenuation for the echolocation of bats (Chiroptera). Anim. Behav. 19:55-61.

Griffin, D.R., J. Friend and F. Webster (1965). Target discrimination by the echolocation of bats. J. Exp. Zool. 158: 155-168.

Grinnell, A.D. (1963). The neurophysiology of audition in bats: Intensity and frequency parameters. J. Physiol. (Lond.) 167:38-66.

Gruschka, H.D., I.U. Borchers and J.G. Coble (1971). Aerodynamic noise produced by a gliding owl. Nature (Lond.) 233: 409-411.

Gunter, R. (1951). The absolute threshold for vision in the cat. J. Physiol. (Lond.) 114: 8-15.

Hebel, R. (1976). Distribution of retinal ganglion cells in five mammalian species (pig, sheep, ox, horse, dog). Anat. Embryol. 150: 45-51.

Heffner, H., R. Ravizza and B. Masterton (1969). Hearing in primitive mammals. III Tree shrew (*Tupaia glis*). J. Aud. Res. 9: 12-18.

Herman, L.M., M.F. Peacock, M.P. Yunker, and C.J. Madsen (1975) Bottlenosed dolphin: Double-slit pupil yields equivalent aerial and underwater diurnal acuity. Science (Wash. D.C.) 189: 650-652.

Jacobs, D.W. (1972). Auditory frequency discrimination in the Atlantic bottlenose dolphin. *Tursiops truncatus* Montague: A preliminary report. J. Acoust. Soc. Am. 52: 696-698.

Jacobs, D.W. and J.D. Hall (1972). Auditory thresholds of a fresh water dolphin, *Inia geoffrensis* Blainville. J. Acoust. Soc. Am. 51: 530-533.

Jamieson, G.S. and H.D. Fisher. 1970. Visual discriminations in the harbour seal, *Phoca vitulina*, above and below water. Vision Res. 10: 1175-1180.

Jamieson, G.S. and H.D. Fisher (1972). The pinniped eye: A review. In: Functional Anatomy of Marine Mammals. R.J. Harrison, editor. Vol. 1, pp. 245-261. Academic Press, N.Y.

Johnson, C.S. (1967). Sound detection thresholds in marine mammals. In: Marine Bio-Acoustics, W.N. Tavolga, editor, Vol. 2. pp. 247-255, Pergamon Press, Oxford.

Johnson, G.L. (1893). Observations on the refraction and vision of the seal's eye. Proc. Zool. Soc. Lond. pp. 719-723.

Johnson, L. (1901). Contributions to the comparative anatomy of the mammalian eye, chiefly based on ophthalmoscopic examination. Philos. Trans. R. Soc. Lond. B. Biol. Sci. 194: 1-82.

Kellogg, W.N. (1961). Porpoises and Sonar. University of Chicago Press, Chicago.

Kellogg, W.N. and C.E. Rice (1964). Visual problem solving in a bottle-nose dolphin. Science (Wash. D.C.) 143: 1052-1055.

Lawrence, B. and W.E. Schevill (1956). The functional anatomy of the delphinid nose. Bull. Mus. Comp. Zool. Harv. Univ. 114: 103-151.

Lay, D.M. (1972). The anatomy, physiology, functional significance and evolution of the specialized hearing organs of gerbilline rodents. J. Morphol. 138: 41-120.

Lay, D.M. (1974). Differential predation on gerbils (*Meriones*) by the Little Owl, *Athene brahma*. J. Mammal. 55: 608-614.

Legouix, J.P. and A. Wisner (1955). Rôle functionnel des bulles tympaniques géantes de certaines rangeurs (*Meriones*). Acustica 5: 209-216.

Lythgoe, J.N. (1972). The adaptation of visual pigments to the photic environment. In: Handbk. Sensory Physiol. H.J.A. Dartnall, editor. Vol. VII/1 Photochemistry of Vision, pp. 529-565. Springer-Verlag, N.Y.

Lythgoe, J.N. and H.J.A. Dartnall (1970). A "deep sea rhodopsin" in a mammal. Nature (Lond.) 227: 955-956.

McCormick, J.G., E.G. Wever, J. Palin and S.H. Ridgway (1970). Sound conduction in the dolphin ear. J. Acoust. Soc. Am. 48: 1418-1428.

McFarland, W.N. (1971). Cetacean visual pigments. Vision Res. 11: 1065-1076.

McFarland, W.N. and F.W. Munz (1975). The visible spectrum during twilight and its implications to vision. In: Light as an Ecological Factor: II, G.C. Evans, R. Bainbridge, and O. Rackham, editors, pp. 249-270. Blackwell Scientific. Oxford.

Manske, U. and U. Schmidt (1976). Visual acuity of the vampire bat, *Desmodus rotundus*, and its dependence upon light intensity. Z. Tierpsychol. 42: 215-221.

Miller, J. (1970). Audibility curve of the chinchilla. J. Acoust. Soc. Am. 48: 513-523.

Møhl, B. (1967). Frequency discrimination in the common seal and a discussion of the concept of upper hearing limit. In: Underwater Acoustics Vol. 2. pp. 43-54.

Møhl, B. (1968). Auditory sensitivity of the common seal in air and water. J. Aud. Res. 8: 27-38.

Moushegian, G. and A.L. Rupert (1970). Response diversity of neurons in ventral cochlear nucleus of kangaroo rat to low-frequency tones. J. Neurophysiol. 33: 351-364.

Muntz, W.R.A. (1972). Inert absorbing and reflecting pigments. In: Handbk. Sensory Physiol. VII/1 Photochemistry of vision. pp. 529-565. H.J.A. Dartnall, editor. Springer-Verlag, N.Y.

Norris, K.S. (1969). The echolocation of marine mammls. In: The Biology of Marine Mammals. H.T. Andersen, editor. pp. 391-423. Academic Press. N.Y.

Norris, K.S., W.E. Evans and R.N. Turner (1967). Echolocation in an Atlantic bottle-nose porpoise during discrimination. In: Animal Sonar Systems: Biology and Bionics. R.G. Busnel, editor. Vol. 2 pp. 409-437. Laboratorie de Physiologie Acoustique, INRA-CNRZ, Jouy-en-Josas, France.

Norris, K.S. and G.W. Harvey (1974). Sound transmission in the porpoise head. J. Acoust. Soc. Am. 56: 659-664.

Ordy, J.M. and T. Samorajski (1968). Visual acuity and ERG-CFF in relation to the morphologic organization of the retina among diurnal and nocturnal primates. Vision Res. 8: 1205-1225.

Payne, R.S. and S. McVay (1971). Songs of humpback whales. Science (Wash. D.C.) 173: 587-597.

Payne, R.S. and B. Webb (1971). Orientation by means of long range acoustic signalling in baleen whales. In: Orientation: Sensory Basis. H.F. Adler, editor. Ann. Acad. Sci. 188: 110-141.

Pedler, C. and R. Tilley (1969). The retina of a fruit bat (*Pteropus giganteus* Brünnich). Vision Res. 9: 909-922.

Pepper, R.L. and J.V. Simmons, Jr. (1973). In-air visual acuity of the bottle-nosed dolphin. Exp. Neurol. 41: 271-276.

Piggins, D.J. (1970) Refraction of the Harp Seal, *Pagophilus groenlandicus* (Erxleban 1777). Nature (Lond.) 227: 78-79.

Poulter, T.C. (1963). Sonar signals of the sea lion. Science (Wash. D.C.),139: 753-755.

Poulter, T.C. (1966). The use of active sonar by the California sea lion (*Zalophus californianus* (L)). J. Aud. Res. 6: 165-173.

Poulter, T.C. (1967). Systems of echolocation. In: Animal Sonar Systems: Biology and Bionics. R-G Busnel editor. Vol. I. pp. 157-186. Laboratorie de Physiologie Acoustique INRA-CNRZ, Jouy-en-Josas, France.

Poulter, T.C. (1969). Sonar of penguins and fur seals. Proc. Calif. Acad. Sci. 36: 363-380.

Poulter, T.C. and R.A. Jennings (1969). Sonar discrimination ability of the California sea lion, *Zalophus californianus*. Proc. Calif. Acad. Sci. 36: 381-389.

Price, L. (1963). Threshold testing with Bekesy audiometer. J. Speech Hearing Res. 6: 64-69.

Purves, P.E. (1966). Anatomy and physiology of the outer and middle ear in cetaceans. In: Whales, Dolphins and Porpoises. K.S. Norris, editor, pp. 320-376. University of California Press Berkeley.

Purves, P.E. (1967). Anatomical and experimental observations on the cetacean sonar system. In: Animal Sonar Systems: Biology and Bionics. R.-G. Busnel, editor. Vol. I. pp. 197-270. Laboratorie de Physiologie Acoustique, INRA-CNRZ, Jouy-en-Josas, France.

Pütter, A. (1902). Die Augen der Wassersäugethiere. Zool. Jahrb. Abt. Allg. Zool. Physiol. Tiere. 99-402.

Ramprashad, F., S. Corey and K. Ronald (1972). Anatomy of the Seal's ear (*Pagophilus groenlandicus*) (Erxleben 1777). In: Functional Anatomy of Marine Mammals. R.J. Harrison, editor. Vol. 1 pp. 263-306. Academic Press, N.Y.

Repenning, C.A. (1972). Underwater hearing in seals: Functional morphology. In: Functional Anatomy of Marine Mammals. R.J. Harrison, Editor. Vol. 1 pp. 307-331. Academic Press, N.Y.

Reysenbach de Haan, F.W. (1957). Hearing in whales. *Acta Oto-Laryngol*. Suppl. 134: 1-114.

Reysenbach de Haan, F.W. (1966). Listening underwater: Thoughts on sound and cetacean hearing. In: Whales, Dolphins and Porpoises. K.S. Norris, editor. pp. 583-595. University of California Press, Berkeley.

Rivamonte, L.A. (1976). Eye model to account for comparable aerial and underwater acuities of the Bottle-nose dolphin. Netherlands J. Sea Res. 10: 491-498.

Rodieck, R.W. (1973). The Vertebrate Retina. Freeman and Co., San Francisco.

Sales, G. and D. Pye (1974). Ultrasonic Communication by Animals. Chapman and Hall, London.

Schevill, W.E. and B. Lawrence (1953). Auditory response of a bottle-nose porpoise, *Tursiops truncatus*, to frequencies above 100 kHz. J. Exp. Zool. 124: 147-165.

Schevill, W.E., W.A. Watkins and R.H. Backus (1964). The 20 cycle signals and *Balaenoptera* (fin whales). In: Marine Bio-Acoustics. W.N. Tavolga, editor. pp. 147-152. Pergamon Press, Oxford.

Schevill, W.E., W.A. Watkins and C. Ray (1963). Underwater sounds of pinnipeds. Science (Wash. D.C.) 141: 50-53.

Schevill, W.E., W.A. Watkins, and C. Ray (1966). Analysis of underwater *Odobenus* calls with remarks on the development and function of the pharyngeal pouches. Zoologica. (N.Y.) 51: 103-111.

Schusterman, R.J. (1967). Perception and determinants of underwater vocalization in the California sea lion. In: Animal Sonar Systems: Biology and Bionics. R-G. Busnel, editor: Vol. I. pp. 535-617. Laboratoire de Physiologie Acoustique, INRA-CNRZ, Jouy-en-Josas. France.

Schusterman, R.J. (1972). Visual acuity in pinnipeds. In: Behavior of Marine Mammals. H.E. Winn and B.L. Olla, editors. Vol. 2, pp. 469-492. Plenum Press, N.Y.

Schusterman, R.J. (1974). Auditory sensitivity of a California sea lion to airborne sound. J. Acoust. Soc. Am. 56: 1248-1251.

Schusterman, R.J. and R.F. Balliet. (1969). Underwater barking by male sea lions (*Zalophus californianus*). Nature (Lond.) 222: 1179-1181.

Schusterman, R.J., and R.F. Balliet (1970). Conditioned vocalizations as a technique for determining visual acuity thresholds in sea lions. Science (Wash. D.C.) 169: 498-501.

Schusterman, R.J. and R.F. Balliet (1971). Aerial and underwater visual acuity in the California sea lion (*Zalophus californianus*) as a function of luminance. Ann. N.Y. Acad. Sci. 188: 37-46.

Schusterman, R.J., R.F. Balliet, and J. Nixon (1972). Underwater audiogram of the California sea lion by the conditioned vocalization technique. J. Exp. Anal. Behav. 17: 339-350.

Schusterman, R.J., R.F. Balliet, and S. St. John (1970). Vocal display underwater by the gray seal, the harbor seal, and the stellar sea lion. Psychon. Sci. Sect. Anim. Physiol. Psychol. 18: 303-305.

Schusterman, R.J. and B. Barrett (1973). Amphibious nature of visual acuity in the Asian "clawless" otter. Nature (Lond.) 244: 518-519.

Schusterman, R.J., R. Gentry, and J. Schmook (1967). Underwater sound production by captive California sea lions, *Zalophus californianus*. Zoologica (N.Y.) 52: 21-24.

Shaver, H.N. and T.C. Poulter (1967). Sea lion echo ranging. J. Acoust. Soc. Am. 42: 428-437.

Shaver, H.N. and T.C. Poulter (1968). Sea lion echo ranging. J. Acoust. Soc. Am. 43: 1459.

Simmons, J.A. (1973). The resolution of target range by echolocating bats. J. Acoust. Soc. Am. 54: 157-173.

Simmons, J.A., W.A. Lavender, B.A. Lavender, C.A. Doroshow, S.W. Kiefer, R. Livingston, and A.C. Scallet (1974). Target structure and echo spectral discrimination by echolocating bats. Science (Wash. D.C.) 186: 1130-1132.

Simmons, J.A., D.J. Howell, N. Suga (1975). Information content of bat sonar echoes. Am. Sci. 63: 204-215.

Slijper, E.J. (1962). Whales. Basic Books Inc. N.Y.

Stone, J. (1965). A quantitative analysis of the distribution of ganglion cells in the cat's retina. J. Comp. Neurol. 124: 337-352.

Suthers, R.A. (1966). Optomotor responses by echolocating bats. Science (Wash. D.C.) 152: 1102-1104.

Suthers, R.A. (1970). Vision, olfaction, taste. In: Biology of Bats. W.A. Wimsatt, editor. Vol. II pp. 265-309. Academic Press. N.Y.

Suthers, R., J. Chase and B. Braford (1969). Visual form discrimination by echolocating bats. Biol. Bull. (Woods Hole) 137: 535-546.

Suthers, R.A., and N. Wallis (1970). The optics of the eyes of echolocating bats. Vision Res. 10: 1165-1173.

Terhune, J.M., and K. Ronald (1971). The harp seal, *Pagophilus groenlandicus* (Erxleben, 1777) X. The air audiogram. Can. J. Zool. 49: 385-390.

Terhune, J.M. and K. Ronald (1972). The harp seal, *Pagophilus groenlandicus* (Erxleben, 1777). III. The underwater audiogram. Can. J. Zool. 50: 565-569.

Terhune, J.M. and K. Ronald (1975a). Underwater hearing sensitivity of two ringed seals (*Pusa hispida*) Can. J. Zool. 53: 227-231.

Terhune, J.M. and K. Ronald (1975b). The upper frequency limit of ringed seal hearing. Can. J. Zool. 54: 1226-1229.

Vakkur, G. and P.O. Bishop. (1963). The schematic eye in the cat. Vision Res. 3: 357-381.

Walls, G.L. (1942). The Vertebrate Eye and Its Adaptive Radiation. Cranbrook Inst. of Science, Michigan.

Walls, G.L. and H.D. Judd (1933). The intraocular color filters of vertebrates. Brit. J. Ophthal. 17: 641-675 and 705-725.

Watkins, W.A. and W.E. Schevill (1977). Sperm whale codas. J. Acoust. Soc. Am. 62: 1485-1490.

Weale, R.A. (1953). The spectral reflectivity of the cat's tapetum measured *in situ*. J. Physiol. (Lond.)119: 30-42.

Weale, R.A. (1966). Why does the human retina possess a fovea? Nature (Lond.) 212:255-256.

Weale, R.A. (1974). Natural history of optics. In: The Eye. H. Davson and L.T. Graham, Jr. editors. Vol. 6. pp. 1-110. Academic Press, N.Y.

Webster, D.B. (1961). The ear apparatus of the kangaroo rat, *Dipodomys*. Am. J. Anat. 108: 123-148.

Webster, D.B. (1962). A function of the enlarged middle ear cavities of the kangaroo rat, *Dipodomys*. Physiol. Zool. 35: 248-255.

Webster, D.B. and M. Webster (1971). Adaptive value of hearing and vision in kangaroo rat predator avoidance. Brain Behav. Evol. 4: 310-322.

Webster, D.B. and M. Webster. (1972). Kangaroo rat auditory thresholds before and after middle ear reduction. Brain Behav. Evol. 5: 41-53.

Webster, D.B. and M. Webster (1975). Auditory systems of Heteromyidae: Functional morphology and evolution of the middle ear. J. Morphol. 146: 343-376.

Webster, D.B. and M. Webster (1977). Auditory systems of heteromyidae: Cochlear diversity. J. Morphol. 152: 153-169.

Wever, E.G. and M. Lawrence (1954). Physiological Acoustics. Princeton University Press, Princeton, N.J.

Wever, E.G., J.G. McCormick, J. Palin, and S. H. Ridgway (1971a). The cochlea of the dolphin, *Tursiops truncatus*: General Morphology. Proc. Natl. Acad. Sci. USA. 68: 2381-2385.

Wever, E.G., J.G. McCormick, J. Palin, and S.H. Ridgway (1971b). Cochlea of the dolphin, *Tursiops truncatus*: The basilar membrane. Proc. Natl. Acad. Sci. USA 68: 2708-2711.

Wever, E.G., J.G. McCormick, J. Palin and S.H. Ridgway (1971c). The cochlea of the dolphin, *Tursiops truncatus*: Hair cells and ganglion cells. Proc. Natl. Acad. Sci. USA 68: 2908-2912.

Wever, E.G., J.G. McCormick, J. Palin, and S.H. Ridgway (1972). Cochlear structure in the dolphin, *Lagenorhynchus obliquidens*. Proc. Natl. Acad. Sci. USA. 69: 657-661.

White, D., N. Cameron, P. Spong, and J. Bradford (1971). Visual acuity of the killer whale (*Orcinus orca*). Exp. Neurol. 32: 230-236.

Whitteridge, D. (1965). Geometrical relations between the retina and the visual cortex. In: Mathematics and Computer Science in Biology and Medicine. John Blackburn, Leeds.

Williams, T.C., L.C. Ireland, and J.M. Williams (1973). High altitude flights of the free-tailed bat, *Tadarida brasiliensis* observed with radar. J. Mammal. 54: 807-821.

Williams, T.C. and J.M. Williams (1970). Radio tracking of homing and feeding flights of a neotropical bat, *Phyllostomus hastatus*. Anim. Behav. 18: 302-309.

Yolton, R.L., D.P. Yolton, J. Renz, and G.H. Jacobs. (1974). Pre-retinal absorbance in sciurid eyes. J. Mammal. 55: 14-20.

ACKNOWLEDGMENT

Original research reported in this chapter was supported by grants from the U.S. National Science Foundation to the author.

3. ADAPTIVE RADIATION OF SENSORY MODALITIES

FUNCTIONAL ADAPTATIONS IN CHEMOSENSORY SYSTEMS

Richard A. GLEESON

Monell Chemical Senses Center, University of Pennsylvania

Philadelphia, Pennsylvania 19104   U.S.A.

INTRODUCTION

The chemical sense is perhaps the most primitive of the sensory modalities utilized by organisms for obtaining information from their environment. Throughout the course of evolution chemosensory systems have evolved to play prominent roles in coordinating both intra- and interspecific interactions as well as monitoring certain abiotic environmental parameters. Within the realm of animal communication, substantial experimental evidence has implicated chemical signaling to be the most important channel of communication for much of the animal kingdom (Shorey, 1976). Indeed, it is probable that the earliest forms of interactions between primordial unicellular organisms were chemically mediated, and that this communication served as an evolutionary substrate for the hormone and neurotransmitter systems of the metazoans (Haldane, 1955; Wilson, 1970).

Although functional adaptations of the chemical senses are quite diverse, these adaptations are not reflected in a comparable number of morphological modifications of the receptors themselves or their accessory structures. This is not to say that elaborate morphologies do not exist, but rather that the functions appear to have primarily evolved through selection pressures acting at the molecular level of the chemoreceptors and/or in the central processing of receptor information (Moulton and Beidler, 1967; Toback, 1971). This chapter, therefore, will primarily be an examination of the functional diversity of chemosensory systems; only those systems involved in monitoring the external environment will be considered. In addition, attention will be given to those environmental parameters which influence chemical transmission; and a brief survey of basic chemoreceptor morphology and mechanisms will be presented. It is not my intention to exhaustively review the literature on the adaptations in chemoreception, but rather to present an overview of the various survival strategies utilized by organisms in different environments.

TERMINOLOGY

Increased research in the chemical senses within the past two decades has generated a corresponding growth in terminology, particularly in the field of chemical ecology. This nomenclature was recently reviewed and expanded by Norlund and Lewis (1976). A *pheromone* is defined as a substance externally secreted by a plant or animal which effects a specific response in another member of the same species (Karlson and Lüscher, 1959). Compounds involved in interspecific interactions, exclusive of food substances as such, are termed *allelochemics* (Whittaker and Feeny, 1971). This latter group is subdivided to *allomones*, which impart adaptive advantage to the emitter (Brown, 1968), and *kairomones*, which are adaptive to the receiver (Brown et al., 1970). Two further subdivisions of allelochemics (*synomone* and *apneumone*) have been proposed by Norlund and Lewis (1976). These are defined:

1) Synomone - "a substance produced or acquired by an organism which, when it contacts an individual of another species in the natural context, evokes in the receiver a behavioral or physiological reaction adaptively favorable for both emitter and receiver."

2) Apneumone - "a substance emitted by a nonliving material that evokes a behavioral or physiological reaction adaptively favorable to a receiving organism, but detrimental to an organism of another species, which may be found in or on the nonliving material."

Both pheromones and allomones may effect immediate behavioral responses (releaser effects) or act over the long term via physiological changes (primer effects) in the receiver organism (Wilson and Bossert, 1963).

Within each of the functional categories single compounds or multicomponent substances may serve as the releasing stimulus. Indeed, multicomponent pheromones may be more common even among the insects (Silverstein and Young, 1976). Moreover, single compounds or mixtures may have multiple functions depending on context. For example, certain secondary plant substances may act as feeding deterrents (allomones) for some organisms, yet simultaneously function as phagostimulants (kairomones) for others (Frankel, 1959).

In mammals, communication via chemical signals is very complex both in a behavioral sense and in the numbers of chemical compounds involved. Consequently, the application of the pheromone concept to mammalian systems may be inappropriate; thus, Beauchamp et al. (1976) have suggested that because of the conceptual implications of the term itself, it should no longer be used for mammals.

PHYSICO-CHEMICAL CONSIDERATIONS

The transmission of chemical information in both aquatic and aerial environments is influenced by several factors. Diffusion, as well as convective and advective circulation patterns, function in dispersion of the stimulus molecules. In those environments in which the medium is dynamic, the convective and advective influences are the more significant distribution parameters. This is particularly so for waterborne compounds; agents of comparable molecular weight have diffusion coefficients about four orders of magnitude lower in water than in air.

Bossert and Wilson (1963) derived mathematical expressions for olfactory communication between animals in air which are also applicable in water. Specifically, they attempted to define the communication distance for a pheromone under various conditions of release: (1) emission as a single pulse into still air; (2) continuous release into still air; (3) emission from a terrestrial trail into still air; and (4) continuous release into moving air. Based on their equations the "active space" of a pheromone released continuously into moving air assumes a semiellipsoid shape extending downwind from the source. The size of this active space is a function of: (1) the amount of pheromone released by the signaling organism; (2) the diffusion rate of the pheromone and the circulation properties of the transporting medium; and (3) the sensitivity of the chemoreceptors in the receiving animal. It should be emphasized, however, that these calculations define an active space which is a time-averaged entity. The actual distribution at a particular point in time would typically describe an irregular, filamentous plume resulting from varying amounts of turbulence (Farkas and Shorey, 1972).

One approach to maximizing the communication potential of a chemical signal is by modulating its release. Various designs have evolved to temporally control pheromone release and increase the efficiency of emission, particularly among the insects. An example of this is seen in several Lepidopteran species which possess eversible scent brushes that facilitate pheromone distribution (Birch, 1974). A similar dispersion function is associated with erectile hair tufts on the tarsal glands in the black-tailed deer, *Odocoileus hemionus columbianus* (Müller-Schwarze, 1971).

Regnier and Goodwin (1977) have presented an experimental and theoretical consideration of the factors involved in pheromone release from scent marks. The Mongolian gerbil (*Meriones unguiculatus*) exhibits scent marking behavior which is characterized by the deposition of secretions from the midventral sebaceous gland onto objects within its immediate environment as well as onto other members of its social group. Thiessen et al. (1974) have demonstrated that one of the behaviorally active components of the midventral secretion is phenylacetic acid. Utilizing this compound in a C^{14}-labeled form, Regnier and Goodwin (1977) demonstrated that release from a scent mark can be modified by: (1) the nature of the surface on which the scent mark is made; (2) the quantitative and qualitative aspects of the other organics contained within the sebum (Fig. 1); (3) the polarity of the agent; and (4) the humidity. From these experimental results they concluded that the volatility of polar odorants is greatly affected by interactions with polar surfaces and sebum. At high humidity the competition of water with these same polar sites substantially increases the evaporation of the odorant. On the other hand, non polar compounds show little interaction with polar surfaces or sebum; consequently their volatility is essentially unaffected by humidity changes. An interesting observation which is relevant to these findings is that most compounds identified as chemical signals in both vertebrates and insects contain fairly polar functional groups (Wheeler, 1977).

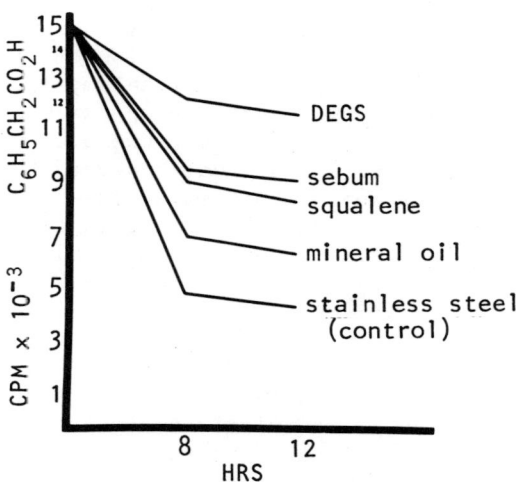

FIG. 1. Evaporation of C^{14}-labeled phenylacetic acid at 20°C and 0% relative humidity from stainless steel and from stainless steel treated with films of sebum and three other liquid materials. From Regnier and Goodwin (1977).

The data of Regnier and Goodwin (1977) clearly demonstrate that several physico-chemical interactions with the environment can significantly modulate the transmission of chemical information from scent marks. It is probable that similar types of interactions occur for airborne (e.g. adsorption to forest vegetation) and waterborne (e.g. adsorption to suspended clay particles) compounds as well. Indeed, it is reasonable to speculate that these factors have importantly influenced the evolution of chemical communication systems; not only of the chemical constituents in the secretions themselves but also of appropriate signal-emission behaviors. The ecological significance of these environmental interactions certainly warrants further study.

CHEMOSENSORY STRUCTURES AND MECHANISMS

The chemical senses are generally divided into three groups, i.e., olfaction, gustation, and the "common chemical sense". These divisions have an anatomical basis in the vertebrates; however, their usefulness in classifying the receptors of other organisms is limited. The first two categories are often distinguished by the operational criteria of "distance" as opposed to "contact" chemoreception or "low-threshold" as opposed to "high-threshold" reception. These distinctions, however, do not always hold in aquatic systems. For example, the dactyl "contact" chemoreceptors in *Gnathophausia ingens*, a bathypelagic crustacean, show sensitivity to very low concentrations of certain amino acids (10^{-8}M) (Fuzessery and Childress, 1975). A similar amino acid sensitivity (10^{-9} to 10^{-11}M) has been reported for the barbel taste buds in the catfish, *Ictalurus punctatus* (Caprio, 1975). Furthermore, catfish have been shown to utilize the gustatory sense alone in following a chemical gradient to locate distant food (Bardach et al., 1967). It is, therefore, puzzling as to why these two distinct chemosensory modalities have evolved. Atema (in press) has suggested that the dichotomy may be related to the functional roles of the two senses. Olfactory systems generally affect higher order behavior patterns (e.g. social interactions), whereas the taste modality is primarily involved in feeding activities and does not show the plasticity seen for olfaction-mediated behavior. Nevertheless, even if such a relationship does exist, the factors leading to this division remain obscure.

The "common chemical sense" is attributed to the nonspecific depolarizing effects of various agents (e.g. certain acids and alkalis) on cell membranes, particularly of excitable tissues. For example, in higher vertebrates this sensitivity is appreciated through nerve endings located in the nasal passages, mouth, and eyelids. Since this type of irritability is a universal characteristic of cells, Moncrieff (1951) has suggested it is the most primitive of the chemical senses.

The information available on the structural characteristics of metazoan chemoreceptors is primarily limited to the molluscs, arthropods, and vertebrates. These reports have been reviewed in detail elsewhere (Ghiradella et al., 1968; Graziadei, 1969, 1977; Laverack, 1974; Moulton and Beidler, 1967; Parsons, 1971; Schneider and Steinbrecht, 1968; Steinbrecht, 1969). A salient feature which emerges is that the sensory elements themselves are generally bipolar primary sensory neurons which exhibit a surprising homogeneity in morphology at the cellular level (Fig. 2) (Vinnikov, 1975). An exception is the development of taste buds in vertebrates which, unlike the olfactory receptors, are secondary sensory receptors (modified epithelial cells interposed between the stimulus and the sensory neuron). However, even these structures have maintained a remarkable similarity throughout the vertebrates (Graziadei, 1974).

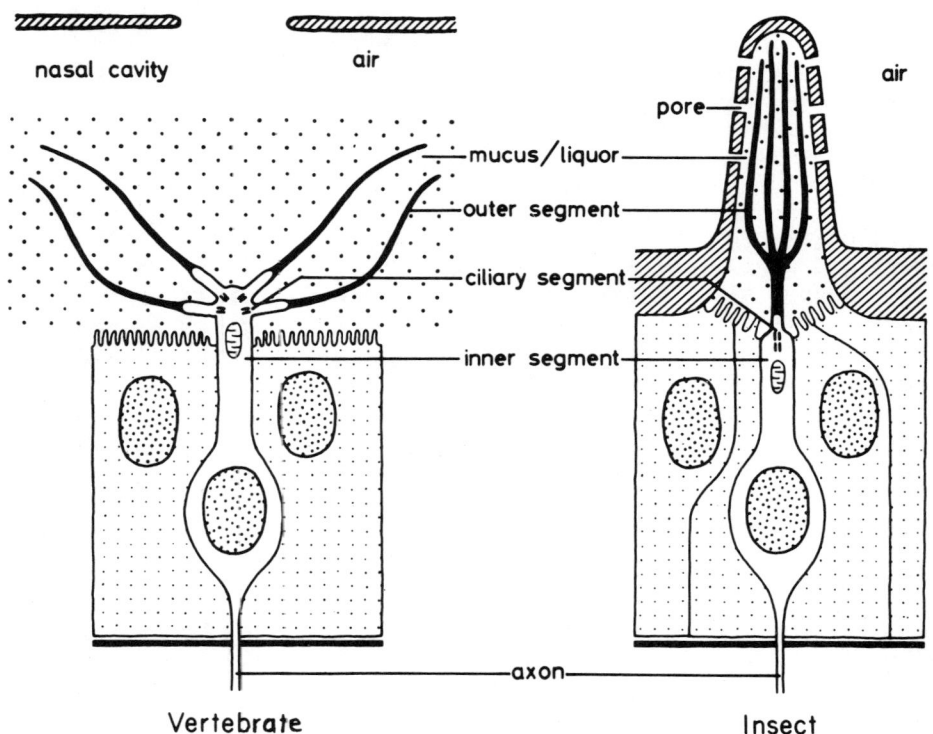

FIG. 2. Schematic comparison between vertebrate and insect olfactory receptors. Possibly analogous structures are drawn similarly. From Steinbrecht (1969).

A feature commonly observed on the outer segments of chemoreceptors is the presence of cilia or microvilli which possibly serve to increase the surface area for membrane-molecule interactions (Moulton and Beidler, 1967). These structures generally extend into a secreted fluid (e.g. mucus) which interfaces the receptor with the external environment (Fig. 2). It has been speculated that the secretions serve as a buffer to maintain constant conditions for the receptor cells (Steinbrecht, 1969; Vinnikov, 1975); however, other possible functions may be served as well (Bannister, 1974).

It is generally assumed that the initial event in chemoreception is an interaction of the stimulating agent with the receptor membrane. Depending on the chemical substance involved and the properties of the receptor, this interaction might be of a nonspecific nature or occur via specialized receptor sites. Several hypotheses have been advanced as to the properties of the interactions involved (Beets, 1971; Beidler, 1954; Davies, 1971); however, much more data is needed before an accurate picture can be formulated. Whatever the nature of these physico-chemical events, the result is a transduction to depolarizing or hyperpolarizing potential changes in the receptor cell which in turn modulates the generation of nerve impulses traveling to the central nervous system.

The mechanisms of quality discrimination in chemoreceptors have been most extensively studied in the insects and vertebrates. Earlier electrophysiological work in insects suggested the presence of two major classes of olfactory receptors; these have been designated "odor specialists" and "odor generalists" (see Kaissling, 1971). Narrow, identical specificities characterize the former group (e.g. certain pheromone receptors), whereas "generalists" have varying specificities with a marked overlap of response from cell to cell. However, this distinct dichotomy has not been supported in more recent studies (e.g. O'Connell, 1975; Schafer, 1977); it appears that the concept of a highly specialized receptor is perhaps an oversimplification. Quality discrimination in receptors with broad and often overlapping specificities presumably occurs centrally via the relative responses of many receptors taken in concert (across-fiber patterning). The on-off patterns of the afferent responses would, therefore, be analogous to the dotted visual image produced on a television screen with each odorant eliciting a different response image.

Studies of single olfactory units in vertebrates have revealed a wide range of responsiveness to test odorants. The majority of cells show rather low specificities (Gesteland et al., 1965; Gesteland et al., 1963; Getchell, 1974; Holley et al., 1974; O'Connell and Mozell, 1969); thus, it would appear that vertebrate receptors conform to the model of the insect generalists. However, before the existence of specialist-like receptors is rejected, more biologically relevant compounds should be examined.

Several studies have been addressed to the properties of spatiotemporal patterning as a quality discrimination mechanism in vertebrate olfactory systems. This hypothesis derives from Adrian's (1950, 1951, 1953, 1954) studies and has been recently reviewed by Moulton (1976) and Mozell (1977). The concept which emerges is that the odorant response patterns of the mucosa, which are in turn projected to the olfactory bulb, result from a combination of: (1) the inherent spatial organization of receptors within the mucosa; and (2) the differential sorption of the odorant molecules onto the mucus sheet. Different odorants would accordingly effect unique patterns or "images" in the bulb as a consequence of these interactions.

As seen for olfactory receptors, responses from nerve fibers innervating taste buds reveal a general lack of stimulus specificity (see Sato, 1971); although in certain mammals, at least, many fibers do appear to display a greater responsiveness to some taste qualities (Pfaffmann, 1974). A somewhat similar situation is found for the contact chemoreceptors of insects (Dethier, 1974). In the blowfly, *Phormia regina*, each of the four labellar chemoreceptors exhibits a spectrum of sensitivities to various natural food stimuli; a spectrum which shows little or no overlap with the response patterns of the other receptors. Thus, in both the insect and vertebrate systems across-fiber patterning has been hypothesized as an integral part in discriminating different taste qualities (Dethier, 1974; Erickson, 1963; Pfaffmann, 1941).

ADAPTIVE FUNCTIONS OF CHEMOSENSORY SYSTEMS

<u>Habitat Selection</u>: Several interacting factors are involved in an organism's selection of a suitable place to live and reproduce, and chemical information is commonly an important parameter in making this decision. In sedentary marine organisms, larval settlement is a critical phase of the life cycle, as selection of an appropriate substrate insures proper conditions for survival and reproduction in the adult form. Studies in several organisms (e.g. bivalves, cirripedes, polychaetes, etc.) have demonstrated that the chemical nature of the substrate is an important factor in settlement induction (Crisp, 1974). Wilson (1968, 1970a, 1970b) found that the substance responsible for gregarious setting in the reef-building polychaete, *Sabellaria*, is a cementing material secreted by adults in the process of tube building. It can only be recognized on contact by the larvae, and, according to Vovelle (1965), is a proteinaceous material. A remarkably similar situation

has been described for the barnacle, *Balanus balanoides* (see Crisp, 1974). The cyprid larvae of this species will preferentially set on surfaces occupied by other barnacles. Again, the settlement factor is a proteinaceous material (arthropodin) which is present in the adult integument, and recognition requires contact. Observations of the setting behavior in the cyprids reveals that during the search phase the antennular organ is repeatedly applied to the substrate; apparently it is this structure which detects the arthropodin.

Homing: Homing behavior is observed in a variety of organisms. For example, several species of limpets (gastropod molluscs) are observed to return after feeding excursions to a particular location within their rocky intertidal habitat. This orientation is so accurate that scars are eventually formed on the rock surface into which the limpet's shell fits precisely. Experimental studies indicate that mucus trails are utilized in effecting this behavior; trails which are specific to individuals, as sites are generally not interchangeable (Cook, A. et al., 1969; Cook, S. B., 1969).

Homing during spawning migrations has been extensively studied in certain fish species. A number of studies on salmonid migration have demonstrated the importance of chemical information in home-stream discrimination (Hasler, 1966). Olfactory imprinting to home-stream odors apparently occurs at the time of, or soon after, smolting (Madison et al., 1973). This was experimentally verified by artificially imprinting salmon during the smolting process using a synthetic chemical (morpholine). The hypothesis that fish so conditioned would return to a morpholine scented stream was confirmed. Tagging programs, tracking studies, and EEG recordings all indicated that the morpholine exposure of juveniles strongly affected subsequent adult migration patterns (Madison et al., 1973; Scholz et al., 1976). Whether imprinting in the natural context is mediated via the minerals and vegetative odors of the home stream (Hasler and Wisby, 1951; Madison et al., 1973) or race-specific pheromones (Døving et al., 1974; Nordeng, 1971; Solomon, 1973) is still a matter of debate.

Recent evidence has implicated chemosensory information as an important parameter in the homing behavior of certain bird species. Grubb (1972, 1974) examined the use of olfaction in navigation and nest location in several Procellariformes. He found that the Leach's storm-petrel (*Oceanodroma leucorrhoa*) returns to its nest burrow at night by flying upwind. A few meters downwind from the entrance it lands and then walks the remaining distance. Nesting material was found to serve as an effective lure in darkness, and Y maze studies revealed preferential selection of the arm containing odors of their own nest material.

Olfaction may also be important in homing pigeon navigation (Papi et al., 1971, 1972, 1973; Benvenuti et al., 1973a,b). Birds deprived of the olfactory sense show disorientation as reflected by the low number of successful returns to the loft. It has been hypothesized that during the early months of life pigeons learn to recognize the loft odors as well as "foreign" odors carried by the wind. Furthermore, they associate these "foreign" odors with the direction from which they come. Accordingly, upon release at a distance, pigeons establish their initial flight orientation by relating the prevailing odor of the release point with its direction relative to the loft. This, of course, assumes that the odor of the release point had been perceived at the loft as one of the "foreign" odors. The home direction is, therefore, opposite to the direction which that odor had normally been sensed in the loft, and the pigeons then presumably use a compass orienting mechanism (e.g. sun compass) to assume the deduced heading.

This rather unusual hypothesis has been supported in more recent studies by Fiaschi and Wagner (1976) as well as in experiments in which pigeons were conditioned to artificial odorous winds (Papi, 1974). However, a serious challenge to the theory has arisen from the results of experiments by Keeton (1974) and Keeton and Brown (1976).

Symbiotic Relationships: Many symbiotic and parasitic associations are chemically-mediated in some manner; particularly in the initial location and recognition of the host or symbiont. Several of these relationships have been examined in aquatic environments. For example, a number of polychaete species exhibit commensal associations with various molluscs and echinoderms, and most display a chemical attraction to their respective hosts (Davenport, 1950; Dimock and Davenport, 1971; Hickok and Davenport, 1957).

In electrophysiological experiments on two commensal shrimps of the genus *Betaeus*, Ache and Case (1969) demonstrated excitation in antennular chemoreceptors when exposed to effluents from their respective hosts. A comparable change was not effected with control stimuli of non-host origin. Behavioral studies demonstrated that both *B. macginnitae* and *B. hartfordi* use chemosensory information in host recognition (Ache and Davenport, 1972).

The clown fish *Amphiprion* is commonly associated with the giant sea anemone *Stoichactis*, and can swim among its tentacles unharmed. Nematocysts are not discharged because of a chemical factor in the mucus of the fish's skin which serves to increase the discharge threshold to mechanical stimulation. Establishment of this association is gradual, entailing progressively longer and more vigorous contact with the tentacles (Davenport and Norris, 1958).

Feeding: The chemical senses are universally important in coordinating the process of food acquisition and ingestion. Lindstedt (1971) reviewed the literature on chemical control of feeding behavior, and modified the classification of stimuli which effect the various phases of feeding. Most of the terms used in this classification have been derived from the numerous studies in insects. The definitions as proposed by Lindstedt (1971) are:

1) Attractant – a stimulus to which an animal responds by orienting toward or becoming receptive to the apparent source. May operate over long distances.
2) Arrestant – a stimulus that causes an animal to cease locomotion when in close contact with the apparent source.
3) Repellent – a stimulus that causes the animal to orient away from or become non-receptive to the apparent source.
4) Incitant – a stimulus that evokes initiation of feeding (tasting).
5) Suppressant – a stimulus that inhibits or prevents initiation of feeding.
6) Stimulant – a stimulus that promotes ingestion and continuation of feeding.
7) Deterrent – a stimulus that prevents continuation of feeding or hastens termination of feeding.

Fig. 3 summarizes the relationships of these stimuli to feeding behavior.

The chemical substances which effect positive feeding responses, at least in carnivorous and omnivorous animals, are generally small molecules (<1000 MW) with a wide distribution in plant and animal tissues (e.g. amino acids) (Ache et al., 1976; Bardach, 1975; Lindstedt, 1971), although there are notable exceptions (e.g. Carr et al., 1974). For example, reduced glutathione (GSH) is characteristically found in living animal tissues, and is a widespread specific feeding incitant in the cnidarians. Interestingly, GSH acts as a suppressant in the blowfly *Phormia regina*, an insect that feeds on carrion (Hodgson, 1968).

Herbivores, on the other hand, have feeding responses cued to both large and small molecules. Stenophagous herbivores tend to respond to only a few compounds (e.g. secondary plant compounds) which are found exclusively in their food plant, whereas euryphagous herbivores show much less specificity in feeding activators (Lindstedt, 1971). The cucumber beetle, *Diabrotica undecimpunctata*, feeds specifically on members of the cucumber family (Cucurbitaceae). Chambliss and Jones (1966) found that one of the several tetracyclic triterpenes, which are commonly associated with this plant family, was a very effective feeding attractant and incitant in *Diabrotica*.

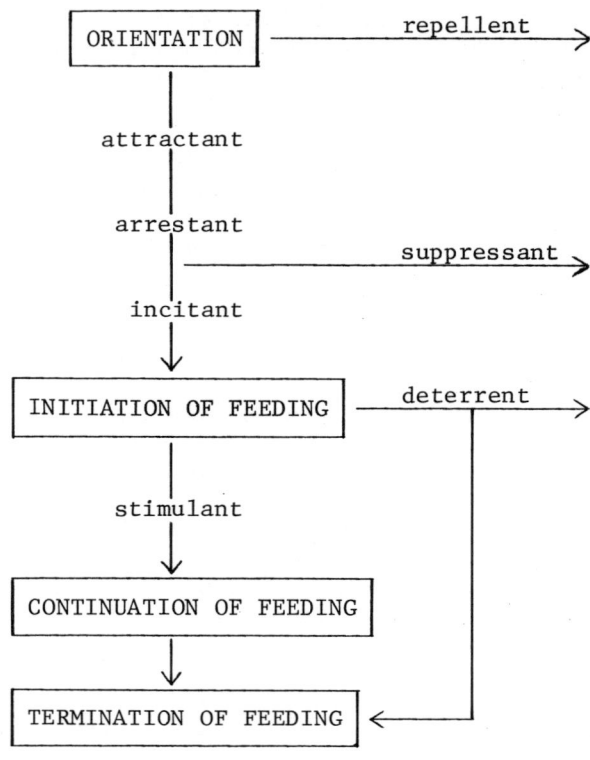

FIG. 3. Relationships of chemical stimuli to feeding behavior. Adapted from Lindstedt (1971).

The regulation of salt and water balance, in terrestrial organisms at least, is partially mediated via the sense of taste; accordingly, these factors will influence appropriate food selection in addition to the organic stimuli (Nachman and Cole, 1971).

In higher organisms it is likely that a combination of substances, particularly as perceived through the olfactory sense, is involved in the release of feeding behavior. It has been speculated that the complex chemical mixtures of prey are recognized as "chemical images" or "fingerprints" much like in visual pattern recognition (Atema, in press; Bardach and Villars, 1974; Polak, 1973). Thus, with the animal in an appropriate motivational state, perception of such an image would serve to trigger the feeding response. Indeed, one could extend this hypothesis to include many other behaviors which presumably involve recognition of chemical combinations (e.g. homing, individual recognition, predator detection, etc.). For each behavior, the processes of establishing and recognizing appropriate images would be differentially influenced by selection pressures depending on survival value; that is, image acquisition might involve a learning or imprinting process or have a more rigidly fixed genetic basis. For example, chemical preference studies in newborn snakes (*Natrix*) have demonstrated a genetic predisposition exists for sympatric species of prey (Burghardt, 1968; Gove and Burghardt, 1975). The concept of chemical images is certainly compatible with the present understanding of quality discrimination in olfactory systems; however, much more experimental support is needed.

Taste stimuli are often effective feeding deterrents when associated with a previous toxicosis. For example, several vertebrates have a demonstrated ability to associate adverse post-ingestional effects with particular tastes (e.g. Braveman, 1975). Possibly related to this is the observation that animals will often display neophobia when presented with a novel food. Garcia and Hankins (1975) noted that aversions to bitter substances are found in a wide variety of animal species; aversions which appear to be innate. They argue that the evolution of this response may be related to the fact that many naturally occurring toxic substances have bitter taste qualities [e.g. certain secondary compounds in plants (Levin, 1976)].

The use of chemosensory information in detecting and orienting to food is often facilitated by structural adaptations of the sensory apparatus. An example of this is found in the kiwi (*Apteryx*), a nocturnal, flightless bird with poor eyesight, which utilizes olfactory cues in locating food (Wenzel, 1968, 1972). In addition to the relatively large olfactory nerves and bulbs in these birds, the external nares are located at the tip of the beak which facilitates foraging for worms. Also, unlike any other bird, kiwis are able to sniff. Another example of structural adaptation is seen in the barbels and modified fins which have developed in several fish species. These structures contain taste cells and have been shown to function importantly in food detection and location (Bardach et al., 1967; Bardach and Case, 1965).

Orientation in feeding commonly involves following a chemical trail. The problems and approaches to this have been summarized by Shorey (1976) with reference to pheromone communication. Because of turbulence it is generally believed that detection of a chemical gradient is not possible in most aerial and aquatic environments beyond a few meters. Consequently, two mechanisms appear to be employed, often simultaneously, in locating a distant odor source. One is by positive anemotaxis or rheotaxis; the other by a form of chemotaxis. The latter is frequently accomplished by a type of chemoklinotaxis (the sequential comparison of concentration differences in space) involving zigzag oscillations across the odorant's active space; a technique commonly observed in terrestrial trail following as well. In combination the two mechanisms complement one another; anemotaxis or rheotaxis confers polarity to the trail, whereas chemotaxis maintains the correct horizontal and vertical orientation.

An interesting example of chemosensory tracking has recently been reported for the planktonic shrimp, *Acetes sibogae australis* (Hamner and Hamner, 1977). This species is able to precisely follow scent trails produced by sinking food particles or bits of paper impregnated with various amino acids. Upon contact with a vertical trail the shrimp will invariably follow it downwards; a behavior particularly well adapted for securing food scraps or wounded prey in its pelagic environment.

Intraspecific Communication: The importance of communication via chemical signals in all major animal phyla has only recently been fully appreciated as increasingly more studies reveal the presence and significance of this mode of organism interaction (Birch, 1974; Cheal and Sprott, 1971; Doty, 1974; Eisenberg and Kleiman, 1972; Johnston et al., 1970; Müller-Schwarze and Mozell, 1977; Otte, 1974; Shorey, 1976). The insects represent the most intensively studied group in this regard; particularly the social insects where chemical messages are extensively utilized to coordinate nearly all facets of their integrated behavior (Blum, 1974; Jacobson, 1972; Shorey, 1973). It has become increasingly evident, however, that for most animals, communication via chemical information is more important than any other communication modality. In Blum's (1974) review he stated, "the road to insect sociality was paved with pheromones." Shorey (1976) has suggested that "although there are many exceptions, Blum's statement should be expanded to include much of the animal kingdom".

Chemical signals are employed for such a variety of functional behaviors that even a superficial consideration of them here would be impractical. Instead it is my intention to examine selected examples of the adaptive functions served by chemical communication systems in an attempt to convey the extent of this diversity. The reader is referred to the literature cited above for a more in-depth survey of this topic.

Reproduction: Pheromones play a prominent role in coordinating various aspects of reproduction in a variety of organisms. A classic example of this is found in the chemical attraction of male silk moths, *Bombyx mori*, to females. Response to the female's sex attractant, Bombykol, occurs at concentrations approaching a single molecule per receptor cell (Schneider et al., 1968; Schneider, 1969). This sensitivity allows detection and location of females which are up to several kilometers away (Bossert and Wilson, 1963).

Sex pheromones also operate in close range interactions for stimulation of courtship behavior and copulation. An example of the former is seen in the reproductive behavior of the blue crab, *Callinectes sapidus* (Gleeson, 1976). Females of this species can only mate during a short period following their maturity molt, which marks the end of the penultimate instar. Males recognize females approaching this critical molt by a pheromone which is present in the female's urine. The pheromone-released courtship behavior of the male leads to the cradle-carry position in which the female is held until she molts. In this configuration the male serves to protect the female from predation during her maturity molt, and is in attendance during the brief period of her life cycle in which she can receive spermatophores for the fertilization of subsequent spawnings.

Reproduction in several aquatic invertebrates involves external fertilization. In many bivalve mollusc and echinoderm species this process is facilitated by spawning inducers which produce localized epidemic spawning of conspecifics. Galtsoff (1940) demonstrated that both sexes of the oyster, *Crassostrea virginica*, are stimulated to release their gametes in response to those of the opposite sex, and that a few individuals could induce simultaneous spawning in large oyster beds. The active material in sperm was found to be extractable with organic solvents (Galtsoff, 1938).

In rodents a number of olfaction-mediated primer effects on reproductive function have been described; some of which are potentially important mechanisms of population control. For example, Whitten demonstrated an acceleration in the estrus cycles of female mice following exposure to the odors of male urine (see Whitten, 1966). Another primer effect described by Ropartz (1966, 1968) involves an increase in adrenal gland size and corticosteroid production resulting in a decrease in reproductive capacity. This effect is produced by the odor of other mice alone. Rogers and Beauchamp (1976) have considered some of the ecological implications of primer stimuli in wild rodent populations. They conclude that the conditions necessary for most of these effects to occur would rarely exist under natural circumstances, and that more studies in free-living populations are necessary before an evaluation of their ecological significance can be made.

Little (1975) demonstrated the importance of chemical information in crayfish maternal behavior. Developing larvae of certain species are observed to seek refuge on the pleopods of their mother's abdomen between feeding excursions. Little (1975) showed that larval orientation to the brooding female is chemically cued, and that attraction is only towards brooding females of the same species. Larvae apparently do not discriminate between different brooding females of the same species.

Sex pheromone detection is frequently associated with specialized sensory structures and/or a sexual dimorphism in size of the existing sensory apparatus. Several examples of the latter situation are known for insects (Kaissling, 1971). In the bathypelagic fishes, *Cyclothone* spp. and the ceratioid angler-fishes, males have an extensively developed olfactory system, whereas the females are relatively microsmatic; this suggests pheromone communication but definitive experiments have not yet been performed (Marshall, 1967). In the amphipod crustacean, *Gammarus duebeni*, pheromone reception apparently occurs via specialized club-shaped structures (calceoli) which are found only on the antennae of males (Dahl et al., 1970).

The vomeronasal system in vertebrates is absent or vestigial in birds, cetaceans, some primates, and most fish, but present in all other classes (Moulton and Beidler, 1967). Its greatest development is found in the snakes and lizards where it functions in monitoring chemical stimuli delivered via the tongue (Burghardt, 1970). Recently it was demonstrated in hamsters that chemical information received via the vomeronasal organ is critical for normal male sexual behavior (Powers and Winans, 1975).

Social Interactions: A variety of intraspecific activities other than those associated with reproduction are coordinated via chemical information. Scent marking behavior is exhibited by numerous mammalian species. The information content of scent marks serves to communicate such things as species group, territory, individuality; including age, sex, reproductive state, and social status (Eisenberg and Kleiman, 1972; Ewer, 1968). Perhaps a major function of scent marking is to maintain familiarity with a territory and/or social group, thereby enhancing "territorial confidence" (e.g. Mykytowycz et al., 1976).

Individual recognition is also chemically transmitted in certain fish species. An elegant example being that of the catfish, *Ictalurus natalis*, which maintains a dominance hierarchy within its social groups. Todd et al. (1967) demonstrated that individual recognition could be altered by subjecting individuals from a stable community to some external stress; for example, by taking the dominant fish from a group and exposing it to shock treatment or aggressive actions from a larger bullhead in another tank. Upon returning this fish to his original community, his status change was chemically recognized by the other community members and he was initially treated as a subordinate. No such change in hierarchy occurred if the dominant individual was only temporarily removed to another tank.

Alarm pheromones released from injured or threatened individuals have been described in several organisms. In mice the odor of a stressed individual causes dispersion of conspecifics (Rottman and

Snowdon, 1972). Von Frisch (1938) demonstrated a fright reaction in minnow schools (*Phoxinus laevis*) following introduction of an injured conspecific. A substance in the skin of the injured minnow effects the response which is prevented if olfaction is disrupted (von Frisch, 1941). Another example is found in the sea anemone, *Anthopleura elegantissima*, which reacts to a chemical released by injured members of its species. The response involves withdrawal of the tentacles and contraction of the column, and is induced by a quaternary ammonium ion, anthopleurine [(3-carboxy - 2,3-dihydroxy - N, N, N-trimethyl) - 1 - propanaminium], which has been isolated from *A. elegantissima* (Howe and Sheikh, 1975).

Terrestrial trail-following is observed in a variety of animals, and is a particularly important aspect of foraging for most ant species (Wilson, 1963). Nestmates are recruited to food by following pheromone trails laid by other ants returning from a food source. Recruits in turn reinforce the trail by releasing additional trail pheromone when coming back to the nest. Once the food source is depleted the pheromone is no longer deposited by returning ants, and the trail dissipates.

Hall (1973) reported intraspecific trail-following in the marsh gastropod *Littorina irrorata*. Individuals of both sexes were observed to track on paths made by either sex and in the proper direction. How the polarity is discerned is not known.

Predator Avoidance: Escape behavior in response to predatory gastropods and starfish is a widespread phenomenon among the molluscs (Bullock, 1953; Feder, 1963; Mackie and Grant, 1974). The reactions are most often triggered by contact, but some occur through distance chemoreception. An example of this behavior was recently reported for two limpets of the genus, *Acmaea* (Phillips, 1975a). *Acmaea limatula* and *Acmaea scutum* both show avoidance responses when exposed to effluents from the predatory starfish *Pisaster ochraceus*. The behavior is characterized by negative rheotaxis and negative geotaxis on horizontal and vertical surfaces, respectively. The chemoreceptors which mediate this avoidance response are located on the mantle margin (Phillips, 1975b).

Idler et al. (1956) noted that salmon of the genus *Oncorhynchus* would temporarily stop their upstream spawning migration in response to mammalian skin rinses. Of the amino acids in the rinses, L-serine was found to elicit the strongest response although it was not as effective as the unfractionated rinse. Presumably this reaction functions in avoiding predators such as bears (Idler et al., 1956).

Another approach to predator avoidance is via the release of repellents which frequently act as irritants to the predator's chemosensory apparatus. A familiar example is the offensive (to some of its predators at least) mixture of mercaptans released by the striped skunk when annoyed (Anderson and Bernstein, 1975). Several insects have likewise evolved a variety of repellents (Eisner, 1970); for example, beetles of the genus *Eleodes* discharge an irritating spray containing benzoquinones which effectively repels many of its predators (Eisner, 1966).

Among the cephalopod molluscs the ink of the octopus reputedly acts to disrupt olfactory function in the moray eel (MacGinitie and MacGinitie, 1968). Apparently the eel requires chemical information to recognize its octopus prey, and the ink serves to block this sensory channel allowing the octopus to escape. It has been hypothesized that this effect is due to the presence of unstable orthoquinones in the ink (Kittredge et al., 1974).

CONCLUSION

Chemical information is utilized by organisms in coordinating many important aspects of their lives. At the species level the diverse functions which have evolved in chemosensory systems are legion, and represent significant adaptations to a variety of environmental conditions. In ecosystems, chemically-mediated interactions are essential features in community organization and niche differentiation. At both levels, however, we are only beginning to understand the complex associations which exist. Furthermore, much more comparative work is needed before an objective assessment of the evolutionary relationships in chemosensory systems can be made.

ACKNOWLEDGMENTS

I wish to thank R. L. Doty, D. G. Moulton and J. Teeter for their helpful comments and criticisms of this manuscript.

REFERENCES

Ache, B.W. and Case, J. (1969). An analysis of antennular chemoreception in two commensal shrimps of the genus *Betaeus*. Physiol. Zool. 42:361-371.
Ache, B.W. and Davenport, D. (1972). The sensory basis of host recognition by symbiotic shrimps, genus *Betaeus*. Biol. Bull. 143:94-111.
Ache, B.W., Fuzessery, Z.M. and Carr, W.E.S. (1976). Antennular chemosensitivity in the spiny lobster *Panulirus argus*: Comparative tests of high and low molecular weight stimulants. Biol. Bull. 151:273-282.

Adrian, E.D. (1950). Sensory discrimination with some recent evidence from the olfactory organ. Brit. Med. Bull. 6: 330-333.

Adrian, E.D. (1951). Olfactory discrimination. Année Psychol. 50:107-113.

Adrian, E.D. (1953). The mechanism of olfactory stimulation in the mammal. Advan. Sci. (London) 9:417-420.

Adrian, E.D. (1954). The basis of sensation - some recent studies of olfaction. Brit. Med. J. 1:287-290.

Anderson, K.K. and Bernstein, D.J. (1975). Some chemical constituents of the scent of the striped skunk (*Mephitis mephitis*). J. Chem. Ecol. 1:493-499.

Atema, J. (in press). Functional separation of smell and taste in fish and crustacea. In: Olfaction and taste VI, edited by P. MacLeod. Information Retrieval Limited, London.

Bannister, L.H. (1974). Possible functions of mucus at gustatory and olfactory surfaces. In: Transduction mechanisms in chemoreception, edited by T.M. Poynder. Information Retrieval Limited, London.

Bardach, J.E. (1975). Chemoreception of aquatic animals. In: Olfaction and taste V, edited by D.A. Denton and J.P. Coghlan. Academic Press, New York.

Bardach, J.E. and Case, J. (1965). Sensory capabilities of the modified fins of squirrel hake (*Urophycis chuss*) and searobins (*Prionotus carolinus* and *P. evolans*). Copeia 1965 (2):194-206.

Bardach, J.E., Todd, J.H. and Crickmer, R. (1967). Orientation by taste in fish of the genus *Ictalurus*. Science 155:1276-1278.

Bardach, J.E. and Villars, T. (1974). The chemical senses of fishes. In: Chemoreception in marine organisms, edited by P.T. Grant and A.M. Mackie. Academic Press, New York.

Beauchamp, G.K., Doty, R.L., Moulton, D.G. and Mugford, R.A. (1976). The pheromone concept in mammalian chemical communication: A critique. In: Mammalian olfaction, reproductive processes, and behavior, edited by R.L. Doty. Academic Press, New York.

Beets, M.G.J. (1971). Olfactory response and molecular structure. In: Handbook of sensory physiology. Olfaction. Vol. 4, Part 1, edited by L.M. Beidler. Springer-Verlag, New York.

Beidler, L.M. (1954). A theory of taste stimulation. J. Gen. Physiol. 38:133-139.

Benvenuti, S., Fiaschi, V., Fiore, L. and Papi, F. (1973a). Homing performances of inexperienced and directionally trained pigeons subjected to olfactory nerve section. J. Comp. Physiol. 83:81-92.

Benvenuti, S., Fiaschi, V., Fiore, L. and Papi, F. (1973b). Disturbances of homing behavior in pigeons experimentally induced by olfactory stimuli. Monit. Zool. Ital. (N.S.) 7:117-128.

Birch, M.C. (1974a). Aphrodisiac pheromones in insects. In: Pheromones, edited by M.C. Birch. North-Holland Pub. Co., Amsterdam.

Birch, M.C. (ed.) (1974b). Pheromones. North-Holland Pub. Co., Amsterdam.

Blum, M.S. (1974). Pheromonal bases of social manifestations in insects. In: Pheromones, edited by M.C. Birch. North-Holland Pub. Co., Amsterdam.

Bossert, W.H. and Wilson, E.O. (1963). The analysis of olfactory communication among animals. J. Theor. Biol. 5:443-469.

Braveman, N.S. (1975). Relative salience of gustatory and visual cues in the formation of poison-based food aversions by guinea pigs (*Cavia porcellus*). Behav. Biol. 14:189-199.

Brown, W.L., Jr. (1968). An hypothesis concerning the function of the metapleural glands in ants. Am. Nat. 102:188-191.

Brown, W.L., Jr., Eisner, T. and Whittaker, R.H. (1970). Allomones and kairomones: Transpecific chemical messengers. Bio. Science 20:21-22.

Bullock, T.H. (1953). Predator recognition and escape responses of some intertidal gastropods in the presence of starfish. Behaviour 5:130-140.

Burghardt, G.M. (1968). Chemical preference studies on newborn snakes of three sympatric species of *Natrix*. Copeia 1968(4):732-737.

Burghardt, G.M. (1970). Chemical perception in reptiles. In: Advances in chemoreception, Vol. 1. Communication by chemical signals, edited by J.W. Johnston, Jr., D.G. Moulton and A. Turk. Appleton-Century-Crofts, New York.

Caprio, J. (1975). High sensitivity of catfish taste receptors to amino acids. Comp. Biochem. Physiol. 52A:247-251.

Carr, W.E.S., Hall, E.R. and Gurin, S. (1974). Chemoreception and the role of proteins: A comparative study. Comp. Biochem. Physiol. 47A:559-566.

Chambliss, O.L. and Jones, C.M. (1966). Cucurbitacins: Specific insect attractants in the Cucurbitaceae. Science 153:1392-1393.

Cheal, M. and Sprott, R.L. (1971). Social olfaction: a review of the role of olfaction in a variety of animal behaviors. Psychol. Rep. 29:195-243.

Cook, A., Bamford, O.S., Freeman, J.D.B. and Teideman, D.J. (1969). A study of the homing habit of the limpet. Anim. Behav. 17:330-339.

Cook, S.B. (1969). Experiments on homing in the limpet *Siphonaria normalis*. Anim. Behav. 17:679-682.

Crisp, D.J. (1974). Factors influencing the settlement of marine invertebrate larvae. In: Chemoreception in marine organisms, edited by P.T. Grant and A.M. Mackie. Academic Press, New York.

Dahl, E., Emanuelsson, H. and von Mecklenburg, C. (1970). Pheromone transport and reception in an amphipod. Science 170:739-740.

Davenport, D. (1950). Studies in the physiology of commensalism. I. The polynoid genus *Arctonoe*. Biol. Bull. 98:81-93.

Davenport, D. and Norris, K.S. (1958). Observations on the symbiosis of the sea anemone *Stoichactus* and the pomacentrid fish, *Amphiprion percula*. Biol. Bull. 115:397-410.
Davies, J.T. (1971). Olfactory theories. In: Handbook of sensory physiology. Olfaction. Vol. 4, Part 1, edited by L.M. Beidler. Springer-Verlag, New York.
Dethier, V.G. (1974). The specificity of the labellar chemoreceptors of the blowfly and the response to natural foods. J. Insect Physiol. 20:1859-1869.
Dimock, R.V. and Davenport, D. (1971). Behavioral specificity and the induction of host recognition in a symbiotic polychaete. Biol. Bull. 141:472-484.
Doty, R.L. (1974). A cry for the liberation of the female rodent: Courtship and copulation in Rodentia. Psychol. Bull. 81:159-172.
Døving, K.B., Nordeng, H. and Oakley, B. (1974). Single unit discrimination of fish odours released by char (*Salmo alpinus* L.) populations. Comp. Biochem. Physiol. 47A:1051-1063.
Eisenberg, J.F. and Kleiman, D.G. (1972). Olfactory communication in mammals. Annu. Rev. Ecol. Syst. 3:1-32.
Eisner, T. (1966). Beetle's spray discourages predators. Nat. Hist. 75:42-47.
Eisner, T. (1970). Chemical defense against predation in arthropods. In: Chemical ecology, edited by E. Sondheimer and J.B. Simeone. Academic Press, New York.
Erickson, R. (1963). Sensory neural patterns and gustation. In: Olfaction and taste I, edited by Y. Zotterman. Pergamon Press, New York.
Ewer, R.F. (1968). Ethology of mammals. Plenum Press, New York.
Farkas, S.R. and Shorey, H.H. (1972). Chemical trail-following by flying insects: A mechanism for orientation to a distant odor source. Science 178:67-68.
Feder, H.M. (1963). Gastropod defensive responses and their effectiveness in reducing predation by starfishes. Ecology 44:505-512.
Fiaschi, V. and Wagner, G. (1976). Pigeons homing - some experiments for testing the olfactory hypothesis. Experientia 32:991-993.
Frankel, G. (1959). The raison d'etre of secondary plant substances. Science 129:1466-1470.
Frisch, K. von (1938). Zur Physiologie des Fischschwarmes. Naturwissenschaften 26:601-606.
Frisch, K. von (1941). Uber einen Schreckstoff der Fischhaut und seine biologische Bedeutung. Z. Vergl. Physiol. 29:46-145.
Fuzessery, Z.M. and Childress, J.J. (1975). Comparative chemosensitivity to amino acids and their role in the feeding activity of bathypelagic and littoral crustaceans. Biol. Bull. 149:522-538.
Galtsoff, P.S. (1938). Physiology of reproduction of *Crassostrea virginica*. II. Stimulation of spawning in the female oyster. Biol. Bull. 75:286-307.

Galtsoff, P.S. (1940). Physiology of reproduction of *Crassostrea virginica*. III. Stimulation of spawning in the male oyster. Biol. Bull. 78:117-135.

Garcia, J. and Hankins, W.G. (1975). The evolution of bitter and the acquisition of toxiphobia. In: Olfaction and taste V, edited by D.A. Denton and J.P. Coghlan. Academic Press, New York.

Gesteland, R.C., Lettvin, J.Y. and Pitts, W.H. (1965). Chemical transmission in the nose of the frog. J. Physiol. (London) 181: 525-559.

Gesteland, R.C., Lettvin, J.Y., Pitts, W.H. and Rojas, A. (1963). Odor specificities of the frogs' olfactory receptors. In: Olfaction and taste I, edited by Y. Zotterman. Pergamon Press, Oxford.

Getchell, T.V. (1974). Unitary responses in frog olfactory epithelium to sterically related molecules at low concentrations. J. Gen. Physiol. 64:241-261.

Ghiradella, H., Case, J.F. and Cronshaw, J. (1968). Structure of aesthetascs in selected marine and terrestrial decapods: Chemoreceptor morphology and environment. Am. Zool. 8:603-621.

Gleeson, R.A. (1976). Pheromone-mediated behavior in the blue crab *Callinectes sapidus*. Am. Zool. 16:197.

Gove, D. and Burghardt, G.M. (1975). Responses of ecologically dissimilar populations of the water snake *Natrix s. sipedon* to chemical cues from prey. J. Chem. Ecol. 1:25-40.

Graziadei, P.P.C. (1969). The ultrastructure of vertebrate taste buds. In: Olfaction and taste III, edited by C. Pfaffmann. Rockefeller University Press, New York.

Graziadei, P.P.C. (1974). The olfactory and taste organs of vertebrates: A dynamic approach to the study of their morphology. In: Transduction mechanisms in chemoreception, edited by T.M. Poynder. Information Retrieval Limited, London.

Graziadei, P.P.C. (1977). Functional anatomy of the mammalian chemoreceptor system. In: Chemical signals in vertebrates, edited by D. Müller-Schwarze and M.M. Mozell. Plenum Press, New York.

Grubb, T.C., Jr. (1972). Smell and foraging in shearwaters and petrels. Nature (London) 237:404-405.

Grubb, T.C., Jr. (1974). Olfactory navigation to the nesting burrow in Leach's petrel (*Oceanodroma leucorrhoa*). Anim. Behav. 22:192-202.

Haldane, J.B.S. (1955). Animal communication and the origin of human language. Sci. Progr. (London) 43:385-401.

Hall, J.R. (1973). Intraspecific trail-following in the marsh periwinkle *Littorina irrorata* Say. Veliger 16:72-75.

Hamner, P. and Hamner, W.M. (1977). Chemosensory tracking of scent trails by the planktonic shrimp *Acetes sibogae australis*. Science 195:886-888.

Hasler, A.D. (1966). Underwater guideposts. University of Wisconsin Press, Madison.

Hasler, A.D. and Wisby, W.J. (1951). Discrimination of stream odors by fishes and its relation to parent stream behavior. Amer. Natur. 85:223-238.

Hickok, J. and Davenport, D. (1957). Further studies in the behavior of commensal polychaetes. Biol. Bull. 113:397-406.

Hodgson, E.S. (1968). Taste receptors of arthropods. In: Invertebrate receptors, edited by J.D. Carthy and G.E. Newell. Symp. Zool. Soc. London 23:269-277.

Holley, A., Duchamp, A., Revial, M.F., Juge, A. and MacLeod, P. (1974). Qualitative and quantitative discrimination in the frog's olfactory receptors: Analysis from electrophysiological data. Ann. N.Y. Acad. Sci. 237:102-114.

Howe, N.R. and Sheikh, Y.M. (1975). Anthopleurine: A sea anemone alarm pheromone. Science 189:386-388.

Idler, D.R., Fagerlund, V.H.M. and Mayoh, H. (1956). Olfactory perception in migrating salmon. I. L-serine, a salmon repellent in mammalian skin. J. Gen. Physiol. 39:889-892.

Jacobson, M. (1972). Insect sex pheromones. Academic Press, New York.

Johnston, J.W., Jr., Moulton, D.G. and Turk, A. (eds.) (1970). Advances in chemoreception, Vol. 1. Communication by chemical signals. Appleton-Century-Crofts, New York.

Kaissling, K.-E. (1971). Insect olfaction. In: Handbook of sensory physiology. Olfaction. Vol. 4, Part 1, edited by L.M. Beidler. Springer-Verlag, New York.

Karlson, P. and Lüscher, M. (1959). "Pheromones" a new term for a class of biologically active substances. Nature (London) 183: 155-156.

Keeton, W.T. (1974). Pigeon homing: No influence of outward-journey detours on initial orientation. Monit. Zool. Ital. (N.S.) 8: 227-234.

Keeton, W.T. and Brown, A.I. (1976). Homing behavior of pigeons not disturbed by application of an olfactory stimulus. J. Comp. Physiol. 105:259-266.

Kittredge, J.S., Takahashi, F.T., Lindsey, J. and Lasker, R. (1974). Chemical signals in the sea: Marine allelochemics and evolution. Fish. Bull. 72:1-11.

Laverack, M.S. (1974). The structure and function of chemoreceptor cells. In: Chemoreception in marine organisms, edited by P.T. Grant and A.M. Mackie. Academic Press, New York.

Levin, D.A. (1976). The chemical defenses of plants to pathogens and herbivores. Annu. Rev. Ecol. Syst. 7:121-159.

Lindstedt, K.J. (1971). Chemical control of feeding behavior. Comp. Biochem. Physiol. 39A:553-581.

Little, E.E. (1975). Chemical communication in maternal behavior of crayfish. Nature (London) 255:400-401.

MacGinitie, G.E. and MacGinitie, N. (1968). Natural history of marine animals. McGraw-Hill, New York.

Mackie, A.M. and Grant, P.T. (1974). Interspecies and intraspecies chemoreception by marine invertebrates. In: Chemoreception in marine organisms, edited by P.T. Grant and A.M. Mackie. Academic Press, London.

Madison, D.M., Scholz, A.T., Cooper, J.C., Horrall, R.M., Hasler, A.D. and Dizon, A.E. (1973). Olfactory hypotheses and salmon migration: A synopsis of recent findings. Fish. Res. Board Can. Tech. Rep. 414:1-35.

Marshall, N.B. (1967). The olfactory organs of bathypelagic fishes. Symp. Zool. Soc. London 19:57-70.

Moncrieff, R.W. (1951). The chemical senses. Leonard Hill, London.

Moulton, D.G. (1976). Spatial patterning of response to odors in the peripheral olfactory system. Physiol. Rev. 56:578-593.

Moulton, D.G. and Beidler, L.M. (1967). Structure and function in the peripheral olfactory system. Physiol. Rev. 47:1-52.

Mozell, M.M. (1977). Processing of olfactory stimuli at peripheral levels. In: Chemical signals in vertebrates, edited by D. Müller-Schwarze and M.M. Mozell. Plenum Press, New York.

Müller-Schwarze, D. (1971). Pheromones in black-tailed deer (*Odocoileus hemionus columbianus*). Anim. Behav. 19:141-152.

Müller-Schwarze, D. and Mozell, M.M. (eds.) (1977). Chemical signals in vertebrates. Plenum Press, New York.

Mykytowycz, R., Hesterman, E.R., Gambale, S. and Dudzinski, M.L. (1976). A comparison of the effectiveness of the odors of rabbits, *Oryctolagus cuniculus*, in enhancing territorial confidence. J. Chem. Ecol. 2:13-24.

Nachman, M. and Cole, L.P. (1971). Role of taste in specific hungers. In: Handbook of sensory physiology. Taste. Vol. 4, Part 2, edited by L.M. Beidler. Springer-Verlag, New York.

Nordeng, H. (1971). Is the local orientation of anadromous fishes determined by pheromones? Nature (London) 233:411-413.

Norlund, D.A. and Lewis, W.J. (1976). Terminology of chemical releasing stimuli in intraspecific and interspecific interactions. J. Chem. Ecol. 2:211-220.

O'Connell, R.J. (1975). Olfactory receptor responses to sex pheromone components in the redbanded leafroller moth. J. Gen. Physiol. 65:179-205.

O'Connell, R.J. and Mozell, M.M. (1969). Quantitative stimulation of frog olfactory receptors. J. Neurophysiol. 32:51-63.

Otte, D. (1974). Effects and functions in the evolution of signaling systems. Annu. Rev. Ecol. Syst. 5:385-417.

Papi, F., Fiore, L., Fiashi, V. and Benvenuti, S. (1971). The influence of olfactory nerve section on the homing capacity of carrier pigeons. Monit. Zool. Ital. (N.S.) 5:265-267.

Papi, F., Fiore, L., Fiashi, V. and Benvenuti, S. (1972). Olfaction and homing in pigeons. Monit. Zool. Ital. (N.S.) 6:85-95.

Papi, F., Fiore, L., Fiashi, V. and Benvenuti, S. (1973). An experiment for testing the hypothesis of olfactory navigation of homing pigeons. J. Comp. Physiol. 83:93-102.

Papi, F., Ioalé, P., Fiaschi, V., Benvenuti, S. and Baldaccini, N.E. (1974). Olfactory navigation of pigeons: The effect of treatment with odorous air currents. J. Comp. Physiol. 94:187-193.

Parsons, T.S. (1971). Anatomy of nasal structures from a comparative viewpoint. In: Handbook of sensory physiology. Olfaction. Vol. 4, Part 1, edited by L.M. Beidler. Springer-Verlag, New York.

Pfaffmann, C. (1941). Gustatory afferent impulses. J. Cell. Comp. Physiol. 17:243-253.

Pfaffmann, C. (1974). Specificity of the sweet receptors of the squirrel monkey. Chem. Senses Flav. 1:61-67.

Phillips, D.W. (1975a). Distance chemoreception-triggered avoidance behavior of the limpets *Acmaea (Collisella) limatula* and *Acmaea (Notoacmea) scutum* to the predatory starfish *Pisaster ochraceus*. J. Exp. Zool. 191:199-210.

Phillips, D.W. (1975b). Localization and electrical activity of the distance chemoreceptors that mediate predator avoidance behavior in *Acmaea limatula* and *Acmaea scutum*. J. Exp. Biol. 63: 403-412.

Polak, E.H. (1973). Multiple profile-multiple receptor site model for vertebrate olfaction. J. Theor. Biol. 40:469-484.

Powers, J.B. and Winans, S.S. (1975). Vomeronasal organ: critical role in mediating sexual behavior of the male hamster. Science 187:961-963.

Regnier, F.E. and Goodwin, M. (1977). Chemical and environmental modulation of pheromone release from vertebrate scent marks. In: Chemical signals in vertebrates, edited by D. Müller-Schwarze and M.M. Mozell. Plenum Press, New York.

Rogers, J.G., Jr. and Beauchamp, G.K. (1976). Some ecological implications of primer chemical stimuli in rodents. In: Mammalian olfaction, reproductive processes, and behavior, edited by R.L. Doty. Academic Press, New York.

Ropartz, P. (1966). Contribution à l'étude du déterminisme d'un effet de groupe chez les souris. Compt. Rend. 262:2070-2072.

Ropartz, P. (1968). Role des communications olfactives dans le comportement social des souris males. Colloq. Intern. Centre Natl. Rech. Sci. (Paris) 173:323-339.

Rottman, S.J. and Snowdon, C.T. (1972). Demonstration and analysis of an alarm pheromone in mice. J. Comp. Physiol. Psych. 81: 483-490.

Sato, M. (1971). Neural coding in taste as seen from recordings from peripheral receptors and nerves. In: Handbook of sensory physiology. Taste. Vol. 4, Part 2, edited by L.M. Beidler. Springer-Verlag, New York.

Schneider, D. (1969). Insect olfaction: deciphering system for chemical messages. Science 163:1031-1037.

Schneider, D., Kasang, G. and Kaissling, K.-E. (1968). Bestimmung der Riechschwelle von *Bombyx mori* mit Tritium-markiertem Bombykol. Naturwissenschaften 55:395.

Schneider, D. and Steinbrecht, R.A. (1968). Checklist of insect olfactory sensilla. Symp. Zool. Soc. London 23:279-297.

Scholz, A.T., Horrall, R.M., Cooper, J.C. and Hasler, A.D. (1976). Imprinting to chemical cues: The basis for home stream selection in salmon. Science 192:1247-1249.

Shafer, R. (1977). The nature and development of sex attractant specificity in cockroaches of the genus *Periplaneta*. IV. Electrophysiological study of attractant specificity and its determination by juvenile hormone. J. Exp. Zool. 199:189-208.

Shorey, H.H. (1973). Behavioral responses to insect pheromones. Annu. Rev. Entomol. 18:349-380.

Shorey, H.H. (1976). Animal communication by pheromones. Academic Press, New York.

Silverstein, R.M. and Young, J.C. (1976). Insects generally use multicomponent pheromones. In: Pest management with insect sex attractants and other behavior-controlling chemicals, edited by M. Beroza. ACS Symposium Series, No. 23, American Chemical Society, Washington, D.C.

Solomon, D.J. (1973). Evidence of pheromone-influenced homing by migrating Atlantic salmon, *Salmo salar* (L.). Nature (London) 244:231-232.

Steinbrecht, R.A. (1969). Comparative morphology of olfactory receptors. In: Olfaction and taste III, edited by C. Pfaffmann. Rockefeller University Press, New York.

Thiessen, D.D., Regnier, F.E., Rice, M., Goodwin, M., Isaacks, N. and Lawson, N. (1974). Identification of a ventral scent marking pheromone in the male Mongolian gerbil (*Meriones unguiculatus*). Science 184:83-85.

Tobach, E. (1971). Photoreception and chemoreception: Questions for the evolution and development of orientation. Ann. N.Y. Acad. Sci. 188:194-201.

Todd, J.H., Atema, J. and Bardach, J.E. (1967). Chemical communication in the social behavior of a fish, the yellow bullhead, *Ictalurus natalis*. Science 158:672-673.

Vinnikov, Y.A. (1975). The evolution of olfaction and taste. In: Olfaction and taste V, edited by D.A. Denton and J.P. Coghlan. Academic Press, New York.

Vovelle, J. (1965). Le tube de *Sabellaria alveolata* (L.) annélide polychète *Hermellidae* et son ciment. Etude ecologique expérimentale, histologique et histochemique. Archs. Zool. Exp. Gén. 106:1-187.

Wenzel, B.M. (1968). The olfactory prowess of the kiwi. Nature (London) 220:1133-1134.

Wenzel, B.M. (1972). Olfactory sensation in the kiwi and other birds. Ann. N.Y. Acad. Sci. 188:183-193.

Wheeler, J.W. (1977). Properties of compounds used as chemical signals in vertebrates, edited by D. Müller-Schwarze and M.M. Mozell. Plenum Press, New York.

Whittacker, R.H. and Feeny, P.P. (1971). Allelochemics: Chemical interaction between species. Science 171:757-770.
Whitten, W.K. (1966). Pheromones and mammalian reproduction. Advan. Reprod. Physiol. 1:155-177.
Wilson, D.P. (1968). The settlement behaviour of the larvae of *Sabellaria alveolata*. J. Mar. Biol. Ass. U.K. 48:387-435.
Wilson, D.P. (1970a). Additional observations on larval growth and settlement in *Sabellaria alveolata*. J. Mar. Biol. Ass. U.K. 50:1-31.
Wilson, D.P. (1970b). The larvae of *Sabellaria spinulosa* and their settlement behaviour. J. Mar. Biol. Ass. U.K. 50:33-52.
Wilson, E.O. (1963). The social biology ants. Annu. Rev. Entomol. 8:345-368.
Wilson, E.O. (1970). Chemical communication within animal species. In: Chemical ecology, edited by E. Sondheimer and J.B. Simeone. Academic Press, New York.
Wilson, E.O. and Bossert, W.H. (1963). Chemical communication among animals. Recent Progr. Horm. Res. 19:673-716.

ADAPTIVE RADIATION OF MECHANORECEPTION

Hubert MARKL

Fachbereich Biologie, Universität Konstanz

D-7750 Konstanz, Fed. Rep. Germany

INTRODUCTION

Sensitivity to mechanical energy is a fundamental characteristic of living systems, from single cells to complex organisms, a sensitivity no less ubiquitous than that to thermal energy or radiation. There are, however, vast differences in reception thresholds and in filtering properties of the mechanoreceptive systems depending on the accessory structures which first accept the stimulus and transmit it to the sensory elements, more often than not transforming it on its way to them, until finally transduction from mechanical stimulus energy to excitation of the sensory cell membrane occurs (Burkhardt, 1960). These processes determine, together with the response characteristics of the sensory cells, to which specific form of mechanical energy a system is sensitive, whether sensitivity is directionally polarized or not and what the difference thresholds are for stimulus changes in the time and in the intensity domains. Central processing of the mechanoreceptive excitation can further affect the mechanoreceptive properties of the whole organism, e.g. by allowing to localize stimulus direction by comparing the input from several spatially distributed sense organs.

TYPOLOGY OF MECHANICAL STIMULI AND OF MECHANORECEPTORS

Whereas mechanical forces basically have always the same and rather simple effects on the living organism - deformation of structures on which the forces impinge and/or setting into motion of the affected body - there is a wide array of discernible ways in which these effects can be brought about, depending on the nature of the mechanical stimuli as well as on the construction of the receptors.

1. According to Newton's law of universal gravitation every organism on earth is under the influence of the force of gravitational attraction from the earth's mass. Sense organs specialized to perceive the direction and magnitude of this force, or of component vectors thereof, are called "gravity receptors" and found in one or the other form in most phyla of the animal kingdom and even in some plants (see reviews: Gordon and Cohen, 1971; Markl, 1974; Lowenstein, 1974; Schöne, 1975). Since there is no difference in the mechanical force derived from mass attraction and that from other ways of accelerating a body (Einstein's equivalence principle) these sense organs are always sensitive to the resultant mechanical force from gravitational and other sources of acceleration of a given mass, e.g. that of a statolith, which is important for analyzing the response characteristics of a graviceptor.

2. Proprioceptive mechanoreceptors are sensitive to mechanical forces within or between body parts resulting from active or passive position of a body or of its parts in the gravitational field or of changes of these positions (motions). They are dealt with in a separate chapter of this volume, (see Wales).

3. While mass attraction applies force without direct contact or without contact transmission through an intervening material medium, the latter forms of stimulus applications are responded to by a number of differently specialized mechanoreceptive mechanisms.

3.1. Touch receptors to be found throughout the animal kingdom react to direct contact with another body in the environment due to pressure or friction caused by motion of one against the other.

3.2. Some "touch" receptors are prepared to monitor the mechanical stimulus derived from the motion of the body relative to its surrounding medium: air-flow or water-flow receptors. They are activated either by outside medium displacement or the active movement of the organism with respect to a resting medium or by any resultant between the two. Others record motion of a fluid, e.g. blood, within body cavities.

3.3. As a derivative from similar medium motion detectors we find mechanoreceptor systems arranged so as to record angular acceleration systems arranged so as to record angular acceleration around the different body axes (or, to look at it differently, in the different planes of space) during rotatory movement.

3.4. There is a continuous transition from DC-flow receptors over low-pass motion detector systems with a low high-cutoff frequency to those with a band pass characteristic making them most sensitive to rhythmical oscillation of medium particles at a given frequency band: *vibration receptors*. Depending on the type of interface between organism and medium we discriminate between substrate -, air - and water-vibration perception (see review: Markl, 1973). Some organisms have developed sensory systems to record vibrations at an air-water interface (surface wave receptors). The oscillatory medium motion can result in different wave-forms depending on the mechanical properties of the medium (e.g. longitudinal, shear, Raleigh waves), opening the possibility for according specializations on the receptor side. Sources of vibration in a medium transmit vibrational energy over distance with medium characteristic velocity and attenuation. There is, however, an important difference between a zone close to the source of excitation (its *near-field*) where source oscillation leads to oscillating flow of medium *around* the source, the amplitude of which can be quite large, depending on source motion amplitude and frequency, source size and geometry and distance from it, and declines much more steeply with distance than the medium oscillations which go along with a pressure wave conducted through the medium over larger distances (*far-field*) (Harris, 1964). Of course, there is no physical difference between vibrational displacement of a medium particle at a given frequency and amplitude associated with the far- or the near-field at any given point in the medium: a displacement-sensitive vibration receptor which is excited by medium oscillation of a given amplitude and frequency in the near-field of a source, will react exactly the same, if the same vibration stimulus arises in the far-field, associated with a pressure wave that is lacking in the near-field. The designation as "near-field" or "far-field" vibration relates to 1. properties of the source, 2. to properties of the medium, and 3. to characteristics of the attenuation of vibration over distance from the source, not to the nature of the vibrational motion itself. The capacity of a source to produce a given vibration amplitude in a given medium at a defined distance in the near- or in the far-field depends above all on source size and vibration frequency, so that transmission efficiency is framed into a well defined frequency-window for each source type of given strength in a given medium (see Markl 1969, 1973). What is most important in that connection, is the often surprisingly large amplitude of oscillatory medium displacements in the near-field of a source which are present even if far-field wave transmission - as for instance monitored by a pressure-sensitive device - is very weak or even not recognizable at a given frequency, because receptors sensitive to rhythmical medium displacement can thus detect the presence of a vibrating source if it is not too far away, even if hardly any far-field sound is radiated by it.

4. There is again a gradual transition from systems which are sensitive to medium vibration to those receptors which react to the rhythmical pressure changes in the surrounding medium which are

usually called "sound", i.e. to the alternating pressure transmitted into the far-field of a vibrating source. For practical reasons it is useful to make a distinction between sound reception, i.e. sensitivity to pressure waves, and vibration reception, i.e. sensitivity to particle displacement of a medium (or a derivative thereof, as velocity or acceleration) even if, as is the case in fishes, the pressure changes may not be perceived by the sense organ directly, but have to be transformed into large amplitude displacements of body tissue by means of an air-filled, compressible swim-bladder, in order to be effective for displacement sensitive receptor mechanisms. This distinction is still useful, if one remains aware of the fact that each and every pressure-sensitive hearing organ in the animal kingdom works on the principle of transforming hydrostatic pressure into displacement of receptive structures, because the distinction between sound and vibration receptors refers not primarily to the level of the receptor cell-where the reception mechanisms may in fact be indistinguishable-but to the level of the whole sensory system with its accessory structures which make the organ receptive to sound pressure or only to medium displacement. For the adaptive radiation of sound receptors see Michelsen (this volume).

5. Finally, medium pressure - be it in the surrounding of the body or within the fluid or air filling of a body cavity-can act as a mechanical stimulus for specialized static pressure receptors (baroreceptors), see Blaxter (this volume).

Some of these different types of receptors are to be classified as static receptors, answering continuously to a maintained mechanical stimulation, others as dynamic or phasic receptors, which react only to arhytmical or rythmical changes in stimulus direction of amplitude, with a defined frequency dependent response.

This can only be regarded as a very rough classification of the kinds of action of mechanical energy on organisms. As is readily seen, it rests rather on aspects relating to stimulus characteristics in environment, to which the stimulus accepting and transmitting accessory structures of the sense organs are specifically adapted, than on differences in the sensory reception processes themselves, which are as a rule much more uniform. For an example, sensory hairs on the body surface of arthropods are stimulated by deflection, reacting in a purely phasic or phasictonic fashion, and have been shown to be involved in almost any of the described mechanoreceptive functions, as 1) tactile receptors, 2) proprioceptors, 3) graviceptors, 4) air- or water-flow receptors, 5) angular acceleration receptors, 6) vibration receptors, 7) sound receptors, and even 8) pressure receptors. Similar versatility can be claimed for other basic forms of mechanoreceptors in other phyla (see below).

If we consider the adaptive radiation of mechanoreceptor mechanisms (with the exception of proprioception, sound reception and baroreception, which are dealt with elsewhere in this volume) it therefore seems useful to restrict the attention in an examplary fashion to the three most wide-spread basic types of mechanoreceptive cells and to follow their adaptive differentiation with the help of elaborated accessory structures. As will become evident, similar functional constraints given by the nature of the surrounding environment and by the physical laws governing generation and transmission of mechanical energy, have led again and again to analogous constructive and functional solutions, more often that not converging to surprising similarities, thus laying open the principle possibilities available to organisms for coping with similar environmental challenges.

THREE LINES OF MECHANORECEPTOR DIVERSIFICATION

1. Ciliated mechanoreceptors in Invertebrates

Although sensitivity of the surface membrane of unicellular organisms to mechanical stimulation is most probably a fundamental constructive feature of even the most primitive forms and can be demonstrated to be present also in protists lacking cilia (e.g. amoeba) it is in ciliates that we have at present the most detailed knowledge of the physiological mechanisms involved (for reviews see Eckert, 1972; Naitoh and Eckert, 1974; Machemer, 1977; Machemer and de Peyer, 1977). One of the most interesting peculiarities in these ciliated cells is that there is no straightforward separation between sensory and motor functions. This holds true not only for protists, where it is of course most evident; mechanical sensitivity of motile cilia may be quite common in the animal kingdom, as was discussed by Thurm (1968, 1969). This is particularly fascinating because numerous sense organs throughout all major phyla have been shown to contain sense cells with either motile or non-motile kinocilia on their outside surface (see below) or with modified ciliary structures in the cell interior (review see Vinnikov, 1974).

The "avoiding reaction" of *Paramecium* was a favorite object of inquiry into the physiological mechanisms of sensitivity to external stimuli in ciliates ever since it was first described by Jennings. It was there that Eckert, Naitoh, Machemer, Kung and others have made decisive progress in elucidating the biophysical basis of a behavioral reaction. A sufficiently strong stimulus to the front side of *Paramecium* releases the stop of the ciliary beat and consecutively the reversal of beating direction in all cilia; after a short backward swim, the animal stops again, rotates to a variable degree around its rear due to asymmetries of ciliary action in restoring their original beat direction, and then swims forward again. Similar stimulation of the rear end leads to no reversal of ciliary beat but rather to increased beating frequency in the normal direction in an accelerated forward lunge, (see also Couillard, this volume).

It has been possible by intracellular recording from *Paramecium* in different ionic environments and by other techniques to correlate specified ciliary activities with bioelectric phenomena, at the cell membrane which also envelopes all cilia, that are caused by changes in membrane conductivity to defined actions on stimulation of the different regions of the cell surface. Mechanical stimulation of the anterior end leads to a depolarization graded according to stimulus strength up to a saturation level. This is caused by an increased Ca^{2+} -conductance; if Ca^{2+} -concentration in the cell is raised above 10^{-6} molar (as compared to 10^{-7} molar in the unexcited cell) the beating direction of the cilia is reversed and their beat frequency increased, resulting in fast backward swim. If stimulation discontinues, Ca^{2+} - conductance returns to normal, the surplus Ca^{2+} is pumped out of the cell, and when its level falls below the threshold, cilia stop to beat in reversed direction and return to normal forward beat. The middle region of the cell is rather insensitive to mechanical stimulation, while stimulation of the rear end leads to a hyperpolarizing membrane response due to increased K^+ -conductance, which results in an increase of ciliary beat frequency in the forward direction. Very gentle touch of the anterior surface can lead to a complete stoppage of all ciliary activity, bringing the animal into thigmotactic rest. The ional mechanism of this response is not yet known.

So far, these are response mechanisms to the most basic form of mechanical stimulation only: touch. Almost nothing is known about whether a ciliate can perceive other forms of mechanical energy, although it seems not impossible that the cilia can also be stimulated by local medium currents, by angular acceleration of the cell, by the action of gravity on the asymmetrical body during swimming or by medium vibrations, e.g. in the near-field of beating cilia or flagella of another organism, as has been demonstrated for *Amoeba* (Kolle-Kralik and Ruff, 1967).

If we move to more highly evolved metazoa, we find among coelenterates examples for most or all of these surmised forms of mechanical sensitivity mediated by ciliated cells. In most cases these cilia are no longer motile.

Sensitivity of ciliated epidermal receptors to medium vibration in the near-field of oscillating objects is widespread in coelenterates and other aquatic invertebrates (review: Markl, 1973). Horridge (1969, 1971) made the suggestion that the different types of gravity receptors of cnidarian medusae could be derived from similar ciliated vibration receptors, with kinocilia being stimulated by the statolith which is typically enclosed in ento- or ectodermal cells of club-like projections of the body surface (lithostyles). By loading the base of the vibration sensitive receptors with calcareous deposits, this base acts as a dense abutment for the motile cilia, thus making them more effective in vibration reception. By protruding these sensory structures into the medium on finger-like tentacles for gaining range in detecting vibrating objects (e.g. prey) these loaded clubs become pendula which are easily displaced by gravity when the animal changes position in space. Sensory ciliated cells surrounding the pendulum become thus part of a mechanoreceptive apparatus of a new quality, able to perceive the direction of gravity (and maybe also of angular acceleration, although experimental evidence is lacking for that). A common role as graviceptor and vibration receptor was also assumed for the apical organ of ctenophores (Krisch, 1973). It is noteworthy that the presumeably mechanoreceptive kinocilia of these receptors are as a rule surrounded by stereocilia, comparable to what we find to be true for vertebrate neuromasts and hair-cells in the stato-acoustic system, and that deflection of the kinocilum in a plane at right angle to the line connecting the two central fibrils is maximally effective as a stimulus (see p. 332). A kinociliar/stereociliar apparatus is also characteristic for the cnidocils, the sensory trigger mechanism of coelenterate cnidocysts (Vinnikov, 1974).

Ciliated tactile receptors on the body surface are common throughout most phyla of aquatic non-arthropod invertebrates. Although sensitivity to water currents and medium vibrations can be suspected for many of these mechanoreceptors, these qualities have only rarely been investigated in detail (see Horridge 1966, Horridge and Boulton, 1967). It is also not particularly difficult to see the graviceptive statocysts of many of these invertebrates derived from similar epidermal mechanoreceptive precursors, as vesicular invaginations from the body surface lined by ciliated sensory cells which record the position of statoliths, particles of higher density than the surrounding fluid, that are either taken up from outside or secreted by the animal itself and that press down on the sensory cilia in the direction of gravitational pull. Annelids, molluscs and echinoderms yield typical examples for different evolutionary stages of this development (review: Markl, 1974). In gastropods and cephalopods they have been studied most extensively both with respect to macro- and micromorphology, to physiological properties and behavioral functions (review: Wolff 1973, 1975; Budelmann, 1976). Interestingly, kinocilia in gastropod statocysts can be continuously or intermittently active as motile organelles. Gastropod statocysts function only as graviceptors, enabling the animals to orient their body axes with respect to the vertical (geotaxis); they convey no sensitivity to angular acceleration. The information about the spatial orientation of the animal is contained in the spatial stimulation pattern of sensory cells by the statolith filling of the cyst rather than by the degree of bending of the cilia of the individual sensory cell.

The cephalopod statocyst system is a culminating pinnacle of mechanoreceptor perfection in the adaptive radiation of ciliated epidermal mechanoreceptors in the non-arthropod, non-chordate phyla. In addition to the statoreceptive apparatus - which is quite extensively differentiated in cephalopods also - further mechanoreceptive functions are served by other parts of the statocyst system in surprising perfection. Structure and functions of this organ complex have recently been reviewed in detail by several authors to which the reader is referred for the complete bibliography (Markl 1974; Vinnikov, 1974; Budelmann, 1975, 1976, 1977). There are considerable differences in structure as well as in physiology between the different cephalopod orders, from a rather snaillike, primitive disposition in the Nautiloidea to the fully differentiated organs in octopods and decapods. The following account relates primarily to the situation in *Octopus*, where the studies are most advanced.

The membraneous statocyst hangs firmly positioned in a lymph-filled cavity of the head cartilage. The sensory cells of the cyst are concentrated primarily into two functionally different regions: the vertical macula (with 5100 hair-cells which are connected with a statolith), and the crista, a greatly elongated ridge, with a course through each of the three planes of space, in which the cilia of several rows of large and small sensory cells are enveloped by a cupula of amorphous and fibrous material, that projects into the sac cavity. The sensory cells of both systems bear numerous (up to over 200) kinocilia of the 9x2+2 type and short microvilli, but no stereocilia, and have recently been shown to be secondary sense cells without axonal fibers, innervated by "second-order" neurons - a most remarkable convergence to vertebrate neuromast hair cells (Budelmann 1977, Budelmann and Thies, 1977).

The sensory cells are polarized due to the arrangement of their kinocilia: the two central filaments of each cilium are in line with the long axis of the elongated group of kinocilia borne by each sense cell, while the basal feet project at right angles to this axis. In the crista the long axis of the kinociliar groups of the sense cells coincides with the long axis of the crista, while on the macula the cells are organized in concentric rings with all the basal feet pointing away from the center. There is efferent in addition to afferent innervation of the sense cells.

Statocyst morphology in the decapod *Sepia* is even more complicated, the macula being divided into three sections in the different spatial planes with altogether 8600 sensory cells, and the four-sectioned crista of ca. 2800 sensory cells forming an elaborately looped system covering all three spatial axes between 11-12 cartilaginous anticristae.

The maculae are principally gravity receptors, the cells being maximally stimulated by shear of the statolith in the direction of the feet of their kinocilia and inhibited by displacement in the opposite direction - which accords well with response polarization in other sense cells bearing kinocilia, from cnidarians (Horridge, 1969, 1971) to vertebrates (Flock, 1965). Depending on the animal's position in space, different groups of macular sensory cells are thus maximally excited. Budelmann (1970) demonstrated in elegant behavioral experiments that, although the sensory cells are clearly able to indicate direction as well as magnitude of the statolith shearing force, the compensatory eye-movements to change of the animal's position in space are only dependent on the direction of that force, not its amplitude. This result - which differs from the outcome of comparable experiments in crustaceans and vertebrates (Schöne, 1959) - has therefore to be ascribed to differences in the central nervous processing of sensory information with respect to positional control in space. As was demonstrated by Wells (1960), postural information from statocysts in cephalopods is not only necessary to control the animal's position in space but also to control the spatial position of the retina in order to allow the animal to discriminate between visual patterns with respect to their orientation in space.

The crista system is primarily a receptor for angular acceleration or deceleration during an animal's rotation in space, with its cupula being deflected by inertial motion of the endolymph with respect to the rotating surrounding head structure, to which the sensory cells are firmly fastened. In addition, Budelmann and Wolff (1973) demonstrated sensitivity to linear acceleration (gravity) in the *Octopus* crista.

In addition to their sensitivity to gravity and angular acceleration the ciliated sensory cells in the cephalopod statocyst have been found to be highly sensitive to vibration, although this seems not to have been followed up quantitatively. Therefore the question of sensitivity to substrate-born and to medium vibration or even to sound can clearly not be regarded as finally answered for the highly evolved cephalopods.

2. Mechanoreceptive hairs in Arthropods

The typical mechanoreceptive exteroreceptor of the arthropod is the *sensory hair*, a developmental unit, in which one or more sensory cells unite with a differentiated apparatus for stimulus transmission, which is the cuticular product of several types of epidermal cells - secreting the elastic basal ring and the hair-shaft - which are also involved in the stimulus transduction process at the sensory dendrite (review: Gnatzy and Schmidt, 1971, Schmidt, 1973, Thurm, 1974, 1977). Most important for the transmission of the mechanical stimulus to the receptor ending are 1) the hair shaft which couples the sensory structure to the environmental stimulus source and the geometry of which determines decisively which form of mechanical energy can excite the receptor; 2) the flexible hair-joint that links the shaft to the surrounding cuticular base, the elasticity and geometry of which is again of profound influence on the stimulus transmitting properties of the hair, and especially on the directionality of its responsiveness. The variability in these cuticular accessory structures much more than differences in response characteristics of the sensory cells themselves are responsible for the large variability of mechanoreceptive function of arthropod hairs, ranging from campaniform receptors of cuticular strain – with no real hair-shaft at all, which is replaced by a cuticular dome – to thread hairs (trichobothria) of enormous length which move with aerial motion in a sound wave, or to mushroomlike hairs in the static organs of some waterbugs which measure the position of an air-water interface and thus enable the animal to infer the direction of an air-bubble's buoyancy, or its reverse: that of gravity. In striking contrast to this diversity, the basic processes of sensory transduction at the level of the dendrite seem to be remarkably similar in these different hair-types (Thurm, 1968, 1969, 1977, Gaffal et al. 1975, Spinola and Chapman, 1975, Moran and Rowley, 1975).

The literature covering structure and function of arthropod mechanoreceptor hairs is very extended and has not been thoroughly reviewed *in toto*. Schwartzkopff (1974) provides a useful overview for the insects, touching many of the diverse aspects which are also valid for the other arthropods. For the receptors for gravity and angular acceleration see Markl (1974), for those for vibration Markl (1973). Since it is impossible to even briefly mention all the different constructive and functional types which have been described, a few well studied cases will have to exemplify the many different ways in which cuticular sensory hairs can serve mechanoreceptive purposes in arthropods.

While the primitive tactile hair with its typically phasic-tonic response characteristic may be mentioned in passing only, notwithstanding its ubiquitous presence in the arthropods, there is again a startling convergence – functional and anatomical – between the statocysts of some aquatic arthropods and the equilibrium apparatus of cephalopods on the one hand, and of vertebrates on the other. In the decapod crustaceans we find again the differentiation between sensory hairs connected to statoliths – usually brought from outside into the invaginated statocyst sac – which act as gravity sensors, and free hairs that project into the fluid filling of the cyst and are stimulated by inertial fluid motion opposite to an accelerated rotation of the animal's body. In crabs, the statocyst wall is shaped so that vertical and horizontal canals for preferential fluid flow are formed, thus enabling the animal to unambigously discriminate between rotations around different body axes (Cohen and Dijkgraaf, 1961; Fraser, 1974; Fraser and Sandeman, 1975; Sandeman, 1975, 1976; Sandeman and Okajima, 1972, 1973 a,b; Silvey et al., 1976). It should also be mentioned that - again comparable to equilibrium organs in other phyla - the decapod statocyst is highly sensitive to substrate vibrations, and that - quite similar to what was recently found in cephalopods - the rotation-sensitive free hairs are also stimulated by linear acceleration, e.g. by gravity.

Sensitivity to vibration in the medium is not at all unusual for arthropod hair receptors. In crayfish, Wiese (1976, Wiese et al., 1976; see also Mellon, 1963; Laverack, 1962 a,b) demonstrated that feathered mechanosensory hairs on the body surface are extremely sensitive to medium displacement in the near-field of oscillating sources, or, if brought close enough to the surface, to surface waves of the water, with a working range between 0.05 and 200 Hz and best thresholds in the low tens of Hertz. Dual innervation - one sensory cell responsive to headward, the other to tailward bending of the hair, to which they are connected - and marked response directionality allow the detection of stimulus direction. The same hairs make the animal even sensitive to sound waves, provided that displacement amplitudes in the wave are sufficiently high (Offutt, 1970). The whole system of these vibration sensitive surface hairs is an almost exact analogue of the lateral line system in fishes and aquatic amphibians (see p. 334).

The use of sensory hairs as receptors for substrate vibrations is also known from arthropods. For instance, they have recently been reported to be involved in the surprising acuity of scorpions to detect burrowing prey from a distance by means of the surface vibrations produced by it (Brownell, 1977). Under somewhat different circumstances, backswimmers (*Notonecta*) use sensory hairs in addition to other mechanoreceptors to detect prey-excited surface waves at the air-water interface (Murphey and Mendenhall, 1973; Lang, 1977).

On land, some insects and spiders make use of highly mobile sensory hairs (trichobothria, thread hairs) for the reception of near-field medium vibration - e.g. as caused by the wing-beat of insects - and evidently also for the detection of sound of sufficient intensity (Görner, 1966; Görner and Andrews, 1969; Drašlar, 1973; Minnich, 1925, 1936; Pumphrey and Rawdon-Smith, 1936; Haskell, 1956; Katsuki and Suga, 1960; Petrovskaya et al., 1970; Edwards and Palka, 1974; Markl and Tautz, 1975; Tautz, 1977 a,b). For one example, the "hearing hairs" of lepidopteran caterpillars are sensitive air-displacement receptors between 40 and 1000 Hz, with best response between 100 and 600 Hz. Although far-field sound is only effective for stimulating them at rather excessive intensities - at best 90 - 100 dB re $2 \cdot 10^{-5}$ N/m^2! - receptor sensitivity to near-field vibration of naturally occurring sources - e.g. a wasp approaching on the wing to attack the caterpillar - is sufficient to detect and to react to the predator from over half a meter distance (Markl and Tautz, 1975; Tautz, 1977 a,b).

As mentioned, hair receptors in crustacean statocysts are used as gravity receptors. There are, however, still other ways, in which cuticular hairs can be transformed into effective sense organs for linear acceleration by gravity. In fact, it is in the arthropods, that we find the richest diversity and ingenuity in the development of graviceptive devices.

Many land insects monitor the position of their body joints by means of proprioceptive fields of hair mechanoreceptors: hair plates. As was most clearly demonstrated in different hymenopterans (bees, ants, wasps) these same proprioceptors function as gravity receptors also, since they indicate precisely not only the position of body parts due to the muscular action of the animal itself, but also any deflections caused by the action of gravity on these mobile body appendages. For example, in an ant that runs at an angle to the lines of gravity on a vertical surface, the gaster, the head and other appendages are deflected downward by gravitational pull. This deviation from the normal position of body appendages is counteracted by a feedback loop in which the proprioceptive hairplates act as sensing devices, the residual deviation being proportional to the sine of the deviation of the animal's long axis from straight upward. Since the same receptors are used to control active motion of body parts and deflections imposed by the influence of gravity - and by the way also by angular acceleration during rotations - the DC-effect of gravity can only be inferred by the animal by comparison of the positional information from several - in ants between 24 and 26 - hair plates, to which, as has been demonstrated in other insects, input from other proprioceptors (e.g. scolopidial organs, campaniform sensilla etc.) is coming. Since gravity affects all of them in the same direction and intensity in contrast to positional changes imposed by other forces, a common-mode-acceptance mechanism in the central nervous processing of proprioceptive positional information, which might be realized in this case in the form of a low-pass filter with a high-cutoff slope in the 0.1-1 Hz-range, allows to extract information about the direction of gravity from proprioceptive input (Lindauer and Nedel, 1959; Markl, 1962, 1964, 1966 a,b, 1971; Bässler, 1965; Wendler, 1964, 1965, 1971, 1972, 1975; Horn, 1970, 1973, 1975 a,b; Jander etal., 1970; Horn and Kessler, 1975).

In crickets, club-shaped hairs on the cerci are deflected by gravity in dependance on the animal's position in space; the response

of the single sensory cell innervating each hair could therefore
reliably indicate to the animal its position with respect to the
vertical, although convincing behavioral evidence is still lacking
that crickets make use of this source of positional information
(Bischof, 1974, 1975).

There is still another way in which hair-receptors can be
involved in gravity reception in insects, again in connection with
a possible function as very low frequency vibration/sound receptors.
Bonke (1975 a,b) studied the structure, the physiological
characteristics and the behavioral function of umbrella-shaped hair
receptors which monitor the position of the air-water interface at
several pairs of transformed abdominal stigmata of the aquatic bug
Nepa. The anatomical arrangement makes these receptors function
as transducer elements in a two-dimensional differential manometer
system which enables the insects to detect the direction of gravity
from the direction of the buoyancy of the air-bubble contained in
their tracheal system, as was first described by Baunacke (1912).
The umbrellar membranes overlap extensively between the approximately
100 hairs situated at each of the six air-water contact zones, the
composite border membrane moving in and out with that border and
thereby pushing in or pulling out the 100 umbrella-sticks of ca.
20 µm length. Different from the typical hair mechanoreceptor in
anthropods, these umbrella-hairs do not respond to bending but to
motion in the direction of their hair-shaft, which relates to a
peculiar $90°$ - bend in the terminal part of the dendrite of the
innervating sensory cell. Although these insects have also tympanal
organs for the reception of sound (Arntz, 1975) as well as trichobothria for the reception of surface waves or near-field water
vibrations caused by prey (Arntz, 1971), the buoyancy-measuring
hair-receptor system would also be able to function efficiently as
a very low frequency vibration or sound detector with an optimum in
the range of a few tens of Hertz.

Summing up this section, it is evident that just as in the
ciliated mechanoreceptors of aquatic non-arthropods the cuticular
hair mechanoreceptor of the arthropods is an extremely versatile
mechanosensory transducer notwithstanding its nearly unvarying basic
construction plan, because the different accessory mechanical
structures and the information processing mechanisms of the central
nervous system, into which the primary informations are fed, adapt
these receptors to a wide variety of mechanoreceptor functions.

3. Mechanoreceptive hair-cell organs in Vertebrates

A third major line of adaptive radiation of mechanosensitive
exteroreceptors is a large number of sense organs in vertebrates
that use another ciliated epidermal cell as their basic receptive
transducer element: the *hair-cell*. Although there are many striking

similarities in the structure of this receptor cell with the cilia-bearing mechanoreceptors of invertebrates (see p. 325) - most prominently the involvement of non-motile kinocilia and of stereocilia in stimulus transmission and possibly also stimulus transduction, which may even be a homologous feature derived from the sensory equipment of a common ancestor, e.g. of coelenterate constitution - there is reason to consider the adaptive differentiation of hair-cell receptors in vertebrates separate from that of the ciliated receptors in invertebrates, since vertebrate hair-cells are always secondary sensory cells lacking an axon of their own, in contrast to what is true for nearly all invertebrate mechanoreceptors (for an exception of this rule, almost certainly arrived at convergently in cephalopods, see p. 326).

Typically the vertebrate hair-cell is dually innervated by both afferent and efferent fibers of the next-order neurons. There can be little doubt that all mechanoreceptive hair-cell organs in vertebrates can be derived from the same ancestral stock: the "neuromast" organ on the body surface of fishes which is sensitive to the relative displacement between its base - the body of the fish - and the surrounding fluid medium. Whereas the response of these hair-cells has in all cases studied been always found to depend on the degree of bending displacement of the hairs by the shearing component of applied mechanical force, the response of the whole sensory organ - i.e. of sense cells plus accessory structures coupling their hairs to the surrounding medium - can not only be dependent on displacement, but also on its change in time - velocity - or the change in time of the latter: acceleration.

Structure and function of the vertebrate hair-cell has been extensively reviewed by Flock (1971), Lowenstein (1974) and Vinnikov (1974), among others. For a most recent experimental study, confirming and extending the general picture of hair-cell function see Hudspeth and Corey (1977). Although some details vary, the basic plan of construction and function is the following.

The epithelial cell is equipped with a bundle of cilia at its outer surface that receive the mechanical deformation from the surrounding fluid through a gelatinous cupula or membraneous cover, into which they are embedded. The cupula projects into the medium motion to hair displacement. In some organs, the cover is loaded by calcareous concretions making the system more sensitive to acceleration. The cilia are usually of two types: a single 9x2+2 type kinocilium is placed at one side of a group of stereocilia, which are the longer the closer they stand to the kinocilium. (The kinocilium is lacking in the organ of Corti of the adult mammal.)

Response of the hair-cell to a shearing force is directionally polarized: displacement of the ciliar bundle in a direction in which

the kinocilium is leading is excitatory: it depolarizes the
hair-cell and results in an increased firing rate of the afferent
fiber, whereas displacement in the opposite direction, from
kinocilium to the shortening stereocilia, hyperpolarizes the cell
and inhibits afferent impulse discharge driving it below the steady
resting discharge level characteristic for these hair-cells. Other
types of mechanical stimuli, e.g. hydrostatic pressure are ineffective on the hair-cell, unless they are transformed by accessory
mechanisms into a form of action which is able to bend the hairs.

Depending on whether all hair-cells in an organ are oriented
morphologically in the same direction or in different ones, the
response directionality *of the organ* will conform to that of the
single sensory cells or differ according to the distribution of the
polarization pattern of the hair-cells. If, as it is the case in
the crista ampullaris of the semi-circular canals, polarization
direction is the same in all cells of an organ, their summated
response (microphonic potential) to sinusoidal cupular displacement
mirrors the stimulus frequency; if, as in the lateral line organ of
fishes, adjacent hair-cells are oriented with their kinocilia facing
into opposite directions, the compound response is the distortion
product between the response potentials of the two sub-populations
of cells, transmitting twice the frequency of the stimulus.

This hair cell unit has become incorporated during vertebrate
evolution into a large number of different mechanoreceptive sense
organs, in adaptation to different mechanoreceptive qualities,
leading again to a pattern of adaptive functional differentiation
of surprising similarity to that described already for the two lines
of development in the invertebrates. It would clearly go beyond
the limits of this chapter to review structure and function of all
these organs in detail, considering the immense body of knowledge
available about them and considering also that excellent treatments
of recent years are at hand, especially in the Handbook of Sensory
Physiology (Springer Verlag: Vol I/1971; III/3/1974; V/1-3/1974-1976;
VI/1-2/1974; see also van Bergeijk, 1967; Cahn, 1967; Schuijf and
Hawkins, 1976). In the present context it seems sufficient to draw
the attention to a few salient features in the major branches of
diversification.

The most primitive hair-cell sense organ is a straightforward
medium displacement detector on the body surface of elasmobranch and
teleost fishes and of aquatic amphibians, called *lateral-line organ*
because of its characteristic distribution alongside their bodies.
In them a number of hair-cells together with their supporting cells
are provided with a common elastic cupula that projects into the
surrounding water. Displacement of this cupula by relative motion
between body and medium stimulates the hair-cells. The mechanical
characteristics of the coupling between water and cupula and the

relation of cupula geometry to hair-cell polarity(ies) determines absolute sensitivity, frequency response, directionality of response and its dependence on water displacement or displacement velocity.

The more highly developed *lateral-line-canal systems* of fishes gain additional directional response precision and guard the cupulae from gross stimulation in running water and during active swim by sinking the neuromast organs into mucus-filled canals below skin level, which communicate with the outside water through narrow pores only. Protection against overstimulation during active locomotion can be quite generally assumed to be one important function of efferent, inhibitory innervation of hair-cells in neuromast organs (Russell, 1971; Russell and Roberts, 1972; Roberts and Russell, 1972). Similar protective feed-forward control has also been assumed for semi-circular canal organs (Klinke and Schmidt, 1970) and fish sound receptors (Piddington 1971 a,b) although other functions of efferent innervation have also to be considered in more advanced sense organs of the acustico-lateralis system (Klinke and Galley, 1974).

Lateral line canal organs respond in a directional manner - depending on canal orientation in space and sense organ polarization - to local water currents, as e.g. caused by approach between the animal and an object in water ("Ferntastsinn": Dijkgraaf, 1934, 1963), and to medium vibrations in the near-field of sources oscillating with low frequency (up to maximally a few hundred Hertz) (Harris and van Bergeijk, 1962; van Bergeijk, 1964) as well as to waves on the water surface (Schwartz, 1965, 1971) thus making it possible for the canal system - not necessarily for the single canal organ - to detect and localize not too distant sources of hydrodynamic motions in the surrounding medium. Since oscillatory medium motion is, of course, connected also with far-field pressure waves in an elastic medium, it is on principle also possible that lateral line organs are able to respond to sound of sufficient intensity and adequate frequency, namely if particle displacement amplitude in the sound wave exceeds displacement threshold of the neuromast organs. For a recent discussion of the question, whether this capacity should be called "hearing" - alive ever since Dijkgraaf (1934) tried to decide it to the negative - see van Bergeijk (1967) Schwartz (1974) and contributions to Cahn (1967) and to Schuijf and Hawkins (1976).

The *semicircular canals* (1 with 2 cristae in *Myxine*, 2 in *Petromyzon*, 3 in the other vertebrates) are generally seen as derived from the lateral-line-canal system as a deeply invaginated and closed-off portion of it, specialized as an inertial navigation system for the detection of angular acceleration during rotatory motion of the animal, the three canals being oriented perpendicularly to each other in the three planes of space (for a lucid discussion of the presumeable evolution of this system, see van Bergeijk, 1967). Hair-cells are concentrated on cristae in dilated portions of the canals (ampullae), with the cupulae connected to the cristal hairs acting as torsion pendula when the endolymph filling of the canal lags behing or overshoots the motion of the canal wall during accelerated or decelerated rotation. Due to physical constraints imposed primarily by canal size, diameter and endolymph viscosity, the semicircular canal system converts angular acceleration to angular velocity that is indicated by the sensory response (Melvill Jones, 1974), which by the way holds also true for the crab canal system (Sandeman, 1976).

The invaginated semicircular canal system is always connected to a second group of hair-cell mechanoreceptor organs: the otolith (statolith) organs. Typically there are three of them with sensory maculae oriented in at least two perpendicular planes of space: the utriculus, the sacculus and the lagena (the latter disappearing in mammals, with the exception of the monotremes). The hair-cells in these organs are coupled to a gelatinous membrane into which one or more aragonite or calcite concretions - the statoliths/otoliths - are embedded, making these organs differential - density gravity sensors and giving them at the same time properties of a vibration receptor which is more or less sharply tuned by the mechanical resonance of the mass-loaded elastic hair-membrane system but highly damped by the viscosity of the surrounding endolymph. In vibration reception frictional fluid drag due to endolymphatic displacement - e.g. in the near-field of the vibrating swim-bladder or even in the far-field of a sound source of sufficient intensity to produce suprathreshold medium displacements - is the main force involved in stimulus transmission to these organs. These properties are the starting point for the development of acute hearing in some fishes, which is, however, beyond the scope of this article, as is consideration of the evolution of the sound-receptive section of the labyrinth in the tetrapods.

As was proven by the elegant centrifuge experiments of von Holst (1950) and the classical electrophysiological studies on these system by Lowenstein and coworkers in fish (see review: Lowenstein, 1974), the sensory epithelia of the otolith organs are sensitive only to the shearing component of gravitational force - endowing the horizontally situated utriculus organ of normally swimming fish with a sinusoidal response characteristic to angular deviation from the resting position - and are also characterized - as is true for hair-cells in the lateralis system and other parts of the labyrinth - by a continuous resting discharge, allowing to modulate response both upward or downward on stimulation in opposite directions. The distribution of hair-cell polarization axes in a sensory macula determines the directionality characteristic of a given macula both to linear acceleration and to vibrational displacement (see reviews by

Lowenstein, 1974; Wersäll and Bagger-Sjöbäck, 1974). While hair-cell maps of different organs in different vertebrate groups differ in characteristic ways, as a general rule sense-cell polarities of all organs taken together cover almost all directions in space, giving the animal excellent possibilities of positional control - and to a degree of localization of sources of vibrations - in three-dimensional space as seems useful for a fish swimming freely in the open water.

This cursory survey of mechanoreceptor organs in vertebrates that make use of the hair-cell as the elementary transducer unit makes it plain again how wide the range of mechanoreceptive functions is that can be covered by one and the same displacement sensitive device by connecting it to particularly adjusted accessory and auxiliary structures that transform inpinging mechanical energy into a quality that is effectively stimulating the transducer mechanism of the sensory cells. It is also evident that strikingly parallel, convergent evolutionary developments can be traced in the vertebrates to what is found in the arthropod and the non-arthropod invertebrates, perfecting again and again the capacity of the different organisms to produce a reliable high-resolution image of the world of mechanical stimuli of different sorts surrounding them.

CONCLUDING REMARKS

Mechanoreceptor organs were treated in this review as systems adaptively radiating in order to serve a multitude of different specific stimulus filtering functions. This view is, however, too restricted, if one wants to fully comprehend the functions of these receptors for an organism. It has been well known for a long time that in many organisms the incapacitation of major sensory organs - especially of those for the perception of light, gravity or sound - has a profound damping influence of the general state of activity and excitability of the animals, making them sluggish and unresponsive, in addition to depriving them of the ability to act to the specific stimuli no longer perceptible to them. It is therefore useful to keep in mind that in addition to supplying the organism with specific information about the environment, receptors serve another important function: to constantly supply driving input to the central nervous system, keeping it at a sufficiently "excited" state for fast action and ready response to new stimulation calling for quick behavioral reactions. Gravity receptors, as well as angular acceleration receptors and medium-flow and -vibration receptors in species which swim constantly throughout their active lives, are particularly apt to provide this "tonic" stimulating input, since they are under continuous stimulation. It may well be that the question, which is often raised about the functionality and adaptedness of very primitve precursor states of receptor organs that seem only fully "useful" for an animal in their completely developed and complex final forms finds a satisfying answer by taking into consideration this tonicizing function of receptors which they can well serve even in their most primordial states of development.

REFERENCES

Arntz, B. (1971). Sinnesphysiologische Untersuchungen beim Beutefangverhalten von *Nepa cinerea* Linné. Diplomarbeit, Math. Naturwiss. Fak. Univ. Bonn.

Arntz, B. (1975). Das Hörvermögen von *Nepa cinerea* L. Zur Funktionsweise der thorakalen Skolopidialorgane. J. Comp. Physiol. 96, 53-72.

Bässler, U. (1965). Proprioreceptoren am Subcoxal und Femur-Tibia-Gelenk der Stabheuschrecke *Carausius morosus* und ihre Rolle bei der Wahrnehmung der Schwerkraftrichtung. Kybernetik 2, 168-193.

Baunacke, W. (1912). Statische Sinnesorgane bei Nepiden. Zool. Jahrb. Anat. 34, 179-342.

Bischof, H.-J. (1974). Verteilung und Bewegungsweise der keulenförmigen Sensillen von *Gryllus bimaculatus* Deg. Biol. Zentralblatt 93, 449-457.

Bischof, H.-J. (1975). Die keulenförmigen Sensillen auf den Cerci der Grille *Gryllus bimaculatus* als Schwererezeptoren. J. Comp. Physiol. 98, 277-288.

Bonke, D. (1975a). Der Bau und die Antwortcharakteristik des Schirmrezeptors aus dem Statoorgansystem von *Nepa cinera* L. (Hemiptera, Rhynchota). Verh. Dtsch. Zool. Ges. 1974, p. 42-45.

Bonke, D. (1975b). Feinstruktur und Funktionsweise der statischen Sinnesorgane von Nepiden (Hemiptera). Diss. Techn. Hochschule Darmstadt.

Brownell, Ph.H. (1977). Compressional and surface waves in sand: used by desert scorpions to locate prey. Science 197, 479-482.

Budelmann, B.-U. (1970). Die Arbeitsweise der Statolithenorgane von *Octopus vulgaris*. Z. Vergl. Physiol. 70, 278-312.

Budelmann, B.-U. (1975). Gravity receptor function in cephalopods with particular reference to *Sepia officinalis*. Fortschritte Zool. 23, 84-96.

Budelmann, B.-U. (1976). Equilibrium receptor systems in molluscs. In: Structure and Function of Proprioceptors in the Invertebrates. P.J. Mill (Ed.), Chapman & Hall, London, p. 529-566.

Budelmann, B.-U. (1977). Structure and function of the angular acceleration receptor systems in the statocysts of cephalopods. Symp. Zool. Soc. London 38, 309-324.

Budelmann, B.-U. and Thies, G. (1977). Secondary sensory cells in the gravity receptor system of the statocyst of *Octopus vulgaris*. Cell. Tiss. Res. 182, 93-98.

Budelmann, B.-U. and Wolff, H.G. (1973). Gravity response from angular acceleration receptors in *Octopus vulgaris*. J. Comp. Physiol. 85, 283-290.

Burkhardt, D. (1960). Die Eigenschaften und Funktionstypen der Sinnesorgane. Ergebnisse Biol. 22, 226-267.

Cahn, Ph.H. (Ed.) (1967). Lateral line detectors. Indiana University Press, Bloomington.

Cohen, M.J. and Dijkgraaf, S. (1961). Mechanoreception. In: T.H. Waterman (Ed.). The Physiology of Crustacea. Vol. 2, 65-108. Academic Press, New York.

Dijkgraaf, S. (1934). Untersuchungen über die Funktion der Seitenorgane an Fischen. Z. Vergl. Physiol. 20, 162-214.

Dijkgraaf, S. (1963). The functioning and significance of the lateral line organs. Biol. Rev. 38, 51-105.

Daršlar, K. (1973). Functional properties of trichobothria in the bug *Pyrrhocoris apterus* (L.). J. Comp. Physiol. 84, 175-184.

Eckert, R. (1972). Bioelectric control of ciliary activity. Science 176, 473-481.

Edwards, J.S. and Palka, J. (1974). The cerci and abdominal giant fibers of the house cricket, *Acheta domesticus*. I. Anatomy and physiology of normal adults. Proc. R. Soc. London B, 185, 83-103.

Flock, Å. (1965). Electron microscopic and electrophysiological studies on the lateral line canal organ. Acta oto-laryng. (Stockh.) Suppl. 199, 1-90.

Flock, Å. (1971). Sensory transduction in hair-cells. In: W.R. Loewenstein (Ed.): Handbook of Sensory Physiology I, p. 396-441, Springer Verlag, Berlin.

Fraser, P.J. (1974). Interneurons in crab connectives (*Carcinus maenas* (L.)): directional statocyst fibers. J. Exp. Biol. 61, 615-628.

Fraser, P.J. and Sandeman, D.C. (1975). Effects of angular and linear accelerations on semicircular canal interneurons of the crab *Scylla serrata*. J. Comp. Physiol. 96, 205-221.

Gaffal, K.P., Tichy, H., Theiss, J. and Seelinger, G. (1975). Structural polarities in mechano-sensitive sensilla and their influence on stimulus transmission (Arthropoda). Zoomorphologie 82, 79-103.

Gnatzy, W. und Schmidt, K. (1971). Die Feinstruktur der Sinneshaare auf den Cerci von *Gryllus bimaculatus* Deg. (Saltatoria, Gryllidae). I. Faden- und Keulenhaare. Zeitschr. Zellforsch. 122, 190-209.

Görner, P. (1966). A proposed transducing mechanism for a multiply-innervated mechanoreceptor (trichobothrium) in spiders. Cold Spring Harbor Symp. Quant. Biol. 30, 69-73.

Görner, P. und Andrews, P. (1969). Trichobothrien, ein Ferntastsinnesorgan bei Webspinnen (Araneen). Z. Vergl. Physiol. 64, 301-317.

Gordon, S.A. and Cohen, M.J. (Eds.) (1971). Gravity and the Organism. University of Chicago Press, Chicago.

Harris, G.G. (1964). Considerations on the physics of sound production by fishes. In: W.N. Tavolga (Ed.): Marine Bio-Acoustics p. 233-247. Pergamon Press, Oxford.

Harris, G.G. and van Bergeijk, W.A. (1962). Evidence that the lateral-line organ responds to near-field displacements of sound sources in water. J. Acoust. Soc. Amer. 34, 1831-1841.

Haskell, P.T. (1956). Hearing in certain Orthoptera. I. II. J. Exp. Biol. 33, 756-766; 767-776.

Horn, E. (1970). Die Schwerkraftreception bei der Geotaxis des laufenden Mehlkäfers (*Tenebrio molitor*). Z. Vergl. Physiol. 66, 343-354.

Horn, E. (1973). Die Verarbeitung des Schwerereizes bei der Geotaxis der höheren Bienen (Apidae). J. Comp. Physiol. 82, 379-406.

Horn, E. (1975). Mechanisms of gravity processing by leg and abdominal gravity receptors in bees. J. Insect. Physiol. 21, 673-679.

Horn, E. (1975). The contribution of different receptors to gravity orientation in insects. Fortschritte Zool. 23, 1-20.

Horn, E. and Kessler, W. (1975). The control of antennal lift movements and its importance on the gravity reception in the walking blowfly, *Calliphora erythrocephala*. J. Comp. Physiol. 97, 189-203.

Horridge, G.A. (1966). Some recently discovered underwater vibration receptors in invertebrates. In: H. Barnes (Ed.): Some Contemporary Studies in Marine Science, p. 395-405. Allen and Unwin, London.

Horridge, G.A. (1969). Statocysts of medusae and evolution of stereocilia. Tissue and Cell 1, 341-353.

Horridge, G.A. (1971). Primitive examples of gravity receptors and their evolution. In: S. Gordon and M.J. Cohen (Eds.): Gravity and the Organism. p. 203-221. University of Chicago Press, Chicago.

Horridge, G.A. and Boulton, P.S. (1967). Prey detection by Chaetognatha via a vibration sense. Proc. Roy. Soc. B. 168, 413-419.

Hudspeth, A.J. and Corey, D.P. (1977). Sensitivity, polarity, and conductance change in the response of vertebrate hair-cells to controlled mechanical stimuli. Proc. Natl. Acad. Sci. USA 74, 2407-2411.

Jander, R., Horn, E. and Hoffmann, M. (1970). Die Bedeutung von Gelenkrezeptoren in den Beinen für die Geotaxis der höheren Insekten (Pterygota). Z. Vergl. Physiol. 66, 326-342.

Katsuki, Y. and Suga, N. (1960). Neural mechanisms of hearing in insects. J. Exp. Biol. 37, 279-290.

Klinke, R. and Galley, N. (1974). Efferent innervation of vestibular and auditory receptors. Physiol. Reviews 54, 316-357.

Klinke, R. and Schmidt, C.L. (1970). Efferent influence of the vestibular organ during active movement of the body. Arch. Ges. Physiol. 318, 325-332.

Kolle-Kralik, U. und Ruff, P.W. (1967). Vibrotaxis von *Amoeba proteus* (Pallas) im Vergleich mit der Cicilienschlagfrequenz

der Beutetiere. Protistologica 3, 319-323.
Krisch, B. (1973). Uber das Apikalorgan (Statocyste) der Ctenophore *Pleurobrachia pileus*. Zeitschr. Zellforsch. 142, 241-262.
Lang, H. (1977). Mechanismen der Beuteerkennung und der intraspezifischen Kommunikation bei der räuberischen Wasserwanze *Notonecta glauca* L. und ihre Rolle bei der Aufrechterhaltung der Populationsstruktur. Diss. Universität Konstanz.
Laverack, M.S. (1962a). Responses of cuticular sense organs of the lobster *Homarus vulgaris* (Crustacea). I. Hair-peg organs as water-current receptors. Comp. Biochem. Physiol. 5, 319-325.
Laverack, M.S. (1962b). Responses of cuticular sense organs of the lobster *Homarus vulgaris* (Crustacea). II. Hair-fan organs as pressure receptors. Comp. Biochem. Physiol. 6, 137-145.
Lindauer, M. und Nedel, J.O. (1959). Ein Schweresinnesorgan der Honigbiene. Z. Vergl. Physiol. 42, 334-364.
Lowenstein, O.E. (1974). Comparative Morphology and Physiology. In: H.H. Kornhuber (Ed.): Handbook of Sensory Physiology Vol. VI/1 p. 75-120, Springer Verlag, Berlin.
Machemer, H. (1977). Motor activity and bioelectric control of cilia. Fortschritte Zool. 24, 195-210.
Machemer, H. and de Peyer, J. (1977). Swimming sensory cells: Electrical membrane parameters, receptor properties, and motor control in ciliated Protozoa. Verh. Dtsch. Zool. Ges. 1977, p. 86-110.
Markl, H. (1962). Borstenfelder an den Gelenken als Schweresinnesorgane bei Ameisen und anderen Hymenopteren. Z. Vergl. Physiol. 45, 475-569.
Markl, H. (1964). Geomenotaktische Fehlorientierung bei *Formica polyctena* Foerster. Z. Vergl. Physiol. 48, 552-586.
Markl, H. (1966a). Schwerkraftdressuren an Honigbienen. I. Die geomenotaktische Fehlorientierung. Z. Vergl. Physiol. 53, 328-352.
Markl, H. (1966b). Schwerkraftdressuren an Honigbienen. II. Die Rolle der schwererezeptorischen Borstenfelder verschiedener Gelenke für die Schwerekompassorientierung. Z. Vergl. Physiol. 53, 353-371.
Markl, H. (1969). Die Verständigung durch Stridulationssignale bei Ameisen. II. Erzeugung und Eigenschaften der Signale. Z. Vergl. Physiol. 60, 103-150.
Markl, H. (1971). Proprioceptive gravity perception in Hymenoptera. In: S. Gordon and M.J. Cohen (Eds.): Gravity and the Organism. p. 185-194. University of Chicago Press, Chicago.
Markl, H. (1973). Leistungen des Vibrationssinnes bei wirbellosen Tieren. Fortschritte Zool. 21, 100-120.
Markl, H. (1974). The perception of gravity and of angular acceleration in invertebrates. In: H.H. Kornhuber (Ed.): Handbook of Sensory Physiology, Vol. VI/1, p. 17-74, Springer Verlag, Berlin.

Markl, H. and Tautz, J. (1975). The sensitivity of hair receptors in caterpillars of *Barathra brassicae* L. (Lepidoptera, Noctuidae) to particle movement in a sound field. J. Comp. Physiol. 99, 79-87.

Mellon, D. (1963). Electrical responses from dually innervated tactile receptors on the thorax of the crayfish. J. Exp. Biol. 40, 137-148.

Melvill Jones, G. (1974). The functional significance of semi-circular canal size. In: H.H. Kornhuber (Ed.): Handbook of Sensory Physiology Vol. VI/1, p. 171-184, Springer Verlag, Berlin.

Minnich, D.E. (1925). The reactions of the larvae of *Vanessa antiopa* L. to sounds. J. Exp. Zool. 42, 443-469.

Minnich, D.E. (1936). The responses of caterpillars to sound. J. Exp. Zool. 72, 439-453.

Moran, D.T. and Carter Rowley III, J. (1975). High voltage and scanning electron microscopy of the site of stimulus reception of an insect mechanoreceptor. J. Ultrastruct. Res. 50, 38-46.

Murphey, R.K. and Mendenhall, B. (1973). Localization of receptors controlling orientation to prey by the back swimmer *Notonecta undulata*. J. Comp. Physiol. 84, 19-30.

Naitoh, Y. and Eckert, R. (1974). The control of ciliary activity in Protozoa. In: M.A. Sleigh (Ed.): Cilia and Flagella. p. 305-352. Acad. Press, London.

Offutt, G.C. (1970). Acoustic stimulus perception by the american lobster, *Homarus americanus* (Decapoda). Experientia 26, 1276-1278.

Petrovskaya, E.D., Rojkova, G.J. and Tokareva, V.S. (1970). Single cercal receptor characteristics in the cricket (*Gryllus domesticus*). (Russ. with Engl. Summary). Biofizika 15, 1112-1119.

Piddington, R.W. (1971a). Central control of auditory input in the goldfish. I. Effect of shocks to the midbrain. J. Exp. Biol. 55, 569-584.

Piddington, R.W. (1971b). Central control of auditory input in the goldfish. II. Evidence of action in the free-swimming animal. J. Exp. Biol. 55, 585-610.

Pumphrey, R.J. and Rawdon-Smith, A.F. (1936). Hearing in insects: the nature of the response of certain receptors to auditory stimuli. Proc. roy. Soc. B 121, 18-27.

Roberts, B.L. and Russell, I.J. (1972). The activity of lateral-line efferent neurones in stationary and swimming dogfish. J. Exp. Biol. 57, 435-448.

Russell, I.J. (1971). The role of efferent fibres in the lateral-line system of *Xenopus laevis*. J. Exp. Biol. 54, 621-641.

Russell, I.J. and Roberts, B.L. (1972). Inhibition of spontaneous lateral-line activity by efferent nerve stimulation. J. Exp. Biol. 57, 77-82.

Sandeman, D.C. (1975). Dynamic receptors in the statocysts of crabs. Fortschritte Zool. 23, 185-191.
Sandeman, D.C. (1976). Spatial equilibrium in the arthropods. In: P.J. Mill (Ed.): Structure and Function of Proprioceptors in the Invertebrates. p. 485-527. Chapman and Hall, London.
Sandeman, D.C. and Okajima A. (1972). Statocyst-induced eye movements in the crab *Scylla serrata*. I. The sensory input from the statocyst. J. Exp. Biol. 57, 187-204.
Sandeman, D.C. and Okajima, A. (1973a). Statocyst-induced eye movements in the crab *Scylla serrata*. II. The responses of the eye muscles. J. Exp. Biol. 58, 197-212.
Sandeman, D.C. and Okajima, A. (1973b). Statocyst-induced eye movements in the crab *Scylla serrata*. III. The anatomical projections of sensory and motor neurones and the responses of the motor neurones. J. Exp. Biol. 59, 17-38.
Schmidt, K. (1973). Vergleichende morphologische Untersuchungen an Mechanorezeptoren der Insekten. Verh. Dtsch. Zool. Ges. 1971, p. 214-219.
Schöne, H. (1959). Die Lageorientierung mit Statolithenorganen und Augen. Ergebn. Biol. 21, 161-209.
Schöne, H. (Ed.) (1975). Mechanisms of spatial perception and orientation as related to gravity. Fortschritte Zool. 23, 1-296, Fischer Verlag, Stuttgart.
Schuijf, A. and Hawkins, A.D. (Eds.) (1976). Sound reception in fish. Developments in Aquaculture and Fisheries Science, 5. Elsevier Sci. Publ. Co., Amsterdam.
Schwartz, E. (1965). Bau und Funktion der Seitenlinie des Streifenhechtlings *Aplocheilus lineatus*. Z. Vergl. Physiol. 50, 55-87.
Schwartz, E. (1971). Die Ortung von Wasserwellen durch Oberflächenfische. Z. Vergl. Physiol. 74, 64-80.
Schwartz, E. (1974). Lateral-line mechano-receptors in fishes and amphibians. In: A. Fessard (Ed.): Handbook of Sensory Physiology Vol. III/3, p. 257-278, Springer Verlag, Berlin.
Schwartzkopff, J. (1974). Mechanoreception. In: M. Rockstein (Ed.): The Physiology of Insecta, Vol. II, p. 273-352, Academic Press, New York.
Silvey, G.E., Dunn, P.A. and Sandeman, D.C. (1976). Integration between statocyst sensory neurons and oculomotor neurons in the crab *Scylla serrata*. II. The thread hair sensory receptors. J. Comp. Physiol. 108, 45-52.
Spinola, S.M. and Chapman, K.M. (1975). Proprioceptive indentation of the campaniform sensilla of cockroach legs. J. Comp. Physiol. 96, 257-272.
Tautz, J. (1977a). Mechanismen und biologische Bedeutung der Luftschallwahrnehmung bei Schmetterlingsraupen. Diss. Universität Konstanz, 188 p.
Tautz, J. (1977b). Reception of medium vibration by thoracal hairs of caterpillars of *Barathra brassicae* L. I. Mechanical properties

of the receptor hairs. J. Comp. Physiol. 118, 13-31.
Thurm, U. (1968). Steps in the transducer process of
 mechanoreceptors. Symp. Zool. Soc. London 23, 199-216,
 J.D. Carthy and G.E. Newell (Eds.): Invertebrate Receptors.
 Acad. Press, London.
Thurm, U. (1969). General organization of sensory receptors.
 Rend. Scuola Intern. Fisica "E. Fermi". XLIII Corso p. 44-68.
Thurm, U. (1974). Mechanisms of electrical membrane responses in
 sensory receptors, illustrated by mechanoreceptors. In:
 L. Jaenicke (Ed.): Biochemistry of Sensory Functions.
 p. 367-390, Springer Verlag, Berlin.
Thurm, U. (1977). Sensorische Transduktion - ein Steuerungsprozess.
 In: W. Hoppe, W. Lohmann, H. Markl, H. Ziegler (Eds.):
 Biophysik p. 391-402. Springer Verlag, Berlin.
van Bergeijk, W.A. (1964). Directional and non-directional hearing
 in fish. In: W.N. Tavolga (Ed.): Marine Bio-Acoustics.
 p. 281-299. Perganon Press, Oxford.
van Bergeijk, W.A. (1967). The evolution of vertebrate hearing.
 In: W.D. Neff (Ed.): Contributions to Sensory Physiology,
 Vol. 2, p. 1-49. Acad. Press, New York.
Vinnikov, Ya.A. (1974). Sensory Reception. Cytology, Molecular
 Mechanisms and Evolution. Springer Verlag, Berlin.
von Holst, E. (1950). Die Arbeitsweise des Statolithenapparates
 bei Fischen. Z. Vergl. Physiol. 32, 60-120.
Wells, M.J. (1960). Proprioception and visual discrimination of
 orientation in *Octopus*. J. Exp. Biol. 37, 489-499.
Wendler, G. (1964). Laufen und Stehen der Stabheuschrecke
 Carausius morosus: Sinnesborstenfelder in den Beingelenken
 als Glieder von Regelkreisen. Z. Vergl. Physiol. 48, 198-250.
Wendler, G. (1965). Über den Anteil der Antennen an der
 Schwererezeption der Stabheuschrecke *Carausius morosus* Br.
 Z. Vergl. Physiol. 51, 60-66.
Wendler, G. (1971). Gravity orientation in insects: the role of
 different mechanoreceptors. In: S. Gordon and M.J. Cohen:
 Gravity and the Organism. p. 195-199. University of Chicago
 Press, Chicago.
Wendler, G. (1972). Körperhaltung bei der Stabheuschrecke:
 ihre Beziehung zur Schwereorientierung und Mechanismen ihrer
 Regelung. Verh. Dtsch. Zool. Ges. 1971, p. 214-219.
Wendler, G. (1975). Physiology and systems analysis of gravity
 orientation in two insect species (*Carausius morosus, Calandra
 granaria*). Fortschritte Zool. 23, 33-48.
Wersäll, J. and Bagger-Sjöböck, D. (1974). Morphology of the
 vestibular sense organ. In: H.H. Kornhuber (Ed.): Handbook
 of Sensory Physiology Vol. VI/1, p. 123-170. Springer Verlag,
 Berlin.
Wiese, K. (1976). Mechanoreceptors for near-field water displacements in crayfish. J. Neurophysiol. 39, 816-833.

Wiese, K., Calabrese, R.L. and Kennedy, D. (1976). Integration of directional mechanosensory input by crayfish interneurons. J. Neurophysiol. 39, 834-843.

Wolff, H.G. (1973). Statische Orientierung bei Mollusken. Fortschritte Zool. 21, 80-99.

Wolff, H.G. (1975). Statocysts and geotactic behavior in gastropod molluscs. Fortschritte Zool. 23, 63-84.

SOUND RECEPTION IN DIFFERENT ENVIRONMENTS

AXEL MICHELSEN

Institute of Biology, Odense University

DK-5230 Odense M, Denmark

INTRODUCTION

The aim of this book is to discuss "sensory ecology", that is the adaptation of sense organs to the properties of the environments. For hearing organs, it is at present only possible to describe such a correlation for hearing in air and water. Some physical parameters are different in these two homogeneous media (see below), and the hearing organs are adapted to the medium in which they are being used (references to the literature on hearing in fish, seals and whales can be found in Schuijf and Hawkins, 1976; Møhl and Ronald, 1975; and Payne and Webb, 1971). The vast majority of hearing animals, however, live in terrestrial environments, and very little is known about the acoustical properties of these environments.

One aim of this chapter is to review our present knowledge of outdoor acoustics, especially the acoustical properties of terrestrial ecosystems. I have tried to point out a number of problems, which should be further investigated, and some of the pitfalls for such investigations. Sensory ecology is an area of research where the investigator is likely to drown himself in data, if he does not manage to keep in mind the order of magnitude of the various phenomena. On the other hand, all the important parameters must be measured simultaneously, if he wishes to analyse, for example, the acoustic properties of vegetation. So far, the majority of the field studies have been performed by acousticians (or biologists), who appear not to be very familiar with the complexity of the problem. In many cases the investigators have ignored important parameters in the environment. The result of such studies are, of course, less useful. (After this chapter was written, I had the opportunity to see the manuscript of a similar review by Wiley and Richards (1978), which is highly recommended. It contains much new material (e.g. on temporal fluctuations in sound pressure), as well as many biological examples, especially from birds. In the present paper the examples have been drawn mainly from insects).

Throughout the chapter a number of examples of acoustic communication in animals are given. I have tried to relate the strategy adapted by each group to known acoustic parameters of the environment. But this correlation is between the environments and the types of *acoustic signals* used by animals living in these environments. Our present knowledge does not allow us to relate the properties of the *hearing organs* to the acoustic properties of the environments. The reason is that we are in most cases ignorant about the nature of the information needed by the listening animal.

The kind of information, which may be of interest to the listening animal, is the identity of the sound producing animal (who is he?), its location (where is he?), and the nature of the message (what is he saying?). In some situations, all this information is important; for example, animals looking for a mate need most of the information mentioned. Other listening animals are only interested in a part of the information. The prey of bats, for example, do not care much about which kind of bat is coming near, but they are very interested to know, how far away, and in which direction the bat is.

This means that there may be more than one reason for the different complexity of the hearing organs in various animals. Animals may need complicated ears and an elaborate system of acoustic neurons in the central nervous system, because they need to extract as much information as possible from the sound signals. In contrast, if the animal happens to live in a very complex acoustical environment, a rather complicated auditory system may be needed to obtain just a small amount of information. Therefore, a meaningful correlation between environments and sensory capacities, is not possible unless the role of sound in the behaviour of the animal is known.

Unfortunately, there is no easy solution to this problem. Some animals have simple calls; others have very complicated calls. We need behavioural studies for determining how much signal redundancy is needed in complex environments, where much filtering takes place as the call propagates to the listening animals. We have very little information on this problem.

SOUND PROPAGATION IN COMPLEX ENVIRONMENTS

There are several reasons why so little is known about the acoustical conditions of relevance to animals in terrestrial environments. One is that most field studies have been performed by acousticians interested in the propagation of noise from traffic or other noisy human activities. The frequency range of most annoying noises is mainly below a few kHz. Most studies have therefore been limited to frequencies below 4 kHz, but biologists are often interested in animal communication at much higher frequencies. Furthermore, humans hearing noises (or rather humans annoyed by noises) normally keep their heads 1.5-2 m from the ground, and therefore most studies have been concerned with the sound propagating to a receiver situated at least 1 m from the ground. So, if one is interested to know, for example, how a grasshopper manages to locate another grasshopper in the vegetation, the majority of the field studies are not of much help.

Another reason for our ignorance is the lack of theoretical understanding of the process of sound propagation in natural environments. A number of theories can be found in the acoustic literature, but very few studies have combined theory and field measurements. Most of these problems have to be studied by biologists or by acousticians working with biologists. The relevant questions about the acoustics of the environment should be asked by scientists who are concerned with the sound parameters carrying the specific information of relevance to the animals.

Sound propagation in complex environments embraces a number of processes of very different natures. It is very important for the investigator to realize, which process is dominant in the particular situation (Lyon,1973). These processes can be divided into 5 groups: geometric spreading, reflection, absorption, refraction and diffraction (scattering).

Geometric spreading. The pressure of sound waves travelling from a sound source will decrease with distance. **This is a simple consequence of the geometry of the air space occupied by the sound energy. Some sound sources (such as pulsating spheres) radiate sound energy evenly in all directions. The geometric spreading from such a *monopole* source will cause a 6 dB (2 times) decrease in sound pressure per doubling of the distance from the origin of the sound wave (Fig. 1,A).**

Other sound sources do not behave in this simple manner. A loudspeaker without the backing enclosure is a *dipole* source. Close to the dipole (in the acoustic *near-field*), the sound pressure decreases 12 dB (four times) with distance doubled (abbreviated dd); far away (in the *far-field*) the decrease is

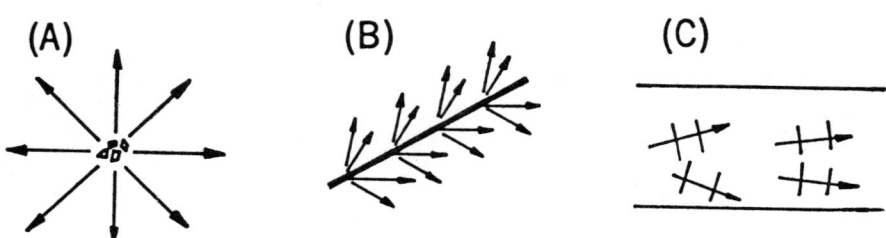

FIG. 1. Geometric divergence of sound waves and resulting attenuation. A. Spherical spreading, attenuation 6 dB/dd (dd = distance doubled). B. Cylindrical spreading, 3 dB/dd. C. Sound in channel, 0 dB/dd. (From Lyon, 1973).

6 dB/dd (as in the monopole source). Furthermore, unlike the sound radiation from a monopole source, the radiation from a dipole source is directional. Singing insects often behave as mixed monopole-dipole sources (Michelsen and Nocke, 1974)

Still other kinds of sound sources exist, which have a different behaviour (see Crocker and Price, 1975). In some bioacoustic problems (e.g. the spreading of low frequency sounds in the oceans, Payne and Webb, 1971) one is dealing with a cylindrical spreading of sound, where the decrease is only 3 dB/dd. Finally, if the sound waves are being reflected without loss in a channel, the attenuation may approach 0 dB/dd (Fig. 1(C)).

It is important to remember that the decrease for other parameters of the sound wave (the oscillation velocity of the molecules, the pressure gradient) is more rapid close to a sound source than is the decrease in sound pressure. This is important for animals with receptors sensitive to such parameters, for example in the near-field communication in fruit flies (Bennet-Clark, 1971).

Reflection. Sound waves are redirected at the surface of another medium if the surface is several wavelengths long and if the acoustic impedance of the two media are different. (The *acoustic impedance* is the density of the medium multiplied by the velocity of sound in the medium). Important examples in a biological context are the reflection from an air-water or water-air interface (where less than 1/1000 of the sound energy is transmitted to the other medium) and reflection from the ground. The magnitude and phase angle of the reflected wave depends upon the ratio of the impedances, Z_1 and Z_2 in Fig. 2, and upon the "grazing" angle ψ (Fig. 2).

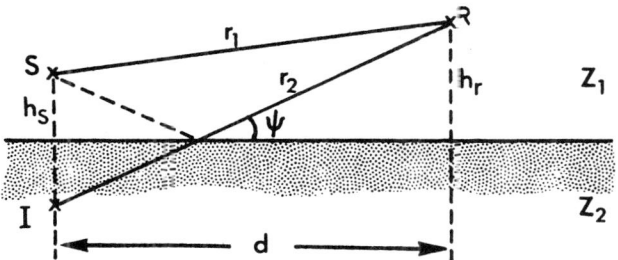

FIG. 2. Reflection from the ground. Location of source (S), and receiver (R) at heights h_s and h_r above flat ground. Z_1 and Z_2 are the impedances of the air and of the ground surface, respectively. The reflected sound wave, which at point R appears to have come a distance r_2 from the sound source I, is interfering with the direct wave (r_1). (From Embleton et al., 1976).

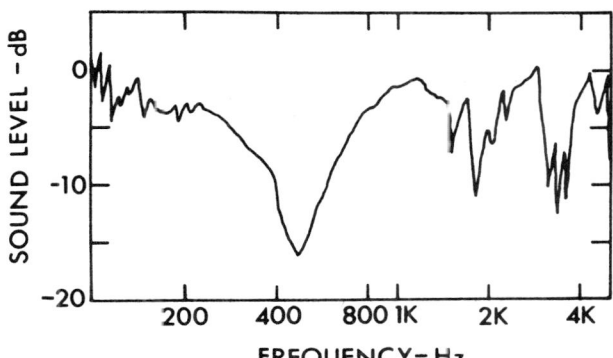

FIG. 3. Reflection from the ground. Frequency filtering of pure tones propagating about 15 m and about 1.2 m above grass. Reference sound level is that due to geometric spreading (6 dB/dd) alone. (From Embleton et al., 1976).

The magnitude of the ground reflection (and ground absorption) is one of the very uncertain parameters, when guesses about the range and filtering of an animal call are made. Two mechanisms are important in causing a filtering of the frequency content of the call: the ground may interfere with the propagation of a sound wave close to the ground (see below), and the reflected wave, r_2

in Fig. 2, may interfere with the direct wave, r_1 in Fig. 2. The latter mechanism may lead to an enhancement (r_1 and r_2 in phase at the receiver), or a destruction (r_1 and r_2 out of phase). The frequency filtering caused by this interference depends very much upon the exact positions of the sender and receiver in space. The magnitude of the filterings may be substantial, when the source and/or receiver are a few metres from the ground. Fig. 3 shows an example of this.

A substantial part of the energy of sound signals in the sea is reflected from the bottom. Also, as previously noted, sound waves are almost entirely reflected by the water-air interface. These reflections, and the small absorption of sound in water (see below), are probably the main reasons why some whales use very short ultrasonic clicks and not longer sound signals for echolocation. But even with short pulses, the reflection from the surface and bottom may cause trouble for the orientation of echolocating animals. This is especially evident in whales, which happen to enter narrow fiords. (By the way, deep and narrow fiords are also ideal hiding places for submarines, which may be impossible to locate by means of sonar or directional listening devices).

Absorption. The absorption of a sound wave includes the dissipation of sound energy to heat, and the transmission of sound into other media. It is, however, often difficult to distinguish between absorption and other causes of attenuation. Good absorbers of sound energy (in air) are porous materials like mineral wool. The air itself also absorbs sound energy. The effect is commonly expressed as a change in sound pressure level over a fixed distance (dB/m).

One can use Fig. 4 for evaluating the effect of absorption on the total decrease in sound pressure from the sound source to the receiver. The sound source is assumed to have an initial level of 0 dB at a distance of 0.1 m from the source, and the sound waves are assumed to spread spherically (attenuation due to geometric spreading: 6 dB/dd). In using the figure, one should subtract the threshold of the receiver from the sound pressure produced by the sound source. If this value is 80 dB, and the attenuation in the path of propagation is 1 dB/m, one finds that the maximum distance, at which the listening animal can hear the sound source, is about 30 m.

Note that *"threshold"* in this context means the smallest sound pressure, which can be heard (and/or "understood") by the listening animal. So, one must allow for the background noise level. For example, the "threshold" may be dependent upon the wind. When the vegetation rustles in the wind, a broadband noise is produced. The sound level of this rustling sound has been measured to about

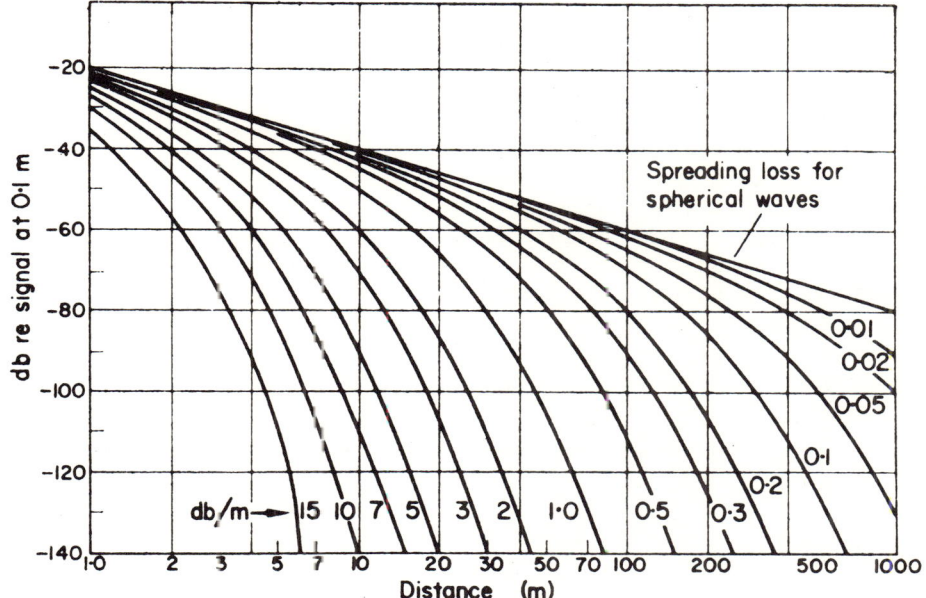

FIG. 4. The relative intensity of a sound wave at various distances from the source. Geometric spreading is assumed to be spherical (6 dB/dd). The sound intensity is arbitrarily assigned an intensity of 0 dB at a distance of 10 cm from the source. Each curve represents a given value of atmospheric attenuation (dB/m); the upper line showing the spreading loss with no atmospheric attenuation added. (From Griffin, 1971).

35 dB(A) at 1 m/s wind speed and to about 60-70 dB(A) at 8 m/s wind speed (Yamada et al., 1977). The presence of such noise is likely to have a masking effect on some animal calls and thus acts to increase the "threshold".

Dissipation in air is caused by several processes: shear viscosity, heat conduction in the air, mass diffusion, thermal diffusion and molecular vibrational relaxation. The attenuation of sound caused by these processes is a fairly complicated function of temperature, relative humidity and sound frequency. The literature and theories have been reviewed recently by Piercy et al. (1977). Rather different estimates of the dissipation can be found in various textbooks, since the contributions from molecular vibration relaxation were not fully understood in some classical papers. The upper curve in Fig. 5 indicates the dissipation in dB/100 m for a pressure of 1 atm, a temperature of 20°C and a relative humidity of 70%.

The effect of *humidity* is not well understood at the molecular level. A fair amount of experimental data is available at various humidities at 20°C, but more data are needed at other temperatures and especially at low humidities. In dry air the dissipation attains a maximum at low frequencies. The position of this maximum is shifted towards higher frequencies when the humidity is increased. (see Piercy et al., 1977).

From the data in Figs. 4 and 5, it is obvious that the use of high frequencies impose a considerable limitation on the range of

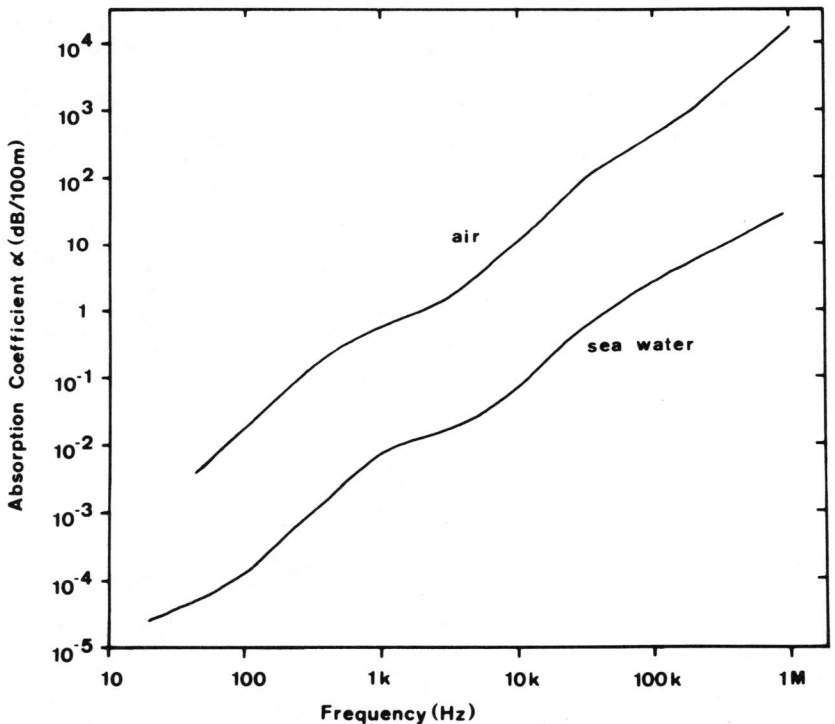

FIG. 5. Dissipation of sound in air (upper curve) and sea water (lower curve) at a pressure of 1 atm. The dissipation in air is for 20°C and a relative humidity of 70% and is redrawn from Piercy et al. (1977). The dissipation in sea water is for 4°C and a salinity of 3.5%. The curve is based on data from Fisher and Simmons (1977) and Payne and Webb (1971).

communication in air. For example, at 100 kHz and at 70% relative humidity the atmospheric attenuation is about 4 dB/m. From Fig. 4, and assuming an 80 dB difference between source pressure and receiver threshold, one finds a maximum distance for communication of about 10 metres. In ecological terms this means a high population density.

I shall not go into the mysteries of bat echolocation in tropical forests where the air is very hot and humid (see Suthers, this volume). The attenuation is enormous under such conditions, and the range of echolocation may be as small as one or two metres (Griffin, 1971). In our part of the world similar unfavourable conditions occur only in foggy weather, and our bats seem to avoid flying in fog (Pye, 1971).

The *dissipation of sound energy in water* (lower curve in Fig. 5) is much less than that in air. At very low frequencies, the absorption is so small that communication over hundreds of kilometres becomes (at least a theoretical) possibility (Payne and Webb, 1971). The absorption in water also increases with increasing frequency, but dissipation never becomes a significant factor within the frequency range used by animals. On the contrary, the absorption is so small that echolocation may be very difficult.

Refraction. The deflection of sound waves known as refraction is caused by an abrupt or gradual change in the velocity of the sound wave. The velocity of sound depends upon the temperature and the density of the medium. The velocity of sound relative to an observer also depends upon the velocity of the medium, the wind. (Water currents are too slow to have any significant effect).

The *velocity of sound in air* depends mainly upon temperature. The velocity is about $(331.4 + 0.607C)$ m/sec, where C is the temperature in degrees Celsius. An increase in temperature of one degree thus causes an increase in sound velocity of about 0.2%. The sound velocity also increases with increasing humidity, except in very dry air (Harris, 1971). At 20°C the sound velocity at 100% relative humidity (R.H.) is about 0.3% larger than at 30% R.H.

Close to the ground or to the surface of objects one often finds *gradients of temperature*, wind and humidity. These parameters vary not only with height above ground, but they may change considerably when the terrain or vegetation change along the path of sound propagation. On sunny days the temperature decreases with distance from the ground (Fig. 6,a) as does the velocity of sound. The sound waves will therefore bend upwards, leaving a *shadow zone* beyond a certain distance from the source (Fig. 7,A). The border of this shadow zone is not sharp. No direct sound can penetrate into the shadow zone, though sound may be scattered into the zone.

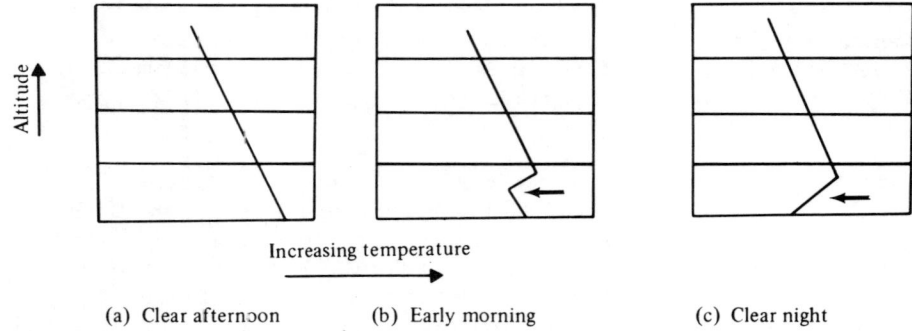

Increasing temperature

(a) Clear afternoon   (b) Early morning   (c) Clear night

FIG. 6. Typical examples of temperature variation with altitude. Arrows indicate that "channeling" of sound is possible. (From Crocker and Price, 1975). (Changed).

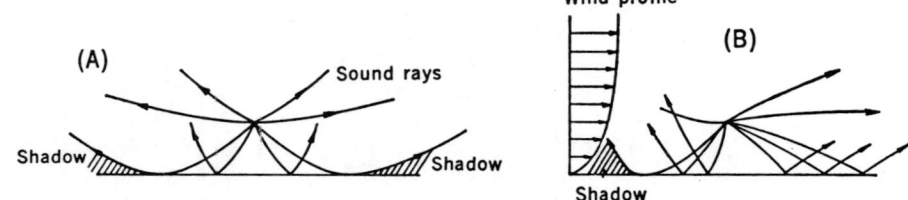

FIG. 7. Refraction of sound by wind and temperature gradients. A. Temperature lapse, no wind. B. Temperature lapse, with wind. (From Lyon, 1973).

If a *wind* is blowing, the wind speed will increase with the distance from the ground, and its effect on sound velocity will add to that caused by temperature. In the upwind direction, the refraction due to temperature and wind cooperate to create an even more pronounced shadow zone (Fig. 7,B). On sunny and windy days the shadow effect for sound propagating upwind may amount to 25 dB (Wiener and Keast, 1959). In contrast, in the downwind direction the effects of temperature and wind will cancel each other, and the result will depend upon the relative strengths of the two parameters.

The situation at other times of the day may be quite different. On a clear night the temperature often increases with distance from the ground, reaching a maximum at a certain altitude (Fig. 6,c). Under these circumstances the sound waves are "caught" in a channel close to the ground, and sounds may be heard at much larger distances from the sound source than during the day. In the early morning, the development of temperature gradients with distance from the ground may be even more complex. Channelling of sound with only little loss is now possible at some distance from the ground (marked with an arrow in Fig. 6,b).

In *water*, the velocity of sound increases with increasing temperature, hydrostatic pressure and salinity. The variation in sound velocity with depth below the surface depends upon the local conditions. In tropical and temperate oceans, the decrease in temperature with increasing depth causes the sound velocity to decrease to a minimum about one to two km from the surface. At larger depths, the velocity again increases due to the hydrostatic pressure. The minimum of sound velocity at a certain depth creates the so-called SOFAR-channel, in which sound communication is particularly favourable (Williams, 1972). In arctic oceans the temperature does not change so much with depth, and the velocity of sound is determined mainly by the hydrostatic pressure (Fig. 8).

Fluctuations of sound pressure. In the previous sections we have seen that geometric spreading, molecular dissipation and refraction caused by wind and temperature gradients all contribute

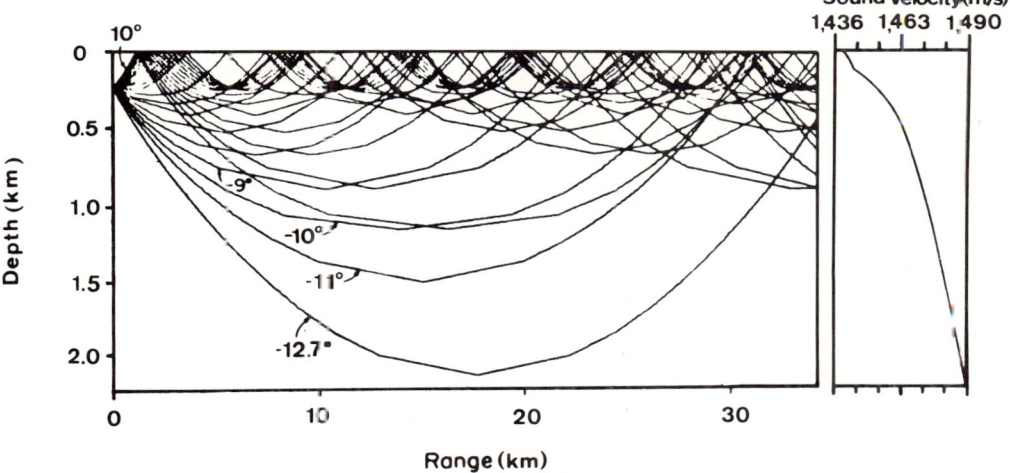

FIG. 8. Propagation of sound waves in Arctic oceans. The velocity profile is shown at the right. The sound source is located about 200 m below the surface. Sound rays are shown at 1° intervals. Rays leaving the source at steep enough angles to reflect from the bottom are omitted. (From Urick, redrawn from Payne and Webb, 1971).

to a smoothly increasing attenuation of sound pressure with distance. The total attenuation depends very much upon the local conditions. It is therefore difficult for an animal to judge the distance to a sound source (another animal) just by measuring the sound pressure. We have also seen that the addition of direct and reflected sound waves may cause local maxima and minima in sound pressure at various distances from the sound source. It is therefore not possible for an animal to locate a sound source just by walking towards higher sound pressure levels.

We noticed before that gradients of wind speed, temperature and humidity are likely to exist close to the ground or to the surface of objects. The mean values of these micrometeorological parameters follow diurnal and seasonal patterns. Furthermore, temperature gradients are only stable, if the cool air is below the warmer air. On sunny days, however, the opposite is true. Due to convection currents, travelling turbulent eddies are formed. They scatter and attenuate the sound waves and cause the sound pressure and phase of the transmitted sound to fluctuate at the receiver. The frequency range of the amplitude fluctuations is about 0.5 to 20 Hz, and their magnitude may be quite large. Wiener and Keast (1959) found peak-to-peak fluctuations of the sound pressure level between 5 and 20 dB during the daytime. At night, the fluctuations are smaller (about 5 dB), but even these fluctuations are very large compared with the change of sound pressure, which an animal experiences when walking towards a distant sound source.

Scattering. The term reflection was used above for describing the redirection of a sound wave impinging upon a "large" surface. If the object is much smaller than the wavelength of sound, it will hardly affect the sound wave. (Physically, the situation is similar to the interaction between light and a virus particle in a light microscope). If the size of the object is of the same order of magnitude as the wave length, the sound (or light) will be redirected. Small objects will scatter the sound in all directions, while large objects will redirect the sound wave almost as a mirror. The redirection of sound from objects of intermediate size may be very complicated, and will depend in part on the shape of the object. The process of redirection will be referred to as scattering in this paper, though it is frequently called diffraction in the literature. The terminology is not quite fixed.

The interference of the scattered sound wave with the sound wave striking the object causes a change in sound pressure around the object. At the surface of a hard object facing the sound source the scattered wave is in phase with the incoming wave, so their pressures will be additive. If the object is large enough compared to the wavelength of sound, the total sound pressure at this region of the surface becomes twice that of the plane wave (6 dB more).

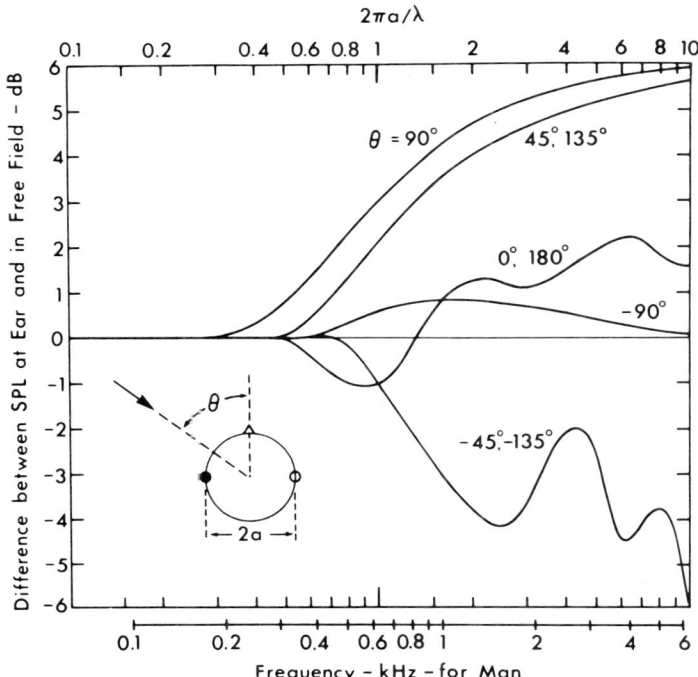

FIG. 9. Calculated transformation of sound pressure level from a free field to a simple ear (point receiver) on the surface of a hard spherical head of radius a as a function of $2\pi a/\lambda$ (where λ = wavelength of sound) for various values of azimuth θ of the incident plane waves. The frequency scale at the bottom is for a sphere of radius a = 8.75 cm. (From Shaw, 1974).

At other parts of the surface or at some distance away from the obstacle, the two sound waves are sometimes in phase and sometimes out of phase. So, the total sound pressure at a point in space may be larger, equal to, or smaller than that in the incoming sound wave. As the sound frequency is varied, a receiver placed at a fixed distance from an object will measure a number of minima and maxima of sound pressure. The fluctuation of the sound pressure with frequency can be calculated for hard objects of simple shape, but such calculations are not realistic for most of the obstacles found outdoors.

The differences in sound pressure caused by scattering around the body (head) is one of the main mechanisms responsible for directionality of hearing (Fig. 9). Scattering also seems to be very important in the attenuation of sound passing obstacles like vegetation (see below). A special case is the scattering of sound from inhomogenities in the medium (like the turbulent eddies mentioned in the previous section).

THE EFFECT OF VEGETATION

The effectiveness of plants for attenuating sound (noise) has been much debated. Several parameters contribute to the total attenuation observed outdoors, and it is very difficult to sort out the contribution of each parameter. In several experimental studies, the attenuation has been observed for sound travelling through a belt of vegetation and compared with the attenuation occuring over an equal distance without vegetation. The ground absorption is, however, not likely to be the same in the two situations. Further, the presence of the vegetation affects the microclimate (temperature, humidity, turbulence of the air, wind) and this again affects the attenuation of sound waves. Most investigators have ignored these parameters in their studies, and only few investigators control the directionality of their sound source (which is important in determining the amount of reflection from the ground). It is therefore not surprising that one can find very different values for the attenuation in a certain kind of vegetation in the literature.

The confusion over the attenuation caused by vegetation becomes even larger, when one compares the physical concepts used by various investigators. One may assume that the attenuation in the vegetation is due mainly to *multiple scattering* by tree trunks, branches and leaves. If this is true, then both theory and experiments demonstrate that the 'excess attenuation' (to be added to the attenuation caused by geometric spreading) should be about 6 *dB per distance doubled* (dd) (Meister and Ruhrberg, 1959; Meister, 1960). In the case of spherical spreading from a monopole source (attenuation 6 dB/dd) the total attenuation should be about 12 dB/dd. In contrast, if the attenuation in the vegetation is due mainly to *absorption*, one would expect the attenuation to be the geometric spreading plus a certain number of *dB per meter*. Most investigators assume that multiple scattering is a major contributor to the excess attenuation in vegetation, but still they express the measured excess attenuation as a certain number of dB per meter. Furthermore, some investigators derive their attenuation values from data collected at only two distances from the sound source. Such experiments do, of course, not allow the investigator to distinguish between attenuation caused by multiple scattering and absorption.

An excess attenuation caused mainly by multiple scattering is found at high frequencies (above 4 kHz) in dense crowns of conifers (Meister and Ruhrberg, 1959). This kind of attenuation is also

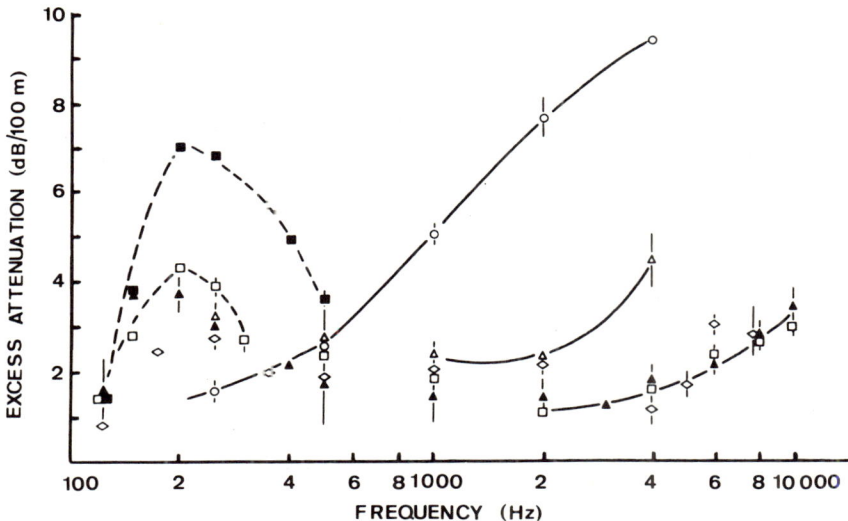

FIG 10. Excess attenuation in decibels/100 m for maize (o), hemlock (◇), pine measured at 30.5 m (■), pine measured at 61 m (□), brush in summer (Δ) and brush in autumn (▲). The source height was 1 m for the maize and 1.5 m for the hemlock, pine and brush. (Redrawn from Aylor, 1971,a).

observed in dense bushes. In contrast, close to the ground, and especially in high grass, these authors found excess attenuations mainly caused by absorption. Still other kinds of vegetation were found to have excess attenuation of a mixed origin.

Some investigators (e.g. Aylor, 1971b) have considered foliage to be a major factor in the attenuation of sound at rather low frequencies (0.5 - 1 kHz). Other authors find, however, that foliage has very little effect at frequencies below 2 kHz (Beck, 1965; Carlson et al., 1977). Beck (1965) investigated the sound attenuation properties of a large number of bushes and trees. He found that a maximum attenuation in the frequency range 2 - 11 kHz was produced by plants with big, hard leaves pointing perpendicular to the direction of the sound.

The total attenuation by bushes and trees reported in the literature varies considerably. Fig. 10 shows the results reported by Aylor (1971a). Most estimates of attenuation are in the range 10 - 30 dB/100m. Most investigators agree that the attenuation increases with frequency. The evidence for a possible minimum of attenuation around 1 - 2 kHz is discussed below.

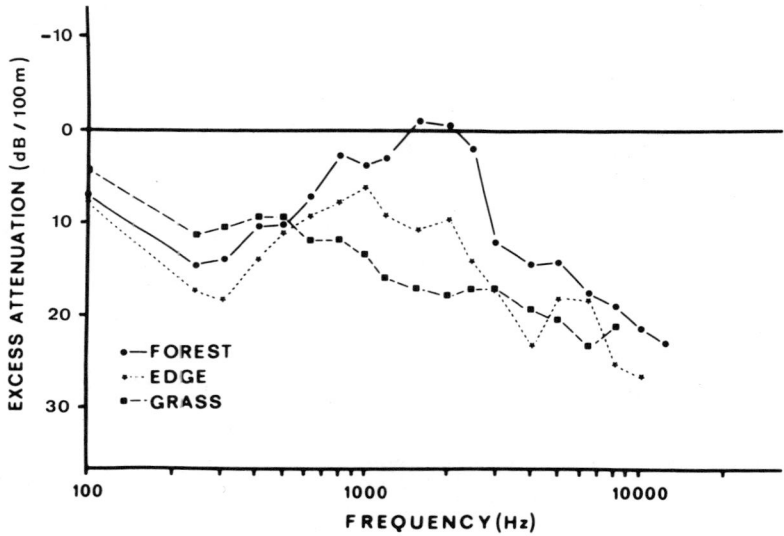

FIG. 11. Excess attenuation levels for pure tone sound propagation about 0.3 m above ground in forest, edge, and grassland habitats. (From Morton, 1975, changed).

BIRD SONG IN DIFFERENT HABITATS

A few attempts have been made to relate the characteristics of animal calls with the acoustical properties of different environments. Morton (1975) investigated the propagation of sound in tropical forest and grassland (Fig. 11) and measured the main frequencies in the songs of the birds living in these habitats (Fig. 12). In the forest he found the transmission to be a maximum around 1.5-2.5 kHz (Fig. 11), if the loudspeaker and microphone were placed about 0.3 to 1.5 m from the ground. At these frequencies the excess attenuation (that is the decrease in sound pressure after correction for spherical geometric spreading) was close to 0 dB. The maximum of transmission around 1.5-2.5 kHz in the forest was, however, not present at about 3 m from the ground, nor was a maximum apparent in the grassland. The edge of the forest appeared to have intermediate properties (Fig. 11). The frequencies emphasized in the song of birds living close to the ground in forests (Fig. 12, "low forest") also showed a maximum around 2 kHz and were significantly different from those of grassland and edge birds (Fig. 12). From these results Morton concluded that the "low forest" birds were well suited for communicating over long distances, while the birds in the other habitats were less capable of long distance communication. The birds in the

SOUND RECEPTION IN DIFFERENT ENVIRONMENTS

"low forest" were found to have predominantly pure tonelike calls, whereas the grassland- and edge birds had both pure tone and modulated tone elements.

One unsolved problem is, why the grassland birds are not equally adapted to long distance communication in their environment. In the grassland the excess attenuation increases with fre-

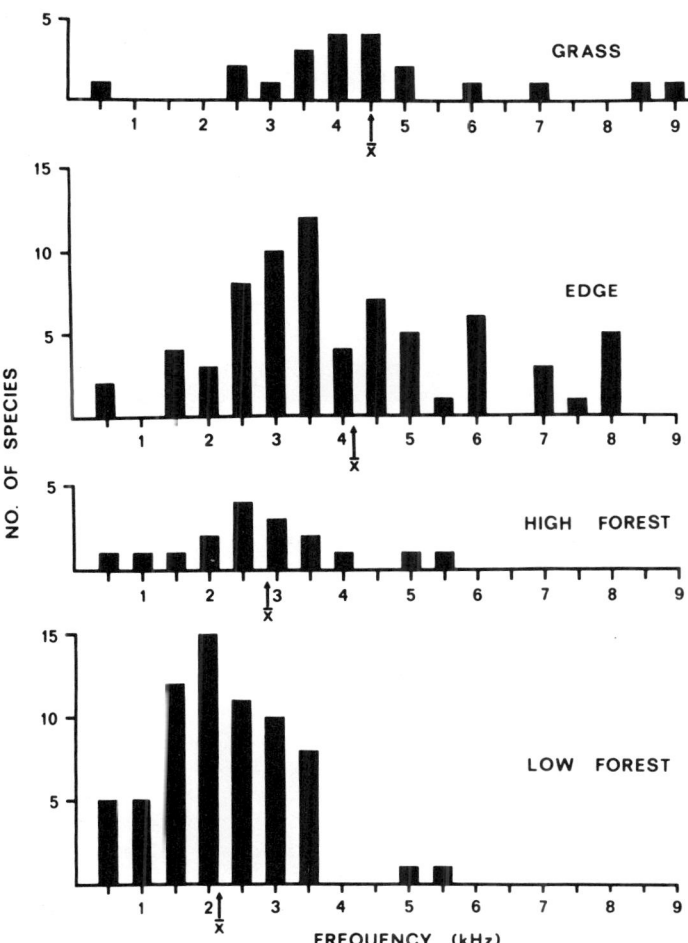

FIG. 12. Histograms showing the frequencies emphasized in the song of bird species living in different habitats. X̄ indicates the mean frequency for each group. From Morton, 1975.

quency (Fig. 11), so one would expect the birds to use very low frequencies. However, this is not the case (Fig. 12). Morton suggests that temporal patterns carried by broad frequency bands provide a more efficient means of communication in open habitats than do pure tone signals, since such signals should be better adapted for preserving the specific information in the presence of shadow-zones and air turbulence. This is a very interesting idea, which may also be applied to grasshopper and bushcricket songs (see below). The validity of this hypothesis should be investigated further.

A central point in Morton's argument is that the selection operating during the evolution of animal calls has favoured communication over as great a distance as possible. This may be true for species which need to have low population densities. But, as pointed out by Marler (1955), several other features of the calls may be important (e.g. the calls should be conspicuous to conspecific animals, but inconspicuous to predators or prey, and the calls should also differ distinctively from the song of other, sympatric, species).

If we now turn to the properties of the auditory system in birds, one might perhaps expect the ear to be most sensitive at the main frequency of the conspecific song. This is, however, not always so (Konishi, 1970). It is hardly surprising that such a connexion does not exist. The sense of hearing is used for many purposes in birds, and the properties of the ear are not determined by just one factor. Probably, the same will appear to be true for the properties of the song of birds. In contrast, in many frogs and insects the main function of the auditory system may be the detection of conspecific calls. So, in these animals one may expect a closer correlation between the acoustical properties of the call and those of the auditory system.

Morton's results are very interesting, but rather different results can be found in the literature. Some studies support the idea of a "sound window" around 1-2 kHz close to the ground (Aylor, 1971a; Carlson et al., 1977), but other studies do not (Eyrings, 1946; Wiener and Keast, 1959). Furthermore, the measurements performed by Embleton (1963) show this maximum only at short distances from the sound source (0-20 m), and not at greater distances. Finally, the "sound window" has been found not only in forests, but also in open habitats (Marten and Marler, 1977; Marten et al., 1977).

The crusial question is, whether this disagreement is due to real differences between the habitats investigated, or to the differences between the apparatus and experimental design in the investigations. In a properly performed investigation the sound power produced by the loudspeakers and their directionality should be known from measurements performed in the laboratory. The attenuation of sound with distance should be measured with at least three microphones placed at different distances from the sound source. Further, the temperature and wind gradients should be measured. So far, no study has been performed, in which all the major sources of error have been controlled.

The directionality of the loudspeaker affects the amount of reflection of sound wave (from the ground or as multiple scattering). Some investigators try to estimate the power output of the loudspeaker by means of a microphone placed several metres away. This is, of course, not possible. For example, in one study (Marten and Marler, 1977) a microphone was placed 2.5 m from the loudspeaker and another microphone about 100 m away. The difference in the reading of the two microphones was then used to calculate the "excess attenuation per 100 m" (by subtracting the attenuation due to geometric spreading). The published attenuation curves clearly demonstrate the presence of fluctuations in sound pressure at the 2.5 m microphone similar to those illustrated in Fig. 3. Furthermore, only two microphones were used, and therefore it is not possible to know, whether the attenuation was due to absorption (dB/m), to multiple scattering (dB/dd) or to something in between (see the section on vegetation above).

THE CRICKET AND THE GROUND IMPEDANCE

The males of field crickets sing their "calling song" while on the ground. The song is heard by other crickets, which are also on the ground. Nocke (1971,1972) measured the sound pressure level produced by a singing male *Gryllus campestris* L. to be around 102 dB at 5 cm distance. The average threshold of hearing at the same frequency (4 kHz) was 44 dB. From these values he calculated a theoretical range of communication of 42 metres, assuming the attenuation to be due only to (spherical) geometric spreading (attenuation 6 dB/distance doubled). We have already seen that the range may be much less, if temperature or wind gradients are present. The acoustical properties of the ground itself may, however, cause further attenuation. In the following we shall examine the propagation of sound waves close to the ground. It will be demonstrated that the actual range is considerably smaller than that calculated by Nocke, even when temperature and wind gradients are absent.

Most of the existing evidence shows that propagation of sound over the ground is adequately described by treating the ground as a locally reacting surface, thus neglecting wave propagation in the ground itself (Piercy et al., 1977). In the habitats of field crickets, the grassy ground surface is behaving as a locally reacting porous medium. In terms of impedance this means a mixed stiffness-resistant element (Embleton et al., 1976).

In the description of reflection (above) we noticed that the magnitude and phase angle of the reflected wave depend upon the impedances of the two media (Z_1 and Z_2) and upon the grazing angle (ψ). For a plane wave the relationship is (Piercy et al., 1977):

$$R_p = \frac{\sin \psi - Z_1/Z_2}{\sin \psi + Z_1/Z_2}$$

where R_p is the amplitude reflection coefficient. Z_1, Z_2 and ψ are shown in Fig. 2. Z_1/Z_2 can be ignored, if the ground is hard, and if the grazing angle is rather large ($Z_2 \gg Z_1$ and $\sin \psi \gg Z_1/Z_2$). The phase change on reflection will then be almost zero. For grounds similar to the habitats of crickets and at 4 kHz (the frequency of the calling song), however, both the resistive and stiffness (real and imaginary) components of Z_2 are only a few times larger than Z_1 (see Fig. 8 in Embleton et al., 1976). This means that at small grazing angles (the sound wave travelling almost parallel to the ground) R_p will be negative, and the reflected wave will be almost out of phase with the incoming wave. So, for crickets on the ground the waves r_1 and r_2 (Fig. 2) will tend to cancel each other, even though their path lengths are about equal.

In practice, this means that for plane waves a path close to the ground represents a *forbidden mode of propagation* (Piercy et al., 1977), except at close distances. The magnitude of the "shadow zone" close to the ground may reach values around 30 dB as the sound propagates outwards from the sound source. Consequently, the range of communication of the field cricket in the absence of temperature- and wind gradients is likely to be about 10 times lower than the value calculated by Nocke, if one allows for the magnitude of the shadow zone. The field measurement of cricket song performed by Popov et al. (1974) show that the sound pressure close to the ground does in fact fall off much faster than expected from geometric spreading alone. Unfortunately, the wind- and temperature gradients were not measured in this study.

The sound pressure likely to reach a listening cricket at a distance of, for example, one metre from a singing cricket is very difficult to calculate. At close range the sound waves are not plane, but spherical. In equations assuming local reaction and homogeneity of the ground, the interaction between a spherical wave and the ground impedance gives rise to a *ground wave*. These assumptions have been questioned by Carlson et al. (1977), but the existence of the ground wave seems well supported by experimental observations (Donato, 1976; Embleton et al., 1976).

SOUND RECEPTION IN DIFFERENT ENVIRONMENTS

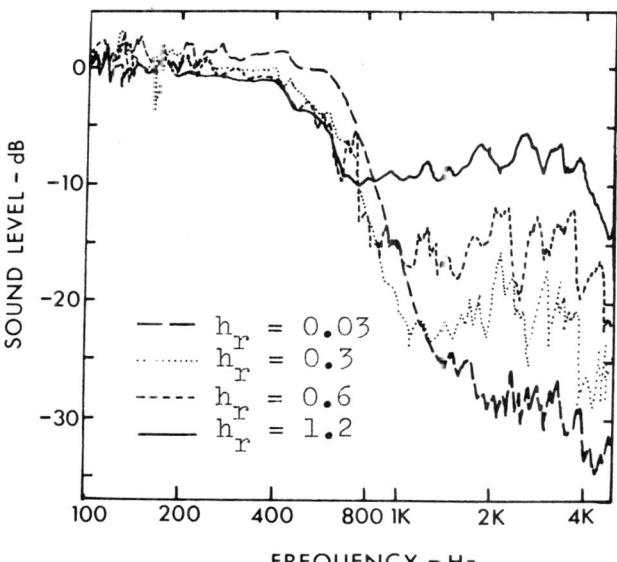

FIG. 13. Propagation of sound over grass. The sound source was 0.03 m from the ground, and the receiver at heights (h_r) between 0.03 and 1.2 metres from the ground. The horizontal distance was about 4.6 m. The reference sound level is that expected from geometric spreading (6 dB/dd) alone. From Embleton et al., 1976 (changed).

The range of the ground wave depends upon the frequency, the grazing angle, and the impedance of the ground. For a constant grazing angle and impedance, the ground wave propagates a certain number of wavelengths away with only the attenuation due to geometric spreading (6 dB/dd). Above this range, however, the ground wave suffers an additional 6 dB/dd attenuation. The ground wave therefore behaves as a *low-pass filter*. Fig. 13 shows the result of measurement over grass, where the distance between the sound source (3 cm over ground) and the receiver (3 to 120 cm over ground) was about 4.6 m. Frequencies below 500 Hz are seen to suffer only the attenuation expected from geometric spreading, whereas higher frequencies sustain an excess attenuation. The attenuation above 500 Hz increases with frequency and decreases with the distance from the ground.

STRATEGIES FOR SOUND COMMUNICATION

It is interesting to compare Fig. 13 with Fig. 3, where both the sound source and the receiver were about 1.2 m above ground.

We can use these results to speculate about the optimum strategy for sound communication in terrestrial animals. From Fig. 13, one can see that animals on the ground may communicate over long distances, if they use low frequencies. This is possible for large animals. Small animals like insects are, however, unable to produce much sound below a few kHz (see Michelsen and Nocke, 1974).

With frequencies of several kHz, the range of communication is limited, if both the singing and the listening animal is on the ground, although the range may be several metres, if the animal produces very loud sounds. The intensity of the sound signals is limited by the muscle power available, but higher intensities may be obtained by means of a resonator. For example, the cricket produces long, pure tones of considerable intensity by driving the sound emitting structure (a part of the wing called the harp) at its resonance frequency. Molecrickets go one step further. They use a horn-shaped burrow for "amplifying" their calls (Bennet-Clark, 1970).

The use of resonators (mechanical oscillators, horns) is an efficient way of producing large sound intensities. The "price" for this solution is that the animals are limited to the resonance frequency (-ies) of the resonator. They cannot signal their message

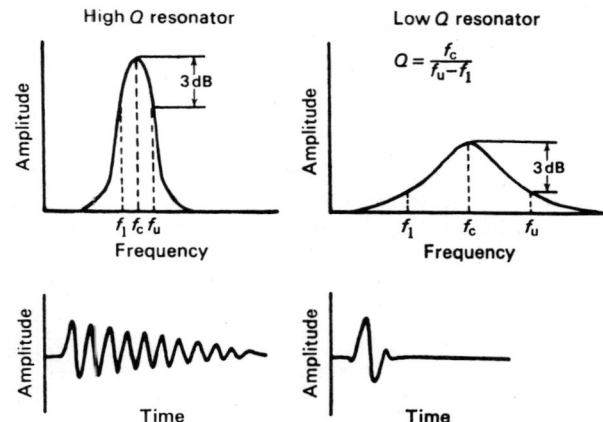

FIG. 14.   The behaviour of a lightly (left) and a heavily (right) damped mechanical oscillator. The upper figures show the frequency response of the two systems. The lower figures show the impulse response (the waveform produced when the oscillator is excited by a sudden shock of broad band energy). The degree of damping is often indicated by the Q-value (a high Q means a small amount of damping). From Sales and Pye, 1974.

by means of, say, frequency sweeps. Furthermore, lightly damped resonators are mechanically "slow" systems (Fig. 14). Such systems cannot be used for transmitting rapid modulations (Michelsen and Nocke, 1974). The message communicated by the calling song of the field cricket is rather simple (I am here, I belong to species X, females are welcome, males are not).

Many species of frogs, toads and birds also produce rather pure tone sounds, which are subjected to a slow amplitude modulation (1-20Hz). The parameters available for carrying the information are limited to frequency and (slow) rhythms. Since the frequency band is limited to one to two decades in small animals (lower frequencies cannot be produced efficiently; higher frequencies do not travel far enough), only a limited number of species can coexist in the same habitat. The use of slow amplitude modulation for carrying information may also be limited by air turbulence, which causes amplitude modulations of similar frequencies (see above).

Some insects (shorthorned grasshoppers, most bushcrickets) are non-resonant singers. The primary vibration is, as in the cricket, produced by the stridulatory organ, where a scraper on one part of the body (leg or wing) hits a series of teeth (the "file") on another appendage. The sound emitting structure (parts of the wings) do not have any lightly damped resonances and generally appear to have a large frictional damping. (The degree of damping can be estimated from the shape of the sound impulses produced by the impacts of the scraper on individual teeth in the file). The sound emitted from such a system is a broad-band noise of rather low intensity. But the system is much faster than the lightly damped, highly tuned sound emitter in the cricket. This means that the animals can use faster rhythms for communication, and thus be undisturbed by the lower-frequency amplitude-variation due to atmospheric turbulence. The "price" is the lower sound intensity, which means a shorter range of communication, and which forces the animals to have higher population densities.

The range of communication may become much greater, if the singing animal and/or the listening animal climbs up the vegetation. Many grasshoppers sing about half a meter above ground, while bushcrickets are often one, two or more metres from the ground. But, as shown in Fig. 3 the price for this longer range of communication may be a large and variable frequency filtering of the sound signal. Unfortunately, very little is known about the magnitude of this frequency filtering in natural habitats.

For many years the insects were considered to be unable to discriminate sound frequencies, and the information was thought to be coded only as amplitude modulations of the carrier sound (Pumphrey, 1940). In 1965 Andrej Popov (Leningrad) and I independently found that the ear of the migratory locust (a short-horned grasshopper) is able to perform some frequency discrimination. Further studies demonstrated that the four anatomical groups of receptor cells have different frequency sensitivities (Michelsen, 1971a; Römer, 1976) and revealed the physical basis for the frequency discrimination (Michelsen, 1971,b). Later, a less detailed discrimination, based on only two groups of receptor cells, has been found in crickets (Nocke, 1972) and molecrickets (Zhantiev and Korsunovskaya, 1973). The physical mechanism has been studied in the latter case (Michelsen, 1978). A much finer frequency discrimination has been found in bushcrickets (Rheinlaender, 1975). Here, 33 receptor cells are situated in a serial arrangement, and their accessory structure become progressively smaller from the proximal to the distal end of the receptor organ. The frequency of maximum sensitivity differs among the cells, probably in a systematic way corresponding to the anatomical arrangement.

Until recently, recordings of insect song were studied (and published) with a time scale, which was suitable only for demonstrating slow amplitude modulations. Finer details in the song did not attract interest, partly because such details are often blurred in tape recordings (either for technical reasons, or because the two sound producing mechanisms in short-horned grasshoppers may be playing slightly different songs, see Elsner, 1974). Elsner's study, however, suggested that specific information may be signalled as changes in the repetition rate for the very short impulses, which are produced during stridulation. This theory will be considered in the following section.

IMPULSE RATE AS CARRIER OF INFORMATION

The song produced by males of the short-horned grasshopper *Omocestus viridulis* L. was studied by Elsner (1974). He noticed that the song of one-legged males contains more details than songs from normal males (in which the two legs do not follow each other precisely, thus leading to the aforementioned blurring). In particular individual impulses lasting about 0.2 msec could be observed (Fig. 15,B). These impulses are caused by the impact of the "scraper" upon a tooth in the "file" of the stridulatory organ. A plot of the time intervals between the impulses demonstrated large and very fast sweeps of impulse frequency (Fig. 15,C). For about half of the song the impulse rate is low (around 300 Hz), but at the end of each chirp the impulse rate reaches 8 kHz. At first sight, such high impulse rates are not likely to be of much interest, since the ears are not fast enough to respond to the individual impulses (the "flicker fusion frequency" for clicks is about 300 Hz; Michelsen, 1966). There is, however, another possibility, which I proposed some years ago (in Elsner, 1974): The impulse frequency could be received by the ear as a frequency, that is, it could be modifying the frequency spectrum of the song, and the grasshoppers could use their ability for frequency discrimination to detect spectral features caused by the impulse rate. This hypothesis has now been confirmed in behavioural experiments (Skovmand and Pedersen, 1978).

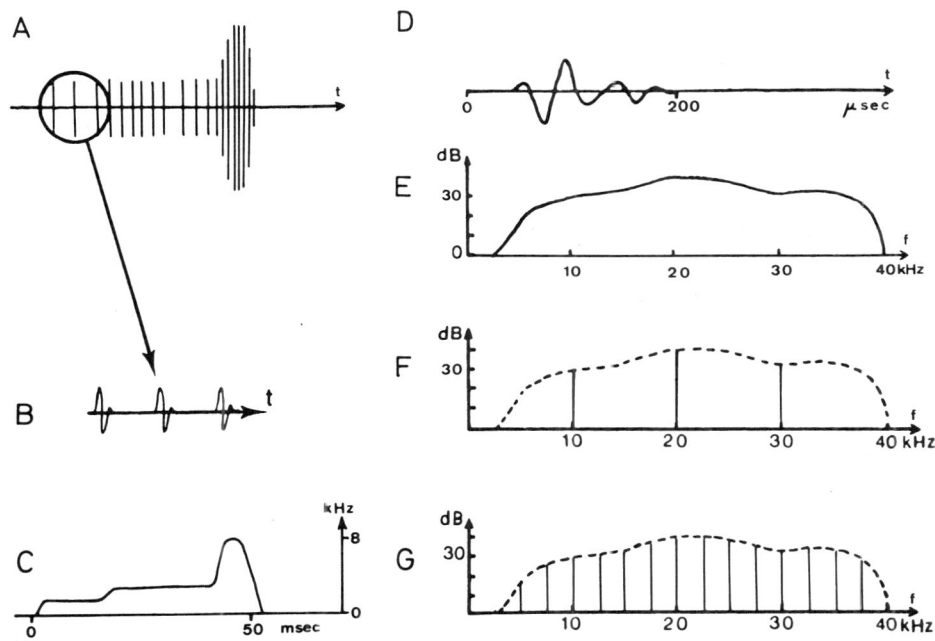

FIG. 15. Analysis of the song of the short-horned grasshopper *Omocestus viridulus*. A. Schematic diagram of one chirp in the song. B. Individual impulses in the chirp. C. The impulse rate during a chirp (schematically). The time scale applies to A and C. D. An impulse shown in more detail. E. One impulse has a continuous frequency spectrum. F and G. The frequency spectrum of a series of impulses with constant repetition rate consists of a number of discrete frequencies (lines). The intensity of each of these frequencies is determined by the spectrum of the individual impulse. The frequency of the lines depends upon the impulse repetition frequency (10 kHz in F, 2.5 kHz in G). From Skovmand and Pedersen, 1978 (changed).

Physically, the situation is rather complicated. The spectrum of a series of impulses (Fig. 15,F-G) differs from the spectrum of one impulse (Fig. 15,E) in being a line spectrum, which is contained in the (broad) spectrum of the single impulse. If the impulse repetition is quite regular, the spectrum of a series of impulses will consist of some discrete frequencies. The lowest of these frequencies is the impulse repetition frequency, and the other frequencies are higher harmonics of this frequency. If the impulse repetition frequency is, for example 5 kHz, the frequency spectrum will have components at 5, 10, 15, 20, -- kHz. When the impulse repetition frequency varies during a chirp, the spectral lines will move.

This is a very complicated pattern of FM-sweeps, but apparently the simple frequency discriminating mechanism of the grasshopper auditory system is able to extract the necessary information. This was shown in experiments, in which behaviourally receptive females were given the choice between various artificial songs (Skovmand and Pedersen, 1973). The songs were generated on a computer in such a way that the individual parameters of the song could be varied and controlled. The variation in (root mean square) sound pressure amplitude appeared to be the most important parameter, but signals with "correct" impulse rate modulation were significantly more attractive than signals without this feature.

These experiments show that frequency information can be coded as the repetition of very short (0.2 msec) impulses. The frequency spectrum of the individual impulses is very broad (1-40 kHz). This means that a considerable part of the impulse will get through the "environmental filter", even in environments with a substantial frequency filtering. No information is presently available about the magnitude of environmental filtering in the habitats of short-horned grasshoppers. From the behaviour one gets the impression of an enormous redundancy in the signal. The males may court a female for many minutes, repeating their chirps several thousand times.

REFERENCES

Aylor, D. (1972a). Noise reduction by vegetation and ground. J. Acoust. Soc. Am. 51: 197-205.
Aylor, D. (1972b). Sound transmission through vegetation in relation to leaf area density, leaf width and breadth of canopy. J. Acoust. Soc. Am. 51: 411-414.
Beck, G. (1965). Pflanzen als Mittel zur Lärmbekämpfung. Patzer Verlag, Hannover, Berlin, Sarstedt.
Bennet-Clark, H.C. (1970). The mechanism and efficiency of sound production in mole crickets. J. Exp. Biol. 52: 619-652.
Bennet-Clark, H.C. (1971). Acoustics of insect song. Nature (Lond.) 234: 255-259.

Carlson, D. E., O.H. McDaniel and G. Reethof (1977). Noise control by forests. Proc. Inter-Noise 77: 576-586.

Crocker, M.J. and A.J. Price (1975). Noise and noise control, vol. 1. CRC Press, Cleveland, Ohio.

Donato, R.J. (1976). Propagation of a spherical wave near a plane boundary with complex impedance. J. Acoust. Soc. Am. 60:34-39.

Elsner, N. (1974). Neuroethology of sound production in gomphocerine grasshoppers (*Orthoptera:Acrididae*). I. Song patterns and stridulatory movements. J. Comp. Physiol. 88: 67-102.

Embleton, T.F.W. (1963). Sound propagation in homogeneous deciduous and evergreen woods. J. Acoust. Soc. Am. 35: 1119-1125.

Embleton, T.F.W., J.E. Piercy and N.Olson (1976). Outdoor sound propagation over ground of finite impedance. J. Acoust. Soc. Am. 59: 267-277.

Eyrings, C. (1946). Jungle acoustics. J. Acoust. Soc. Am. 18: 257-270.

Fisher, F.H. and V.P. Simmons (1977). Sound in sea water. J. Acoust. Soc. Am. 62: 558-564.

Griffin, D.R. (1971). The importance of atmospheric attenuation for the echolocation of bats (*Chiroptera*). Anim. Behav. 19: 55-61.

Harris, C.M. (1971). Effects of humidity on the velocity of sound in air. J. Acoust. Soc. Am. 49: 890-893.

Konishi, M. (1970). Comparative neurophysiological studies of hearing and vocalizations in songbirds. Z. vergl. Physiol. 66: 257-272.

Lyon, R.H. (1973). Propagation of environmental noise. Science 179: 1083-1090.

Marler, P. (1955). Characteristics of some animal calls. Nature (Lond.) 176: 6.

Marten, K. and P. Marler (1977). Sound transmission and its significance for general vocalization. I. Temperate habitats. Behav. Ecol. Sociobiol. 2: 271-290.

Meister, F.J. (1960). Über einige Besonderheiten der Schallausbreitung auf natürlich bewachsenen Flächen. Frequenz 14: 211-217.

Meister, F.J. and W. Ruhrberg (1959). Der Einfluss von Grünanlagen auf die Ausbreitung von Geräuschen. Lärmbekämpfung. 1: 5-11.

Michelsen, A. (1966). Pitch discrimination in the locust ear: observations on single sense cells. J. Insect. Physiol. 12: 1119-1131.

Michelsen, A. (1971a). The physiology of the locust ear. I. Frequency sensitivity of single cells in the isolated ear. Z. vergl. Physiol. 71: 49-62.

Michelsen, A. (1971b). The physiology of the locust ear. II. Frequency discrimination based upon resonances in the tympanum. Z. vergl. Physiol. 71: 63-101.

Michelsen, A. (1978). Comparative biophysics of hearing : insect ears as mechanical systems. Am. Sci. (submitted).

Michelsen, A. and H. Nocke (1974). Biophysical aspects of sound communication in insects. Adv. Insect Physiol. 10: 247-296.

Morton, E.S. (1975). Ecological sources of selection on avian sounds. Am. Nat. 109: 17-34.

Møhl, B. and K. Ronald (1975). The peripheral auditory system of the harp seal, *Pagophilus groenlandicus*, (Erxleben, 1777). Rapp. P.-v.Réun. Cons. int. Explor. Mer. 169: 516-523.

Nocke, H. (1971). Biophysik der Schallerzeugung durch die Vorderflügel der Grillen. Z. vergl. Physiol. 74: 272-314.

Nocke, H. (1972). Physiological aspects of sound communication in crickets (*Gryllus campestris* L.). J. Comp. Physiol. 80: 141-162.

Payne, R. and D. Webb (1971). Orientation by means of long range acousitc signalling in baleen whales. Ann. N. Y. Acad. Sci. 188: 110-141.

Piercy, J.E., T.F.W. Embleton and L.C. Sutherland (1977). Review of noise propagation in the atmosphere. J. Acoust. Soc. Am. 61: 1403-1418.

Popov, A.V. (1965). Electrophysiological studies on peripheral auditory neurons in the locust. J. Evol. Biochem. Physiol. (in Russian).

Popov, A.V., V.F. Shuvalov, I.D. Svetlogorskaya and A.M. Markovich (1974). Acoustic behaviour and auditory system in insects. In: Mechanoreception (J. Schwartzkopff,ed.). Westdeutscher Verlag, Opladen p. 281-306.

Pumphrey, R.J. (1940). Hearing in insects. Biol. Rev. 15: 107-132.

Pye, J.D. (1971). Bats and fog. Nature (Lond.) 229: 572-574.

Rheinlaender, J. (1975). Transmission of acoustic information at three neuronal levels in the auditory system of *Decticus verrucivorus (Tettigoniidae,Orthoptera)*. J. Comp. Physiol. 97: 1-53.

Römer, H. (1976). Die Informationsverarbeitung tympanaler Rezeptorelemente von *Locusta migratoria (Acrididae,Orthoptera)*. J. Comp. Physiol. 109: 101-122.

Sales, G. and D. Pye (1974). Ultrasonic communication by animals. Chapman and Hall, London.

Schuijf, A. and A.D. Hawkins (ed., 1976). Sound reception in fish. Developments in aquaculture and fisheries science, vol. 5. Elsevier, Amsterdam, Oxford, New York.

Shaw, E.A.G. (1974). The external ear. In: Handbook of sensory physiology, vol. V/1: Auditory system, anatomy, physiology (ear). Springer Verlag, Berlin. p. 455-490.

Skovmand, O. and S. Boel Pedersen (1978). Tooth impact rate in the song of a shorthorned grasshopper : a parameter carrying specific behavioural information. J. Comp. Physiol. (In press).

Wiener, F.M. and D.N. Keast (1959). Experimental study of the propagation of sound over ground. J. Acoust. Soc. Am. 31: 724-733.

Wiley, R.H. and D.G. Richards (1978). Physical constraints on acoustic communication in the atmosphere : Implications for the evolution of animal vocalizations. Behav.Ecol.Sociobiol. (In press).

Williams, A.O. (1972). Propagation. J. Acoust. Soc. Am. 51: 1041-1048.

Yamada, S., T. Watanabe, S. Nakamura, H. Yokoyama and S. Takeoka (1977). Noise reduction by vegetation. Proc. Inter-Noise 77: 599-606.

Zhantiev, R.D. and O.S. Korsunovskaya (1973). Sound communication and some characteristics of the auditory system in mole crickets *(Orthoptera,Gryllotalpidae)*. Zool. J. 52: 1789-1801. (In Russian).

Figures 6 and 7 are reprinted with permission and are the copyright of the Chemical Rubber Company, C.R.C. Press, Inc. and the American Association for the Advancement of Science, respectively.

BARORECEPTION

J.H.S. Blaxter

Dunstaffnage Marine Research Laboratory, P.O. Box 3

Oban, Argyll, PA34 4AD, Scotland

INTRODUCTION

It is possible to make a fairly clear distinction between the effects of pressure on organisms in the aquatic and in the terrestrial environment. In water, pressure increases with depth roughly at the rate of 0.1 atm/m (Fig. 1) so that in the deep ocean pressures near the sea bed may be as high as 1000 atm. In more shallow areas pressures will be less but the effect of tidal changes, which are often 10 m or more, or waves, which may be several metres high, will impose greater percentage changes in pressure than they will over deep water.

Atmospheric weather changes of pressure of the order 0.01 atm will be a negligible component in water but may be more significant on land. For birds a change in altitude will involve a decrease in pressure of only about 0.0001 atm/m. Altitude discrimination will therefore require much better sensitivity than depth discrimination. In addition the concentration of oxygen decreases with height which might limit the altitude at which birds could fly or to which mountainous areas could be colonized. In the sea there may be a so-called "oxygen minimum" layer between 100 and 1000 m or regions of anoxia as in the depths of the Black Sea, but in general the oxygen concen-

---

Footnote

Kinne (1972) gives an elaborate account of units used in pressure studies and their interconversions. For the purpose of this paper pressures are given above and below atmospheric pressure (1 ATA) in atmospheres (atm.) 1 atm $\equiv$ 1000 mb $\equiv$ 1033 cm H_2O $\equiv$ 76 cm Hg $\equiv$ 14.7 lbs /in^2 $\equiv$ 10^5 N/m^2

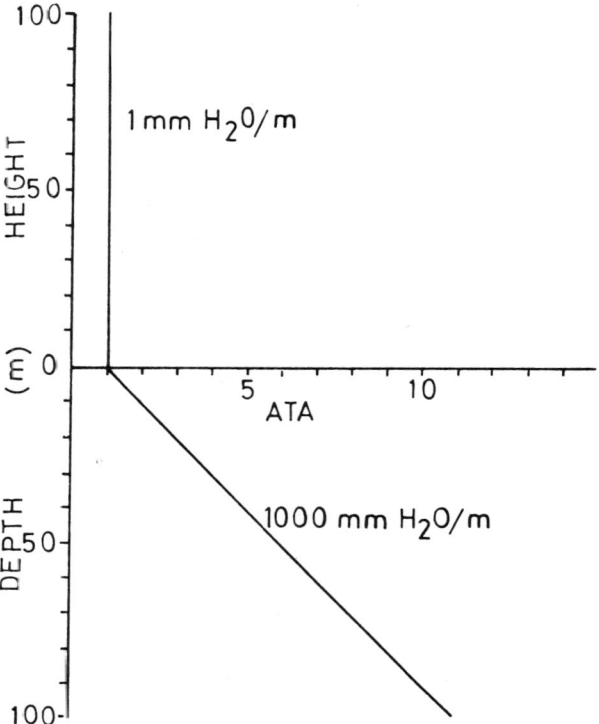

Fig. 1. Changes in pressure at different depths and altitudes.

tration of the water remains similar to that at the surface regardless of depth.

Aquatic plants have the problem of obtaining sufficient light to photosynthesise. The euphotic zone will rarely extend beyond a pressure of 20 atm and may often be much less in turbid water. High altitude plants, like high altitude animals or airborne birds, may adapt to the low oxygen concentrations found at great heights. Oxygen supply may well be much more limiting for them than it is for deep sea organisms.

Two quite distinct approaches have been made to the pressure responses of organisms and their systems. Very high pressures of the orders of hundreds of atmospheres influence enzyme systems, protein and RNA synthesis, cell division, ciliary action, respiration rate and general activity; they also inhibit bacterial reproduction and affect photosynthesis. Many organisms die at high pressures. For current descriptions of this type of work the results of the 1971 Society

for Experimental Biology Symposium on pressure may be consulted
(Sleigh and Macdonald, 1972). A series of papers from Florida State
University were also published in 1974 in the Internationale Revue
der gesamten Hydrobiologie Vol. 59. In 1973 the United States
Research Vessel "Alpha Helix" was used to conduct a multidisciplinary
investigation of the effects of pressure on deep sea fish and the
resulting papers are collected in Comparative Biochemistry and
Physiology Vol. 52B No. 1 (1975). Finally may be mentioned a symposium on high pressure aquarium systems organized by Brauer and others
in the United States and published in book form (Brauer, 1972) and a
multi-author chapter in Marine Ecology, Vol. 13 (Kinne, 1972).

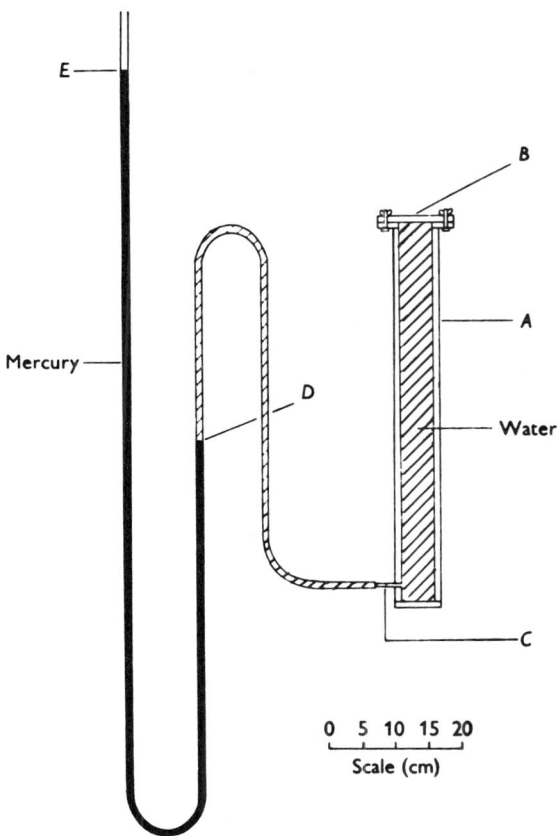

Fig. 2. Simple apparatus to investigate the response of marine
animals to pressure changes. A-observation tank; B-removable end-
plate; C-inlet tube; D,E-ends of mercury manometer (from Rice, 1964).

In the context of ecology it may be assumed that organisms living deep in the sea have their biochemical and other systems adapted to high pressures. A great volume of work has been done on subjecting deep sea organisms to decompression and compressing surface-living and terrestrial organisms to very high pressures which are ecologically meaningless. What is of more interest for the present volume is how organisms respond to pressure changes within their normal range, how this may limit or influence their distribution and how the sensory system appreciates either absolute pressure or pressure change.

METHODS

Many aquatic organisms have diel rhythms, rising to the surface at dusk and sinking at dawn. This usually occurs within the top 1000 m. Other aquatic organisms apparently respond to tides or waves and terrestrial organisms to changes in weather. It is rarely clear whether the animals are responding to pressure or using it to monitor their movements. In many instances changes in pressure are accompanied by changes in the quality or intensity of light or with other types of stimulus.

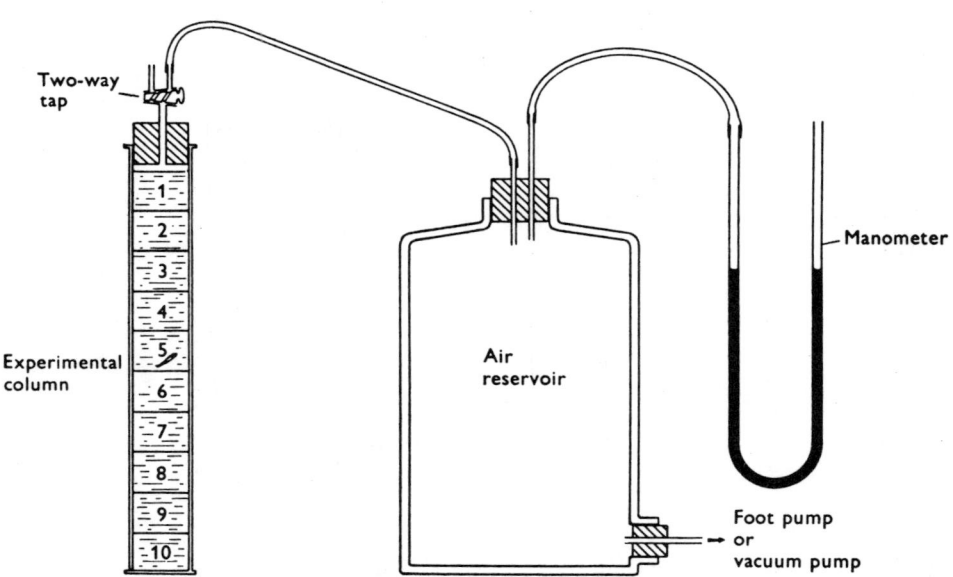

Fig. 3. Apparatus for "scoring" pressure responses (from Blaxter and Denton, 1976).

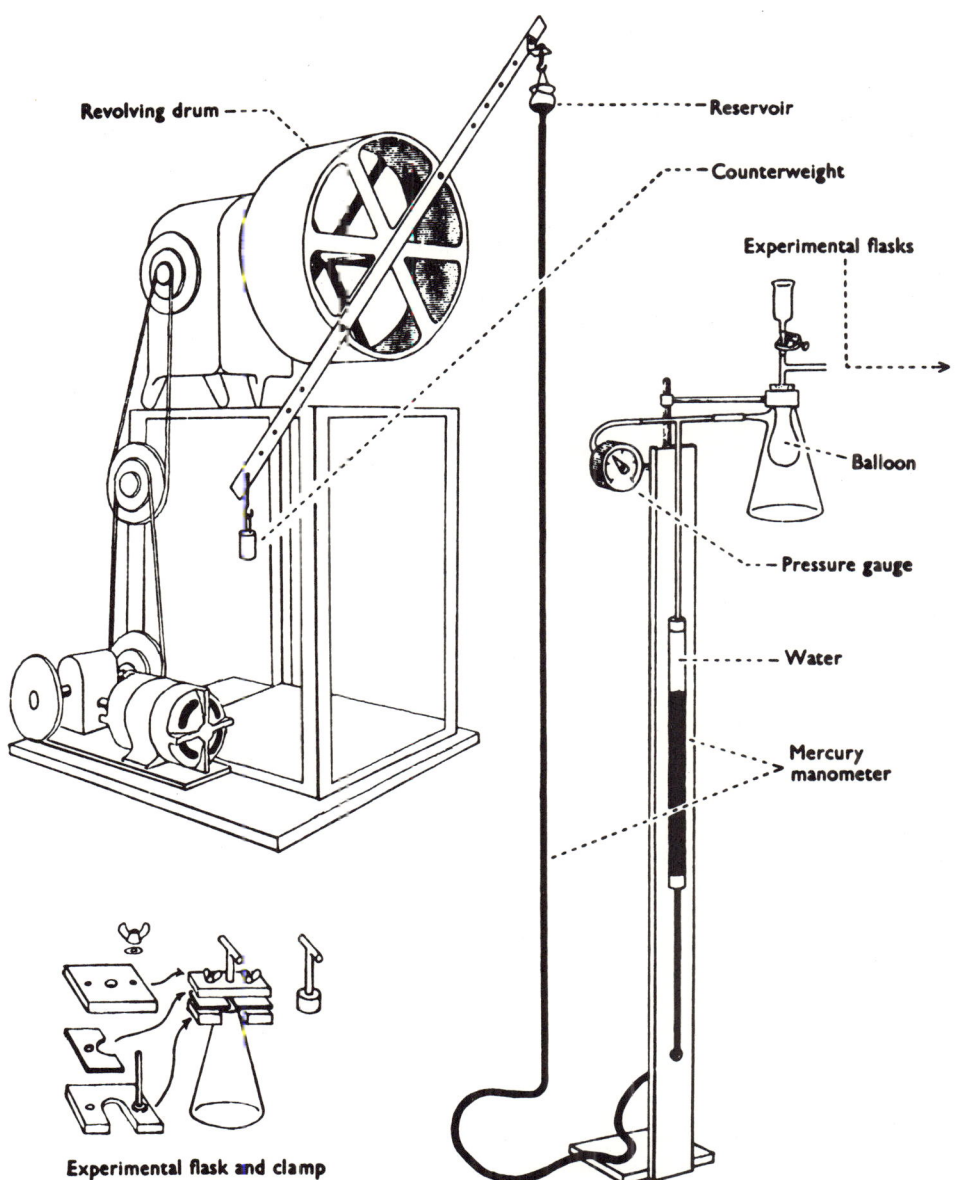

Fig. 4. Apparatus for producing cyclical changes of pressure (from Morgan et al. 1964).

Pressure responses then must be investigated by imposing pressure changes on organisms and observing either responses in behaviour or in some physiological function such as nerve impulse frequency or heart rate. Three types of apparatus for observing behaviour are shown in Figs. 2, 3 and 4. In Figs. 2 and 3 quick pressure changes were applied directly through the water or through an air phase over the surface. The apparatus shown in Fig. 4 was used to apply cyclical pressure changes and so to simulate tides. The tank shown in Fig. 5 could be inclined at different angles to give varying pressure differentials during swimming and was used in a study of buoyancy.

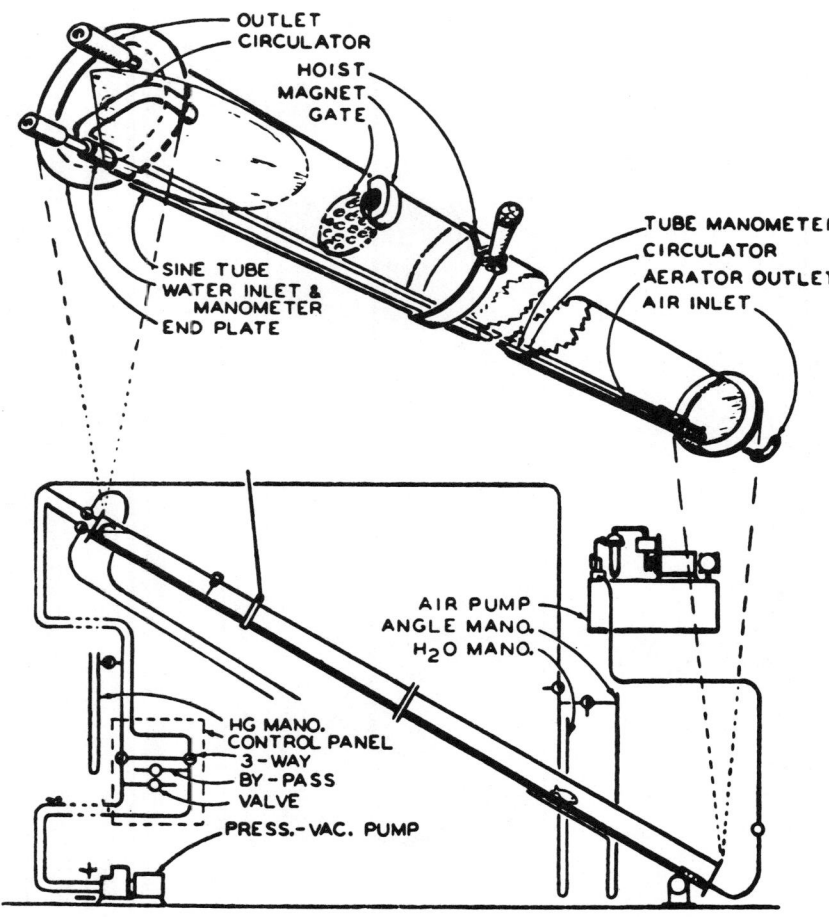

Fig. 5. Inclined tank to investigate buoyancy responses (from McCutcheon, 1958).

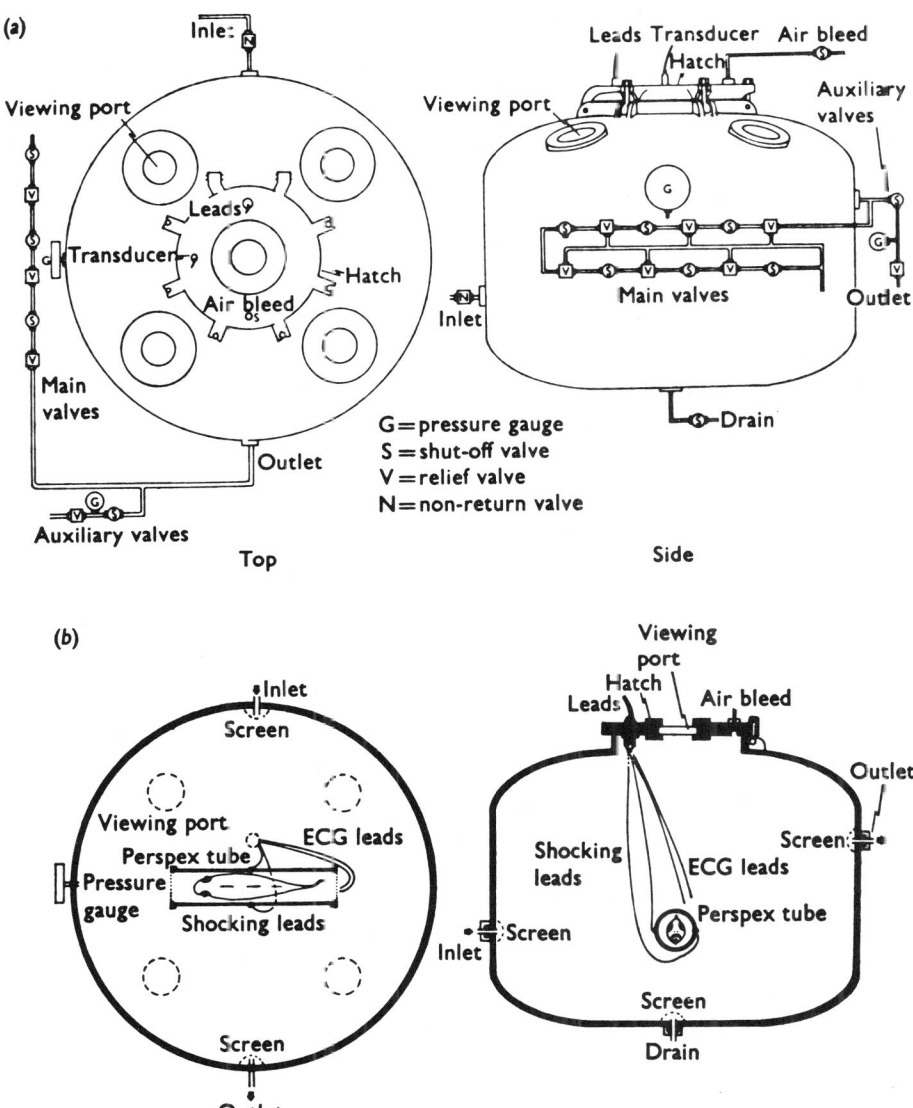

Fig. 6. Views of pressure vessel for cardiac conditioning of fish in relation to pressure up to 20 atm (from Blaxter and Tytler, 1972).

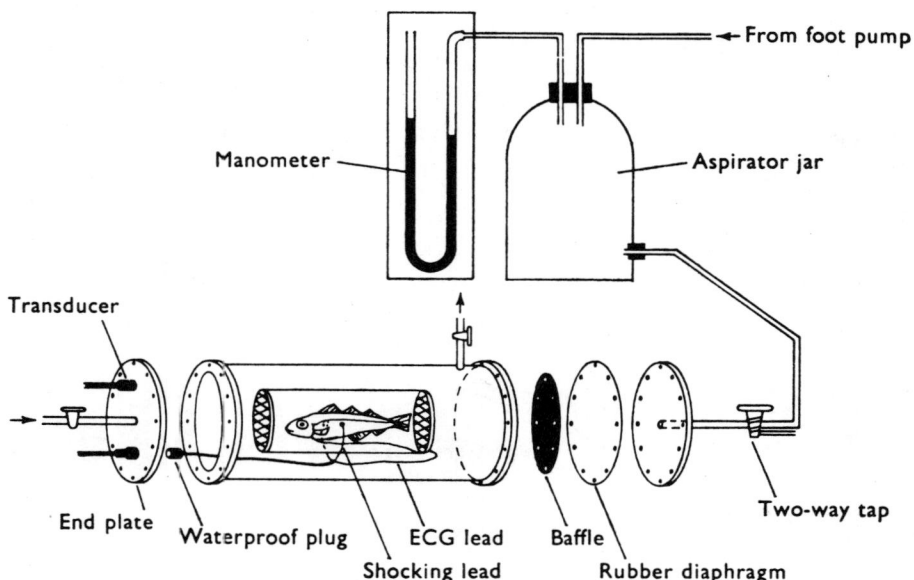

Fig. 7. Diagram of pressure vessel for cardiac conditioning of fish up to 0.5 atm (from Blaxter and Tytler, 1972).

In Figs. 6 and 7 apparatus is shown for a cardiac conditioning technique where fish are "trained" to vary the heart rate in response to pressure changes, so permitting pressure thresholds to be measured. A similar technique is seen in the air-filled pressure tank used by Kreithen and Keeton (1974) to measure pressure thresholds in pigeons.

RESPONSES OF AQUATIC ANIMALS

Behavioural responses: Pressure sensitivity was originally thought to exist only in animals with a gas phase such as a swimbladder or buoyancy float. It has become increasingly obvious in the last twenty years that many planktonic invertebrates and fish larvae without gas respond to pressure (see reviews by Knight-Jones and Morgan (1966) and Flügel (1972). These responses seem to be associated with two important rhythmical behaviour patterns, vertical migration and tidal activity. Examples of these will be taken rather than attempting a comprehensive review.

Rice (1964) described pressure responses in 41 out of 53 plankton organisms he studied. Quick pressure changes of the order of 1 atm were given over 2 or 3s and the three main responses were:

1. Pressure change causes movement orientated entirely with respect to gravity, pressure increase causing an upward movement, pressure decrease a downward movement (e.g. ctenophores, crab megalopa, *Loligo* larvae.
2. As 1 but in horizontal light increased pressure causes movement towards the light, decreased pressure causes reduced activity or movement away from the light (e.g. adult mysids, plaice larvae).
3. Pressure increase causes or enchances movement towards the light whether this involves upward, sideways or downward movement (e.g. decapod zoeae, adult copepods, barnacle nauplii).

No response to pressure was observed in some hydromedusae, polychaetes, chaetognaths and amphipods, nor in *Sepiola* larvae.

Upward movement in response to increased pressure and downward movement in response to decreased pressure is clearly a depth regulating response. It is usually associated with a geo- or phototaxis, the pressure change only acting as a trigger for movement; there is no evidence that organisms swim up or down a pressure gradient. Subsequent interpretations of these sorts of responses show how they may be involved in controlling distribution. For instance Naylor and Isaacs (1973) using the megalopa larvae of the decapods *Callinectes sapidus* and *Macropipus holsatus* showed that they have quite high thresholds which would keep them away from the surface waters of estuaries where net flow would be seaward and away from the deeper bottom water where net flow would be landward (Fig. 8). The megalopas would therefore tend to stay in the estuaries but avoid being swept ashore.

It should be noted that these organisms are all devoid of a gas phase which would be the obvious site of a pressure receptor. In larvae of herring *Clupea harengus* (see Fig. 9) the development of a gas phase in the bony otic bulla was associated with an improvement in pressure sensitivity so that they made depth changes in the apparatus shown in Fig. 3 to pressure changes of less than 13 cm H_2O (Blaxter and Denton, 1976). Unlike most other investigators these authors used positive and negative pressures and checked to see whether movement compensated for the pressure change; it did not.

Another type of result is typified by Morgan's (1965) work on the amphipod *Corophium volutator* which has no gas-filled structures. This amphipod, when brought into the laboratory, displayed an endogenous rhythm of activity with a 12.4 h cycle, the maximum being correlated with the early ebb of the tide in the original environment. The rhythm persisted for three days (see Fig. 10). *Corophium* collected from non-tidal pools was arhythmic but in such animals a tidal cycle could be imposed by subjecting the animals to cycles of pressure

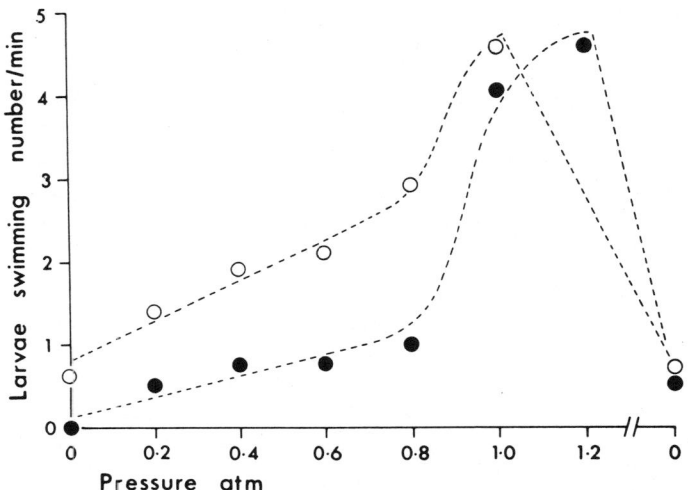

Fig. 8A. Percentage of *Callinectes sapidus* megalopas in upper half of a pressure vessel subjected to step-wise 1 min pressure increments above ambient. Results based on 4 experiments each with 10 1st day megalopas.

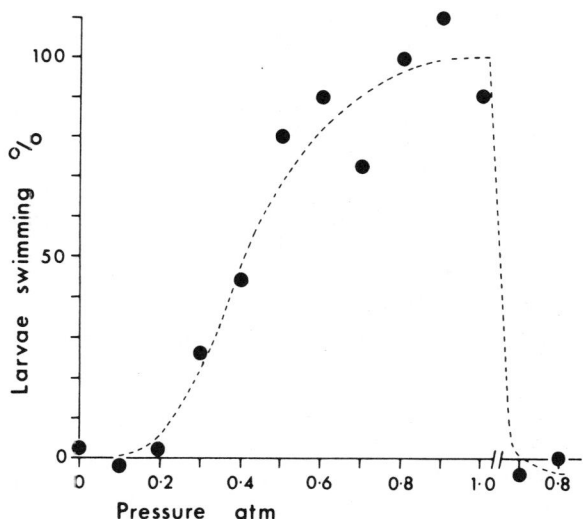

Fig. 8B. Average numbers of *Macropipus holsatus* megalopas in upper half of a pressure vessel subjected to step-wise 1 min pressure increments above ambient. O mean of 8 experiments each with 5 larvae; ● mean of 4 experiments each with 5 larvae (both figures from Naylor and Isaac, 1973).

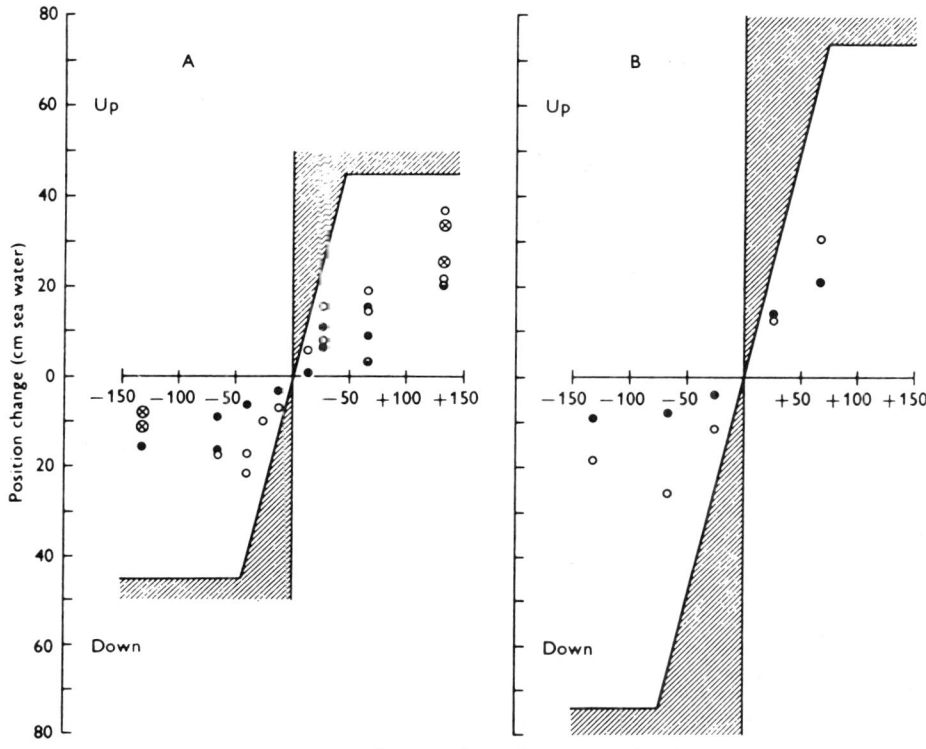

Fig. 9.

Effect of positive and negative pressures on change of larval position (all values given in cm H_2O). A. 50 cm high cylinder (45 cm depth). B. 80 cm cylinder (74 cm depth). ● otic bulla full of liquid, ◑ bulla with gas bubbles, O bulla full of gas, ⊗ bulla full of gas and pressure applied slowly. The oblique line shows the position where the larvae would have compensated precisely for the pressure change, the shading shows where there would have been over-compensation (from Blaxter and Denton, 1976).

change. The behaviour of *Corophium* - increased activity on the ebb (i.e. with falling pressure) - apparently prevented the animals being stranded as the tide went out. In the pycnogonid *Nymphon gracile* a similar reverse response was found by Morgan et al. (1964) with pressure increases inhibiting swimming. Imposed cyclical pressure changes with an amplitude of 0.8 atm caused most active swimming at late ebb or low water of the artificial cycle.

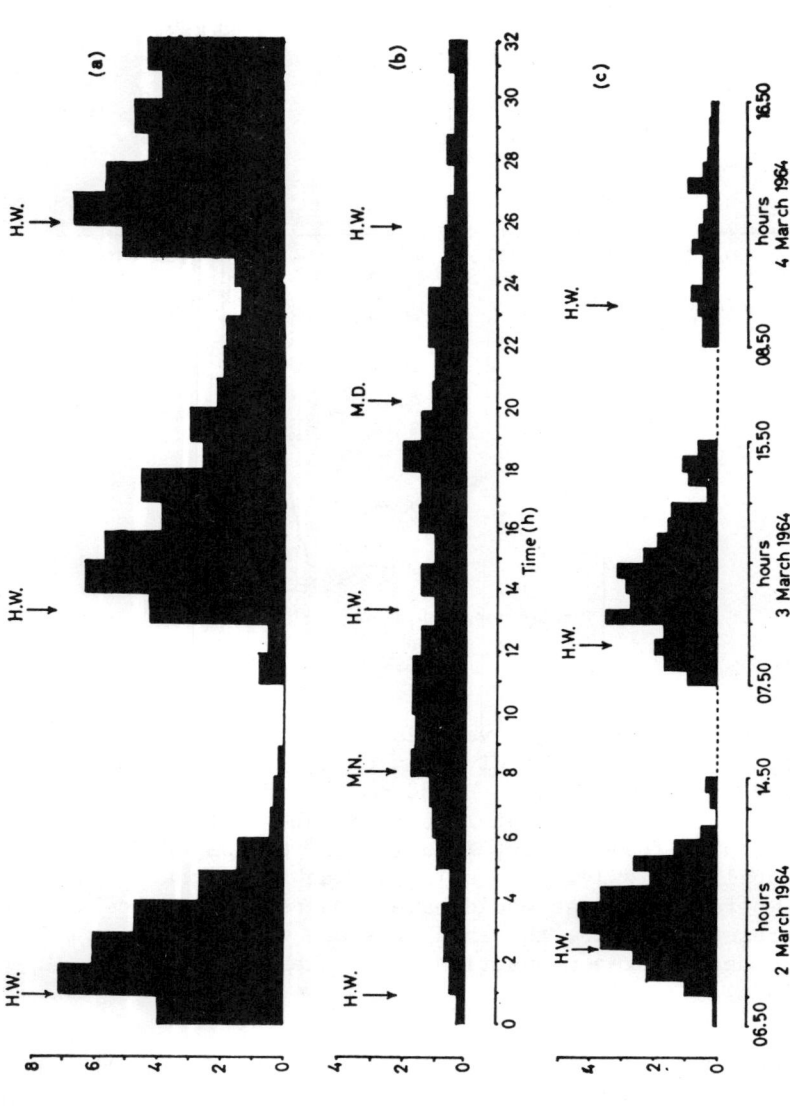

Fig. 10. Number of *Corophium* swimming above half-depth in a vessel 18 cm deep (a) collected intertidally, (b) collected generally in non-tidal pools, (c) collected intertidally and observed over a longer period, H.W. time of high water in natural environment (from Morgan, 1965).

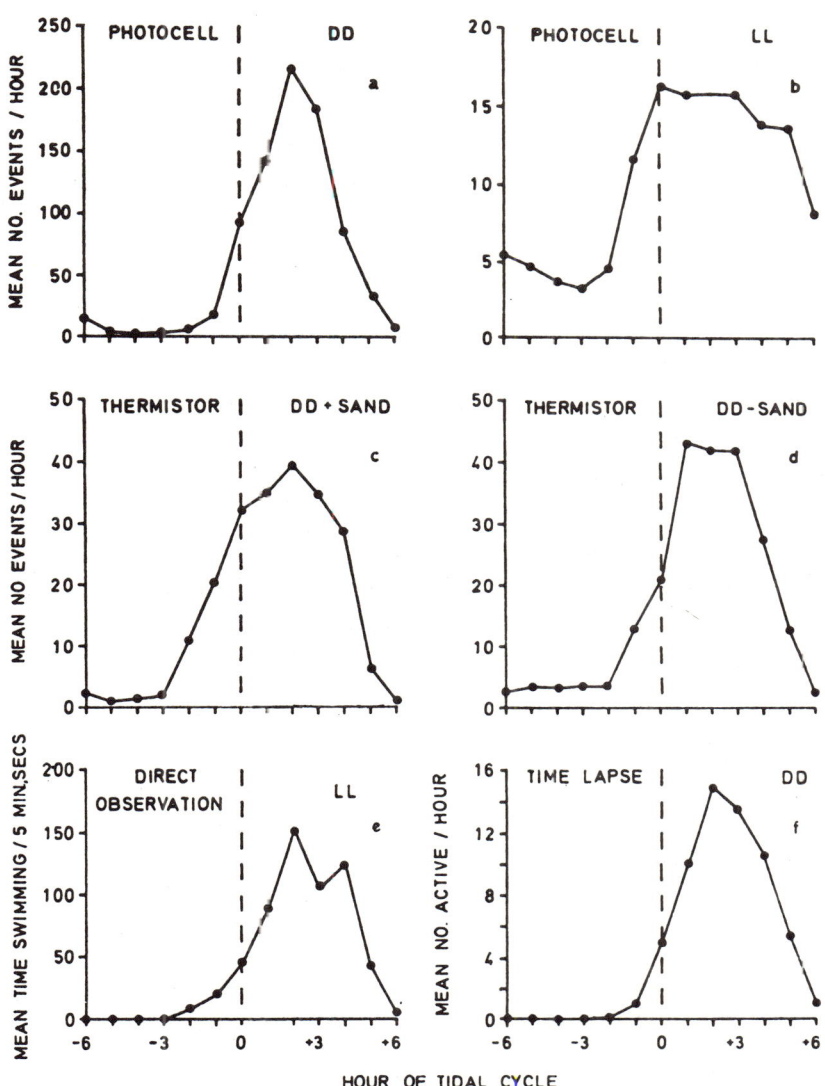

Fig. 11. Activity peaks of young plaice measured in various ways. (a) photocell and infra-red light in darkness (DD); (b) as in (a) but in continuous light (LL); (c) using thermistors in darkness with a sand substratum; (d) as (c) but without sand; (e) direct observation; (f) using a time lapse technique. High tide in natural environment is shown by the vertical dashed line (from Gibson, 1975).

Tidal cycles are also shown by flatfish such as plaice *Pleuronectes platessa* without swimbladders (Gibson, 1975; Gibson *et al.*, 1978). These fish are most active on the "ebb" (Fig. 11); if this behaviour occurs in the sea it might prevent stranding as the tide recedes. The threshold for pressure responses was of the order of 100 cm H_2O.

Spontaneous behaviour of a different sort was used by McCutcheon (1966) in the pinfish *Lagodon rhomboides* and sea bass *Centropristus striatus* and in 9 other physoclists (with closed swimbladders) and the goldfish, a physostome (with pneumatic duct from the swimbladder to the gut). All species showed yawning behaviour if the pressure was reduced by 0.2 cm H_2O for a minute or more. A cyclical swimming activity around a position in the tank (Fig. 5) to which the fish was adapted was observed when the pressure was changed by ±1.0 cm H_2O with the first movement being downwards for a pressure decrease and upwards for an increase.

Rather different and non-spontaneous behavioural criteria have been used on other fish. Thus Dijkgraff (1941) used an operant conditioning technique with minnows *Phoxinus laevis*. They were trained to associate food with small changes of pressure and showed clear searching movements after training with pressure changes of only 0.5 - 1.0 cm H_2O. These fish possess Weberian ossicles connecting the swimbladder to the inner ear; when the ossicles were extirpated the fish would not respond to 40 cm H_2O.

Physiological responses: It is noteworthy that almost all experiments on pressure sensitivity in invertebrates have involved observing changes in behaviour while in fish physiological criteria for pressure sensitivity have been used. This is, of course, partly explained by the difficulty of making physiological measurements on small invertebrates and the difficulty of observing the behaviour of fish in rather large tanks which need to be pressurized. Some statements in the literature on invertebrates e.g. Laverack (1976) or Clarac (1976) that pressure could stimulate the hair sensillae of crustacea may be misleading since these organs are more likely to respond to the particle displacement stimulus of sound waves, or to water currents, and not to DC or AC pressure changes.

Qutob (1962) reviewed the sparse literature on the use of physiological responses to determine pressure sensitivity in fish. Vasilenko and Livanov (1936) recorded activity in the pneumogastric branch of the vagus in isolated preparations of the swimbladder of the carp *Cyprinus carpio* and found a change in the frequency of impulses as the internal pressure of the swimbladder was changed by a cannula. Qutob retained the swimbladder of roach *Leuciscus rutilus* and rudd *Scardinius erythrophthalmus* in situ. Tactile stimulation of the swimbladder wall of roach and increases of internal pressure of 53 cm H_2O caused an increase in discharge rate of the pneumogastric nerve. Release of internal pressure from 79 cm H_2O to ambient relaxed the wall and caused depression or loss of the impulses. In the rudd external pressure decreases of 16 cm H_2O also caused distension of the swimbladder wall and increased nervous activity. Increase in external pressure of 23 cm H_2O relaxed the wall and was inhibitory.

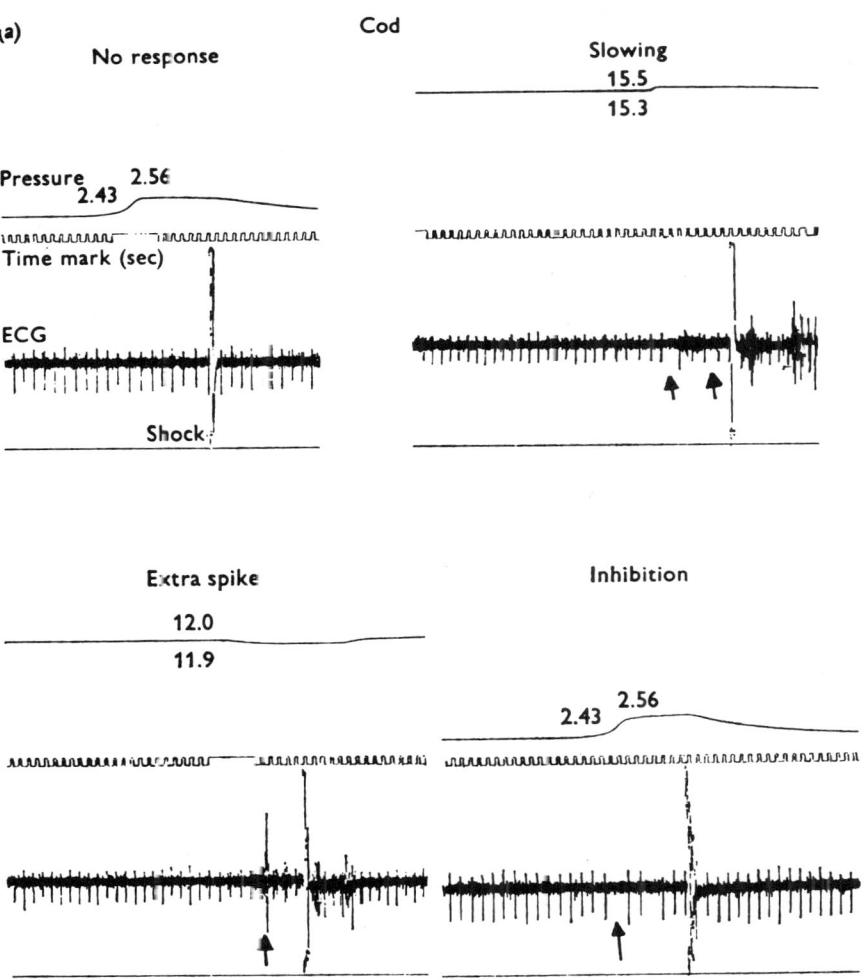

Fig. 12. Cardiac responses of cod shown as ECG traces from a Devices M2 pen recorder. Pressure changes are given in atm. The arrows indicate cardiac responses before an electric shock, which is shown as a large spike on the ECG record (from Blaxter and Tytler, 1972).

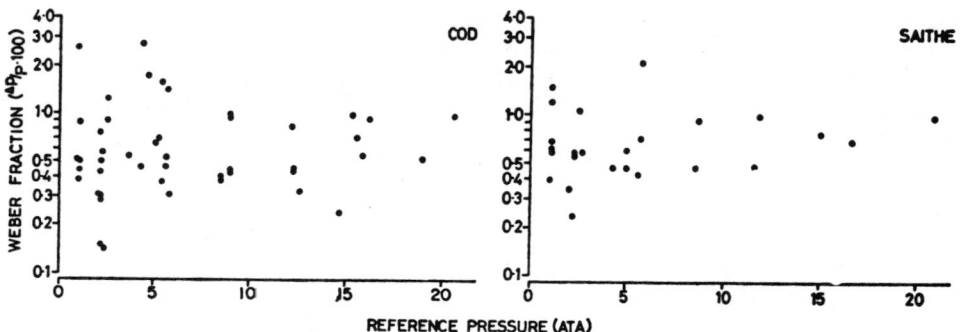

Fig. 13. Thresholds of pressure discrimination, at different ambient pressures in cod and saithe (from Tytler and Blaxter, 1973).

These three species have a pneumatic duct to the gut and Weberian ossicles which link the swimbladder to the inner ear and which could enhance pressure sensitivity. The gadoid fishes have a simpler closed swimbladder. A cardiac conditioning technique was used by Blaxter and Tytler (1972) and Tytler and Blaxter (1973) to investigate their pressure sensitivity. Some of the gadoid species are known to make substantial diel vertical migrations. The apparatus used is shown in Figs. 6 and 7, a wide range of ambient pressure up to 21 atm being used, equivalent to the pressure on the sea bed at the edge of the continental shelf. In cardiac conditioning the fish is "trained" to respond to a stimulus by slowing the heart beat (see Fig. 12). Thresholds of perception can be easily established in this way. In cod *Gadus morhua* and saithe *Pollachius virens* the threshold was 5 - 10 cm H_2O when adapted to atmospheric pressure and remained at about 0.5-1% of the pressure when the fish were adapted to higher pressures (see Fig. 13). In the haddock *Melanogrammus aeglefinus* with a swimbladder and in the dab *Limanda limanda* without a swimbladder the threshold was about 2% of the adapted pressure.

RESPONSES OF TERRESTRIAL ANIMALS

Cardiac conditioning has also been used in a very different species, the pigeon *Columbia livia*, by Kreithen and Keeton (1974). They found a threshold of about 0.001 atm equivalent to about 1 cm H_2O or 10 m of altitude (see Fig. 14). It is interesting that the threshold corresponds closely to that of the most sensitive aquatic organisms with a swimbladder. The conditioned response was, however, a cardiac acceleration rather than slowing.

RESPONSES OF PLANTS

Membrane permeability, photosynthesis, growth and other plant functions are inhibited by very high pressures outside the ecological range (Vidaver, 1972). Much lower pressures of the order of 1- 2 atm can, however, influence differentiation and morphogenesis.

Once of the most interesting self-regulating pressure responses has been described by Walsby (1972). Some photosynthetic blue-green algae contain very small gas vesicles. When transferred from a low light to a high light intensity some of the vesicles are lost with an associated reduction in buoyancy. This effect seems to be due to an increase in turgor pressure of the cell as a result of photosynthetic products. As the cell sinks the light intensity falls, photosynthesis becomes reduced, these products start to be metabolised and the vesicles reappear and help the cell to rise again in the water (see Fig. 15).

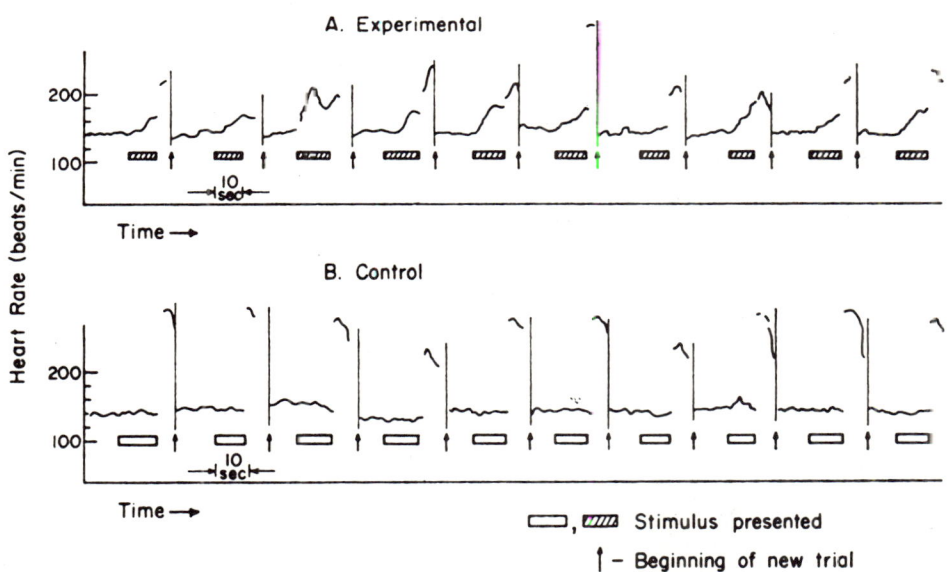

Fig. 14. Cardiac responses of pigeons. In A 10 consecutive trials were done with ± 2 cm H_2O pressure change as shown by the hatched block. The shock followed the end of the pressure change. In 3 10 control trials are done without a pressure change. Arrows show where the chart is stopped. Note how the heart rate rises during the pressure stimulus but before the shock. The shock itself also causes a massive acceleration, (from Kreithen and Keeton, 1974).

QUANTITATIVE ASPECTS OF THE STIMULUS

Most authors have used a sudden square wave or DC hydrostatic pressure change to evoke responses. Almost all the thresholds for response are based on this type of stimulus which is ecologically abnormal. The pressures from sound waves are simusoidal in form as are changes in tidal height. There is unlikely to be a square wave front in any type of stimulus except perhaps in front of a towed obstacle such as a trawl door.

From Table 1 it can be seen that many animals with different types of putative receptor have a threshold between 1 and 10 cm H_2O (0.001 - 0.01 atm) or 0.1 - 1% above atmospheric pressure. Where gas-filled structures are present sensitivity seems to be enhanced with a threshold of 0.1 - 1.0%. Where no gas is present the thresholds tend to be in the range 1.0 - 10.0%. Further results (Table 2) from Tsvetkov (1969) on a number of freshwater fish with swimbladders also show thresholds equivalent to a pressure change of 0.1%, different techniques tending to give the same results.

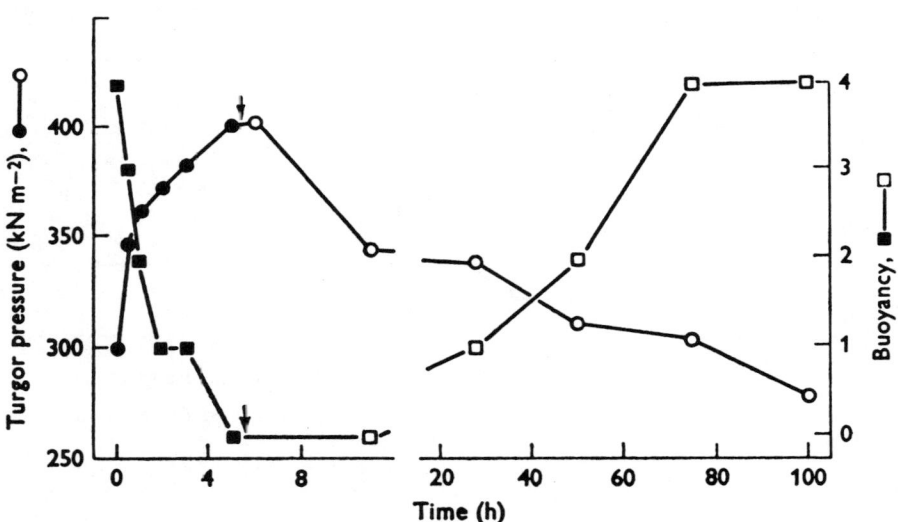

Fig. 15. Changes in turgor pressure and buoyancy of blue-green algae *Anabaena flos-aqua* ● turgor pressure rising following transfer to bright light ■ loss of buoyancy, ○ turgor pressure falling, □ increase of buoyancy following return to dim light (arrowed) (from Walsby, 1972).

TABLE 1. THRESHOLDS FOR PRESSURE DISCRIMINATION

SPECIES	THRESHOLD CM H_2O	AUTHOR(S)
NO GAS PHASE		
Synchelidium sp. (Amphipod)	5 – 10	Enright (1962)
Callinectes sapidus (Both decapod megalopa)	400	Rice (1964)
Macropipus holsatus	800 – 1000	
Coelenterate medusae	300 – 500	
Polychaete adult and larvae	20 – 50	
Temora longicornis (Copepod)	10	Knight-Jones and Morgan (1966)
Cirripede nauplii	< 100	
Cirripede cyprids	10	
Decapod larvae	10 – 25	
Calanus helgolandicus (Copepod)	700	Lincoln (1971)
Daphnia magna (Cladoceran)	3500	
Carcinus maenas (Shore crab)	< 100	Naylor and Atkinson (1972)
Centronotus gunnellus (Larval blenny)	25	Qasim *et al.* (1963)
Limanda limanda (Adult dab)	20	Blaxter and Tytler (1972)
Pleuronectes platessa (Juvenile plaice)	100	Gibson *et al.* (1978)

TABLE 1 (CONTINUED)

SPECIES	THRESHOLD CM H_2O	AUTHOR(S)
WITH GAS PHASE		
Phoxinus laevis (Minnow)	0.5 – 1.0	Dijkgraaf (1941)
"	5	Qutob (1962)
Lagodon rhomboides (Pinfish) }	0.5 – 2	McCutcheon (1966)
Centropristus striatus (Sea bass) }		
Scardinius erythrophthalmus (Rudd)	20	Qutob (1962)
Blennius pholis (Larval blenny)	5	Qasim *et al.* (1963)
Gadus morhua (Cod) }	5 – 10 }	Blaxter and Tytler (1972)
Pollachius virens (Saithe) }		
Melanogrammus aeglefinus (Haddock)	20	
Clupea harengus (Herring larvae and juveniles)	< 13	Blaxter and Denton (1976)

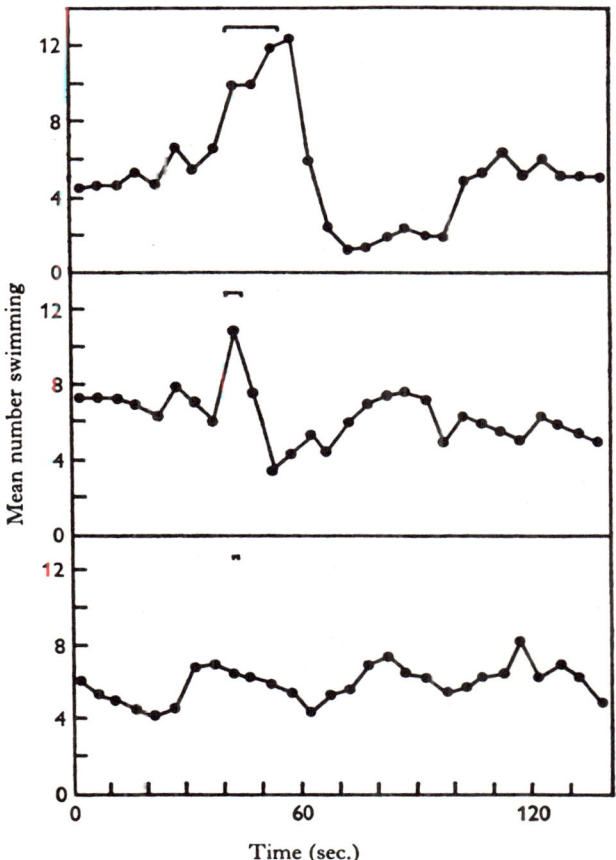

Fig. 16. Number out of 20 *Nephtys* swimming at a pressure of 67 cm H$_2$O applied for 1.5 and 15s as shown by the horizontal bar (from Morgan, 1969).

Enright (1962) considered the liminal time (the minimum stimulus duration for response) for a pressure stimulus of 53 cm H$_2$O using the amphipod *Synchelidium* and found a value of 1s. Morgan (1969), however, found a liminal time between 1 and 5s for the polychaete *Nepthys* (Fig. 16). Enright also showed that the resulting activity followed the Fechner principle and increased linearly as the logarithm of the pressure increment (Fig. 17). Sulkin (1973), however, found a linear relationship between the pressure increment and swimming speed of xanthid megalopas (Fig. 18).

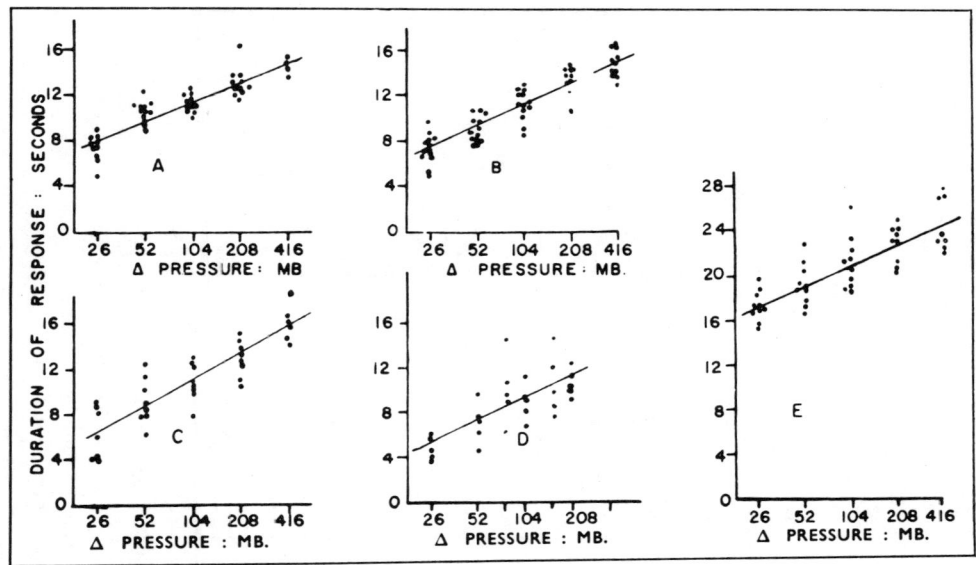

Fig. 17. Duration of activity of *Synchelidium* related to log pressure increment in mb (1 mb = 1 cm H_2O). A-D inactive phase of tidal rhythm. E active phase (from Enright, 1962).

Very few workers have considered the effect of adapting or holding organisms at pressure above atmospheric and then re-testing the response. Tytler and Blaxter (1973) measured the pressure sensitivity of saithe at adaptation pressures up to 20 atm. Weber's Law, that the relative sensitivity is independent of absolute stimulus intensity, seemed to apply, the sensitivity being about 0.5% of adaptation pressure (Fig. 13).

ECOLOGICAL IMPLICATIONS OF BARORECEPTION

Knight-Jones and Morgan (1966) and Naylor and Atkinson (1972) have reviewed the ecological importance of pressure responses. Diel vertical migration is a widespread phenomenon but it seems to be predominantly directed by responses to light. Light, however, while varying in a regular way to signal dusk and dawn over a 24 h period, is much less satisfactory than pressure as a depth-sensing device because of the wide range of turbidity in adjacent areas. It is quite possible for light penetration to vary by over two orders of magnitude in the same area depending on the influence of run-off from the land or turbulence induced by bad weather.

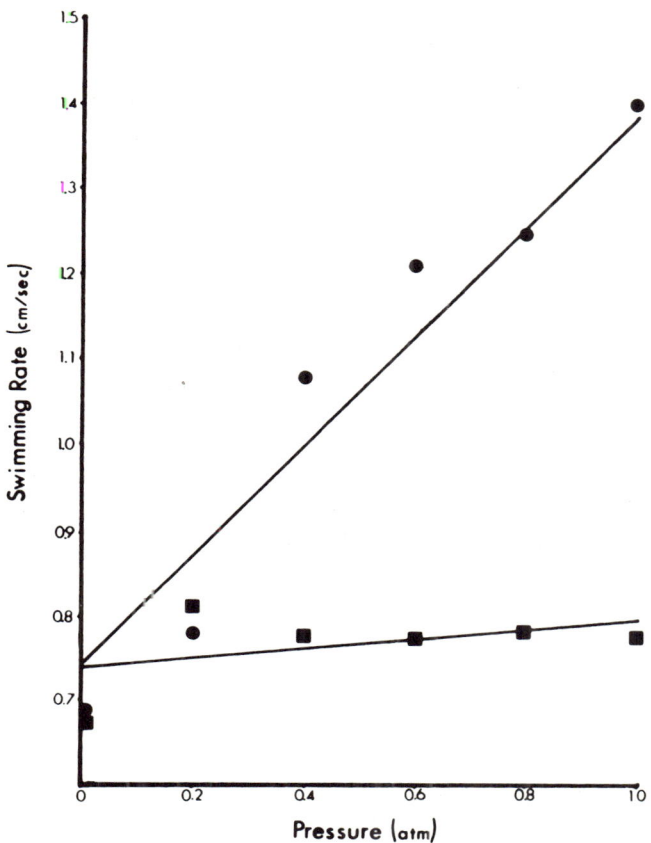

Fig. 18. Regression of swimming speed on pressure in Stage I (●) and Stage IV (■) soeae of the crab *Leptodius floridanus* (from Sulkin, 1973).

The tidal rhythms of inshore invertebrates described by many workers seem much better candidates for a control by pressure. The organisms are near the surface so that given increments or decrements of depth will give much bigger percentage pressure changes. The tidal areas will be especially subject to changes of turbidity which would make light an unpredictable stimulus. Some of the earlier examples show how tidal rhythms may prevent stranding or retain organisms in the coastal belt. Endogenous rhythms have been demonstrated especially in surf plankton animals which enable them to undergo tidal flood transport to return to estuaries or inshore settling sites, or ebb tide transport to prevent stranding (see Naylor and Atkinson, 1972).

TABLE 2. THRESHOLDS FOR PRESSURE
DISCRIMINATION USING DIFFERENT
TECHNIQUES (FROM TSVETKOV, 1969)

SPECIES	NUMBER USED	THRESHOLD IN CM H_2O		
		FROM BUOYANCY	FROM FRIGHT RESPONSE	FROM CONDITIONED REFLEX
Cichlasoma biocellatum	2	± 1.6	± 2.0	–
Acerina cernua (Ruffe)	2	–	± 1.4 – 1.5	–
Corydoras paleatus	2	–	± 0.4	± 0.4 – 0.5
Perca fluviatilis (Perch)	5	± 1.1 – 1.7	± 1.1 – 1.7	± 1.4
Macropodus opercularis	2	–	–	± 1.2
Esox lucius (Pike)	6	± 0.5 – 0.6	–	–

Pressure sensitivity thresholds of the order of 0.1 – 1% of adapted pressure (see Tables 1 and 2) are commonplace, especially in organisms with gas-filled structures. The ability to appreciate changes in depth will be much enhanced near the surface as shown in Fig. 19; a downward movement of 1 – 10 cm would, for example, be appreciated at the surface but at 90 m only a downward movement of 10 – 100 cm would be sensed.

In birds like the pigeon, a 0.1% threshold would enable them to monitor a change in altitude of 10 m and to respond to changes in atmospheric pressure of 1 mb. Nisbet and Drury (1968) found that migration density of birds was correlated with low and falling atmospheric pressure. Although it is difficult in field observation to eliminate multiple variables it seems that the sensory system of birds has the potential to perceive weather change.

In an analagous way the possibility that fish might respond to waves passing overhead and therefore respond to bad weather indirectly is worth exploring. A change in atmospheric pressure of 10 mb would alter the pressure at the sea surface by 1% which would be within the sensitivity range of many fish but only as tested for rapid pressure changes. Waves would cause much more rapid pressure changes but only

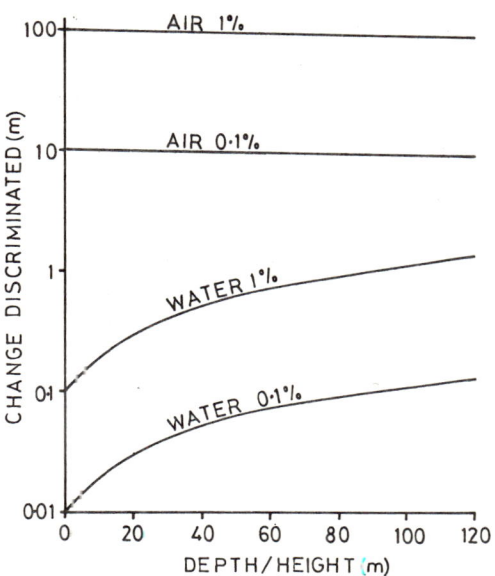

Fig. 19. Vertical excursions which would be discriminated assuming a 1% or 0,1% threshold change in pressure at different depths/altitudes.

to fish or other organisms on the sea bed. Off the bottom circular movements of water below waves tend to compensate for pressure changes (Blaxter and Tytler, 1972). In Fig. 20 the sea bed depths at which waves of various height or length could be perceived are plotted on the assumption that the threshold for pressure sensitivity is 1%.

Implicit in this discussion is that animals can sense gradual changes of pressure or pressure gradients, but almost all experiments have been done using square wave (DC) or sudden pressure changes. A few results e.g. Enright (1962) indicate that thresholds are less when the pressure stimulus changes slowly. Thus the threshold for *Synchelidium* was 25 - 50 cm H_2O when the pressure was raised at 2.8 cm H_2O/s but was 5 - 10 cm H_2O with a rapid rise. Morgan (1969) found the swimming activity of *Nephtys* depended on rate of change of pressure between ± 2.5 cm/sec (see Fig. 21). Tidal ranges of 5 - 10 m are commonplace giving average rises or falls of pressure of only 0.023 - 0.046 cm H_2O/s. For non-aquatic animals atmospheric pressure changes of 1 mb in 30 min could mark the passage of a cold front (Kreithen and Keeton, 1974), but this is equivalent to only 0.0006 cm H_2O/s. If aquatic or terrestrial animals are to perceive such slow changes an absolute sense of pressure giving a fixed reference pressure seems desirable.

It is far from clear whether animals possess an absolute pressure sense. The swimbladder is often an actively secreting or resorbing organ adapting to changes of depth and is quite unsuited to absolute perception (see Blaxter and Tytler, 1978 for a full discussion of this). Other possible pressure receptors, as will be described in the last section, or a non-adapting lung may well be better suited to absolute pressure perception.

THE RECEPTOR

Gas-containing structures: Baroreceptors containing gas might be expected to have the greatest sensitivity since the volume changes for given pressure changes will be greatest. The data set out in Tables 1 and 2 tend to confirm this expectation. Amongst the aquatic invertebrates gas-filled structures are found in siphonophores,

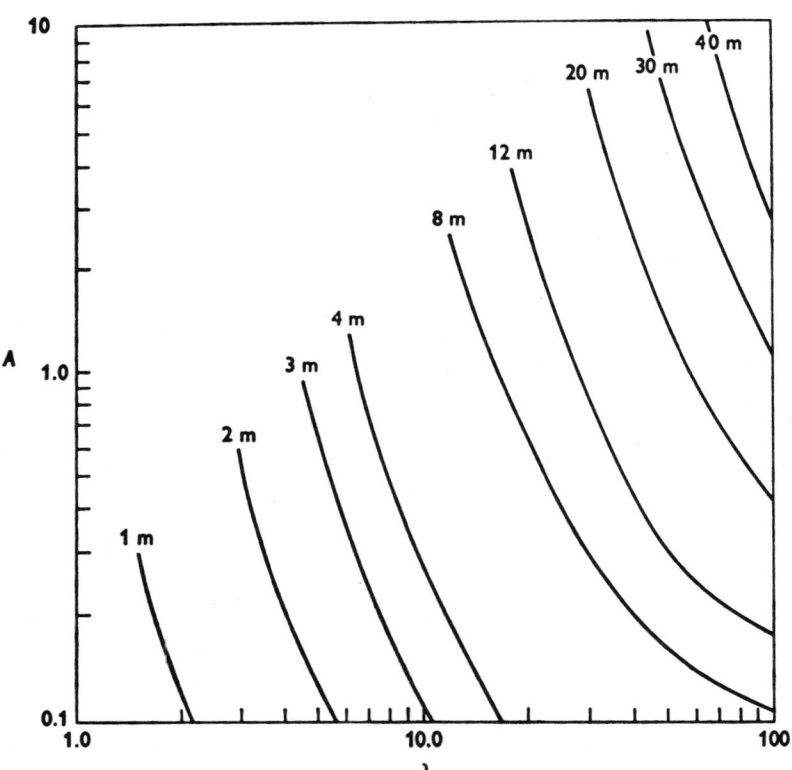

Fig. 20. The sea bed depth at which a pressure changes of 1% would be experienced for different wave heights (A) and lengths (λ) expressed in metres (from Blaxter and Tytler, 1972).

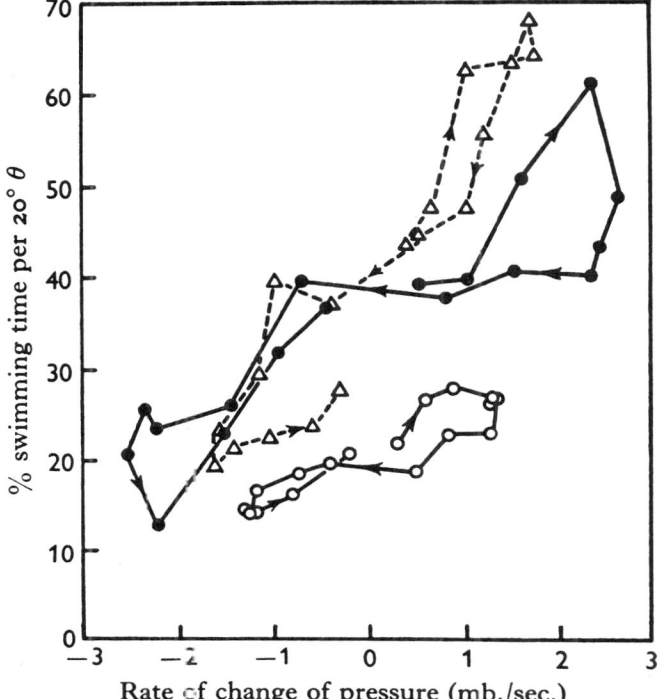

Fig. 21. Relationship between swimming activity and rate of pressure change in *Nephtys*. The activity is expressed as % of total possible swimming time related to the phase of the pressure change, i.e. $0°$ = atmospheric pressure, $180°$ ≡ maximum pressure over 3 consecutive cycles each frequency. ● cycles of 415 cm H_2O over 480s, O cycles of 415 cm H_2O over 860s, Δ cycles of 200 cm H_2O over 360s (from Morgan, 1969).

cephalopods and aquatic insects and in vertebrates in the swimbladders or teleosts and the lungs of aquatic reptiles, diving birds and mammals.

A review of the physiology of the swimbladder by Blaxter and Tytler (1978) shows that it may contain gas at a somewhat higher pressure than ambient but generally it is compliant and changes volume with depth according to Boyle's Law. This would also be true of the unsupported gas bubbles of siphonophores but not of the chambered buoyancy shells of cephalopods which implode at high pressures as a result of their rigid envelopes. The cuttlebone of *Sepia officinalis* imploded at 24 atm (Denton *et al.* 1961) and the shell of *Spirula spirula*, a deepwater form, at 130 - 230 atm (Denton and Gilpin-Brown, 1971).

There is some direct evidence that the swimbladder is acting as a baroreceptor. Qutob (1962) found an increase in the firing rate of the pneumogastric nerve in cyprinids when the swimbladder wall was distended by increased internal pressure or decreased external pressure. Relaxation of the swimbladder wall was inhibitory. Tytler and Blaxter (1977) found that the pressure threshold for saithe was greatly decreased after withdrawal of the swimbladder gas but the threshold was partly or wholly restored when gas had been re-secreted (see Table 3).

TABLE 3. PRESSURE DISCRIMINATION THRESHOLDS OF SAITHE AFTER DEFLATING THE SWIMBLADDER AND ALLOWING RECOVERY. THRESHOLDS IN CM H_2O. NUMBER OF FISH IN BRACKETS (FROM TYTLER AND BLAXTER, 1977).

CONTROL	SHAM OPERATION	1ST DEFLATION	1ST RECOVERY	2ND DEFLATION	2ND RECOVERY
12.6 (10)	10.6 (7)	86.6 (8)	44.6 (8)	131.0 (2)	9.0 (2)

Although there is some evidence of innervation of the swimbladder wall (see Qutob, 1962) it is not clear whether the nerve endings are stretch receptors or whether they have a function in controlling gas secretion and resorption as a part of buoyancy regulation. Only Qutob's own electrophysiological work suggests that some of them have a role in baroreception. It is also quite possible that changes in volume of the swimbladder are stimulating receptors in the surrounding body wall.

Although stretch receptors have been described in fish (e.g. Bone and Chubb, 1975) they have been associated with fin movements and there are no quantitative data on the degree of stretch required to elicit a response. A simple calculation shows that the *percentage* linear stretch of the wall of a gas-filled structure is independent of its size (Fig. 22). If the response of a stretch receptor depends on its change in length relative to the "resting" length the size of the associated gas-filled structure is therefore unimportant.

A gas-filled baroreceptor has also been described in the water bug *Aphelocheirus aestivalis* by Thorpe and Crisp (1947). This organ is found in shallow depressions in the cuticle covered with backwardly

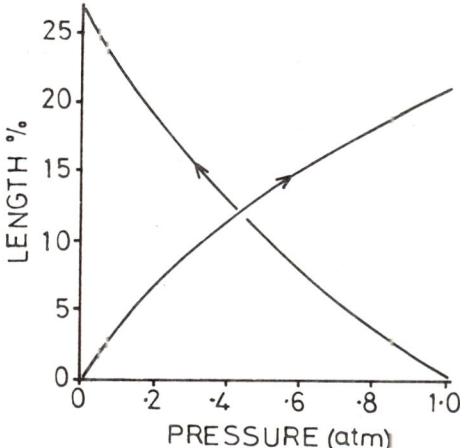

Fig. 22. % change in length of a hypothetical stretch receptor activated by a gas-filled structure "obeying" Boyle's Law as the pressure changes from 0 atm (1 ATA) to 1 atm and back again.

pointing hairs, some hydrofuge and some sensory, overlain by a film of air. When this film is compressed the hydrofuge hairs are pressed down and move the sensory hairs. Adjacent to this structure is a small air sac which may damp out fluctuations of pressure caused by muscle or body movements. *Nepa*, another aquatic bug, has pressure receptors linked into a differential manometer system which should permit the animal to orientate its body axes with respect to the horizontal plane (Fig. 23).

Charged membranes: Digby (1972) reviews his well-known hypothesis that pressure influences the potential across the cuticle of decapod crustacea. Because of their high sensitivity to pressure change he postulated that there must be an ultra-thin layer of gas over the cuticle, probably of hydrogen. Pressure-sensitive behaviour is reversibly reduced or abolished by oxygen or hydrogen peroxide suggesting the presence of nascent hydrogen and is also reduced by strong salt solutions which would decrease the potential across the cuticle (Fig. 24). Digby also showed that current flow through a pair of metallic electrodes immersed in fresh or salt water is sensitive to small changes in hydrostatic pressure. This sensitivity is also reduced by oxygen and hydrogen peroxide. The crustacean cuticle is semi-conducting showing an electrolytic effect when potentials are applied across it, hydrogen being produced at the cathode. These electrolytic effects may well be involved in the deposition of calcium.

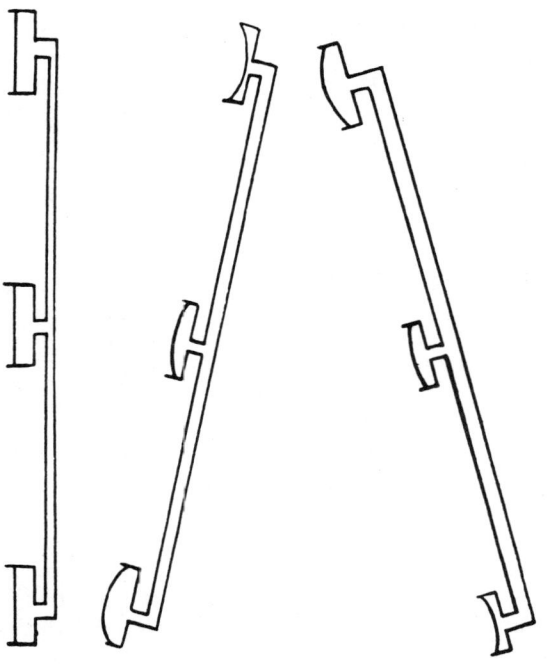

Fig. 23. Showing the principle of action of the linked pressure receptors in *Nepa*. The system consists of three distensible membranes enclosing an air space and interconnected by a gas-filled tube. As the body tilts the membranes are affected differentially by the pressure (from Thorpe and Crisp, 1947).

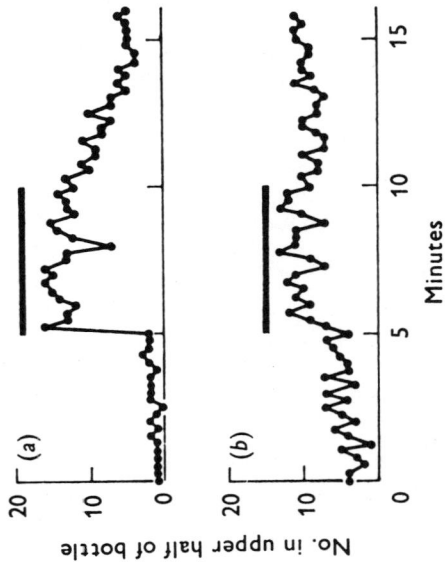

Fig. 24. Effect of substances on pressure sensitivity of *Praunus flexuosus*. Effect of H_2O_2 (a) control in sea water (b) experimental groups 0.2 vol H_2O_2 in sea water. Heavy lines show pressure increase of 1 atm above ambient (from Digby, 1972).

Much of the evidence for this hypothesis is circumstantial or based on physical experiments with electrodes and it is far from clear how a layer or hydrogen, perhaps a few angström units thick and changing in thickness with pressure, could stimulate the nervous system. Enright (1963) calculated that at threshold the surface displacement of a gas layer would be only 2°A. Nor is it clear how such a thin layer would be maintained. Being volatile and very thin the layer would not be visible by any know microscopic techniques, so that proof of the existence of the layer is lacking.

Tissue compressibility: Shearing might occur between adjacent tissues of different compressibility. Such shear could be monitored by hair sensillae or other receptors. Enright (1963) found that the compressibility of *Euphausia pacifica*, an oceanic euphausiid, and of *Excirolana chiltoni*, an intertidal isopod, was 15 – 40% lower than sea water. Some of the more watery tissues are likely to have the same compressibility as sea water and presumably the least compressible structures will be endo or exo-skeleton. The potential structures for differential compressibility creating shear are therefore available.

It is possible that the pressure receptors reported in fish gills work on this principle. Laurent and Rouzeau (1972) analysed the electric activity in the pseudobranch nerve of the rainbow trout *Salmo gairdnerii*. They found an increase in activity when the pressure of the perfusion fluid was pulsed. This is, of course, internal hydrostatic pressure and there is no evidence that such a system would respond to external changes in hydrostatic pressure.

Piezo-electric effects: Crystalline structures like bone and otoliths may have piezo-electric properties. Morris and Kittleman (1967) found that crushed otoliths of the Antarctic nototheneid *Trematomus bernacchii* and Pacific flatfish *Parophrys vetulus*, neither with a swimbladder, were strongly piezo-electric with a magnitude similar to Rochelle salt. The otolith of *Trematomus* oscillated at 300 – 800 Hz and *Parophrys* at 1 – 15 kHz. About one third of the 32 known classes of crystals respond to hydrostatic pressure, changing the output frequency when a pressure stimulus is applied. This clearly could provide a mechanism of pressure reception in fish, especially as the otoliths are in such close contact with the highly sensitive maculae of the inner ear.

CONCLUSIONS

Many animals seem to be sensitive to pressure changes of the order of 1% of the adapted pressure, the thresholds tending to be lower where a gas-filled structure is present. There is, however, a surprising gap in our knowledge of whether such structures are, indeed, pressure sensors or what other structures may act as sensors in pressure-sensitive animals without a gas phase.

High sensitivities of the order of 0.1 - 1% would enable animals, especially in water, to be aware of rather small changes in depth. The problem of detecting such changes against the background "noise" created by the animals own movements seems to be great and it is not at all clear why such high sensitivity seems to have been developed.

A considerable limitation of many experiments to date has been the use of a sudden change of pressure. It is not always certain whether the animals have responded to a change of hydrostatic pressure or to some transient stimulus associated with the pressure front. An experiment needs to be devised where aquatic animals impose pressure changes on themselves by their vertical movements and at the same time some sort of physiological changes are measured.

Gas-filled structures like the swimbladder are often in a dynamic state, adjusting for depth changes by gas secretion or resorption. Gas will also tend to diffuse out of such structures especially at high ambient pressures. If an absolute pressure sense exists as a reference for slow pressure changes then it might be better situated in an all-fluid structure. In fact, we do not know whether animals have an absolute pressure sense, which would enable them to detect slow pressure changes incurred during vertical migration or tidal changes in water, or during the passage of weather fronts on land.

REFERENCES

Blaxter, J.H.S. and Denton, E.J. (1976). Function of the swimbladder-inner ear-lateral line system of herring in the young stages. J. mar. biol. Ass. U.K. 56: 487-502.
Blaxter, J.H.S. and Tytler, P. (1972). Pressure discrimination in teleost fish. Symp. Soc. exp. Biol. 26: 417-443.
Blaxter, J.H.S. and Tytler, P. (1978). Physiology and function of the swimbladder. Adv. comp. Physiol. Biochem. 7:
Bone, Q. and Chubb, A.D. (1975). The structure of stretch receptor endings in the fin muscles of rays. J. mar. biol. Ass. U.K. 55: 939-943.
Brauer, R.W. (1972). Barobiology and the experimental biology of the deep sea. North Carolina Sea Grant Programme pp. 428.
Clarac, F. (1976). Crustacean cuticular stress detectors. In "Structure and Function of Proprioceptors in the Invertebrates". 299-321, Ed. P.J. Mill, Chapman & Hall: London.
Denton, E.J. and Gilpin-Brown, J.B. (1971). Further observations on the buoyancy of *Spirula*. J. mar. biol. Ass. U.K. 51: 363-373.
Denton, E.J., Gilpin-Brown, J.B. and Howarth, J.V. (1961). The osmotic mechanism of the cuttlebone. J. mar. biol. Ass. U.K. 41: 351-364.

Digby, P.S.B. (1972). Detection of small changes in hydrostatic pressure by Crustacea and its relation to electrode action in the cuticle. Symp. Soc. exp. Biol. 26: 445-471.

Dijkgraaf, S. (1941). Über die Bedeutung der Weberschen Knöchel für die Wahrnehmung von Schwankungen des hydrostatischen Druckes. Z. vergl. Physiol. 28: 389-401.

Enright, J.T. (1962). Responses of an amphipod to pressure changes. Comp. Biochem. Physiol. 7: 131-145.

Enright, J.T. (1963). Estimates of the compressibility of some marine crustaceans. Limnol. Oceanogr. 8: 382-387.

Flügel, H. (1972). 8. Pressure 8.3 Animals. In Marine Ecology Vol.1, Part 3; 1407-1437. Ed. O. Kinne, Wiley Interscience: London, New York.

Gibson, R.N. (1975). A comparison of field and laboratory activity patterns of juvenile plaice. Proc. 9th Europ. Mar. Biol. Symp. 13-28. Ed. H. Barnes, Aberdeen University Press.

Gibson, R.N., Blaxter, J.H.S. and de Groot, S.J. (1978). Developmental changes in the activity rhythms of the plaice. Proc. Symp. "The Rhythmic Activity of Fish" held at University of Stirling, Scotland in July 1977. Academic Press; London and New York.

Kinne, O. (1972). 8. Pressure (1) General aspects. In Marine Ecology Vol 1, Pt 3. 1323-1360. Ed. O. Kinne, Wiley Interscience; London, New York.

Knight-Jones, E.W. and Morgan, E. (1966). Responses of marine animals to changes in hydrostatic pressure. Oceanogr. mar. Biol. Ann. Rev. 4: 267-299.

Kreithen, M.L. and Keeton, W.T. (1974). Detection of changes in atmospheric pressure by the homing pigeon *Columba livia*. J. comp. Physiol. 89: 73-82.

Laurent, P. and Rouzeau, J.-D. (1972). Afferent neural activity from pseudobranch of teleosts. Effects of PO_2, pH, osmotic pressure and Na^+ ions. Resp. Physiol. 14: 307-331.

Laverack, M.S. (1976). External proprioceptors. In "Structure and Function of Proprioceptors in the Invertebrates". 1-63 Ed. P.J. Mill, Chapman & Hall: London.

Lincoln, R.J. (1971). Observations of the effects of changes in hydrostatic pressure and illumination on the behaviour of some planktonic crustaceans. J. exp. Biol. 54: 677-688.

McCutcheon, F.H. (1966). Pressure sensitivity, reflexes and buoyancy responses of teleosts. Anim. Behav. 14: 204-217.

Morgan, E. (1965). The activity rhythm of the amphipod *Corophium volutator* (Pallas) and its possible relationship to change in hydrostatic pressure associated with the tides. J. anim. Ecol. 34: 731-746.

Morgan, E. (1969). The responses of *Nephtys* (Polychaeta: Annelida) to changes in hydrostatic pressure. J. exp. Biol. 50: 501-513.

Morgan, E., Nelson-Smith, A. and Knight-Jones, E.W. (1964). Responses of *Nymphon gracile* (Pycnogonida) to pressure cycles of tidal frequency. J.exp. Biol. 41: 825-836.

Morris, R.W. and Kittleman, L.R. (1967). Piezoelectric property of otoliths. Science 158: 368-370.

Naylor, E. and Atkinson, R.J.A. (1972). Pressure and the rhythmic behaviour of inshore marine animals. Symp. Soc. exp. Biol. 26: 395-415.

Naylor, E. and Isaac, M.J. (1973). Behavioural significance of pressure responses in megalopa larvae of *Callinectes sapidus* and *Macropipus* sp. Mar. Behav. Physiol. 1: 341-350.

Nisbet, I.C.T. and Drury, W.H. (1968). Short-term effects of weather on bird migration: A field study using multivariate statistics. Anim. Behav. 16: 496-530.

Qasim, S.Z., Rice, A.L. and Knight-Jones, E.W. (1963). Sensitivity to pressure changes in teleosts lacking swimbladders. J. mar. biol. Ass. India 5: 289-293.

Qutob, Z. (1962). The swimbladder of fishes as a pressure receptor. Arch. neerl. Zool. 15: 1-67.

Rice, A.L. (1964). Observations on the effects of changes of hydrostatic pressure on the behaviour of some marine animals. J. mar. biol. Ass. U.K. 44: 163-175.

Sleigh, M.A. and Macdonald, A.G. (editors). (1972). The effects of pressure on organisms. Symp. Soc. exp. Biol. 26: 516 pp.

Sulkin, S.D. (1973). Depth regulation of crab larvae in the absence of light. J. exp. mar. Biol. Ecol. 13: 73-82.

Thorpe, W.H. and Crisp, D.J. (1947). Studies on plaston respiration III. J. exp. Biol. 24: 310-328.

Tsvetkov, V.I. (1969). Sensitivity of certain freshwater fish to quick pressure changes (in Russian). Voprosy Ikhtiol. 9: 928-935.

Tytler, P. and Blaxter, J.H.S. (1973). Adaptation by cod and saithe to pressure changes. Neth. J. Sea Res. 7: 31-45.

Tytler, P. and Blaxter, J.H.S. (1977). The effect of swimbladder deflation on pressure sensitivity in the saithe *Pollachius virens*. J. mar. Biol. Ass. U.K. 57: 1057-1064.

Vasilenko, T.R.D. & Livanov, M.N. (1936). Oscillographic studies of the reflex function of the swimming bladder in fish. Bull. Biol. Med. Exp. U.S.S.R., 2: 264-266.

Vidaver, W. (1972). 8 Pressure 8.2 Plants. In Marine Ecology Vol. 1 Pt 3, 1389-1405. Ed. O. Kinne, Wiley Interscience: London, New York.

Walsby, A.E. (1972). Gas-filled structures providing buoyancy in photosynthetic organisms. Symp. Soc. exp. Biol. 26: 233-250.

ACKNOWLEDGEMENT

The author acknowledges the following for permission to reproduce figures: Company of Biologists Figs 4, 16, 21, 23; Society for Experimental Biology Figs 6, 7, 12, 15, 20, 24; Gordon and Breach Fig. 8; Blackwell Scientific Publications Fig. 10; Dr. Gibson/ Dr. Barnes Fig. 11; Netherlands Journal of Sea Research Fig. 13;

Springer-Verlag Fig. 14; Pergamon Press Fig. 17; North Holland Publishing Co. Fig. 18; Cambridge University Press Figs 2, 3, 9; **Baillière** Tindall Fig. 5.

THE ADAPTIVE RADIATION OF PROPRIOCEPTORS

W. WALES, Biology Dept. U. of Stirling and

Dunstaffnage Marine Research Lab., P.O. Box 3

Oban, Argyll, Scotland

INTRODUCTION

The term "proprioception", coined by Sherrington (1906) has become an accepted and frequently used word in the vocabulary of sensory physiologists. However its popular use has been accompanied by a loss of precision in its meaning. It is, therefore not inappropriate in 1977 to look again at the definition of proprioception as it was originally conceived.

Sherrington considered that in terms of its mechanoreceptive input the body could be divided into three receptive fields.

1) <u>The exteroceptive field</u> consisting of the total external covering of the animal.

2) <u>The interoceptive field</u> lining the alimentary tract.

3) <u>The deep field</u> consisting of all subepidermal tissues.

The deep field mechanoreceptors are separated from the others as they respond primarily to parameters of the internal environment, that is to stimuli which result from the actions of the animal, much more so than the receptors of the surface fields. Sherrington, therefore, coined "proprioceptive" from the Latin <u>propius</u> meaning "one's own" to describe these self stimulated mechanoreceptors. Sherrington's definition was intended as a functional one and use of the word in Sherringtonian terms implies that the function of the receptor is known. The term is none-the-less currently used to describe receptors when their function is unknown and in some instances where even the adequate stimulus is yet to be described. It must, however, be admitted that there is a need for a term to describe presumptive proprioceptors where direct evidence of function is not available and yet, by homology with known proprioceptors, the function can be implied.

In an attempt to overcome the limitations imposed by Sherrington's definition, "proprioceptor" has been redefined a number of times. These definitions have ranged from the simple:-

"A mechanoreceptor which normally signals movement of position of the parts of the body" (Bullock and Horridge, 1965).

To the complex:-

"Sense organs capable or registering continuously deformations (change in length) and stress (tensions, compressions) in the body. These can arise from the animals own movements or may be due to its weight or to other external mechanical forces" (Lissmann, 1950).

It is perhaps this definition which has come closest to meeting universal approval although it is not without its inadequacies. Pringle (1961) points out that it excludes "vibration", the lower frequencies of which are considered by many to fall within the field of proprioceptive stimuli.

The need for a more flexible definition of proprioception arises from the functional implications of the original definition. We now have knowledge of many mechanoreceptors, throughout the animal kingdom, which conform to Lissman's definition but for which we lack knowledge of their reflex roles and an equally large number known only anatomically but which are probably proprioceptive. It would thus seem advisable to include these presumptive proprioceptors in any comprehensive review of proprioception as has been done by Mill (1976a).

The solution is perhaps to use the term "presumptive proprioceptors" for possible proprioceptors where we lack knowledge of the adequate stimulus or function. The adequacy of any definition will, however, depend much upon the needs of the author and it is sufficient that he states clearly which definition he is conforming to in his usage of the word.

A second problem associated with the use of the term "proprioception" lies in the definition of the boundaries covered by the field. There are in fact three important boundaries to be considered.

1) <u>Relative magnitude of body part</u>. We would agree that a receptor recording limb movement is a proprioceptor whereas a sensory cell monitoring the movement of a hair on that limb would normally be excluded. Yet if that hair is deflected by another part of the body then the hair receptor is providing information pertaining to the relative position of those body parts and the information would be

of a proprioceptive nature e.g. hair plates of insects (Pringle, 1938) and arthropod mouthpart sensillae (Wales, 1976).

On the other hand, if we conclude that, because the hair projects from the external surface and responds to environmental changes, it is not proprioceptive, then must we exclude Johnston's organ for the same reasons? Johnston's organ monitors movement of the antennal flagellum which is deflected by wind in flight (Bässler, 1958; Hollick, 1940).

2) <u>Rate and amplitude of movement</u>. Where a part of the body is undergoing slow, large amplitude oscillations we would accept that receptors monitoring this movement are proprioceptors. Where the amplitude decreases to very low levels the receptor remains proprioceptive but if the frequency increases to a high level at low amplitudes, such as a propagated sound wave, then we would accept that the information carried by the receptor is no longer proprioceptive. There are exemples of receptors, particularly in the anthropods, which traverse this boundary e.g. Johnston's organ of insects (Burkhardt and Gewecke, 1966; Burkhardt and Schneider, 1957) and the myochordotonal organ in the walking leg of crabs (Clarac, 1968a,b; Horch, 1971).

3) <u>Source of mechanical stimulus</u>. Sherrington (1906) made it clear that he included not only those mechanoreceptors responding to body movement but also those mechanoreceptors responding to orientation in space, the gravity detectors or, as he described them, "Labyrinthine proprioceptors". This view appears to be generally accepted by other authors (Bullock and Horridge, 1965; Lissmann, 1950; Mill, 1976a). However, if orientation in space is within the field of proprioception then it is not such a large step to include movement in space as Laverack (1976) has done.

The boundaries between "proprioception" and other forms of mechanoreception are not sharply defined (Fig. 1). Clearly some mechanoreceptors are proprioceptive only part of the time depending on the nature of the stimulus to which they respond and the only definition which will separate proprioceptive input from non proprioceptive input is a functional one.

My brief is to deal with the "adaptive radiation of proprioceptors" and it will be clear to anyone who knows this field that it is far too large to cope with in a single chapter. I have, therefore, elected to examine a single group of proprioceptors which I have some personal experience of, and which admirably suit the purpose of this chapter, having undergone adaptive radiation both at a cellular level and at the organ level. These are the connective chordotonal organs of Arthropods.

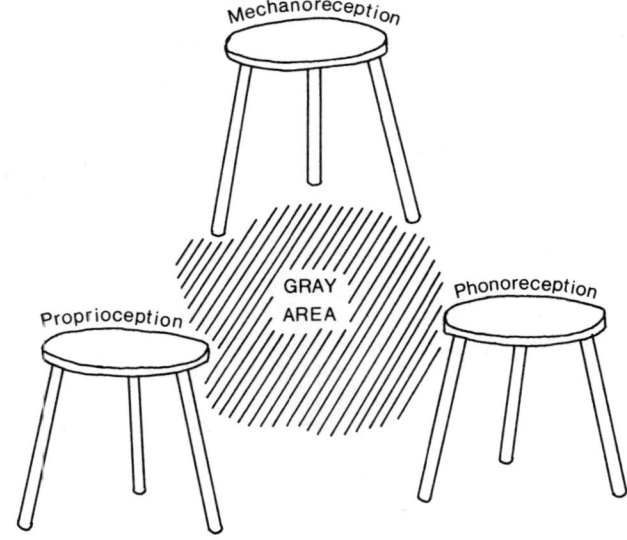

Fig. 1. There are no distinct boundaries between proprioception, phonoreception and other forms of mechanoreception. These senses are thus separated by a considerable grey area.

CHORDOTONAL ORGANS

Before looking specifically at the connective chordotonal organs it will be helpful to look at the basic structure of chordotonal organs. Chordotonal organs, which were first described in insects by Graber (1882), are a morphological group of mechanoreceptors characterised by a special sensilla, the scolopidium. They occur in two classes of the Arthropods, the Crustacea and the Insecta but are not known in the remaining classes. Our fine structural knowledge of chordotonal organs relates mainly to the Insecta whereas functional data is available largely for the Crustacea. This is primarily due to the size of the receptors, the small insect chordotonal organs being ideally suited to electron micrograph studies whereas the larger crustacean organs have facilitated some of the more elegant functional studies.

Chordotonal organs are primarily "deep field" mechanoreceptors and are therefore mainly proprioceptive but not exclusively so. The scolopidium is also sensitive to vibration and the organs of hearing in the Insecta and Crustacea are based on this type of sensillium.

Morphology of chordotonal organs

Chordotonal organs consist of one or more sensory units called scolopidia. These are associated with a support system which may be a cuticular structure, or a cellular or connective tissue strand. Figure 2 shows a simple chordotonal organ from the leg of the cockroach (Young, 1970). This organ which occurs at the tibio-tarsal joint of the legs is a typical connective chordotonal organ (Howse, 1968). That is one in which the support system is a tissue strand. In this figure the essential organization of a scolopidium can be clearly seen.

A number of bipolar uniterminal neurones can be seen proximally on the strand (Fig. 2) with their dendrites running distally, parallel to the strand. These primary sensory cells have processes running directly to the central nervous system. The neurones insert into a scolopidium singly (distal main strand only) or in pairs.

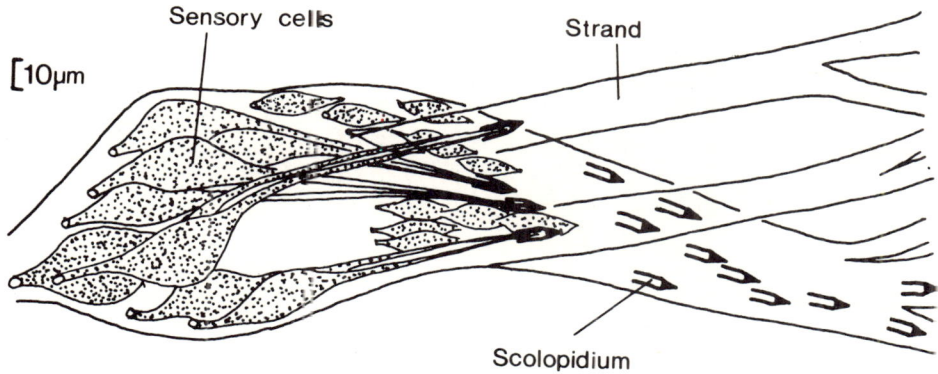

Fig. 2. A reconstruction of the tibio-tarsal connective chordotonal organ of the cockroach showing the relationship between the sensory cells, scolopidia and the strand. The scolopidia lie distal to the sensory cells (After Young, 1970).

The cells of the scolopidium are a morphological unit derived from a single epidermal cell (Jägers-Röhr, 1968). The scolopidial unit consists of one to three uniterminal cells and usually two accessory cells which are the scolopale cell and the attachment cell, but larger numbers of cells may be present (see below). Fig. 3 shows a scolopidium with two sensory cells from the cockroach tibio-tarsal connective chordotonal organ which displays all of the typical features of the scolopidium. The terminal processes of the sensory cells exhibit a ciliary centriolar derivative. This typically consists of two basal bodies (cen. and cil.b. fig. 3), a ciliary root system and a ciliary segment containing an axoneme of 9+0 structure, lacking the central filaments of motile cilia.

Sensory cells make contact with one another inside the scolopidium as the cells are not separately ensheathed and therefore are in close contact with one another. Specialised contact between neurones also occurs in the form of a belt desmosome in the apical region between sensory cells and between them and the scolopale cell. The latter junction may also occur where a single cell inserts in the scolopidium and this most probably serves as a structural support (Satir and Gilula, 1973).

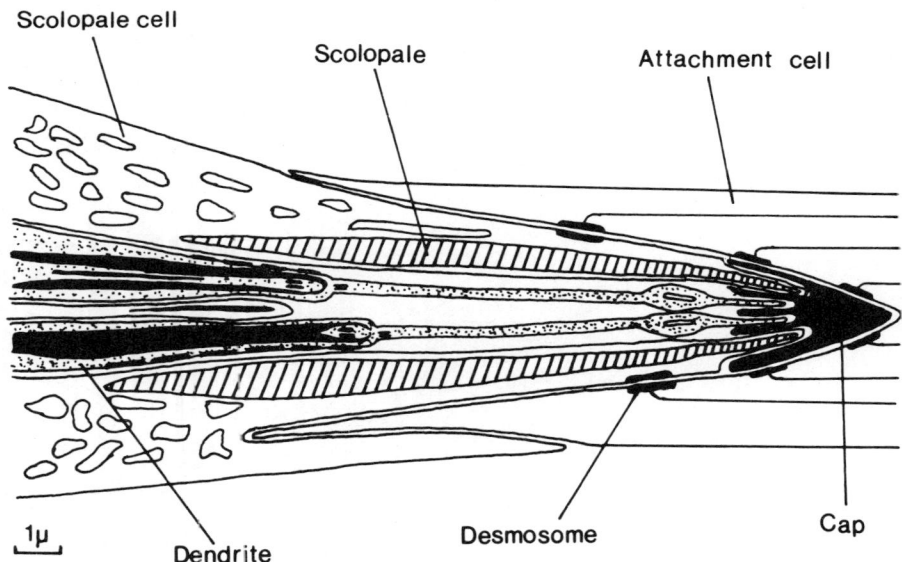

Fig. 3. The scolopale unit of the tibio-tarsal connective chordotonal organs showing the insertion of the dendrites into the scolopale and the structural relationships between the component cells (After Young, 1970).

The terminal regions of the sensory cells are ensheathed by the scolopale cell and the attachment cell. The scolopale, located in the scolopale cell, is an intracellular matrix which may be fenestrated to form a labyrinth continuous with the extracellular spaces. It consists of a cylinder of electron dense material deposited round longitudinally orientated microtubules (Mill and Lowe, 1973). It is this intracellular scolopale which is the prime diagnostic feature separating the chordotonal organ from the other types of arthropod sensillae (Moulins, 1976).

The terminal processes of the sensory cell are enclosed in an amorphous extracellular structure described as a tube or cap according to its morphology and composition. This structure along with other amorphous extracellular material found in the scolopale space is believed to be secreted by the scolopale cell which frequently has the vacuolated appearance of a secretory cell (Moulins, 1976).

The final unit of the scolopidium is the attachment cell. This cell is difficult to identify in chordotonal organs with a connective tissue strand and may be absent (Whitear, 1962), but is usually present when the strand is cellular (Young, 1970). Where no strand is present, the attachment cell is found and may form the apical anchorage for the scolopidium. Some scolopidia in crustacean chordotonal organs have a canal cell (Whitear, 1962) which may be equivalent to the attachment cell.

The extracellular tube which envelopes the apical sensory processes may penetrate the attachment cell and be continuous with the cuticle. Where this occurs the tube is lost at ecdysis (Richard, 1957). In insects the tube may be attached to a distinct invagination of the cuticle and in cerambycid larvae a complex structure, termed the pleural disc, marks externally the location of an abdominal chordotonal organ (Hess, 1917). It is, however, in the crustacea that the most interesting cuticular structures are found. These are considered at length below.

The functional significance of the different morphological features are not clearly understood and authors are at varience regarding the adequate stimulus of chordotonal organs. This point is considered below with respect to a particular class of chordotonal organs.

CONNECTIVE CHORDOTONAL ORGANS

Connective chordotonal organs, CCO, (Howse, 1968) are found in the thoracic appendages of decapod crustaceans (Whitear, 1962; Wales et al. 1970) and in some insect limbs. At least one of these organs is found at each articulation of the walking legs and chelae.

Similar receptors also occur in some of the mouthparts (Wales et al. 1970) and the antennae (Wyse and Maynard, 1965; Hartman and Austin, 1972) but these will not be considered here.

The terminology of decapod walking limb (pereiopod) articulations and proprioceptors can be a little confusing but conforms to a simple pattern. The segments of the appendage are named, as shown in Fig. 4 and, where two segments have fused, they are identified by the combined name i.e. the basi-ischiopodite. The limb joints are named for the segments which they articulate i.e. the joint between the meropodite and carpopodite is the meropodite-carpopodite or M-C joint. Likewise the chordotonal organs are known by the initials

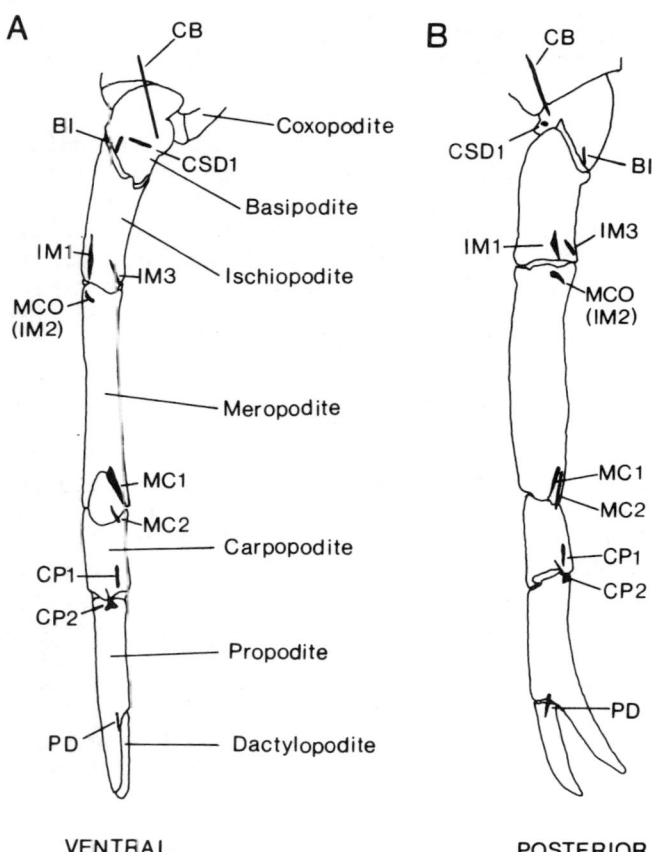

Fig. 4. The terminology of decapod chordotonal organs (CCO) may at first seem a little confusing but the CCOs are named for the articulation at which they occur. This is further explained in the text (After Wales et al. 1970).

of the articulation where they are located. Where more than one
occurs they are also allocated a number i.e. the CCO's at the C-P
joint are CP1 and CP2. However not all CCO's occur at an articulation
(see CSD1 below) nor are they necessarily monitoring movement of the
articulation where they are located (see MCO1 and MCO2 below).

The connective chordotonal organs have a strand with an outer
layer of amorphous connective tissue and an inner structure of
connective tissue, collagen and strand cells (Whitear, 1962; Mill
and Lowe, 1973). In some cases the strand may be almost completely
cellular (Young, 1970; Moulins and Clarac, 1972). The concentration
of these elements may vary from one part of the strand to another
and this is of possible functional significance. The shape of the
connective strand varies greatly from a simple cylindrical structure
to a flat sheet of tissue. The structure is frequently complicated
by accessory strands connecting the receptor strand to other nearby
structures. Some of these accessory strands can be extremely
important to the function of the organ by altering its effective
sensory field (see MCO2 below).

The connective tissue support system is elastic, it can be
stretched and will shorten if allowed to. It lies between two points
usually associated with two leg segments and thus functionally spans
the joint. Only two receptors physically span the articular
membrane from one segment to the other, these are the TC and CB CCO's.
Several lie proximal to the articulation but are stretched between
the proximal segment and the tendon of a muscle, which lies in the
proximal segment but moves the distal segment. As the tendon
primarily moves in sympathy with the distal segment, this may be
considered functional spanning of the articulation. Another group
lie distal to the articulation being stretched between the distal
segment and a muscle which moves this segment. The mechanics of
strand movement is less obvious in these examples. The PD CCO has
been examined by Lowe and Mill (1972) who found that strand length
increases with an approximately linear relationship to joint angle
from the fully closed position through about two thirds of its arc
of movement. So, although the PD organ does not cross the joint
morphologically it is still a proprioceptive organ of the propopodite-
dactylopodite articulation. Likewise MC2 and CP2 have been shown to
respond to movement of their respective articulations (Bush, 1965a,c),
though the mechanical relationship is in these cases not known.

The remaining CCO's which do not conform to the above pattern,
CSD1, CSD2, MCO1 and MCO2 are considered in the appropriate sections
below.

The CCO's are proprioceptors in the Sherringtonian sense of the
word. They lie below the body surface (though they may retain
connections with the external cuticle) and monitor position, change

of position and rate of change of position of the body parts which
they span. They respond to movement of the respective articulations
on passive movement and on active contraction of the muscles moving
these articulations (Burke, 1954). The proprioceptive input has
been shown to elicit resistance reflexes in static preparations
(Bush, 1962a,b, 1963, 1965c) and during walking (Barnes, Spiroto and
Evoy, 1972).

Functional Adaptation

The best known connective chordotonal organ is that at the most
distal articulation of the walking leg, the PD organ. This CCO has
been chosen partly for historical reasons that it was the first
examined functionally by Burke (1954) but also because it is one of
the least complicated. The PD joint has only one CCO, it is moved
by a single pair of antagonistic muscles and its movement is restricted to a single plane. The location and structure of the receptor is
shown in figure 5. The receptor strand, which is usually oval in
section, is anchored to the tendon of the dactylopodite closer
(flexor) muscle and distally to the cuticle of the dactylopodite.
The receptor strand is relaxed on action of the opener (extensor)
muscle and stretched by joint movement produced by the closer muscle.
The strand is under constant tension, being stretched even in its
most relaxed position Burke (1954). The PD receptor has some 80
sensory cells in two groups, a proximal group of large cells and a
distal group of small cells (Wiersma and Boettiger, 1959).

Electrophysiological recordings were first made by Burke (1954)
who noted that the receptor responded to change in length of the
receptor strand both in situ and when isolated. There is a resting
discharge from the static receptor which increases on movement of
the joint (and hence change of strand length) produced passively,
or by the action of leg muscles. Burke concluded that the organ
signals the rate and extent of movement and he considered that the
receptor could probably signal the position of the joint. He,
moreover, noted that the receptor was sensitive to small amplitude
vibrations up to 1000 HZ.

Burke had known of the presence of one other CCO located at
the M-C joint but gives no indication of anticipating these receptors
to be widespread throughout crustacean appendages. He may therefore
have considered the PD organ to be more unique than we now know it to
be and this would have coloured his views on receptor function,
particularly with regard to vibration. Burke noted the large size
of some of the axons and considered that this may be an important
proprioceptive input which was probably involved in rapid reflex
action. This point was confirmed by Bush (1962b).

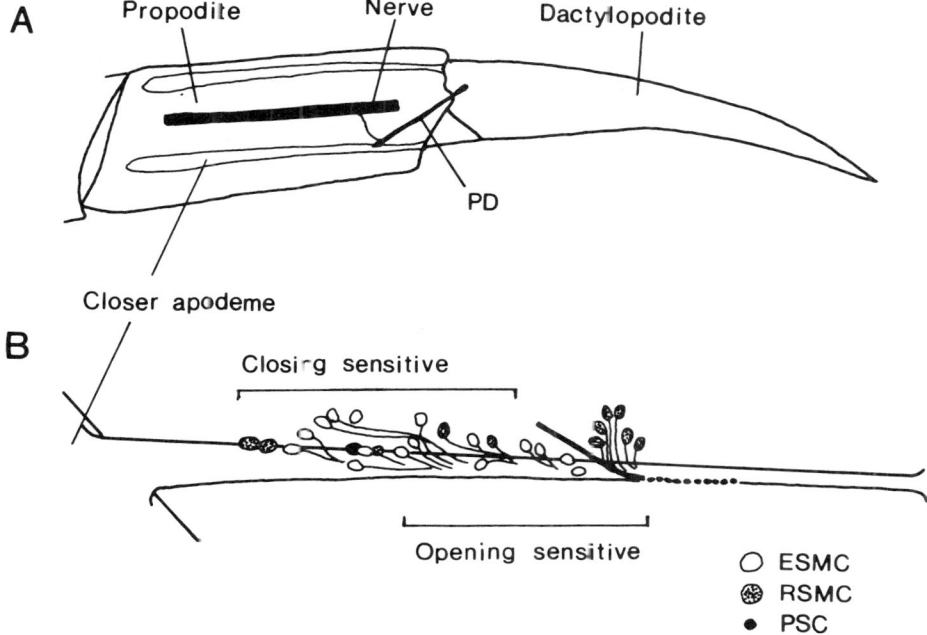

Fig. 5. The propodite-dactylopodite CCO (PD) in the crab leg.

A. Shows the location of the PD organ (seen in trans-transparency) relative to the leg segments and the closer muscle apodeme (After Burke, 1954).
B. The distribution of the different classes of sensory cells on the strand (After Hartman and Boettiger, 1967).

The PD organ is composed of a large number of sensory cells each of which is functionally unique (Wiersma and Boettiger, 1959; Hartman and Boettiger, 1967; Mendelson, 1963; Mill and Lowe, 1972; Mill, 1976). All of these authors are agreed that the sensory responses fall into three categories 1) movement cells 2) position cells and 3) intermediate cells (Fig. 6). In fact the responses of the cells form a complete spectrum from purely phasic (movement sensitive) to purely tonic (position sensitive) units with phase tonic intermediate units which convey information concerning both movement (change in strand length) and position (strand length).

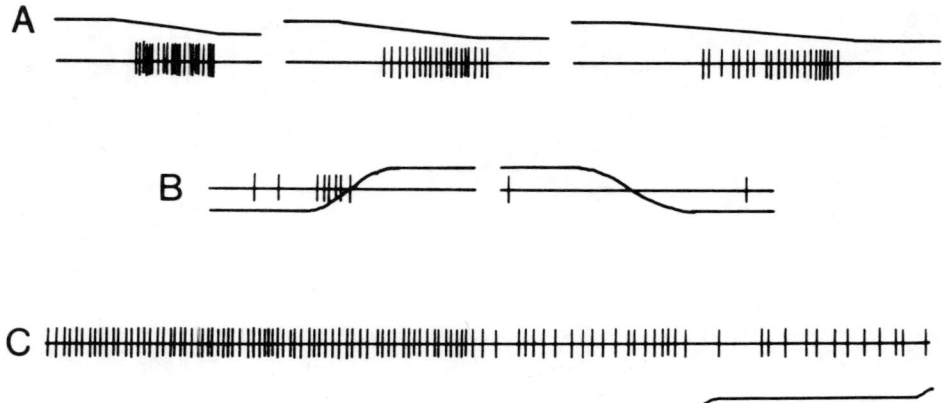

Fig. 6. Typical responses from CCO sensory units, A. an elongation sensitive movement cell. B. A relaxation sensitive movement cell, C. A position sensitive cell. (After Mill, 1976).

If we look first at the two ends of the spectrum we can better understand the potential value of this type of proprioceptor. Firstly the pure phasic units effectively respond only during change in strand length and their frequency is dependent upon the rate of change, i.e. velocity. According to Mill and Lowe (1972) the output of these "movement" units depends upon three criteria.

1) the sensitivity of the fibre

2) the rate of movement (velocity)

3) the position (strand length) within the sensitive range at which the movement occurs

Movement units are unidirectional in that they respond during strand elongation or relaxation. The units are referred to respectively as "elongation sensitive movement cells" (ESMC) or "relaxation sensitive movement cells". The units may be active over the total range of movement or their activity may be restricted to a portion of the total arc of movement. Some of the sensory cells are extremely sensitive to movement, responding to movements of less than $1°$/sec at which velocity the total movement would take two minutes. One fibre reached its maximum response at the low rate of 1.2°/sec (Wiersma and Boettiger, 1959). As most of the joint movements during locomotion will be much faster than this, the receptor cannot measure the velocity and therefore will only indicate when movement occurs with an output which is not related to velocity and, as such is a pure movement receptor. Many of the velocity sensitive units have upper and lower thresholds within the normal range of stretch and therefore only respond for part of the arc of movement. The sensitivity of movement cells may vary throughout the range of strand length over which they respond. These cells will convey information regarding position as well as velocity and movement. Some velocity sensitive units do not reach saturation frequency within the physiological range.

The purely tonic units tend to show their maximum response either on relaxation or elongation of the strand and may therefore be classified as "elongation sensitive position cells" (ESPC) or "relaxation sensitive position cells" (RSPC). The response of position cells is frequently asymmetrical in that the response of the cell at a given strand length will depend upon whether that length was attained during relaxation or elongation i.e. they exhibit hysteresis. No single cell monitors the total range of positions but the position cells as a group cover the total range of *in situ* strand length.

Between these extremes the units are responsive to both movement and position but have their distinctive sensitivity ranges. The phasic and tonic elements may vary considerably from unit. It is more difficult to understand the function of these units as they do not respond to a single physical parameter and we have no knowledge of the information that the CNS can extract from this input.

Wiersma and Boettiger (1959) and Hartman and Boettiger (1967) examined sensory cell function relative to the location of the cell on the strand and found that the main functional cell types occur in specific regions of the strand. The larger more proximal cells are movement sensitive whereas the smaller distal cells are position sensitive. Of the large movement cells, those on the dorsal surface of the strand are relaxation sensitive and those on the anterior surface are elongation sensitive. It is possible to split the strand to divide the two groups of movement cells without destroying their function. Of those movement cells whose response has a position component, the more distal cells in each group exhibit increased sensitivity to strand relaxation whereas the more proximal cells are more sensitive cells on the distal strand can be separated as a distinct bundle but no one has been able to demonstrate what the correlation of cell location and function is for these small sensory cells.

The importance of the above work is that it opened the way for others to correlate fine structure and function in a CCO. All of the above sensory units are based on the same morphological unit and the study of adaptive radiation necessarily has to be performed at a fine structure level.

The fine structure of crustacean CCO's was first performed by Whitear (1962) who adopted a comparative approach and in doing so failed to take full advantage of the findings of Wiersma and Boettiger (1959). A significant finding by Whitear was the different structure of the two sensory endings inserting into the same scolopidia. This led to the hypothesis that these two dendritic types, the ciliary and paraciliary might explain the presence of relaxation and elongation sensitive neurones. Whitear found that MC1 and CP1 had poorly developed paraciliary endings which correlated with the findings of Wiersma (1959) that these proprioceptors were mainly relaxation sensitive. Thus ciliary endings could be relaxation sensitive and paraciliary endings elongation sensitive. Thus by differential bending (proposed by Mendelson (1963) and Bush (1965b) as the adequate stimulus) causing a shearing action of the tube-scolopale junction, the activity of the sensory cells could depend upon the direction of the bending.

Whitear's hypothesis however is not accepted for two reasons. Firstly other CCO's do not correspond to this pattern, for example CB has sensitivity to relaxation and elongation of the strand yet the scolopidia only have sensory endings of the ciliary type (Whitear, 1962) moreover the work of Hartman and Boettiger (1967) later showed that the two endings in each scolopidium are of the same response type.

With this knowledge Mill and his colleagues re-examined the fine structure of the PD organ of *Cancer pagurus* (Mill and Low, 1973; Low, Mill and Knapp, 1973). It was now clear that the functional differences had to be found in the region of the strand, or the properties of the scolopidium, for each response type. Using the PD organ topography described above, Mill selected out known elongation sensitive movement cells, relaxation sensitive movement cells, and position sensitive cells for examination. It must be borne in mind however that the sensory cells are each unique and they were therefore looking for gross differences between groups of sensilla rather than minute ones between neighbouring sensillae which would be difficult to explain.

This study showed that the strand varies in composition from region to region with few clear trends. The "position cell" strand region does not differ significantly from the "movement cell" region but that in which the RSMC embed tends to have fewer cells and more collagen fibres than the EMSC region. The main differences observed occur in the structure of the scolopidia (Fig. 7). The ESMC scolopidia have the appearance of a more rigid structure. They possess a "canal cell" round the distal region and this is absent in the RSMC. The canal cell is probably equivalent to the attachment cell (Moulins, 1976). In the PD organ of *Cancer pagurus* the scolopale cell is surrounded by enveloping cells which may exhibit cytoplasmic continuity so that the dendrites are ensheathed by 1 to 3 cells. However to avoid confusion these structures will be referred to as individual cells. The enveloping cells of the ESMC contain scolopale like material at the tube – scolopale junction and there are desmosomes between enveloping cells and between enveloping cells and the scolopale cell. The RSMC lack the scolopale like material in the enveloping cells and have fewer desmosomes. All of the scolopidia of the PD organ in *C. pagurus*

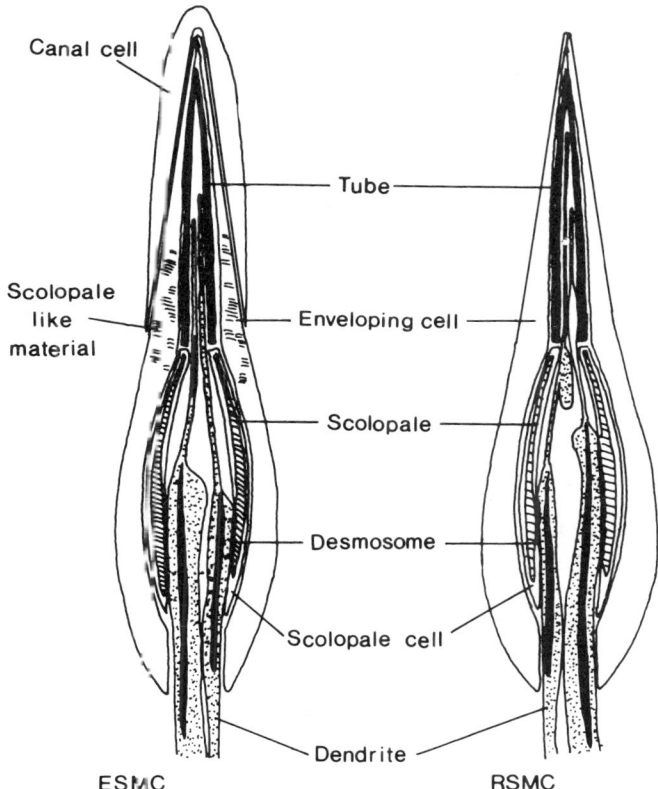

Fig. 7. The structure of an elongation sensitive (ESMC) and relaxation sensitive (RSMC) movement cell from the PD CCO of *Cancer pagurus* (After Mill and Lowe, 1973).

have ciliary and paraciliary dendrites but small differences in structure occur. Both the ciliary and paraciliary dendrites of RSMC scolopidia have dilations which are absent in the ESMC and the ESMC has a paraciliary rod formed from the ciliary tubules and microtubules. This rod again occurs at the region of the tube scolopale junction.

Mendelson (1966) has shown that the origin of the transduced sensory response lies within the scolopale space and, as the scolopale and tube will be relatively rigid structures, one can assume that if the adequate stimulus is bending as proposed by Bush (1965b), Howse (1968) and Taylor (1967) then this must occur at the tube-scolopale junction. Mill and Lowe (1973) point out that this is unlikely for two reasons. Firstly the bending theory requires that the scolopidia cross the boundary between materials of different mechanical properties but this is not supported by fine structural studies (Whitear, 1962; Mill and Lowe, 1973). Secondly the structure of the ESMC seems to be specifically evolved to reduce bending in that the occurance of desmosomes, the scolopale material in enveloping cells, the presence of the canal cell, and the paracilliary rod will all contribute to make the structure more rigid.

Mill and Lowe (1973) propose that longitudinal stretch is the adequate stimulus. The development of the ESMC is such that the cellular components of the scolopidia are well anchored to each other and to the strand cells. Thus stretching the strand will result in stretching of the scolopidium, and movement of the tube away from the scolopale will stretch the dendrites. Young (1970) proposes a similar mechanism for the connective chordotonal organ of the cockroach leg.

The RSMC scolopidium does not exhibit the same rigid structure or close association with strand cells, and the collagen fibres which surround them retain their length even when the strand is stretched. Thus the strand elongation will not be conveyed so readily to the RSMC scolopidia. Mill & Lowe propose that due to the tapered shape of the scolopidium the collagen molecules will make contact with the enveloping cells on relaxation and contact will be greater at the proximal end thus stretching the scolopidium and hence the dendrites.

The structure of the position sensitive cell (PSC) is described by Lowe, Mill and Knapp (1973) in *C. pagurus* who found its structure to be intermediate between the ESMC and RSMC types. The canal cell is absent but the scolopale like material occurs in the enveloping cells and the paraciliary rod is present. The distal ends of the dendrites inside the tube is often set in an amorphous matrix "glue" of Whitear (1962). Only very small amounts of this material occur in movement sensitive scolopidia. The authors again propose that longitudinal stretch is the adequate stimulus but can find little structural basis for the difference between movement and position cells. They suggest that position cell dendrites may return to their resting length very slowly on deformation which may correlate with the presence of amorphous material round the dendritic endings. The distribution of elongation and relaxation sensitive position cells is not yet known and Mill and his colleagues were unable to identify the appropriate cells to explain the known physiological differences.

The differences in sensitivity between units and the rate of adaptation may both depend considerably on the membrane properties of the dendrites, but it is clear that the many sensory cells in the connective chordotonal organ have become adapted structurally and physiologically to provide the CNS with varied data concerning limb joint position and movement.

THE EVOLUTION OF CONNECTIVE CHORDOTONAL ORGANS

Two trends are apparent in the evolutionary development of connective chordotonal organs in the decapod limb. These are the association of the organs firstly with the external cuticle and secondly with the limb muscles.

CONNECTIVE CHORDOTONAL ORGANS AND CAP SENSILLAE

The connective strand of CCO's is normally suspended between two parts of the internal cuticular surface. These points of insertion may be unspecialised areas of cuticle, the apodemes forming the tendons of muscles, internal cuticular projections (TC and CSD1, below), areas of specialised relatively uncalcified cuticle (CSD1 and MCO of *Ocypode*) or externally visable sensillae. It is the latter category which are of primary interest in an evolutionary context.

In his original description of the myochordotonal organ of crabs, Barth (1934) observed that the sensory cells of this elaborate connective chordotonal organ were associated with blind ending pits in the cuticle. In other groups of the decapod crustaceans (Alexandrowicz, 1972) the sensory processes of MCO1 are associated with cuticular pores closed by a thin cuticular membrane bearing a small cuticular spike, the length of which varies with species. These sensillae, named cuticular articulated pegs (CAP) by Wales et al. (1970) are found distal to the 1-M, M-C and P-D joints of the pereiopods of macruran decapods. They occur at the same location of some limb joints of some anomurans but are absent from the brachyurans (Alexandrowicz, 1972; Laverack, 1976).

The CAP are usually formed into lines approximately parallel to the limb axis and thus perpendicular to the joint axis. Their proximity to the joint is such that in most cases the sensillae are covered by the articular membrane on flexion of the joint. In their location they are similar to the hair plate organs of insects (Pringle, 1938).

Laverack (1976) has examined the CAP organs of a number of decapods and found a wide variation in number of sensillae with species and with age. In the natantian *Pasiphaea* he found bunches of hairs where the CAP would normally be located. In *Crangon* the sensilla form a well ordered line (Fig. 8) and are still distinct hairs though somewhat shortened. In the Astacrura the well ordered line of sensillae is replaced by several lines forming a fan shaped group and the cuticular projection is reduced to a short peg. The sensillae are present in the Palinuridae where the shortened peg becomes extremely small relative to the size of the limb (Wales et al. 1970).

Laverack (1976) further demonstrated that the number of sensillae in each group increase in number during the larval development of the common european lobster *Homarus gammarus*. One may therefore conclude that these are functional sensillae which serve some purpose in adult life. They would appear to be mechanoreceptive hairs. Their location relative to the limb joint further suggests that they are proprioceptive.

At each of the three joints where CAP sensillae are found, there is a CCO located internally (CP2, MC2 and MCO1) and these proprioceptors are directly associated with the sensillae. Wales et al. (1970), in their description of the proprioceptors of the walking legs of *Homarus*, described the CAP sensillae as being innervated by some of the cells previously thought to be part of the chordotonal organ. They concluded that the sensory cells at the CCO consisted of two groups, one of which innervated the CCO strand and the other innervating the CAP sensillae. As the number of sensory cells associated with the CAP was less than the total number of sensillae it was considered that only a portion of the neurones were visible due to capricious staining by methylene blue.

In his final paper, Alexandrowicz made a more detailed examination of the CCO's at the three joints where CAP occur and his observations led him to conclude that it was the sensory cells of the CCO which were associated with the CAP. As described above, the sensory cells insert in pairs into the scolopidium and the scolopidia are attached to the base of the CAP spine by a non-nervous prolongation which is presumably an extension of the tube. If so then the attachment may be by a cuticular thread similar to

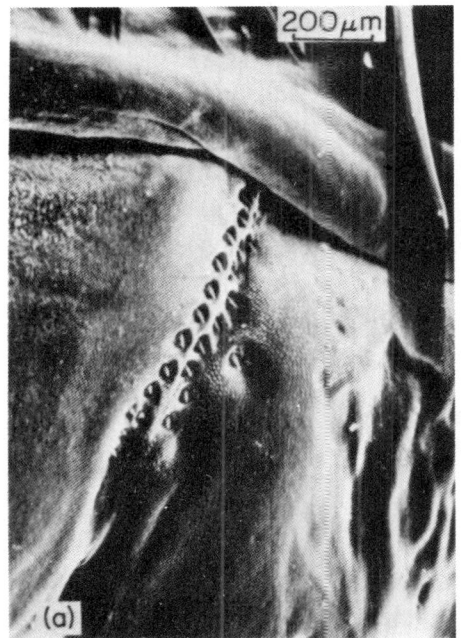 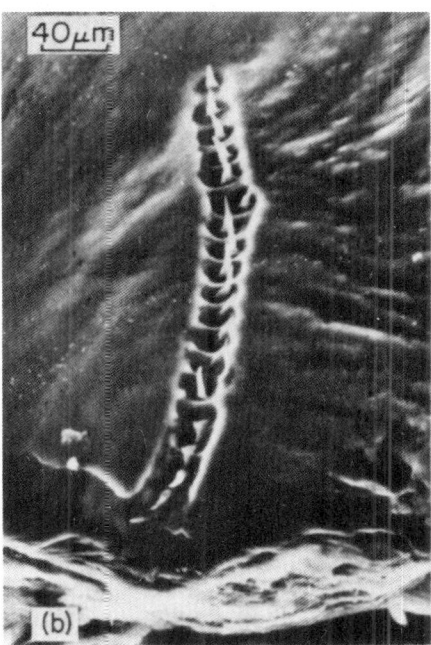

Fig. 8. Typical arrangements of CAP organs at, (a) the 2nd maxilliped of *Panulirus argus* and (b) the MC joint of a leg joint of *Pandalus borealis*. (From Laverack, 1976).

those found in some insect non-connective chordotonal organs (Moulins, 1976).

It is still not clear whether the CAP are functional sensillae or vestigal structures anchoring the scolopidia to provide more effective stretch. Wiersma (1959) recorded responses from CAP sensillae on touching the external cuticular projections of CP2. This can be achieved by recording from a separate sub-bundle of the main nerve not the nerve to the CCO which is contrary to the findings of Alexandrowicz (1972). Wales et al. (1970) were unable to get any response from the chordotonal organ nerve which they could with certainty attribute to the CAP's. The function of the CAP's, therefore, remains unproven.

One point on which most authors seem to agree is that the CAP sensillae are probably a primitive feature. It is highly probable that the chordotonal organs are derived from external receptors which responded to similar parameters, i.e. joint movement. The advantage of evolving an external hair plate type organ into an internal chordotonal organ is obvious from a consideration of the physiological responses above. The external sensilla will provide positional information but data on movement, particularly relating to velocity can only be obtained in a limited form by sophisticated processing of information in the CNS whereas the connective chordotonal organ provides data on position, movement and velocity.

The number of external sensillae receiving a stimulus will depend on the position of the articular membrane whereas all of the internal sensillae are stimulated all the time.

The above study has provided some information on the evolution of CCO's from the primitive organ and I now want to consider some specialisations exhibited by CCO's.

CONNECTIVE CHORDOTONAL ORGANS AND SOFT CUTICLE

In the basipodite and ischiopodite of the pereiopods of macuran decapods and the fused basi-ischiopodite of anomuran and brachyuran decapods lie two CCOs which do not correspond to the normal pattern in that they neither span an articulation nor are attached to a muscle or its tendon. These CCO's, named "Cuticular Stress Detectors" (CSD1 and CSD2) by Wales et al. (1971) have a highly specialised association with discrete areas of soft cuticle. Although these receptors have some differences in their organization and CSD1 appears to be more highly specialised, they are functionally similar. In the following description I am, therefore, going to consider only the better known CSD1.

As described by Wales et al. (1971), CSD1 lies in the basi-ischiopodite proximal to the performed breakage plane at which autotomy occurs. In *Carcinus maenas* it consists of a cellular strand (Moulins and Clarac, 1972) innervated by more than 40 bipolar neurones. The strand is stretched between a flexible cuticular peg and a discrete area of soft cuticle on the anterior surface of the appendage. The strand (Fig. 9) lies parallel to the aututomy plane and is thus orientated approximately perpendicular to the other CCO's which are more or less parallel to the limb axis. The autotomy plane, which forms a line circumscribing the basi-ischiopodite, is a rigid structure of considerable strength and the only visible movements associated with it occur at autotomy.

CSD1 is a typical CCO consisting of a wide range of units varying from pure phasic to pure tonic. The adequate stimulus is

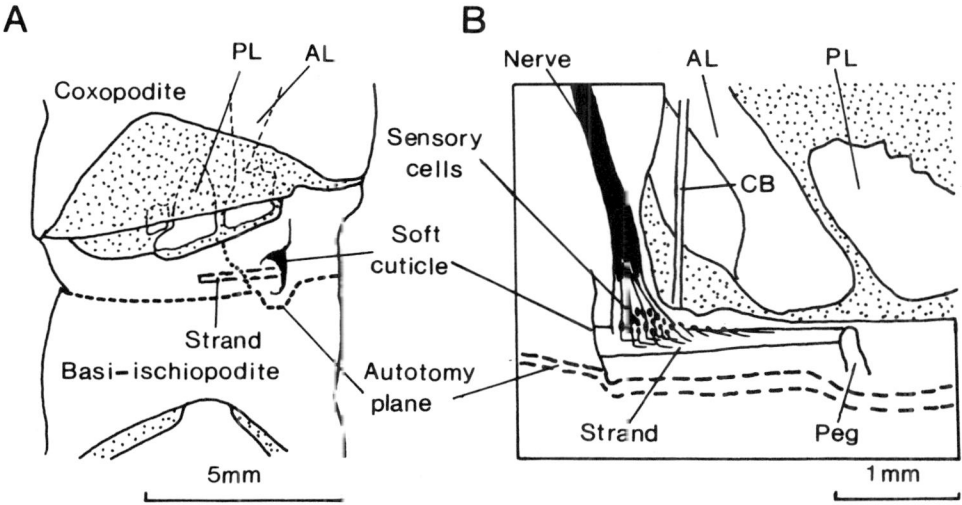

Fig. 9. CSD1 from the walking leg of the crab *Carcinus maenas*, showing; A. the position of the receptor relative to the levator muscles and the autotomy plane and, B. the general disposition of the receptor. AL, Anterior levator; PL, Posterior levator (After Wales et al. 1972).

most probably change in strand length though this will be small as the two ends of the receptor strand are attached to cuticle between which there is no articulation. Clarac et al. (1971) demonstrated that the receptor responded to any stimulus which might be expected to stress the basi-ischiopodite. Pressure applied generally to the basi-ischiopodite produces a marked response but external manipulation of the area of soft cuticle onto which the strand inserts proved a particularly potent stimulus. This proved a convenient stimulus to examine the component units as pressure applied externally to the soft cuticle has a known effect on the strand i.e. it shortens it. As the CSD strand is suspended between two fixed structures changes in strand length occurring *in vivo* will normally be minimal.

CSD1 exhibits tonic activity in the absence of any apparent stimuli. Shortening of the strand produces activity in phasic units and change of frequency in tonic units, thus the two basic classes of unit, corresponding to "movement" and "position" units, are present. Removal of the stimulus causes the strand to be restretched to its original resting length and elicits phasic responses from a smaller population of units indicating that both "relaxation sensitive" and "elongation sensitive" movement units are present. However, as the strand length alters over such a small range, the terms "relaxation sensitive" and "elongation sensitive" position units may no longer be appropriate. The nature of our experiments did not allow us to measure the sensitivity of the units to changes of strand length but otherwise CSD1 seems to have the same basic unit types as other CCO's.

Clarac et al. (1971) also measured the response of CSD1 to a range of more natural stimuli which might be expected to stress the basi-ischiopodite by 1) pulling the tendons of the C-B joint muscles, 2) flexing the I-M joint, and 3) flexing the B-I joint in *Homarus* where these segments are unfused. All these stimuli produced good responses from the receptor. Because of the proximity of the receptor to the autotomy plane, we also monitored CSD1 activity during artificially produced autotomy, in an isolated limb, and found that the receptor responded both during fracture of the cuticle and rupture of the soft tissues. Although some damage may occur to the receptor during autotomy it is still responsive following autotomy.

CSD1 input is known to reflexly modify the activity of C-B joint muscle motoneurones, exciting the levators and inhibiting the depressors. Interpretation of CSD1's possible role in autotomy is confused by the lack of agreement by authors on the function of one of the levators, the posterior levator muscle. According to McVean (1974) this muscle plays an essential role in the prodution of autotomy and it must contract prior to the anterior levator muscle for successful autotomy to occur. This conflicts with the findings of Clarac and Wales (1970) who have shown the posterior levator to be active during normal walking, and with Moffett (1975) who demonstrated that this muscle is inhibited prior to anterior levator activity at autotomy. It is not possible to ascertain the true role of CSD1 in autotomy until the function of the muscles, which it reflexly controls, is clarified.

Although the proximity of CSD1 to the autotomy plane obviously implicates it in autotomy, this may not be its only, or even its prime function. Clarac (1976) has shown that CSD1 input excites not only the motoneurones of the C-B joint muscles but also those of muscles at the M-C and P-D joint. These leg articulations, along with the C-B joint are important in raising the limb and lateral stepping when the animals walk. This suggests that CSD1 is used in control of limb posture or gait. The value of having these receptors at the C-B joint may be that it is the depressor muscle at this articulation which produces the main lifting force to raise the animal from the substrate and the CSD's are ideally situated to monitor the load borne by each limb during locomotion.

CONNECTIVE CHORDOTONAL ORGANS AND LIMB MUSCLES

Alexandrowicz (1972), in a comparative study of decapod leg proprioceptors, observed that whereas the CCOs of the macruran decapods tend not to be directly associated with limb muscles, those of the higher decapods are. The connection between the CCO and the muscle occurs distally and usually involves the attachment of the receptor strand to the connective tissue ensheathing the muscle tendon. As mentioned above, the muscle tendons largely move in sympathy with the distal segment of each limb joint and the receptor will still monitor joint movement but in a modified form due to any elasticity in the articular membrane, onto which the tendon inserts. These are however two notable exceptions in the sensory cell groups which form the myochordotonal organ (Fig. 10) where the cells are associated with the proximal insertion of a limb muscle. Although it has been shown that these receptors do respond to movement of the M-C joint (Clarac, 1970) the mechanical relationship between joint movement and the stimulus at the receptor will be complex.

With other proprioceptor types, the development of a sensory cell group in conjunction with a specific set of muscle fibres has produced some elaborate proprioceptors e.g. the muscle spindles of vertebrates and the abdominal muscle receptor organs (MRO), thoracico-coxal MROs and mandibular MRO of decapod crustaceans. In each of these organs the receptor cells are associated with groups of muscle fibres, the relative bulk of which is small with respect to the total muscle at that articulation. This muscle provides a means of central adjustment of excitatory level of sensory cell input (Evoy and Cohen, 1969). Thus, the most interesting question to ask of the CCO is "has the association with limb muscles endowed the properties of an MRO on any CCO?"

To answer this question we must first determine whether any of the CCO's are associated with a small group of muscle fibres. Only one receptor system comes close to meeting this requirement and that is the myochordotonal organ (MCO) which is associated with the accessory flexor muscle (AFM) of the M-C joint. The size of this muscle is however relatively larger than those of known MROs and the function of the AFM is not yet fully understood. Alexandrowicz (1972) points out that the name of this organ arose largely because it was one of the first CCOs described in the decapod leg and Barth (1934) may not have used this name had he known of the widespread occurrence of CCOs in association with limb muscles. Because of the name attributed to this receptor, most authors have worked on the premise that the MCO is a type of MRO but none have set out without

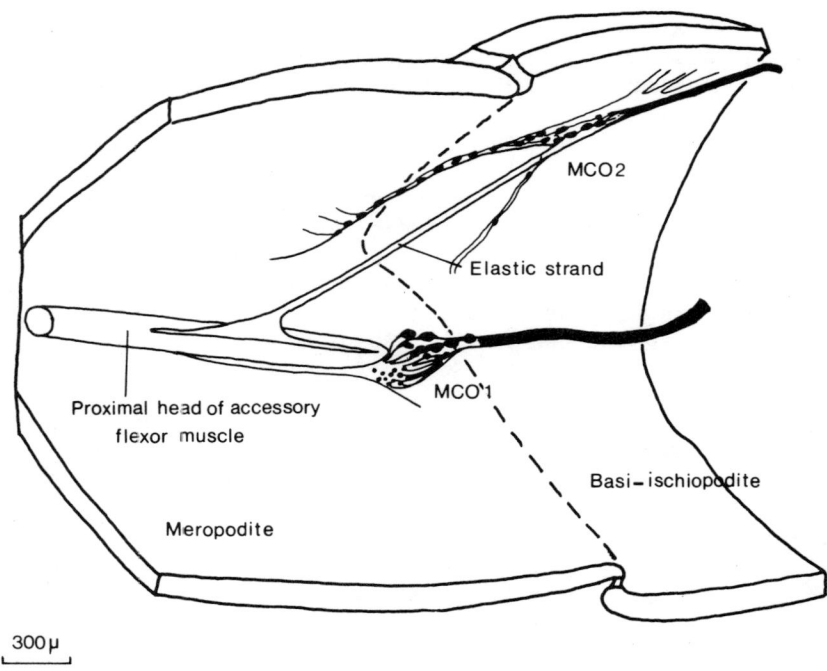

Fig. 10. The myochordatonal organ in the walking leg of the crab *Carcinus maenas* showing the two sense organs (MCO1, MCO2.) associated with the proximal head of the accessory flexor muscle (After Clarac, 1968a).

preconceived ideas to determine the function of the AFM. It should be pointed out that the M-C joint is not the only articulation with more than two muscles, having a small muscle of unknown function (Wales, et al. 1970). Therefore the function of the AFM need not be primarily correlated with MCO function and indeed the association could be no more significant than that between other CCOs and leg muscles (Alexandrowicz, 1972).

The AFM has two muscle heads, one proximal and one distal connected by a long filamentous tendon. The distal head is not directly associated with a CCO and it is much larger than the proximal head, especially in brachyurans. The two muscle heads share the same innervation consisting of an excitatory axon functionally linked in its discharge with one of the four flexor motoneurones and an inhibitory motoneuron common to the extensor muscles (Angaut-Petit et al. 1974). Thus the AFM will tend to contract with the flexors but be inhibited with the extensors. Exactly how this simple control system will act is not clear, particularly as common inhibitor neurone function is not at all understood (Ferrero and Wales, 1976). The basic questions regarding the control and function of the AFM is therefore as yet unanswered and indeed the association between CCOs and muscles in general is not understood. It may prove that the prime role of the MCO is to involve the CCOs of the I-M joint in proprioception of the M-C joint which is much more important in locomotion, particularly in sideways walking.

CONCLUSION

The study of CCOs has revealed that these organs may have evolved from externally located sensillae without loss of function but with increased sensitivity to a greater range of stimuli. This has been achieved by the adaptive radiation of the sensory units, scolopidia, to respond unidirectionally to fractions of the range of length changes and movements experienced by the receptor strand.

On an organ level the CCOs have also undergone adaptation to enable them to respond not only to joint movement but also muscle contraction. Some of the CCOs have become adapted to respond to joint movement of articulation other than the one at which they occur. Further CCOs have become associated with the cuticle, but without sensillae, to provide sensitivity to cuticular stress. It could therefore be construed that adaptive radiation has also occurred at an organ level.

REFERENCES

Alexandrowicz, J.S. (1972). The comparative anatomy of leg proprioceptors in some decapod crustacea. J. mar. biol. Assoc. U.K. 52: 605-634.
Angaut-Petit, D., Clarac, F. and Vedel, J.P. (1974). Excitatory and inhibitory innervation of a crustacean muscle associated with a sensory organ. Brain Res. 70: 148-152.
Barnes, W.J.P., Spirito, C.P. and Evoy, W.H. (1972). Nervous control of walking in the crab Cardisoma quanhumi. II Role of resistance reflexes in walking. Z. Vgl. Physiol. 76: 16-31.
Barth, G. (1934). Untersuchungen über Myochordotonalorgane bei Dekapoden Crustacean. Z. Wiss. Zool. 145: 576-624.
Bässler, U. (1958). Versuche zue Orientierund der Stechmücken: Die Schwarmbildung und die Bedeutung des Johnstonschen Organs. Z. Vgl. Physiol. 41: 300-330.
Bullock, T.H. and Horridge, G.A. (1965). Structure and function in the nervous system of invertebrates. Freeman, New York.
Burke, W. (1954). An organ for proprioception and vibration sense in Carcinus maenas (L). J. Exp. Biol. 31: 127-137.
Burkhardt, D. and Gewecke, M. (1966). Mechanoreception in Arthropoda: the chain from stimulus to behaviour pattern. Cold spring Harbor Symp. Quant. Biol. 30: 601-614.

Burkhardt, D. and Schneider, G. (1957). Die Antennen von *Caliphora* als Anzeiger der Fluggeschwingigkeit. Z. Naturforsch. Teil. B. 12: 139-143.

Bush, B.M.H. (1962a). Peripheral reflex inhibition in the claw of the crab, *Carcinus maenas* (L) J. Exp. Biol. 39: 71-88.

Bush, B.M.H. (1962b). Proprioceptive reflexes in the legs of *Carcinus maenas* (L). J. Exp. Biol. 39: 89-105.

Bush, B.M.H. (1963). A comparative study of certain limb reflexes in decapod crustaceans. Comp. Biochem. Physiol. 10: 273-290.

Bush, B.M.H. (1965a). Proprioception by chordotonal organs in the merocarpopcdite and carpo-popodite joints of *Carcinus maenas* legs. Comp. Biochem. Physiol. 14: 185-199.

Bush, B.M.H. (1965b). Proprioception by the coxo-basal chordotonal organ, CB, in legs of the crab, *Carcinus maenas*. J. Exp. Biol. 42: 285-297.

Bush, B.M.H. (1965c). Leg reflexes from chordotonal organs in the crab, *Carcinus maenas*. Comp. Biochem. Physiol. 15: 567-587.

Clarac, F. (1968a). Proprioceptor anatomy of the ischio-meropodite region in legs of the crab *Carcinus mediterraneus* C. Z. Vgl. Physiol. 61: 203-223.

Clarac, F. (1968b). Proprioception by the ischio-meropodite region in legs of the crab, *Carcinus mediterraneus* C. Z. Vgl. Physiol. 61: 224-245.

Clarac, F. (1976). Crustacean cuticular stress detectors. In: Structure and Function of Proprioceptors in the Invertebrates, Ed. P.G. Mill. Chapman & Hall, London.

Clarac, F. and Wales, W. (1970). Contrôle sensoriel des muscles élévateurs au cours de la marche et de l'autotomie chez certains crustacés, decapods. C.R. Hebd. Séances Acad. Sci. D.271: 2163-2166.

Clarac, F., Wales, W. and Laverack, M.S. (1971). Stress detection at the autotomy plane in the decapod crustacea. II. The function of receptors associated with the cuticle of the basi-ischiopodite. Z. Vgl. Physiol. 73: 383-407.

Evoy, W.H. and Cohen, M.J. (1969). Sensory and motor interaction of the locomotor reflexes of crabs. J. Exp. Biol. 51: 151-169.

Ferrero, E. and Wales, W. (1976). The mandibular common inhibitor system. I. Axon topography and the nature of coupling. J. Comp. Physiol. 109: 123-134.

Graber, V. (1882). Dier Chordotonalen Sinnesorgan und das Gehür der Insekten. Arch. Mikrosk. Anat. Entw. Mech. 20L: 506-640.

Hartman, H.B. and Austin, W.D. (1972). Proprioceptor organs in the antennae of decapod Crustacea. 1. Physiology of a chordotonal organ spanning two joints in the spiny lobster *Panulirus interruptus* (Randall). J. Comp. Physiol. 81: 187-202.

Hartman, H.B. and Boettiger, E.G. (1967). The functional organization of the propus-dactylus organ in *Cancer inoratus* Say. Comp. Biochem. Physiol. 22: 651-663.

Hess, W.N. (1917). The chordotonal organs and pleural discs of cerambycid larvae. Ann. Entomol. Soc. Am. 10: 63-74.

Hollick, F.S.J. (1940). The flight of the dipterous fly *Muscina stabilans* Fallen. Philos. Trans. R. Soc. Lond. B. Biol. Sci. 230: 357-390.

Horch, K. (1971). An organ for hearing and vibration sense in the ghost crab *Ocypode*. Z. Vgl. Physiol. 73: 1-21.

Howse, P.E. (1968). The fine structure and functional organisation of chordotonal organs. Symp. Zool. Soc. Lond. 23: 167-198.

Jägers-Rühr, E. (1968). Untersuchungen zur Morphologie und Entwicklung der Scolopidial Organe bei der Stabheuschrecke *Carausius morosus Br*. Biol. Zentralbl. 87: 393-409.

Laverack, M.S. (1976). External proprioceptors. In: Structure and Function of Proprioceptors in the Invertebrates. Ed. P.J. Mill. Chapman & Hall, London.

Lissmann, H.W. (1950). Proprioceptors. Symp. Soc. Exp. Biol. 4: 34-59.

Lowe, D.A. and Mill, P.J. (1972). The relationship between the PD proprioceptor, the propodite-dactylopodite joint and the dactylopodite flexor muscle in the walking legs of *Cancer pagurus*. Mar. Behav. Physiol. 1: 157-170.

Lowe, D.A., Mill, P.J. and Knapp, M.F. (1973). The fine structure of the PD proprioceptor of *Cancer pagurus* II The position sensitive cells. Proc. R. Soc. Lond. B. Biol. Sci. 184: 199-205.

Mendelson, M. (1963). Some factors in the activation of crab movement receptors. J. Exp. Biol. 40: 157-170.

Mendelson, M. (1966). The site of impulse initiation in bipolar receptor neurons of *Calinectes sapidus L*. J. Exp. Biol. 45: 411-420.

Mill, P.J. (1976). Structure and Function of Proprioceptors in the Invertebrates, Chapman & Hall, London.

Mill, P.J. (1976). Chordotonal organs of crustacean appendages. In: Structure and Function of Proprioceptors in the Invertebrates, Ed. P.J. Mill. Chapman & Hall, London.

Mill, P.J. and Lowe, D.A. (1972). An analysis of the types of sensory unit present in the PD proprioceptor of decapod crustaceans. J. Exp. Biol. 56: 509-525.

Mill, P.J. and Lowe, D.A. (1973). The fine structure of the PD proprioceptor of *Cancer pagurus* 1. The receptor strand and the movement sensitive cells. Proc. Soc. Lond. B. Biol. Sci. 184: 179-197.

Moffett, S.B. (1975). Motor patterns and structural interactions of basi-ischopodite levator muscles in routine limb elevation and production of autotomy in the land crab *Cardisoma quarihumi*. J. Comp. Physiol. 96: 285-305.

Moulins, M. (1976). Ultrastructure of chordotonal organs. In: Structure and Function of Proprioceptors in the Invertebrates. Ed. P.J. Mill. Chapman & Hall, London.

Mouuns, M. and Clarac, F. (1972). Ultrastructure d'un organe chordotonal associé à la cuticule dans les appendices de l'Esrevisse. C.R. Hebd. Séances, Acad. Sci. D. 274: 2189-2192.

Pringle, J.W.S. (1938). Proprioception in insects. III. The function of the hair sensilla at the joints. J. Exp. Biol. 15: 467-473.

Pringle, J.W.S. (1961). Proprioception in arthropods. In: The Cell and the Organism. Edit. J.A. Ramsay and V.B. Wigglesworth. University Press, Canbridge.

Richard, G. (1957). L'ontogenese des organes chordotonaux antennaires de *Calotermes falvicollis* (Fab). Insectes Sociaux, 4: 106-111.

Satir, P. and Gilula, N.B. (1973). The fine structure of membranes and intracellular communications in insects. Annu. Rev. Entomol. 18: 143-166.

Sherrington, C.S. (1906). On the proprioceptive system, especially in its reflex aspect. Brain. 29: 467-482.

Taylor, R.C. (1967). The anatomy and adequate stimulation of a chordotonal organ in the antennae of a hermit crab. Comp. Biochem. Physiol. 20: 709-717.

Wales, W. (1976). Receptors of the mouth parts and gut of anthropods. In: Structure and Function of Proprioceptors in the Invertebrates. Ed. P.J. Mill. Chapman & Hall, London.

Wales, W., Clarac, F., Dando, M.R. and Laverack, M.S. (1970). Innervation of the receptors present at the various joints of the periopods and third maxilliped of *Homarus gammarus* (L) and other Macruran Decapods (crustacea). Z. Vgl. Physiol. 68: 345-384.

Wales, W., Clarac, F. and Laverack, M.S. (1971). Stress detection at the autotomy plane in the decapod crustacea. 1. Comparative anatomy of the receptors of the basi-ischiopodite region. Z. Vgl. Physiol. 73: 357-382.

Whitear, M. (1962). The fine structure of crustacean proprioceptors. 1. The Chordotonal organs in the legs of the shore crab, *Carcinus maenas*. Philos. Trans. R. Soc. Lond. B. Biol. Sci. 245: 291-325.

Wiersma, C.A.G. (1959). Movement receptors in decapod crustacea. J. Mar. Biol. Assoc. U.K. 38: 143-152.

Wiersma, C.A.G. and Boettiger, E.G. (1959). Unidirectional movement fibers from a proprioceptive organ of the crab, *Carcinus maenas*. J. Exp. Biol. 36: 102-112.

Wyse, G.A. and Maynard, D.M. (1965). Joint receptors in the antennule of *Panulirus argus* (Latreille). J. Exp. Biol. 42: 521-535.

Young, D. (1970). The structure and function of a connective chordotonal organ in the cockroach leg. Philos. Trans. R. Soc. Lond. B. Biol. Sci. 256: 401-428.

PERIPHERAL THERMAL RECEPTORS

R. Loftus, S.J.

Institut für Zoologie, Universität Regensburg

D - 8400 Regensburg, Germany

I. INTRODUCTION

Thermal receptors as a rule do not provide an organism with the kind of detailed information one is accustomed to expect from sense organs for vision, hearing, or smell. The location of nest, brood, mate, foraging areas, and predators within a biotope is mediated largely by sensory systems other than thermal, though there are notable exceptions. The facial pits of crotalid vipers (Bullock and Diecke, 1956; Goris and Nomoto, 1967) enable their possessors to strike successfully at small warm objects in the dark. Similar pits line the mouth region of the Boidae (Haris and Gamow, 1971; Hensel, 1947b). *Melanophila*, a bupestrid beetle, has sense organs at the rim of the depressions where its middle legs join its mesothorax. Apparently the beetle employs these organs to locate the scene of forest fires (Evans, 1964, 1966a,b). There it seeks fairly intact stumps of freshly injured trees for its grubs to feed in. Fire damage affords access. Mosquitoes are another example. They are equipped with antennal thermal receptors which in combination with chemoreceptors presumably assist them in the search for their warm-blooded prey (Davis and Sokolove, 1975). But on the whole, thermal receptors tend to provide background information such as temperature internal to the organism, of its periphery and of the medium in which it is immersed, and also on the direction and rate with which these temperatures are changing. Such information is as important as it is general. At least in indirect form, it is prerequisite for thermal regulation.

The range of body temperature at which an animal functions effectively is quite narrow, often including little more than $4^\circ C$. Thus at least during those periods in which survival demands active and successful competition, it becomes incumbant on the organism to match heat flow with heat production precisely in order to maintain its internal temperature within the range in face of variation in external temperature and its own activity. To make the match, control can be exerted on heat production, on heat flow, on peripheral or even on external temperature, and of course, on combinations of these parameters. Heat production is no great problem for desert jackrabbits forced to run in the cool early morning. But it the heat of the day locomotor activity can drive their internal temperature as high as 42 or $43°C$. Then their running periods are much shorter. They must stop and do so, preferably in the shade (Shoemaker et al., 1976). Similarly on a hot day the thorax temperature of a bee may exceed $38°C$ during flight. But when the air streaming past is no longer enough to offset the rate of temperature increase, a drop of fluid appears in the mouthparts of the bee which it proceeds to smear over its thorax (Esch, 1976). Evaporating water is an effective coolant even when it is not pure. Moreover, the same heat production which becomes a problem at high ambient temperatures can be a solution when they are low. Sphinx moths employ prolonged bursts of muscular activity to elevate their thorax to flight temperature (McCrea and Heath, 1971; Heinrich, 1974). Shivering is commonplace in the cold, as it is also in animals emerging from hibernation. A more economical form of thermoregulation, however, is adjustable insulation. Birds ruff their feathers when they sit in the cold, and when it gets too hot pigeons increase the air gap between wings and body even to the point of extending a wing, the better to expose otherwise densely covered surfaces. Another alternative is to enhance or curtail peripheral blood flow, depending on the desirability in the circumstances of bringing core temperature blood to the surface where it can function as a heat exchange medium. Still another is to regulate external temperature by avoiding areas with extremes, as fish do in the neighborhood of a power plant dissipating waste heat into the water (Gift, 1977). An animal can move down the temperature gradient into a more suitable environment. When avoidance is impossible, however, a further alternative is to stop fighting to maintain a large temperature difference towards the outside. An animal may wait out a season in hibernation, in the inactivity of lowered temperature. Or in the extreme, it might leave the waiting to embryonic progeny in its eggs, and no longer able itself to regulate, simply die.

II. ADAPTIVE RADIATION

The preceding examples illustrate that thermoregulation is commonplace in the animal kingdom and that there is a variety of mechanisms which animals can make use of to achieve independence of ambient temperature. Wherever thermoregulation is accomplished through nervous control, the existence of thermal sense organs is strongly suggested. In many instances their existence has been demonstrated by direct experiment. But the animals in which they are found display considerable diversity and so do the biotopes they live in. Since this double diversity reflects the demands the biotope places on the organism as well as those which the organism places on the biotope, each animal represents a possible solution to the network of problems posed by a given environment in terms of the physiological, morphological, and behavioral systems which the animal incorporates. Therefore it would not seem unreasonable to expect thermal sense organs to manifest to some degree not only the sameness which permits classification of many animals into a single group despite their diversity, but also differences whose development and distribution reflect environmental pressures in varying ways. Shouldn't one expect to find adaptive radiation of thermal receptors similar to that of other functional structures of the animals possessing them?

Adaptive radiation would mean: 1) that thermal receptors should be encountered in widely separate species within a taxonomic group, 2) that at least within the group these receptors should be in principle the same, and 3) their variations should reflect the physiology and behavior of the animal they are found in and help fit it into its biotope. Ideally one would find a single basic type throughout the animal kingdom varying according to the use it is put to and helping make each different species a match for its environment. Unfortunately experimental results hardly begin as yet to equal the richness of the expectation.

The number of animals whose behavior is demonstrably influenced by temperature is much greater than the number in which specifically thermal sense organs have been identified electrophysiologically, the latter being restricted so far to insects and vertebrates. The vertebrates include man (Hensel and Bowman, 1960), monkeys (Iggo, 1969; Darian-Smith et al., 1973), rats (Iggo, 1969; Hellon et al., 1975), dogs (Iggo, 1969), cats (Hensel and Zotterman, 1951; Benzing et al., 1969; Hensel and Kenshalo, 1969), pigeons (Necker, 1973), snakes (Bullock and Diecke, 1956; Bullock and Fox, 1957; Goris and Nomoto, 1967; Goris and Terashima, 1976; Harris and Gamow, 1971), and frogs (Spray, 1974). The areas examined, though not the same in all species, are usually such as one might suspect as having thermoregulatory relevance: tongue, face, extremities, and scrotum in mammals, the beak of pigeons, and the body skin of frogs. Insects include bees (Lacher, 1964), moths (Schoonhoven, 1967; Becker, in prep.), predaceous bugs (Bernard, 1974), sap-sucking bugs (Čokl, 1972), grasshoppers (Waldow, 1970), cockroaches (Kerkut and Taylor, 1957; Loftus, 1968, 1969), walking sticks (Tichy, in prep.). (The references mentioned above by no means constitute a complete list. For further references the lists in the literature cited at the end of the chapter are recommended, especially in the reviews.)

Research so far has revealed great similarity in the functional characteristics of thermal receptors, and the problems arising from them are still to a large extent common. For this reason the material will be treated mainly from the standpoint of functional characteristics rather than taxonomy or environment, and exemplified by the spectrum of animals available. The resulting juxtaposition of thermal receptors from widely separate groups reflects functional similarity and is intended to facilitate comparison, not to prejudice the issue of whether thermal receptor types are homologous wherever found. Even the scope of this chapter limits treatment; there is no pretense at an exhaustive survey.

III. UNIMODAL FUNCTION

From their physiology, thermal receptors can be divided into two antagonistic categories: the warm receptors which respond with a rise in impulse frequency to increasing temperature, and the cold receptors which respond with a rise in impulse frequency to decreasing temperature. Thus defined, the designations of warm and cold refer simply and directly to the physiological responses just described. They are not meant to bias the question of whether either of the two alone or even both in combination are sufficient to account for an unambiguous behavioral response to temperature or for the sensation of warmth or cold in a given situation.

A. INFRARED RECEPTORS

Warm receptors have not been found as often as cold receptors, but what is probably the most famous of all thermal receptors belongs to the warm category: the facial pit organs of crotaline vipers (Fig. 1). These pit organs, which rattlesnake and their relatives share with the family, Boidae, consist of a richly innervated membrane in places less than 15 μm thick. The membrane is supplied by trunks from all three branches of the trigeminus nerve. Individual fibers within the membrane undergo repeated branching (Bullock and Fox, 1957) and form a terminal nerve mass about 10 μm thick and 40 μm in diameter (Terashima, Goris, Katsuki, 1970). In crotaline vipers the membrane itself lies at the bottom of a pit on either side of the head above the upper lip and between the eye and nostril, and is suspended all along its perimeter. Beneath it is an air chamber with an opening to the surface. Thus insulated from other tissues, this thin membrane can be quickly influenced by the temperature of objects outside it, either warming up slightly when presented with an object warmer than its surroundings, or losing heat when its field of view is directed at cooler objects. Because it lies at the bottom of a pit, it is somewhat shielded from fluctuating temperatures of air currents. Location in the pit also provides directionality. For even though infrared radiation corresponding to the temperatures which the membrane detects is of wavelengths

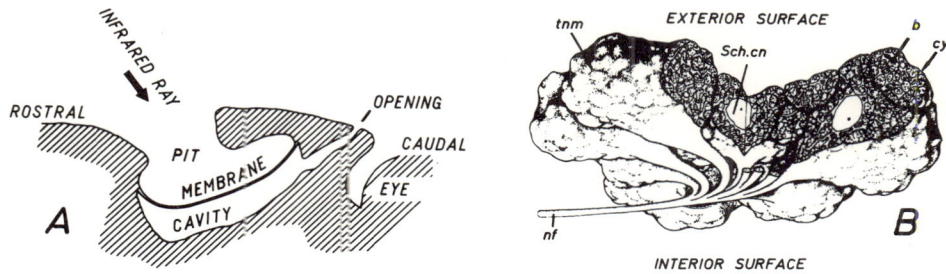

Fig. 1. A. Facial pit organ of crotalid viper in schematic sagittal section. Infrared-sensitive thermoreceptors dispersed throughout membrane. Membrane often thinner than 15 μm is suspended over air-filled cavity.
B. Schematic drawing of terminal nerve masses in pit membrane. b, branchlet; cy, cytoplasm of Schwann cell; nf, nerve fiber; Sch.cn, Schwann cell nucleus; tnm, terminal nerve mass. (A. and B. from Goris and Terashima, 1976, slightly altered.)

10 or more times greater than that of visible light, its 1 to 20 μm is still short enough to cast a fairly sharp heat shadow across the membrane when the heat source is sufficiently off center. Presumably the directionality is enhanced by deepening the pit and thus increasing the ratio of its depth to the diameter of the membrane. Such would appear to be the case for *Trimerosurus flavoviridis*. The directionality itself is a fact is clear. Goris and Terashima (1976) describe experiments in which the scales covering the snake's eyes were blackened. Thus blinded, the snake not only strikes successfully at a warm object. When the object suddenly moves in midstrike, the snake can still hit it. This behavior corresponds to the snake's nocturnal hunting habits and fits in with recordings from single units in the optic tectum where neurons were found with overlapping fields of 50 or 60°. This relationship does not yet of course define the precise number or location of individual "nerve masses" in the membrane which the brain is drawing on at a given instant as the animal strikes at a moving object.

That such an organ is quite sensitive to warm or cool objects in its surroundings is clear enough, but to determine the threshold of individual units is not easy. The membrane can be irradiated with a calibrated incandescent heat source whose distance from the pit can be varied arbitrarily. But because the thickness of the membrane can be under 15 μm and the distance from its outer surface to the receptors themselves much less, direct measurements of the temperature course within the membrane during the first few critical milliseconds of exposure are hardly possible and assumptions have to be made regarding heat absorption and transfer during this period. Bullock and Diecke (1956) made a conservative calculation of 0.02°C within the membrane, or a heat flux slightly over 1 mW/cm^2 of irradiated surface. Using generator potential as an indicator, Terashima, Goris and Katsuki (1968) were able to lower this estimate to about 100 μW/cm^2. Another approach made by Bullock and Diecke (1956) was to let warm water at constant temperature flow gently across the membrane and then, after the flow as such had ceased to stimulate, to warm the water by about 0.1°C and observe the response during the transition. Impulse frequency rose 50% during the first 60 msec (Fig. 2A). As evidenced by a small thermocouple in the stream, water temperature during this period increased by only 0.003°C. The high heat capacity of water suggests a similar rise time for both thermocouple and membrane under these conditions and thus provides an estimate of 0.003°C as the minimum rise in temperature (ΔT) needed to stimulate the organ clearly. It could be, however, that the rate of temperature change (dT/dt) rather than the temperature difference (ΔT) is the significant stimulus parameter. Unfortunately its value is almost impossible to determine for fast temperature changes with sufficient accuracy, since the time course of temperature recorded is that of the thermocouple, not of the organ, except indirectly by way of inference.

The results of the above experiments can be readily interpreted as meaning that what the membrane is really responding to is its own changes in temperature as brought about by infrared (IR) radiation from the objects it faces. There is an alternative hypothesis, how-

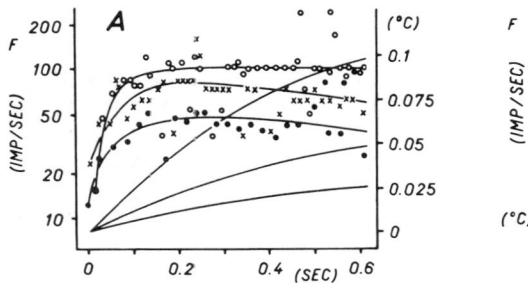 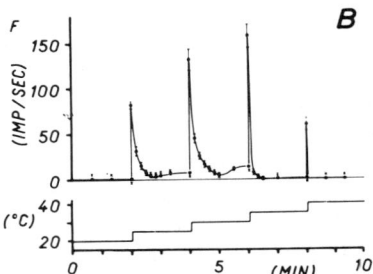

Fig. 2. A. Responses of single units from rattlesnake pit membrane to sudden warming brought about by raising membrane temperature with gentle stream of water. As temperature rises 0.003°C (lower 3 traces) impulse frequency increases by 50%. (From Bullock and Diecke, 1956, modified.)
B. Response of units from *Boa* pit organ to 5°C step functions. Maximum impulse frequency varies with initial temperature of step. (From Hensel, 1976, slightly altered.)

ever. The slight changes in temperature might be only a concommitant phenomenon and the low energy IR photons may trigger photochemical processes in the membrane analogous to those in photoreceptors. Either hypothesis is beset with the difficulty to explaining the detection of very small changes, whether in temperature or in IR radiation as such. But regardless of the nature of their primary processes, the pit organs are obviously stimulated by warm objects, as the behavior of the snakes demonstrated, and the response of the organs bears a clear relationship to temperature changes in the membrane in any case. Consequently they can be classed at least in a broad sense as thermal. (For evidence in this regard -- against the infrared photoreceptor hypothesis, as it turns out -- see the experiments of Harris and Gamow (1971) on the pit organs of the family Boidae.)

Hensel (1974b, 1976) found that more goes into determining the response of the facial pits of *Boa* than only the magnitude of the temperature jump (ΔT). Beginning at $20°C$, he stimulated the organ with a series of rapid rises in temperature, each of $5°C$. As Fig. 2B shows, response magnitude also depends on initial temperature; the temperature, that is, from which the jump begins. The consequences of these experiments are important. First, they place these receptors in a framework not so very different from that of other thermal receptors, since the response of other thermal receptors to rapid temperature change is also dependent on initial temperature -- cold fibers in the lingual nerve of the cat, for example, observed by Hensel and Zotterman as early as 1951. Second, they indicate that a given response is ambiguous. The same temperature jump produced different responses at different initial temperature. Presumably then, the same response could be produced by a variety of combination of temperature jumps and initial temperatures. As will be seen, such a situation has been demonstrated for insect cold receptors.

In the case of pit organs though, the ambiguity may not be very important. When individual receptors in the membrane are suddenly confronted with an object several degrees warmer than its surroundings, their impulse frequency shoots to high values capable of communicating the presence of a warmer object to the central nervous system. But how necessary or even advantageous it would be for the snake to be able to determine that an object was very close to the surface temperature of a mouse and not merely warmer than its surroundings may be questionable. The position of the object together with its direction and manner of motion (smooth, jerky, undulating, erratic, still) might offer a predatory animal better clues regarding suitability than a precise measurement of its temperature, particularly in the dark. But how spacial and temporal resolution of these organs or their ability to discriminate temperatures is correlated with the hunting habits of their possessors remains to be examined.

There appears to be another infrared sense organ with a function closely resembling that of facial pit organs of vipers. Evans (1964, 1966a,b) describes it in the case of the bupestrid beetle, *Melanophila accuminata*. Active during the heat of the day, bupestrids in northern climes are attracted to tree trunks warmed in the sun. There they lay their eggs. But *M. acuminata* even flies to the scene of forest fires and lays its eggs near root level in trunks of trees which fresh fire damage has exposed to attack (Palm, 1962). The organ claimed responsible is a field of sensilla covered with wax filaments and situated in a depression at the edge of both the right and the left middle coxal cavities. A short flash of infrared radiation directed at either was immediately followed by a twitch of the antenna on the same side as the field, a twitch whose amplitude increased with the intensity of the IR source. When other modalities were tested, this behavior failed to develop, and it ceased altogether when the fields were coated with bronze paint. Maximum response (twitch, that is) to equal energy wavelengths occurred just a bit below 3.5 µm, right in the window of the IR absorption spectrum for carbon dioxide and water vapor. This window represents a narrow band of wavelengths relatively intense at the temperatures of fire and which is not filtered out by the main products of combustion. Threshold energy was calculated at 60 $\mu W/cm^2$, a value close to the estimates for crotaline facial pits. Such a threshold should enable the beetle to detect a 4 hectar fire 3 km away and at an angle of $45°$. These bupestrids therefore would appear to be an excellent example of an animal's adaptation to a limited biotope by reason of a specialized thermoreceptor. Unfortunately there seems to have been no electrophysiological confirmation of these results. This method would not only eliminate any lingering doubt as to whether the sensilla grouped in this pair of small fields in fact represent a concentration of thermal receptors. Comparison at the receptor level should also indicate how the spectral sensitivity and threshold of individual receptive units fit in with the behavioral response. It should also indicate the role played by other modalities, if any.

B. WARM AND COLD RECEPTORS

As was pointed out earlier, receptors demonstrated electrophysiologically to be thermal fall into two categories: warm and cold. This classification does not always reflect the temperature range in which either category responds (Hensel, 1976) but rather the antagonism of their responses to temperature change. The antagonism can be illustrated by recordings from the antenna of the yellow fever mosquito, *Aedes aegypti* (Fig. 3). Close proximity of the two receptor types there permitted Davis and Sokolove (1975) to pick up impulses from both with the same electrode simultaneously. (Parenthetically, the data in this paper are the most extensive published on any invertebrate warm receptor so far. Čokl found the same basic type of response earlier in the antenna of the sap-sucking bug, *Oncopeltus* (1972). The recordings of Van Weel and McDonald (1971) are also similar, but do not exclude the possibility of thermosensitive mechanoreceptors as a source.) As the temperature trace in Fig. 3 shows, both types react to the same temperature changes, but in opposite direction. The same cooling lowers impulse frequency in the warm receptor and raises it in the cold receptor. Correspondingly contrary effects are produced by warming. Closer inspection of Fig. 3 also suggests that just as in the case of the pit organs of *Boa*, more is involved in determining response than just temperature change. This "more" is a source of ambiguity, important for the limitation it can place on a receptor's usefulness and for the resulting role of the receptor in adapting an animal to its environment.

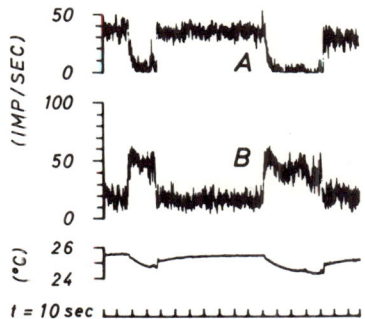

Fig. 3. Time course of antagonistic responses to temperature changes (third line). A, warm receptor; B, cold receptor, together on antenna of yellow fever mosquito, *Aedes aegypti*. (From Davis and Sokolove, 1975, slightly altered.)

1. RESPONSE TO STEADY TEMPERATURE

In general thermal receptors tend to respond both to stationary temperature and to temperature change. The first type of response resembles that of a thermometer. Just as each temperature in the range is assigned a position on the thermometer scale, so is each temperature indicated by a definite impulse frequency. This simple sounding relationship is probably more or less what one would expect from a thermal receptor, but it is in fact acceptable only with reservation. As exemplified by the warm and cold receptors on the antenna of the yellow fever mosquito (Davis and Sokolove, 1975) to be seen if Fig. 4, two differences from a thermometer scale are obvious. The first is a very high degree of variation in response, which was also observed in antennal cold receptors of the cockroach, *Periplaneta* (Loftus, 1968). Each steady temperature can occupy such a large segment of the frequency range available that very little distinction in temperature could be made on the basis of the stationary frequencies they elicit. In the case of *Periplaneta*, the response to temperature change is quite accurate, but the spread in response of

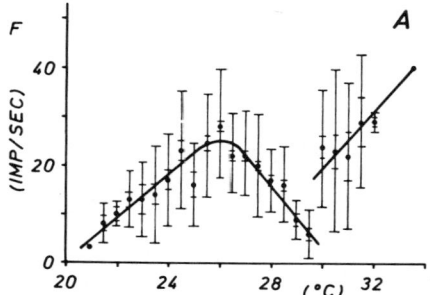 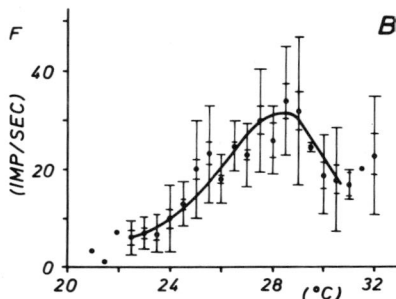

Fig. 4. Responses of warm (A) and cold (B) receptors of yellow fever mosquito to steady temperature. Similar large variance was also found in the antennal cold receptor of the cockroach *Periplaneta* (Loftus, 1968) and is by no means unique. Outer brackets, standard deviation; inner brackets, standard error of mean. (From Davis and Sokolove, 1975, slightly altered.)

even single receptors to temperatures held constant within a few tenths of a degree for several hours was so great that these single receptors could be classed at best only as very poor thermometers. Yet *Periplaneta* displays very precise temperature preference (Coenen-Staß, 1976). The corresponding variation in cold receptor response of bees (Lacher, 1964) and migratory locusts (Waldow, 1970) seems smaller, but the data are not so extensive. To judge by a fairly large number of response curves from the cat's tongue (Benzing et al, 1969) and nose (Hensel and Kenshalo, 1969), there can also be large variation in the response of mammalian cold and warm receptors. On the other hand response curves from individual receptors often tend to be quite smooth and when many are pooled, a curve of similar shape emerges also smooth and of much greater amplitude (Hensel and Zotterman, 1951). Pooling should tend to even out irregularities and to compensate for the low rate of information transfer implicit in the low frequencies of individual receptors.

The second and even more important difference from a thermometer is shared with vertebrate thermal receptors (Figs. 5 and 6): impulse frequency as a function of temperature passes through a maximum. In the case of the mosquito the curve appears to reverse its course even a second time. This kind of non-linearity means that a condition exists in thermoreceptors which would be intolerable in a thermometer: each value on the scale except the highest (that is, each impulse frequency in the range except the maximum) designates not one but a pair of temperatures. Such ambiguity would not be eliminated even if response variation should be reduced to zero. If cold and/or warm receptors are to provide unequivocal data on their temperature and therefore on that of their immediate environment, something beyond the stationary frequency of just one or the other type is needed. Several avenues are open. Iggo (1969), for example, has suggested

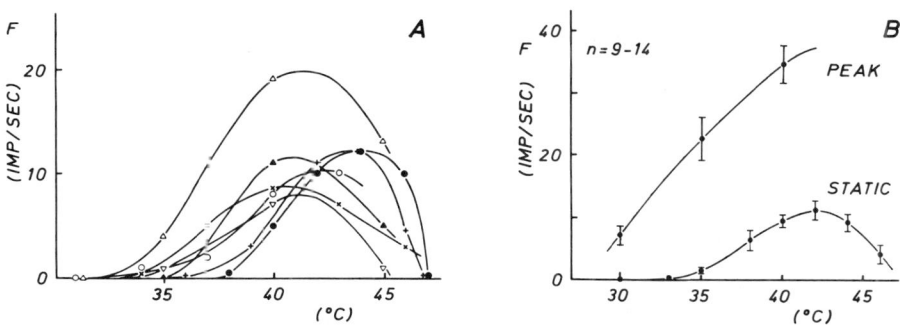

Fig. 5. A. Stationary frequency from 7 warm fibers in rat scrotal nerve. Response to steady temperature passes through maximum and is thus ambiguous even if variance were zero.
B. Lower curve: stationary frequency averaged from 9-14 single fibers. Brackets indicate standard error of mean. Upper curve: responses to 5°C step functions resembling those of *Boa* in their dependence on initial temperature (Fig. 2B). (From Hellon et al., 1975, slightly altered.)

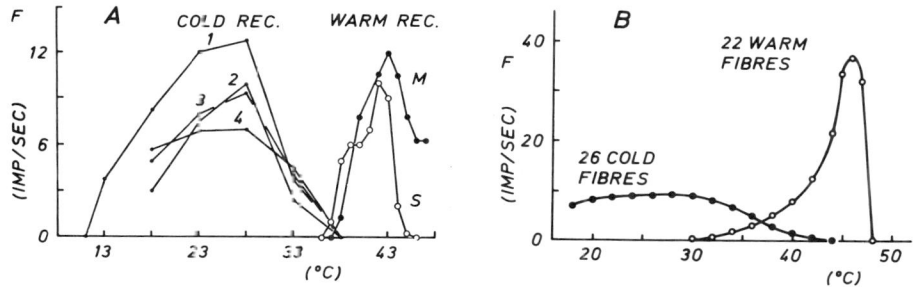

Fig. 6. A. Stationary frequency from single fibers in rat scrotal nerve. Responses from both cold and warm receptors pass through maximum with corresponding ambiguity and suggest only slight overlapping in their respective temperature ranges. S, single fiber; M, multifiber preparation. (From Iggo, 1969, slightly altered.)
B. Average stationary frequencies in response to steady temperature from warm and cold receptors in nasal region of cats. Ambiguity inherent in each steady-temperature response curve may be eliminated in range of overlapping. (From Hensel and Kenshalo, 1969, slightly altered.)

differences in bursty patterns developing in the sequence of impulses at certain temperatures. In *Periplaneta* (Loftus) and in the walking stick, *Carausius* (Tichy, in prep.), however, burstiness often appeared rather independently of temperature, after lengthy experimentation and by no means always. Another possibility would be to place the maximum outside the range of normal operation temperatures. Such a solution could be envisioned for receptors monitoring core temperature in birds and non-hibernating mammals, where the range is narrow anyway. It may turn out to be the case in *Carausius* where the cold receptor response maximum is at quite a low temperature, lower perhaps than those which the insect usually encounters (Tichy, in prep.).

A third possibility, suggested by Kenshalo et al. (1971) and Järvilehto (1973), would be the interaction of warm and cold receptors. A comparison of the simultaneous outputs of both could prove a means of discrimination analogous to that of photoreceptors. The response of different photoreceptors also passes through a maximum, but depending on the type, at different wavelengths whose counterpart here would be different temperatures. A proper combination of the responses could actually enhance sensitivity, but the curves should cross below their maxima yet still at a fairly high frequency value. When they meet close to zero as in Fig.6A, one of the two would be silent at most temperatures (Iggo, 1969). The outlook would be better for fibers from the nasal region of cats (Hensel and Kenshalo, 1969), where overlapping is more extensive (Fig. 6B). Between $30°$ and $45°C$ and within the limits of "random" variability, the combination of stationary impulse frequencies from these warm and cold receptors is unique at each temperature and might therefore be used as a basis for discrimination. A very similar situation is reported for warm- and cold-sensitive neurones in the raphé, once more of rats (Dickenson, 1977). Overlapping is displayed here between $29°$ and $38°C$. The activity of these neurones is unimodal and depends only on skin, not on body temperature.

2. RESPONSE TO TEMPERATURE CHANGE

A third characteristic of thermal receptors distinguishes them more profoundly still from thermometers: their response is not only affected by the temperature they happen to have. A portion of their total frequency spectrum is partially reserved for temperature change. Of course a thermometer can be used to determine the direction and extent of temperature change by making readings at different times, but there is no part of its scale which refers to anything but temperature. (The physical states, e.g. expansion or deformation, which are a function of temperature and which termometers directly indicate, are not meant in this connection.) Though as will be seen, there appears to be no impulse frequency which refers simply to temperature, there are two ranges of the frequency scale whose values the receptor's firing rate never assumes unless temperature has just changed or is still doing so: those values, namely, which are either above or below the segment which stationary temperature elicits. As soon as the frequency exceeds the maximum for stationary temperature, a change is signalled: falling temperature if cold receptor exceeds its maximum and a rise if the warm receptor exceeds its. Frequencies below the minimum for stationary temperature are elicited by the correspondingly opposite changes in temperature.

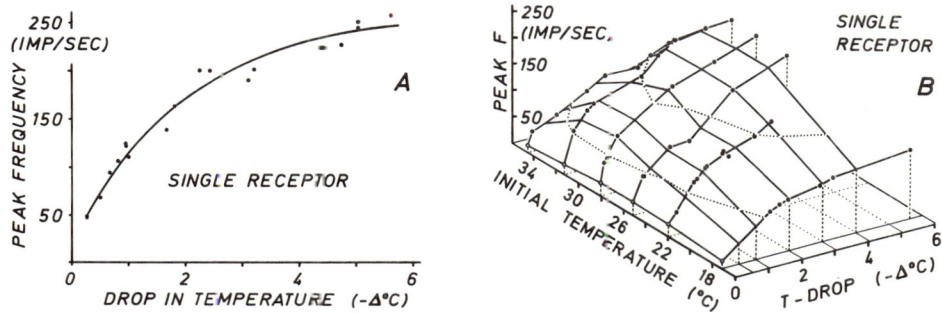

Fig. 7. A. Peak frequency response of antennal cold receptor of cockroach, *Periplaneta americana*, to rapid drop in temperature (step function).
B. Characteristic curves as in A depend on temperature from which drops were initiated. Dotted equal frequency curves across irregular surface display ambiguity resulting from this double dependence. Each peak frequency corresponds not to a single combination of a temperature drop and an initial temperature, but to a set of such combinations. (From Loftus, 1966, 1968, slightly altered.)

As might be expected, the greater the temperature change, the greater the magnitude of the response, depending on course on direction and receptor type. Such a relationship is illustrated by the cold receptor of *Periplaneta* in Fig. 7A (Loftus, 1966). The curve approximating this function tends to be steeper where the drops are smaller and to flatten as they become greater, with the consequence that the receptor is relatively more sensitive to smaller drops than to larger ones. Furthermore and very important, this function is not be same at all temperatures. As was seen in the case of Bca (Fig. 2B) and the rat scrotum (upper curve in Fig. 5B), the response to the same change in temperature varies with the temperature from which the change initiated. In Fig. 7B (Loftus, 1968) it is also clear that each response can be elicited by a variety of combinations of initial temperatures and temperature drops, a situation quite similar to that observed by Darian-Smith, Johnson, and Dykes in their recordings from cold fibers of the palmar and digital skin of rhesus monkeys at temperatures between 31° and 41°C (1973). The response then is not to a single combination of these parameters but to a set of them. Ambiguity is intensified. To be sure, high impulse frequencies are produced only by big changes, but how big cannot be decided on the basis of a single response, or even from the responses of a group of receptors of exactly the same type to the same stimulus situation.

3. RESPONSE TO LOW RATE OF TEMPERATURE CHANGE

It may be, however, that neither *Boa*, rhesus monkeys, nor *Periplaneta* require precision in this regard. Or if one or the other does, one might envision a comparison of successive response magnitudes. When the response is stationary, indicating constant temperature, its value might be used as a setting with which to compare ensuing high frequency responses. A solution along this line is regarded difficult, though, not only by reason of the ambiguity involved in stationary responses. In *Periplaneta* the high sensitivity of antennal cold receptors to low rates of change would have an additional effect. Once the animal ventured from under cover and into moving air, its superficial receptors would be confronted constantly with at least minute changes in temperature. Under these circumstances frequency varies continually, reflecting from instant to instant not only a succession of temperatures but also the rate with which temperature happens to be changing. Even frequencies in the range indicating steady temperature become additionally ambiguous during temperature change by reason of the simultaneous effect of the rate of change on the receptor's output. The same situation has also been observed by Tichy in the antennal cold receptors of *Carausius* (in prep.). An example in the case of *Periplaneta* can be seen in Figs. 8 and 9.

When receptor temperature moves slowly and steadily up and down at varying speeds and amplitudes, impulse frequency is not in step with temperature but ahead of it (Fig. 8B). Frequency has its counterpart mainly in the rate of temperature change, though to some extent also in the actual temperature at each instant. As shown in Fig. 8A (Loftus, 1969), where frequencies during a single "oscillation" are plotted as a function of the temperatures passed through, ambiguity is massive. Not only is the same response elicited at more than one temperature, more than one response is elicited at the same temperature -- many in fact, since the temperature and its rate of change are independent parameters. But the two basic relationships are maintained (Fig. 9A). As a function of temperature frequency passes through a maximum, while it is at the same time a steadily decreasing function of the rate of temperature change.

In itself this source of ambiguity does not come as a surprise. Impulse frequency of both warm and cold receptors in general is driven by sudden changes in temperature to heights which no steady

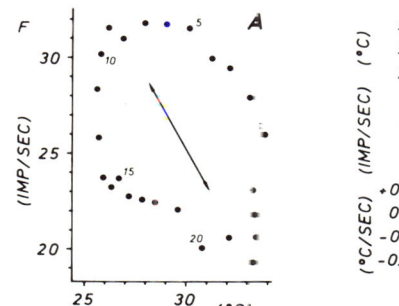
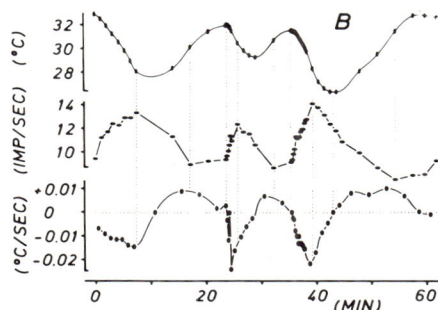

Fig. 8. A. Sequence of responses from single cold receptor of *Periplaneta* during slowly changing temperature. Ambiguity is heightened still further here where more than one frequency corresponds to a given temperature and where each frequency corresponds to more than one temperature. Lissajous figure. B. Frequencies during slow changes in temperature are in close step with the rate of temperature change but also to a lesser extent dependent on the temperature at which the change is occuring. (Loftus, 1969, slightly altered.)

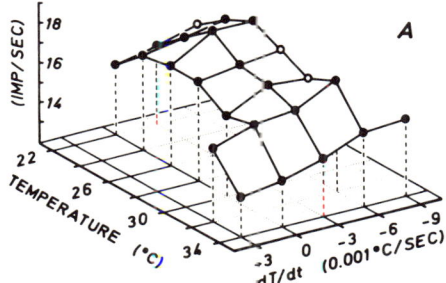
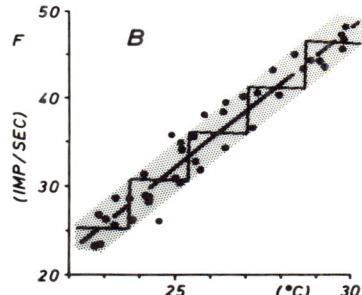

Fig. 9. A. Dependence of impulse frequency both on the rate of temperature change and on instantaneous temperature as in Fig. 8B. The dependence on temperature passes through a maximum here and is at the same time a decreasing function of the rate of change. (Loftus, 1969, slightly altered.) B. Response of antennal dry receptor of *Periplaneta* to slowly changing temperature at constant partial pressure of water vapor. The linearity of this dependence during temperature change contrasts with that of the cold receptor. The darkened band encloses 90% of the responses of this single receptor and suggests that it could distinguish several different temperatures in the $7^{\circ}C$ range with high propability if the dependence on humidity were eliminated from the response (Loftus, in prep.).

temperature elicits. But if such large frequency displacements occur, smaller ones should be possible too, so small that frequency stays in the range elicited by steady temperature but with values also being influenced by low rates of temperature change. The important thing regarding insects is that the rates of temperature change producing such ambiguity are so small (a few degrees per hour maintained only for seconds) that they become unavoidable in open spaces by reason of the small size and superficial location of the sensilla (see Fig. 12). Nevertheless, comparsion of the fluctuating antagonistic responses of warm and cold receptors should inform the insect at least as to when temperature is changing and could thus be a help in the search for shelter. Moreover, since both types respond differently to the same two parameters, proper central processing of both might yield unambiguous information concerning either. Unfortunately such considerations are for the present quite hypothetical. The number of insect species in which thermal receptors have been identified electrophysiologically is still quite small (7-10 for cold, 2-3 for warm), and concerning central processing very little is known beyond the existence of neurons in *Periplaneta* whose activity is influenced by several modalities at the antenna, including temperature changes (Waldow, 1975).

IV. BIMODAL FUNCTION

A. HYGRO- THERMAL

Multimodal neurons such as those described by Waldow (1975) suggest that sense organs other than those exclusively thermal may be involved in thermal perception and orientation. Hygroreceptors in insects are among the probable candidates. Often, though not exclusively, found in the same sensillum with cold receptors of honey bees (Lacher, 1964), locusts (Waldow, 1970), predaceous bugs (Bernard, 1974), cockroaches (Yokohari and Tateda, 1976; Loftus, 1976), moths (Becker, in prep.), and walking sticks (Tichy, in prep.), these receptors display considerable dependence on temperature, which at least in the case of the dry receptor of *Periplaneta* cannot simply be attributed to the effect of relative humidity. This receptor which responds to decreasing humidity with increasing impulse frequency also responds to slowly changing temperature, but differently from the cold receptor. Rate of change here has little if any influence on impulse frequency, but the temperatures encountered during the changes do. The dependence is quite linear and reliability appears sufficient to permit the distinction of several different temperatures in a $10^{\circ}C$ range with high probability (Fig. 9B) (Loftus, in prep.). If such a possibility were to be exploited, though, the relationship to humidity would have to be eliminated from the function. For this purpose the antagonistic response of the moist receptor may also be essential in the processing.

B. MECHANO- THERMAL

Another possibility is mechanoreceptors. Their thermosensitivity, long recognized, has been reported e.g., in crustacean stretch receptors (Burkhardt, 1959), in mechanoreceptive hairs of bees (Thurm, 1963), and in the trichobothria of the sap-sucking bug, *Pyrrhocorix* (Draslar, in prep.) but much more frequently in vertebrates (for references and treatment, see Hensel, 1952, 1974a, 1976; Järvilehto, 1973). The oral and facial regions of squirrel monkeys provide an example which permits comparison of cold receptor responses with those of cold-sensitive mechanoreceptors both to sudden temperature change. Recording from trigeminal neurons, Poulos and Lende (1970) found that either type reacts to sudden temperature drops with sharp rises in frequency and that both depend on the extent of the drop, but with pronounced differences in range and sensitivity (Fig. 10A). Cold unit response is maximal once drops reach 8-10°C, whereas response of mechano-cold units continues to rise for a full 20°C, though much less steeply. Within the first 8°C, therefore, the extent of the drop should be more effectively distinguished

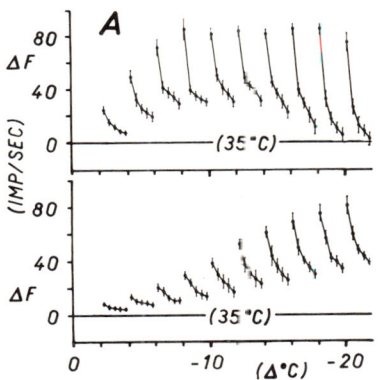 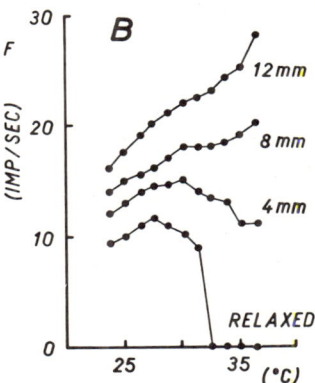

Fig. 10. A. Comparison of cold receptor (above) with mechano-cold (below), both in oral-facial region of squirrel monkey, *Saimiri sciureus*. Both respond to sudden drops in temperature (Δ°C) from 35°C with an increase in impulse frequency (ΔF). One offers the advantages of greater sensitivity and the other of greater range. (From Poulos and Lende, 1970, modified.)

B. Response of muscle spindle endings in cat gastrocnemius to constant muscle stretch during gradual cooling from 37° to 25°C. Where muscle is relaxed, spindle response to cooling resembles that of cold receptor. (From Michalski and Séguin, 1974, slightly altered.)

by means of cold unit response, but thereafter the mechano-cold units still continue to offer a basis for discrimination. That mechano-thermal receptors are probably much less effective in this regard than purely thermal receptors has been pointed out by Johnson, Darian-Smith and LaMotte (1973). The first two authors and Dykes (1973) found in psychophysical experiments that human subjects can distinguish two successive temperature drops with 75% probability even if the difference between the drops is significantly less than $0.1°C$. These findings hold for trained subjects in the $29°$ - $41°C$ range and for drops between $1°$ and $8°C$. Calculation from recordings of cold fibers in the hand of rhesus monkeys indicate that the responses of 16 independent fibers would, when properly pooled, convey sufficient information for comparable discrimination. For the same results full 2000 mechano-thermal units would be required on the same basis. If less resolving power were demanded though, the range might be extended from $8°$ to $20°C$, but in steps much larger. Depending on the need, range can be as important as precision.

Fig. 10B is another instance of biomodal mechanoreceptor response, illustrated by one of the three types of muscle spindle which Michalski and Séguin (1975) have found in cats. Not only does impulse frequency depend on the amount of constant stretch to which the muscle is subjected. Frequency also varies when the muscle is cooled gradually from $37°C$ down to $25°C$ at a rate of about $0.9°C/min$ (i.e., $0.015°C/sec$). On the one hand, differences in response are greatest towards $37°C$, a relationship suggesting the spindles would be most useful in distinguishing the degree of stretch at temperatures where the muscle is active. On the other hand, the other two types of spindle are silent during cooling when the muscle is relaxed. A comparison of their responses in this condition could indicate the temperature of the relaxed muscle, but as the lower curve of Fig. 10B illustrates, with some of the same ambiguity which cold receptors are beset with. Such thermal dependence might lead one to wonder just how intimate the relationship between thermal and mechanoreceptors may be as regards utilization, function, and origin.

Still another example is the mechanoreceptors in the skin of bony fish. Späth and Grocki (1976) succeeded in training goldfish to press laterally against a thermode while being fed. Both temperature and pre-tension of the thermode could be varied, and the pressure exerted by the fish against the thermode grew stronger as the magnitude of either parameter was increased. When both were raised, the reaction was stronger still. Späth had recorded from receptors in cyprinids with just this kind of dual sensitivity earlier (1967), but the search for purely thermal receptors in the periphery of bony fish is still in vain.

C. ELECTRO- THERMAL

The search for unimodal thermal receptors in elasmobranchs has not been any more successful than in bony fish. The ampullae of

Lorenzini, long renowned for their cold-receptor response (see Hensel, 1955), react to electric currents so minute (in the vicinity of 10pA, depending on temperature (Bromm, Hensel, Tagmat, 1976)), that they should be quite effective electroreceptors. But since both modalities influence the responses, unambiguous information regarding either will not be automatically available. If within this class of receptors there are sufficient differences in the type of response to each of the modalities in the same stimulus situation, the CNS may with proper processing be able to extract unambiguous information on temperature and the other modality too. But whether it does so or to what extent may well vary from species to species. A clear correlation of receptor response with behavior in an environment is still a long way off.

V. STRUCTURE

Because of the possibilities it opens and of the limits it places on the function of any organ, a consideration of structure is essential to the understanding of thermal as well as of other receptors. Furthermore, from the standpoint of adaptive radiation, structure assumes additional importance. Aside from the fact that no researcher ever dug up a paleontological function but only concluded to functions however unique by reason of the structures he found, even among contemporary animals structure is presumably a better indicator of evolutionary pathways than function is, and a discussion of adaptive radiation is concerned precisely with this kind organic relationship. One might expect a morphological correlate of function common to all thermal receptors which could be used to identify them as such and which would perhaps betray their origin. Actually though, knowledge of their fine structure is on at least two counts still quite meager. The first is that fine structure has been examined only in a small number of the receptors whose function has been proven directly and clearly to be thermal. The second is that the connection which can be established between function and observable structure is still largely descriptive, though not without interest.

In the pit organs of crotalid vipers the trigeminal nerve undergoes repeated branching (Bullock and Fox, 1957) and ends without synapses or tight junctions in clusters of "nerve masses" (Terashima, Goris, and Katsuki, 1970). Extensions of Schwann cells are dispersed among the terminals. Clusters are about 40 µm in diameter and thin, 10 µm or less, (see Fig. 1B) and are scattered without overlapping throughout the pit membrane. The membrane's large surface relative to its thickness and slight mass should allow radiant heat to influence its temperature very effectively. Thus if there is an obvious connection between morphology and function, it could perhaps be characterized as high terminal density extended over a large surface by a very thin membrane. It would appear to be the number of the terminals and the membrane they are imbedded in

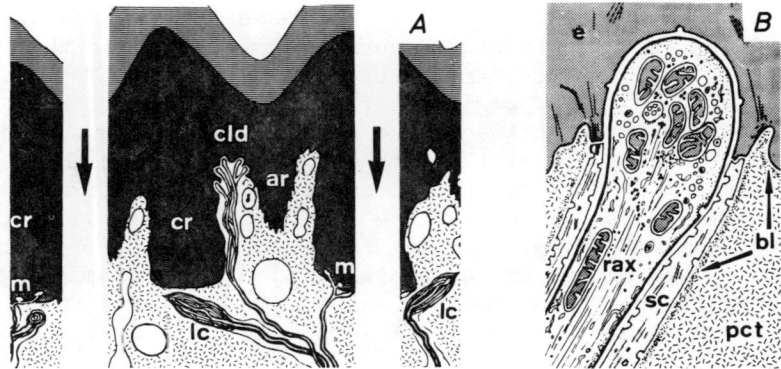

Fig. 11. A. Schematic section through glabrous nose skin of cat, showing cutaneous receptors and holes from wires (large arrows) marking cold-sensitive point. Adhesive ridge (ar), axon terminal reported to be cold receptive (cld), columnar ridge (cr), lamellated receptors (lc), Merkel cell complexes (m).
B. Detail of "cld" in Fig. 11A. Basal lamina (bl), epidermis (e), papillary connective tissue (pct), Schwann cell (sc). (Both A und B from Hensel, Andres, and Düring, 1974; slightly altered.)

rather than the structure within each terminal which indicates the modality to which the organ responds.

Thermal receptors in other vertebrates have been much harder to identify. Since there are usually several sensory structures dispersed through a given section of skin, there is the problem of narrowing the possibilities down to the precise receptor responsible for the impulses recorded a few centimeters further up the axon. The closest approach so far is that of Hensel, Andres, and Düring (1974). They used a wire tip thermode (diameter, 60 μm) similar to that of Kenshalo et al. (1971) to localize cold spots in the nasal region of cats, then marked the area with a steel needle on two sides of the spots (Fig. 11A), and sectioned it after perfusing. Only axon terminals of the type in Fig. 11B were common to all eleven successfully marked and sectioned areas. These were therefore considered to be the cold receptors they were seeking.

The argument from a single common structure would seem quite conclusive unless one can question the method used to zero in on the cold spot or is willing to entertain the possibility that one or other of the three structures near dermal papillae of this region is not modality-specific. The branched, unmyelinated, cold-receptive terminals as shown in Fig. 11 are different from any known mechanoreceptor (Hensel et al., 1974) but lack unique accessory structures. From the standpoint of stimulus-conducting mechanisms, this lack hardly comes as a great surprise. Since the thin (0.1 mm) layer of skin separating the receptor from the surface is anything but perfect insulation, it necessarily conducts heat in its entirety as long as a difference in temperature persists between the surface and level of the receptor. A specialized stimulus-conducting mechanism or structure is hardly needed for this purpose when all the surrounding tissues fulfill the requirement. However, even though heat conductance is inescapable, heat capacity can and does vary. The mere mass of tissue both above and behind thermal receptors must influence the speed with which their temperature changes and therefore on their sensitivity to changes in the temperature of the surrounding medium. The lower the heat capacity of the medium and the greater the mass of the tissue in which the receptor is imbedded, especially that above the receptor, the more the effect of temperature changes in medium will be smoothed out. The slight depth of these cold receptors can hardly be without significance for receptor function and presumably also for use. Furthermore, when temperature changes in the organ are to be produced by IR radiation from bodies whose temperature is close to that of the organ, low organ mass becomes increasingly advantageous, as would appear to be the case in crotalid pit membranes. If on the other hand, transparency for certain IR wavelengths turns out to be essential for receptor function, the result might be a characteristic structure of outer layers beyond mere thinness. But whether there are still other structures specific to some group of thermal receptors either immediately external to them or attached to their membranes has yet to be determined. It is noteworthy in this regard that although warm receptors eluded identification by the above methods, perhaps by reason of greater depth, the structure shown in Fig. 11B was not found in the vicinity of warm spots. Therefore at least two types of mammalian thermoreceptors can be distinguished: the cold receptor just discussed and something else. It is to be hoped that morphological examination of cold and warm receptors in other mammals will indicate structural relationship at least within each of the functional groups and perhaps to receptors of other modalities.

In contrast to cutaneous receptors of vertebrates, the superficial cuticular receptors of insects offer the advantage of fairly easy accessibility. (For recent treatment of insect sensillum structure in relation to function and in quite a general framework, see the survey of Altner, 1977). Fig. 12A,B (Schaller, 1978) shows the two types of sensilla in *Periplaneta* in which cold receptors have been found. Both are double-walled. The type in Fig. 12A, first described by Yokohari et al. (1975) is rigid and often contains dendrites of two antagonistic hygroreceptors in combination with the cold (Yokohari and Tateda, 1976; Loftus, 1976; Altner, Sass, and Altner, 1977). All dendrites are enclosed in an inner cuticular wall

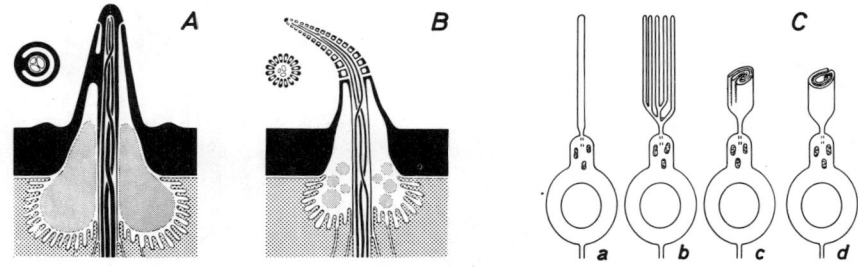

Fig. 12. A. and B. Two antennal sensilla (pegs) of *Periplaneta*, each with a cold receptor. Both are double-walled. Dendrites are neither branched nor folded in either, though they are intertwined with dendrites of other receptors. Base diameter about 4.5 μm. (Schaller, 1978, modified.) C. Dendritic forms in insect sensilla: a, simple and unbranched; b, simple and branched; c, flattened and branched with interlocking folds; d, flattened and rolled; e, flattened with complex folding and/or rolling. (Courtesy of Altner, unpublished.)

without lateral pores. Communication with the outside therefore can hardly occur unless the wall is permeable to critical substances, or unless the moulting pore at the tip were to provide a channel between moults, a function for which there is as yet no persuasive evidence. A sensillum containing the same triad of receptors on the antenna of the walking stick, *Carausius*, also lacks lateral pores (Altner, Tichy, and Altner, in prep.), and according to McIver (1973) such is also the case for the coeloconic sensilla of *Aedes* in which Davis and Sokolove (1974) found both warm and cold receptors. The above triads also appear to be sensilla without lateral pores in the predaceous bug, Triatoma (Bernard, 1974; Steinbrecht and Müller, 1976) and in the honey bee (Slifer and Sekhon, 1961; Lacher, 1964). Cold receptors do exist, however, in perforated sensilla. One such case in *Locusta* (Waldow, 1970; Altner, Sass, and Altner, 1977) where the triad is present. The other is *Periplaneta* (Fig 12B) where the cold receptor is in combination with a fatty-acid receptor (Altner, Sass, and Altner, 1977). Both of these two sensilla are double-walled and neither possesses pore tubules.

In insects, therefore, thermal receptors are not tied down to a single sensillum type. Even the two basic forms indicated in Fig. 12 as well as their differences may be more the result of the modalities of the receptors with which the thermal are combined rather than of the thermal receptors themselves. For again, the structures surrounding the dendrites can hardly be classed as effective insulation, whatever their form.

Another structure conceivably characteristic of thermal receptors is the dendrite of the receptive cell. (In this context dendrite is the equivalent of vertebrate axon terminals above.) As Fig. 12C (courtesy of Altner, unpublished) indicates, dendrites take on many forms. They can approximate wires fairly straight or crooked, simple or branched. They can also be flattened and rolled or folded into lamellae sometimes of highly complex arrangement. In *Periplaneta* it is clear that thermoreceptive dendrites can resemble unbranched wires (Fig. 12Ca), for all the dendrites of the sensilla in Fig. 12A and B are of this type. Even when the hexanoic-acid receptor is present, the branching often found in chemoreceptive dendrites (Fig. 12Cb) is lacking (Altner, Sass, and Altner, 1977). The situation with *Carausius* is more complicated, for although the same triad is present as in *Periplaneta* one of the dendrites is folded (Altner, Tichy, and Altner, in prep.). Which is which? A warm and a cold receptor were found in the pair of sensilla at the tip of the antenna of *Aedes*. Each sensillum has three dendrites and one in each is highly lamellated (McIver, 1973; Davis and Sokolove, 1975). Is it thermal? A similar dendrite even more highly lamellated is to be found on the tip of the antenna of the crabhole mosquito, *Deinocerites cancer* (McIver and Siemicki, 1976), but the lamellation becomes extravagant in an antennal dendrite of the cave beetle, *Speophyes lucidulus* (Corbière-Tichané and Bermond, 1972). Because the latter dendrite responds to prolonged incubation in osmium tetroxide like retinal outer segments do, Corbière-Tichané (1974) feels it might be an infrared receptor.

Although quite limited, the above data nevertheless do permit a few inferences. As *Periplaneta* has demonstrated, simple unbranched dendrites can be chemo-, hygro-, and thermosensitive. This structure therefore is not exclusively thermal, and its presence does not permit the conclusion that a given sensillum houses a thermoreceptor. But if on the basis of branching and lamellation some dendrite does turn out to be exclusively thermal, the type will not be common to all thermal receptors. Its absence alone will therefore permit no conclusion as to be absence of a thermoreceptive dendrite in a given sensillum. For an inventory of sensory function on the above basis, or analogously, on the basis of sensillum type, a physiological check will remain essential.

As for substructures, the number of microtubules packed into the dendritic outer segments of the sensillum in Fig. 12A is remarkably large (Altner, 1977). The packing is not so dense as that of the tubular bodies of mechanoreceptors (Thurm, 1965; Gaffal et al., 1975), but the large number of tubules does provoke the question of whether thermo- and/or hygroreceptors may have their origin in mechanoreceptors (Gaffal, Tichy, discussion). As desirable, however, as it would be to find such a connection (even the rattlesnake pit membrane can be stimulated by a wisp of cotton (Bullock and Diecke, 1956) and cold spots on the cat's nose by excessive pressure

(Hensel, Andres, and Düring, 1974)), until more hard evidence is available, the speculation may suggest that one should look, but the question is where.

In conclusion, therefore, thermal receptors have been demonstrated in two widely divergent groups: arthropods and vertebrates. Their location and degree of concentration at definite points near the surface of the body are sometimes clearly strategic and represent differences significant for the physiology and behavior of the animal. Measurements of their threshold, sensitivity, and range indicate they could be put to effective use, and despite extensive ambiguity in both temperature and its rate of change, behavior, where observed, at least suggests and in some instances even proves that they are. Moreover, their physiology indicates a great deal of similarity. But does physiological similarity automatically mean homology; variation, that is, on a single theme whose unity is the result of organic relationship; and not convergence? Or even from the standpoint of physiology, just how different is the thermosensory equipment of animals clearly adapted to divergent climates: arctic or tropic, desert or swamp? As yet the number of thermal receptors whose physiology has been even cursorily examined is still quite small. What is known concerning their primary processes can hardly be compared with our knowledge of the chain of events initiated in photoreceptors by a flash of light. For the study of thermal receptors is in many ways still in its infancy, and the well-rounded picture of their adaptive radiation is a goal, an interesting hypothesis awaiting experimentation to show where it must be rejected or under what limitations it can or must be accepted as a fact.

The author wishes to express his indebtedness to Prof. Dr. Helmut Altner for supporting this study, to Drs. Manfred Späth, Harald Tichy, and Peter Gaffal for their very helpful discussion, to Ms. Christel Danner and Ms. Hildegard Hallmer for their assistance at the drawing board and in the darkroom in reworking the text figures, and to Ms. Renate Rochelt for skilled and patient typing of the manuscript.

REFERENCES

Altner, H. (1977). Insect sensilla: Principles of structure and function. Verh. Dtsch. Zool. Ges. 1977, 139-153.
Altner, H., H. Sass, and I. Altner (1977). Relationship between structure and function of antennal chemo-, hygro-, and thermoreceptive sensilla in *Periplaneta americana*. Cell Tiss. Res. 176, 389-405.
Altner, H., H. Tichy, and I. Altner (1978). Folded outer dendritic segments of a sensory cell within a thermo- and hygroreceptive sensillum of the insect, *Carausius morosus*. In prep.

Becker, D. Styloconic sensilla at Mamestra. Doctoral Dissertation. In prep.

Benzing, H., H. Hensel, and R. Wurster (1969). Integrated static activity of lingual cold receptors. Pflügers Arch. ges. Physiol. 311, 50-54.

Bernard, J. (1974). Etude électrophysiologique de récepteurs impliqués dans l'orientation vers l'hôte et dans l'acte hématophage chez un Hémiptère: *Triatoma infestans*. Thèse Doct. Sci. Nat. Rennes, pp. 1-285.

Bromm, B., H. Hensel, and A.T. Tagmat (1976). The electrosensitivity of the isolated ampulla of Lorenzini in the dogfish. J. comp. Physiol. 111, 127-136.

Bullock, T.H. and F.P.J. Diecke (1956). Properties of an infra-red receptor. J. Physiol. (London) 134, 47-87.

Bullock, T.H. and W. Fox (1957). The anatomy of the infrared sense organ in the facial pits of pit vipers. Quart. J. Microsc. Sci. 98, 219-234.

Burkhardt, D. (1959). Die Erregungsvorgänge sensibler Ganglienzellen in Abhängigkeit von der Temperatur. Biol. Zbl. 78, 22-62.

Coenen-Staß, D. (1976). Vorzugstemperatur und Vorzugsluftfeuchtigkeit der beiden Schabenarten *Periplaneta americana* und *Blaberus trapezoideus* (Insecta: Blattaria). Ent. Exp. & Appl. 20, 143-153.

Corbière-Tichané, G. (1974). Sur la presence possible d'un pigment visuel dans un récepteur sensoriel de l'antenne d'un coléoptère cavernicole *Speophyes lucidulus* Delar. (Bathysciinae). Vision Res. 14, 819-822.

Corbière-Tichané, G. and N. Bermond (1972). Sensilles énigmatiques de l'antenne de certains Coléoptères. Z. Zellforsch. 127, 9-33.

Čokl, A. (1972). Thermorecepcija pri stenici vrste Oncopeltus fasciatus. Bioloski vestnik 20, 39-45.

Darian-Smith, J., K.O. Johnson, and R. Dykes (1973). "Cold" fiber population innervating palmar and digital skin of the monkey: responses to cooling pulses. J. Neurophysiol. 36, 325-346.

Davis, E.E. and P.G. Sokolove (1975). Temperature responses of anntenal receptors of the mosquito, *Aedes aegypti*. J. Comp. Physiol. 96, 223-236.

Dickenson, S.H. (1977). Specific responses of rat raphé neurones to skin temperature. J. Physiol. (London) 273, 277-293.

Drašlar, K. Temperature dependence of trichobothria response in *Pyrrhocoris*. In prep.

Esch, H. (1976). Body temperature and flight performance of honey bees in a servo-mechanically controlled wind tunnel. J. Comp. Physiol. 109, 265-277.

Evans, W.G. (1964). Infrared receptors in *Melanophila acuminata* De Geer. Nature 202, 211.

Evans, W.G. (1966a). Perception of infrared radiation from forest fires by *Melanophila acuminata* de Geer (Buprestidae, Coleoptera). Ecology 47, 1061-1065.

Evans, W.G. (1966b). Morphology of the infrared sense organs of

Melanophila acuminata (Buprestidae: Coleoptera). Ann. Entomol. Soc. Amer. 59, 873-877.

Gaffal, K.P., H. Tichy, J. Theiß, and G. Seelinger (1975). Structural polarities in mechanosensitive sensilla and their influence on stimulus transmission. Zoomorph. 82, 79-103.

Gift, J.J. (1977). Application of temperature preference studies to environmental impact assessment. J. Fish. Res. Board Can. 34, 746-749.

Goris, R.C. and M. Nomoto (1967). Infrared receptopn in oriental crotaline snakes. Comp. Biochem. Physiol. 23, 879-892.

Goris, R.C. and S. Terashima (1976). The structure and function of infrared receptors of snakes. Progress in brain research, vol. 43. Somatosensory and visceral receptor mechanisms. A. Iggo and O.B. Ilyinski (Edit.). Elsevier Scientific Publ. Co. Amsterdam, Oxford, N.Y.

Harris, J.F. and R.I. Gamow (1971). Snake infrared receptors: thermal or photochemical mechanism. Science 172, 1252-1253.

Heinrich, B. (1974). Thermoregulation in endothermic insects. Science 185, 747-756.

Hellon, R.F., H. Hensel, and K. Schäfer (1975). Thermal receptors in the scrotum of the rat. J. Physiol. (London) 248, 349-357.

Hensel, H. (1952). Physiologie der Thermoreception. Ergebn. Physiol. 47, 166-368.

Hensel, H. (1955). Quantitative Beziehungen zwischen Temperaturreiz und Aktionspotentialen der Lorenzinischen Ampullen. Z. Vergl. Physiol. 37, 509-526.

Hensel, H. (1974a). Thermoreceptors. Ann. Rev. Physiol. 36, 233-249.

Hensel, H. (1974b). Properties of warm receptors in *Boa constrictor*. Naturwissenschaften 61, 369.

Hensel, H. (1976). Functional and structural basis of thermoreception. Progress in brain research, 43. Iggo, A. and O.B. Ilynski (Edit.). Elsevier Scientific Publ. Co., Amsterdam, Oxford, N.Y.

Hensel, H., K.H. Andres, and M. v. Düring (1974). Structure and function of cold receptors. Pflügers Arch. Ges. Physiol. 352, 1-10.

Hensel, H. and K. Bowman (1960). Afferent impulses in cutaneous sensory nerves in human subjects. J. Neurophysiol. 23, 564-578.

Hensel, H. and D.R. Kenshalo (1969). Warm receptors in the nasal region of cats. J. Physiol. (London) 204, 99-112.

Hensel, H. and Y. Zotterman (1951). Quantitative Beziehungen zwischen der Entladung einzelner Kältefasern und der Temperatur. Acta Physiol. Scand. 23, 291-319.

Iggo, A. (1969). Cutaneous thermoreceptors in primates and subprimates J. Physiol. (London) 200, 403-430.

Järvilehto, T. (1973). Neural coding in the temperature sense. Ann. Acad. Sci. Fenn. B, 184, 1-71.

Johnson, K.O., I. Darian-Smith, and C. LaMotte (1973). Peripheral neural determinants of temperature discrimination in man: A correlative study of responses to cooling skin. J. Neurophysiology 36, 347-370.

Kenshalo, D.R., H. Hensel, P. Graziadei, and H. Fruhstorfer (1971). On the anatomy, physiology, and psychophysics of the cat's temperature-sensing system. In: Dubner, R. and Y. Kawamura (Eds.), Oral-Facial Sensory and Motor Mechanism. New York: Appleton-Century-Crofts.

Kerkut, G.A. and B.J.R. Taylor (1957). A temperature receptor in the tarsus of the cockroach, *Periplaneta americana*. J. Exp. Biol. 34, 486-493.

Lacher, V. (1964). Elektrophysiologische Untersuchungen an einzelnen Rezeptoren für Geruch, Kohlendioxyd, Luftfeuchtigkeit und Temperatur auf den Antennen der Arbeitsbiene und der Drohne. Z. Vergl. Physiol. 48, 587-623.

Loftus, R. (1966). Cold receptor on the antenna of *Periplaneta amercana*. Z. Vergl. Physiol. 52, 380-385.

Loftus, R. (1968). The response of the antennal cold receptor of *Periplaneta americana* to rapid temperature changes and to steady temperature. Z. Vergl. Physiol. 59, 413-455.

Loftus, R. (1969). Differential thermal components in the response of the antennal cold receptor of *Periplaneta americana* to slowly changing temperature. Z. Vergl. Physiol. 63, 415-433.

Loftus, R. (1976). Temperature-dependent dry receptor on antenna of Periplaneta. Tonic response. J. Comp. Physiol. 111, 153-170.

McCrea, M.J. and J.E. Heath (1971). Dependence of flight on temperature regulation in the moth, *Mandula sexta*. J. Exp. Biol. 54, 415-435.

McIver, S.B. (1973). Fine structure of antennal sensilla coeloconica of culicine mosquitos. Tissue and Cell 5, 105-112.

McIver, S.B. and R. Siemicki (1976). Fine structure of the antennal tip of the crabhole mosquito, *Deinocerites cancer* (Diptera, Culicidae). Int. J. Insect Morphol. Embryol. 5, 319-334.

Michalski, W.J. and J.J. Séguin (1975). The effects of muscle cooling and stretch on muscle spindle secondary endings in the cat. J. Physiol. (London) 253, 341-356.

Necker, R. (1973). Temperature sensitivity of thermoreceptors and mechanoreceptors on the beak of pigeons. J. Comp. Physiol. 87, 379-391.

Palm, T. (1962). Zur Kenntnis der früheren Entwicklungsstadien schwedischer Käfer. 2. Bupestriden-Larven, die in Bäumen leben. Opuscula Entomolog. 27, 65-78.

Poulos, D.A. and R.A. Lende (1970). Response of trigeminal ganglion neurons to thermal stimulation of oral-facial regions. II. Temperature change response. J. Neurophysiol. 33, 518-526.

Schaller, D. (1978). Sensory system of *Periplaneta americana* L.: Distribution and frequency of morphologically different sensillum types and their sex-specific changes during postembryonic development. In prep.

Schoonhoven, L.M. (1967). Some cold receptors in larvae of three lepidoptera species. J. Insect Physiol. 13, 821-826.

Shoemaker, V.H., K.A. Nagy, and W.R. Costa (1976). Energy utilization and temperature regulation by jackrabbits (*Lepus califorincus*) in the Mojave Desert. Physiol. Zool. 49, 364-375.

Slifer, E.H. and S.S. Sekhon (1961). Fine structure of the sense organs on the antennal flagellum of the honey bee, *Apis mellifera*) Linnaeus. J. Morph. 109, 351-381.

Späth, M. (1967). Die Wirkung der Temperatur auf die Mechanorezeptoren des Knochenfisches *Leuciscus rutilis* L. Z. Vergl. Physiol. 56, 431-462.

Späth, M. and K. Grocki (1976). Reactions of the goldfish (*Carassius auratus auratus* L.) to quantified mechanical and thermal stimuli. Experientia (Basel) 32, 1253-1254.

Spray, D.C. (1974). Characteristics, specificity and efferent control of frog cutaneous cold receptors. J. Physiol. (London) 237, 15-38.

Steinbrecht, R.A. and B. Müller (1976). Fine structure of the antennal receptors of the bedbug *Cimex lectularius* L. Tissue and Cell 8, 615-636.

Terashima, S.-I., R.C. Goris, and Y. Katsuki (1968). Generator potential of crotaline snake infrared receptor. J. Neurophysiol. 31, 282-288.

Terashima, S.-I., R.C. Goris, and Y. Katsuki (1970). Structure of warm fiber terminals in the pit membrane of vipers. J. Ultrastr. Res. 31, 494-506.

Thurm, U. (1963). Die Beziehungen zwischen mechanischen Reizgrößen und stationären Erregungszuständen bei Borstenfeldsensillen von Bienen. Z. Vergl. Physiol. 46, 351-382.

Thurm, U. (1965). An insect mechanoreceptor. I. Fine structure and adequate stimulus. Cold Spring Harb. Symp. Quant. Biol. 30, 75-82.

Tichy, H. Antennal thermoreceptors of *Carausius morosus*. In prep.

Waldow, U. (1970). Elektrophysiologische Untersuchungen an Feuchte-, Trocken- und Kälterezeptoren auf der Antenne der Wanderheuschrecke Locusta. Z. Vergl. Physiol. 69, 249-283.

Waldow, U. (1975). Multimodale Neurone im Deutocerebrum von Periplaneta americana. J. Comp. Physiol. 101, 329-341.

Van Weel, P.B. and D.L. McDonald (1971). A preliminary report on heat perception in cockroaches. Netherlands J. Zool. 21 (4); 487-491.

Yokohari, F. and H. Tateda (1976). Moist and dry hygroreceptors for relative humidity on the cockroach *Periplaneta americana* L. J. Comp. Physiol. 106, 137-152.

Yokohari, F., Y. Tominaga, M. Ando, and H. Tateda (1975). An antennal hygroreceptive sensillum of the cockroach. J. Electron Micr. 24, 291-293.

PHOTORECEPTION

M.A. ALI, M. ANCTIL and L. CERVETTO

Dépt. Biol. Univ. Montréal, C.P. 6128, Montréal, Canada

Lab. di Neurofisiologia del CNR, Pisa, Italy

> Art thou not, fatal vision, sensible
> to feeling as to sight? Or art thou but
> A dagger of the mind, a false creation
> Proceeding from the heat-oppressed brain?
>
> Macbeth, Act II, Sc. 2.

INTRODUCTION

Sensitivity to solar radiation is a widespread phenomenon among living forms and is commonly expressed in phototropic, phototactic or visual responses. Whenever sufficiently detailed investigations have been made, a molecular mechanism involving photolabile pigments and associated membrane specialisations was found to mediate this photosensitivity. We find this applicable to the bacteriorhodopsin of some bacteria (Oesterhelt and Stoeckenius, 1971), the flavin-type and carotenoid pigments of the phototropic fungi (Wolken, 1975), the chloroplast-based chlorophyl of green plants and, of course, the photopigments of animal photoreceptors. Photoreception, especially vision, has a prominent role among the sensory modalities used by many animal forms to provide information on the nature of the environment. It is reasonable to assert that, in the most organised visual systems, the sensory information contained in the visual messages is one of the richest, both in quality and detail.

Processing of photic or visual information begins in the photoreceptors. These are cells in which structures associated with cilia or rhabdomeric membranes (Eakin, 1972) are specialised for absorbing *light* and transforming the absorbed energy into a signal

capable of influencing neurones in sensory centres or central
nervous systems (CNS). These processes transmit to integration
centres, when present, information on the amount of light falling
upon the eyes, the spatial distribution of the illumination, the
chromaticity of photic stimuli and sometimes even on their movements.
The end-result of these processes and the response mechanisms may
differ in a number of important details, depending on the survival
strategies of animals exposed to different conditions and environments. We shall discuss these adaptive features shortly, but first
let us examine a few general principles concerning the photic
environment and the functional properties of photoreceptors.

THE PHYSICAL NATURE OF THE STIMULUS

Before discussing the functional properties of photoreceptors,
it may be useful to review some general notions on the physical
nature of the stimulus and on the processes of its interaction with
matter.

Light is a form of radiant energy. The complete spectrum of
the electromagnetic energy contains about seventy *octaves*. It
begins at one end with the long wavelengths of radio-waves, proceeds
through the infra-red rays, visible light, ultraviolet, X and gamma
rays to the other extreme represented by the cosmic rays with an
exceedingly short wavelength. As it is apparent from Fig. 1,
visible light occupies a negligible fraction of this broad spectrum.

The phenomena involved in the emission, propagation and
absorption of the radiant energy are explained by the quantistic
theory of electromagnetic waves. According to the Planck hypothesis,
the radiant energy is emitted and propagated in small discrete units
called *quanta* (*photons* in the case of visible light). The energy
of a single quantum is given by $h\nu$ where h is the universal Planck's
constant and ν is the frequency of the radiation which may be
expressed in terms of wavelength λ by the relation.

$$\nu = \frac{c}{\lambda} \quad (1)$$

where c is the velocity of propagation of light. Interaction of the
radiation field with matter involves several phenomena such as
emission, absorption and *scattering* of energy.

When a system passes from a state of Energy E_1 to a state of
lower Energy E_2, the frequency of the emitted radiation is such that
the quantal energy is given by

$$n \; h\nu = E_1 - E_2 \quad (2)$$

Likewise, the same frequency may be absorbed by a system in the state
of Energy E_2 which is thereby raised to a state of Energy E_1. In

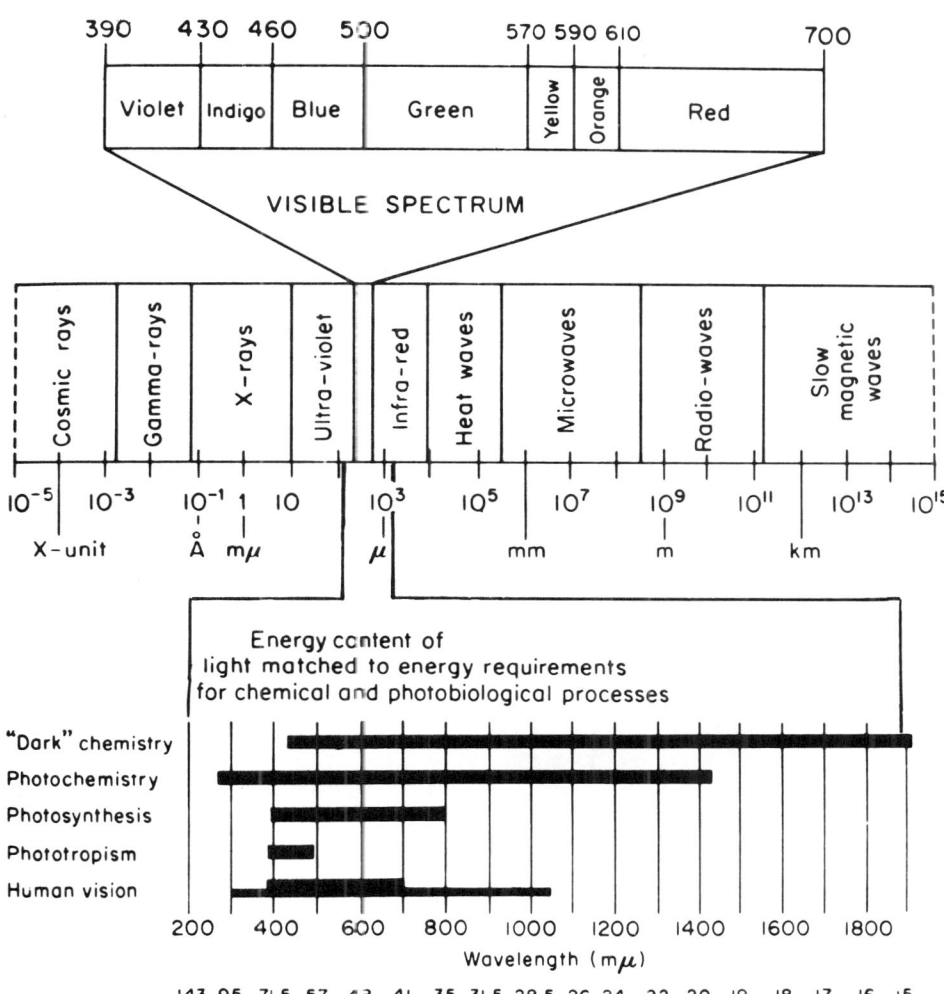

Fig. 1. Representation of the electromagnetic spectrum (middle). The visible spectrum is enlarged above, and its energy content is expressed in relation to requirements of chemical and photobiological processes (bottom). Note that the visible spectrum forms a very small part of the electromagnetic spectrum. It is also noteworthy that the region between 400 to 500 nm, especially around 500 nm, is common to all processes. (After Hoar, W.S., General and Comparative Physiology, 2nd ed., Prentice-Hall, 1975).

general radiations at a given frequency may be absorbed only if the energy of the corresponding quantum is sufficient to rise an atom or molecule to a higher quantal energy level. Thus the degree of absorption depends on both the physical properties of the matter and the frequency of the radiating energy.

The energy of a beam radiating through a body is attenuated by an amount equivalent to the energy transferred to the molecules of the body. If the body medium is homogeneous, the energy attenuation can be calculated from the Lambert-Bouguer law:

$$\ln \frac{E_t}{E_i} = -\alpha x \qquad (3)$$

where E_t is the energy of the transmitted radiation, E_i the energy of the incident monochromatic radiation, α the coefficient of absorption of the medium and x is its thickness.

In non-homogeneous media, transmission is complicated by scattering. This phenomenon occurs when the refractive index is not constant and the radiating energy deviates from the original direction. Energy is thus lost to the beam and re-irradiated from each scattering centre.

The first quantitative study of the phenomenon was performed in the 19th century by Lord Rayleigh who showed that scattering of radiant energy by small independent particles, as the molecules of a gas, is proportional to the reciprocal of the forth power of the corresponding wavelength. This relation can be expressed in simplified form by:

$$E_s = K_R \frac{E_i}{\lambda^4} \qquad (4)$$

where E_s is the amount of the scattered energy, E_i that of the incident energy and K_R a constant depending on a number of factors like: the refractive index of the medium, the number of scattering centres per unit of volume, the angle formed by the incident and scattered beams and the distance from the scattering centres. One interesting consequence emerging from this relation is that the amount of energy scattered from a beam of ultraviolet rays is substantially greater than that scattered from a beam of infrared rays. One can easily calculate that the energy scattered at wavelengths of 400 nm is about 10 times greater than the energy scattered at 700 nm. An important restriction of this theory is that it applies only when the diameter of the scattering particles is much smaller than the wavelength of the incident beam.

These phenomena partially explain the colour of the sky. The blue colour of the sky and its dependence on the size of the scattering centres was studied by Tyndall. Rayleigh scattering accounts also for the fact that much ambiant light in the atmosphere and underwater is plane-polarised.

For all the organisms living on our planet, by far the most important source of visible light is the sun. Sunlight reaches the earth through the atmosphere which is known to absorb most of the short wavelengths. The degree and the selectivity of the atmosphere filtering out solar radiations may change substantially in different conditions: the presence of clouds, for instance, extends the spectrum of absorbance towards the infrared rays which are otherwise transmitted through a clear atmosphere (for details on light transmission in the atmosphere and under water, see Lythgoe, 1972).

Although photons might in principle interact with the molecules in many different ways, their absorption always involves electronic transitions. After a photon is absorbed, several processes are available but the most relevant to the present discussion are the chemical changes that the absorbing molecules undergo.

PHOTOCHEMICAL PROCESSES

The interpretation of the phenomena involved in photochemical reactions rests upon two postulates: i) only the light actually absorbed is effective; ii) only one *quantum* is involved in a primary reaction.

The efficiency of an absorbed *quantum* inducing a chemical transformation is called "*quantum yeild*" and it is defined as the number of the affected molecules divided by the number of the absorbed photons. According to the second postulate mentioned above, the *quantum yield* should not exceed the unity.

Quantum yields greatly in excess of the unity, however, are not uncommon, but they are interpreted as the result of chain reactions initiated by the molecules which have originally absorbed a single quantum.

The absorption of a quantum of ultraviolet light by a molecule of chlorine provides sufficient energy to dissociate it into atoms and, in presence of hydrogen, an atomic chain reaction is then started which produces large amounts of HCl. The reaction will proceed until the probability that the activated molecules (radicals) react among themselves becomes very high. This will produce the reactions leading to the extinction of the process.

In nature, most of the molecules react to ultraviolet radiations. The relative stability of the matter to the visible radiation may be explained by considering that the planet has been irradiated by sunlight for something like 200 billions of years, thus leading to a complete dissociation of the light reacting molecules. Important exceptions are represented in living organisms by the photosynthetic pigments of plants and bacteria and by the visual pigments of animals. This occurs because living organisms are capable of reforming the molecules dissociated by light. While in plants the light absorbing molecules are used to build up energy, animals utilise their visual pigments for obtaining information on the environment. For the effects that light may produce on the protoplasm, see Couillard (this volume).

Photoreceptors may assume a great variety of structural features and organisation (Eakin, 1972; Crescitelli, 1972). In spite of the many differences, all photoreceptors have in common a morphology based on protrusions or infoldings of their membrane or cilium. These structures may assume the forms of microvilli, discs or lamellae (Fig. 2). Studies performed during the last decade and including microspectrophotometry, X-rays diffraction, cristallography, immunology, electrophysiology, have established that the visual pigment monolayer is the major proteinic constituent of these membranes (for references, see Rodieck, 1973).

From all these observations, the general conclusion emerges that orderly arrangement of pigment molecules in the specialised structures of visual cells is required for the production of visual responses. In addition, the great expansion of the membrane area occurring at the infoldings is probably important to accomodate a large number of pigment molecules. A great light sensitivity may be achieved if the cell is specialised to absorb light with high efficiency. The maximum light absorbing power of a single electronic transition is limited, and a photoreceptor can efficiently absorb an incident beam if it contains a large number of pigment molecules. It has been estimated that a visual cell, to properly perform should contain at least 10^7 molecules of pigment (Hagins, 1972). Other estimates based on experimental measures suggest for turtle cones values of about 10^8 (Liebman and Granda, 1971; Hodgkin and O'Bryan, 1977).

The pigment consists of a conjugated protein in which the vitamin A aldehyde is bound as a chromophore to a proteinic component called opsin. The vitamin A aldehyde (retinal) conjugated to the protein gives origin to a pigment known as *rhodopsin* while the dehydrated form of *retinal*, when conjugated to the opsin gives a pigment known as *porphyropsin*. *Rhodopsin* absorbs maximally at about 500 nm, while the spectral maxima of porphyropsin range between 520-540 nm.

Rhodopsin is generally found in both marine and terrestrial animals, while *porphyropsin* is found in fresh water vertebrates, including fishes that spawn in freshwater (Crescitelli, 1972; Goldsmith, 1972). This rule, however, is flexible and exceptions will be described later.

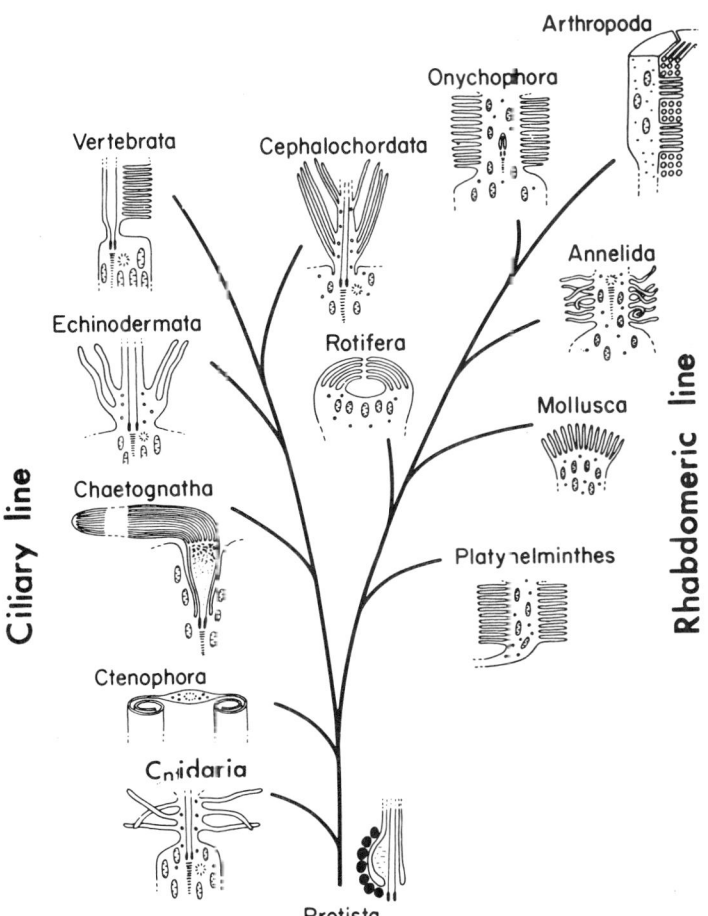

Fig. 2. Schematic representation of the phylogeny of photoreceptor organelles in representative groups of animals along "ciliary" and "rhabdomeric" lines of evolution. The protist's eyespot (see Couillard, this volume) is a cup of pigment granules (carotenoid?) which lies outside the flagellum and may be an inert shading device. The nature of the photopigment in the paraflagellar body, the cnidarian and ctenophoran receptors is unknown. One may observe that typically, evolutionary trend is an increase in the surface area of the cell membrane rather than the development of an intracellular structure analogous to the paraflagellar body. The distal retina of the scallop (Mollusca), a protostome, is ciliary, thus an exception to the phylogenetic generalisation depicted. (After Dobzhansky, T. et al., eds., Evolutionary Biology, Vol. 2, Appleton-Century-Crofts, Meredith Corporation, 1968).

The structure of the opsin is not entirely understood. Retinal is known to be formed by a chain of carbon atoms connected by alternating single and double bonds. In darkness, the chain is twisted to the so-called 11-cisform. Following absorption of a photon, a single molecule is isomerised and changed into all-trans configuration. A sequence of reactions then follows, wherein light is no more necessary (dark reactions). The final step involves the separation of the chromophore from the opsin.

ELECTRICAL PROPERTIES OF PHOTORECEPTORS

Pioneering investigations on the electric signals generated by visual cells were performed by Hartline (1934). He recorded impulses originating from single fibres in the lateral eye of the horseshoe crab *Limulus* and observed that the photic stimulus evokes a discharge of pulses whose frequency is related to the intensity of the stimulation, confirming a notion derived from studies performed on other receptor functions by Adrian (1928) and Matthews (1931). Later, the properties of these responses were investigated by intracellular recordings (Hartline et al., 1952). It was shown that as other nerve cells, invertebrate photoreceptors generally have an internal negativity at resting. Light was shown to produce a depolarisation of the membrane, which in turn is responsible for the nerve impulses. Depolarisation is roughly proportional to the log of the light intensity, while the frequency of firing is linearly related to the depolarisation. More important, it was shown by Fuortes (1959) that injected current produces across the membrane, smaller potential drops in light than in darkness, thus indicating that the *membrane resistance is decreased by illumination*.

In addition, it has been shown that the changes in conductance occur *at the site where light is absorbed*. Therefore, it seems safe to conclude that photon absorption sets off a sequence of intracellular events which ultimately lead to a change of the ionic permeability responsible for the electric signals.

Studies on vertebrate photoreceptors, in particular on the cones of the turtle *Pseudemys scripta elegans*, have shown that their membrane potential in darkness is unusually low (-30 to -40 mV). Following a flash, they do not generate action potentials but a graded hyperpolarising wave which may reach an amplitude of 30 mV. Thus at the peak of the response the membrane potential approaches -60 mV, a value close to the resting potential of other nerve cells. On the other hand, in darkness the cone is depolarised as other cells do during activity, including invertebrate photoreceptors. Current-voltage and ionic exchange measurements have led to the conclusion that in *darkness the membrane is leaky because of a relatively high sodium permeability*. Light decreases the leak, thereby shifting the membrane potential toward the equilibrium for potassium (Baylor and

Fuortes, 1970; Korenbrot and Cone, 1972; Cervetto, 1973).

Developing the general notion that light activates an internal transmitter which is capable of acting on the membrane permeability, Baylor et al. (1974 a,b) have worked out a kinetic model which accounts for the observed responses of turtle cones. A model has also been proposed, which describes as well the behaviour of rods in response to a variety of stimuli (Cervetto et al., 1977).

In both cones and rods it has been observed that the *electrical response is proportional to the number of photons absorbed only in a narrow range*. In particular, linearity applies only to responses whose mean amplitude does not exceed 1-2 mV. The ability of a photoreceptor to signal the number of absorbed quanta independently of wavelength is called *spectral univariance*. This property is usually approximated only in conditions wherein the illuminated area of the retina is extremely small (less than 100 µm). Univariance implies that a flash of intensity I_1 and wavelength λ_1 has effects which are identical in all respects to those produced by a flash of intensity KI_1 and wavelength λ_2, where K is the relative spectral sensitivity and is function of λ_1 and λ_2 but not of I_1.

Electrophysiological measurements of the spectral sensitivity of photoreceptors (Tomita, 1965; Baylor and Hodgkin, 1973) are in good agreement with the microspectrophotometric measurements (Marks et al., 1964; Liebman, 1972) and have indicated the existence of visual cells which absorb preferentially at different wavelengths.

In the turtle, three types of cones have been described: red (630 nm), green (550 nm), and blue sensitive (460 nm), and one class of rods which absorbs maximally at 520 nm. It is suggested that the turtle rods and the green cones possess the same pigment and that the drop of sensitivity observed in the cones at short wavelengths is due to the presence of an oil droplet (Liebman, 1972; Baylor and Hodgkin, 1974).

Using as a stimulus a monochromatic light at the optimal wavelength within the range over which the intensity is proportional to the response, the ratio between the peak hyperpolarisation to the intensity of light measured in photons per μ^2 per sec gives the absolute sensitivity according to the relation

$$\sigma = \frac{V_{peak}}{IA_c \Delta t} \quad (5)$$

where A_c is the actual collecting area calculated from the geometric area A_g of the outer segment of a photoreceptor by the equation

$$A_c = A_g \emptyset [1 - T(\lambda)] \quad (6)$$

where ∅ is the quantum efficiency and T (λ) the transmittance which is obtained from equation (3). In turtle cones it has been estimated that a single photoisomerisation produces ca. 25 µV. In the rods of a toad, we find values 20 times larger (ca. 500 µV.)

Psychophysical experiments as well as electrical recordings from cat ganglion cells have estimated that threshold for excitation of the visual system is very low and, in dark-adapted conditions it is reached when a small number of photoreceptors absorb a single photon each (*single quantum detection*). Single quantum detection has been measured also in the photoreceptors of *Limulus*.

Such a high sensitivity may be explained by assuming the existence of mechanisms of amplification in the synaptic pathways. Consistent with this hypothesis are the observations that the ratio of the sensitivity of bipolar cells to that of rods (dynamic gain) measured in the dogfish is about 50 (Ashmore and Falk, 1976). In turtle, however, it has been recently shown that at least 130 photoisomerisations are necessary to excite the most sensitive paths from cones and rods to ganglion cells (Baylor and Fettiplace, 1977). This seems to rule out the single photon detection in the visual system of the turtle. Observations on rods of *Bufo marinus*, showing a sensitivity substantially greater than in the cones of turtle (Fain, 1976; Cervetto et al., 1977), strongly suggest that the relatively high sensitivity of visual system of the toad is already operating at the receptor level.

Until recently, the prevailing idea concerning the organisation of retinal receptors was that they form a *mosaic of independent units*, the response of each cell being controlled only by the light impinging upon it. Visual recognition in fact requires a transformation of the optical images produced by the lens on the retina in a composite electrical signal generated by the photoreceptors. A faithful transcription of these messages would be obtained if the response generated by each receptor were uniquely related to the light falling upon it. Interactions can only complicate the process and may be expected to cause distortion. This notion implies that the response of a photoreceptor should be the same whether or not other receptors are activated by the stimulus. Therefore, provided that light intensity is the same, the response of a single cell should be identical for small or large fields of illumination.

Independence of photoreceptors appears especially desirable for maximal spatial resolution of the image. Contrary to this prediction, experiments performed on the retina of the turtle have shown that the response of a cone is increased by the activity of nearby receptors. Thus, turtle cones seem to be subjected to interactions opposite to those described in the eye of *Limulus*, namely "lateral facilitation". *Summative interactions have been* reported also between rods of the toad. *The interactions occurring in the cones of the turtle have been shown, however, to occur only among cones with the same pigment.*

In essence, these experiments lead to the conclusion that interactions among visual cells do not impair discrimination of colours, even though they may decrease spatial resolution. The question that we may ask concerns the biological significance that the emphasis on chromaticity rather than on sensitivity or spatial resolution has for the survival of an animal species. The development of the ability to better perform a certain discriminative function of the visual system certainly reflects the adaptational problems solved during evolution. It may be instructive to look and see how far in the details, experimental findings, obtained in the extreme conditions of a laboratory, can be correlated with the *biology* of animals living freely in their natural environment. The hope is that reciprocal extrapolations are applicable from the two approaches.

PHOTORECEPTOR ORGANISATION AND THE ENVIRONMENT

Diversity in the photic environment and the modes of life (see Fig. 3) has probably been a major evolutive factor in the spectacular development of ocular adaptations in the animal world. This diversity is diagrammatically illustrated in Fig. 4, and summarised in Table 1. We shall first review some morphological and optical structures of the photoreceptors which appear to have adaptive value in relation to environmental conditions.

Simple photosensitive devices are tracked down the phylogenetic tree wherever one finds cell membranes which provide large surface areas for light capture by a visual pigment. As cells and multicellular organisms evolved, motile systems using cilia and flagella ensured versatile locomotory habits in the aquatic environment. In the protozoan *Euglena*, a structure associated with the flagellum, the paraflagellar body, contains a photopigment and thus provides a simple receptor system for sensing the photic environment (Wolken, 1975). Then, in the course of evolution, ciliary adaptations may have evolved into a variety of specialised receptor organs, including photoreceptors. Ciliary photoreceptors are present among the cnidarians, ctenophores, chaetognaths, echinoderms, protochordates and vertebrates (Eakin, 1972) and therefore are characteristic of the deuterostome line of evolution. Photoreceptors formed by more or less elaborate evaginations of the plasma membrane probably appeared later, and are typical of the protostome (annelid-arthropod) line. This is a flexible rule, since annelid worms, for example, may possess ciliary or rhabdomeric-type photoreceptors, or even both in the same animal. To this diversity of membrane specialisations are added the impressive dioptric accessories of many forms. Beyond the obvious or

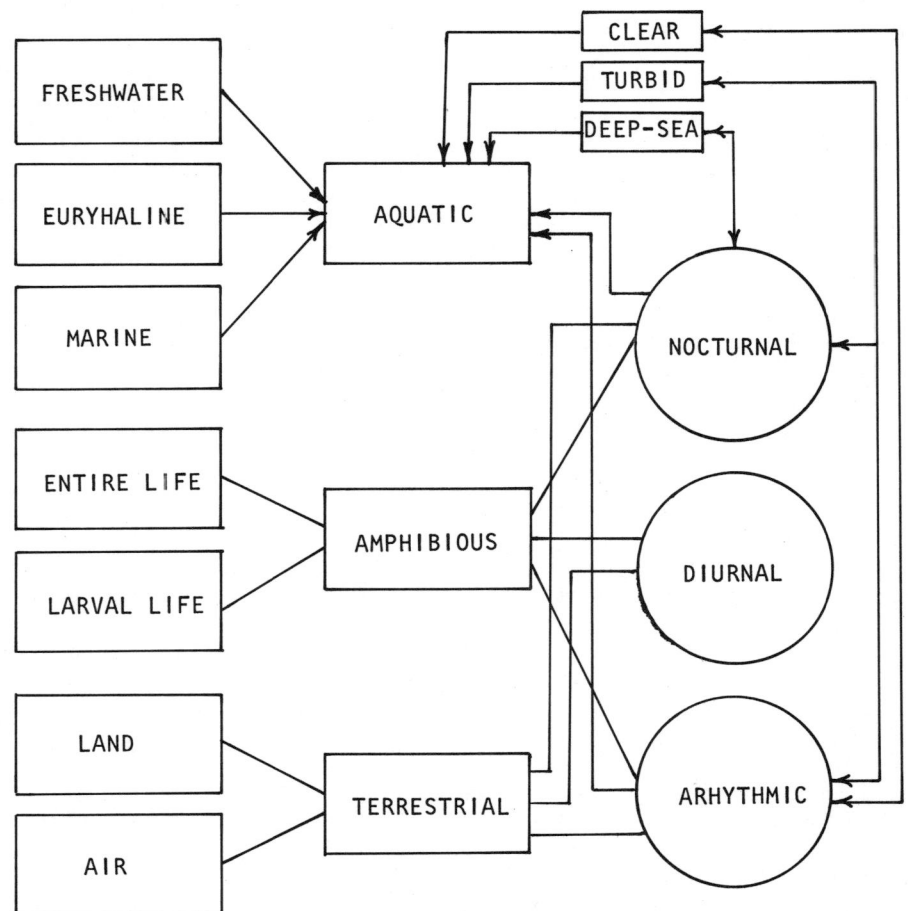

PHOTORECEPTION : ADAPTIVE RADIATION

Fig. 3. Diagram showing the relationship among habitats, modes of life and photic environmental conditions. The reader is referred to the phylogenetic survey of sensory functions by Ali, Croll and Jaeger (this volume). It is noteworthy that a purely diurnal mode of life is not to be found in the aquatic environment. This may be due to the greater variation in photic quality that one encounters there. The deep-sea and turbid environments have often been compared, but the only similarity is the paucity or attenuation of light.

Table 1. Summary of visual adaptations to photic environments and habits

	SCOTOPIC →			←	
Nocturnal Crepuscular	Deep-Sea	Turbid waters	Caves Oceanic Abyss		
Pure-rod retinas	—Pure-rod retinas—		—Pure-rod retinas		
Rod-rich retinas	—Rod-rich retinas—	—Rod-rich retinas—	—Eyes degenerated		
Tapeta lucida	—Tapeta lucida—	—Tapeta lucida	—Eyes absent		
Optical guides (crystalline cones)	—Optical guides— (rod myoids)	—Optical guides (cone myoids)			
Superimposed layers of rhabdomeres	—Superimposed layers of rods				
Closed-type rhabdoms	—Large mass of rod outer segments				
Superposition eyes	Grouped rods Tubular eyes	—Grouped rods and cones			
Color vision	—Colour vision—	—Colour vision—	—Colour vision		
Coloured intraocular filters			—Coloured intraocular filters		
Cone-rich retinas	—Cone-rich retinas—	—Cone-rich retinas—	—Cone-rich retinas		
Pure-cone retinas	Colourless cone droplets	Polarotaxis	—Polarotaxis		
Apposition eyes	Double optical axes				
Open-type rhabdoms					
Diurnal	Water surface	Clear shallow waters	Terrestrial Aerial		
		PHOTOPIC			

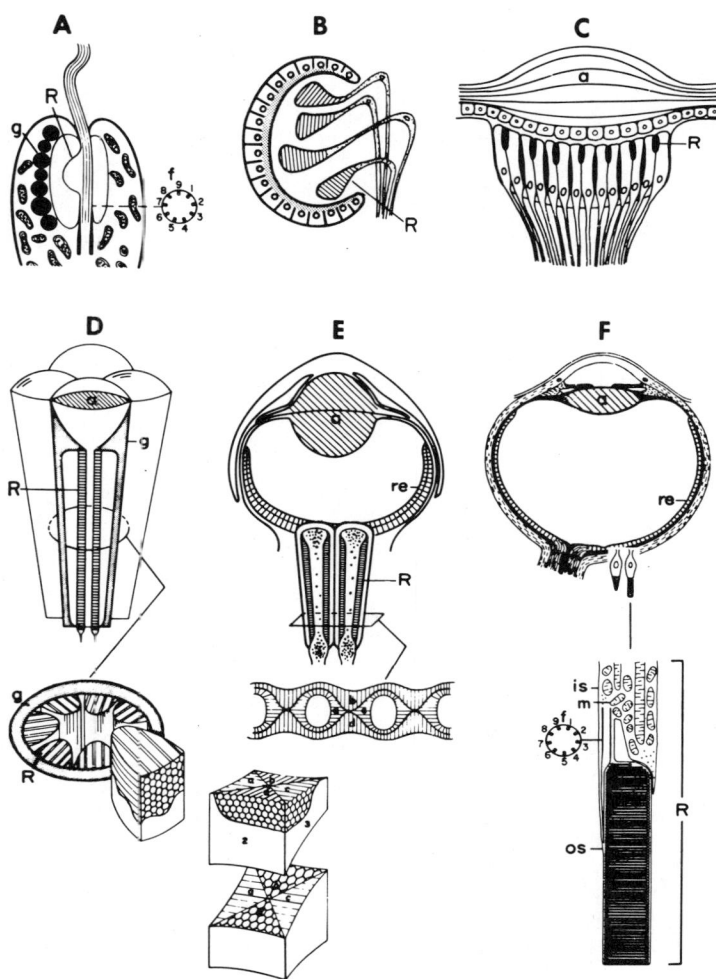

Fig. 4. Schematic representation of basic types of photoreceptor organs from various organisms. A, eyespot-flagellum of protozoans; B, ocellus of flatworms; C, ocellus of insects; D, compound eye of arthropods; E, eye of cephalopod molluscs; F, vertebrate eye. a, lens; f, flagellum; g, pigment granules; is, inner segment; m, mitochondria; os, outer segment; re, retina; R, photoreceptor (rod or rhabdomere). Note that the basic optic design is similar in the ocellus of insects and the eyes of the cephalopods or vertebrates, although the disposition of the light sensitive parts is entirely different in the three cases. The lens is not flattened in all vertebrates (see Sivak, this volume). It is circular in fishes, less so in reptiles and quite flattened in mammals. (From Wolken, 1975, Academic Press).

proposed light sensitivity of these structures, one may ask what else is involved to ensure that the mediators of this sensitivity reflect the sensory demands of the animal in a given set of environmental conditions.

The dermal light sense of molluscs and echinoderms is mediated, not by ocular specialisations, but by diffusely distributed neurons which happen to contain photopigments (Millott, 1968). These sensory units not only provide information on the presence of light, but also on its direction and changes of intensity. These sedentary animals make use of neuron properties, especially "off-responses" (Millott, 1968), for defensive shadow responses. A similar, but apparently less sophisticated, diffuse light sense is present in some fishes (Steven, 1963; Wales, 1975). It is known to mediate phototactic responses and to monitor vertical migrations in fish larvae.

When we turn our attention to relatively complex imaging eyes, we find a certain degree of convergence among invertebrates and vertebrates in the development of adaptive features to scotopic *vs* photopic conditions. The most obvious convergence is that of the cephalopod and vertebrate eyes. Both have refracting, camera-like eyes, although the cephalopod photoreceptors differ in being rhabdomeric and exposed directly to the incident light (Yamamoto *et al.*, 1965). Both groups have been considered highly successful at exploring their environment visually. However, it is more revealing for the purpose of this discussion to compare arthropod compound eyes with the refracting eyes of vertebrates. Insects and teleost fishes are, within their respective phyla, very diversified and successful groups. In both groups nocturnal or deep-sea species tend to include a larger volume of photosensitive membranes in their photoreceptors than in the diurnal or shallow-water species. For example, insect compound eyes comprise a number of ommatidia in which there may be several grouped retinula cells with membranes (rhabdomeres) extending towards the centre of the bundle. The light sensitive elements, rhabdomeres, of all adjacent retinula cells form an entity called the rhabdom. In diurnal insects each rhabdomere is relatively small and the rhabdom is of the open type, that is, the rhabdomeres are distant one from another. In nocturnal insects, on the other hand, the rhabdomeres tend to be larger and the rhabdom is of the closed type, that is, all adjacent rhabdomeres are fused together. The latter appears to be an adaptation for increasing the overall sensitivity of the eye, by providing a larger area for light capture (Wolken, 1975). Among fishes, the best comparison can be made between shallow-water and deep-sea teleosts (Fig. 5 a,f). Shallow-water species tend to possess many rods of moderate length, as well as cones, but deep-sea species have usually pure-rod retinas (Ali and Anctil, 1976). Moreover, the rod outer segments, enclosing the membrane-bound scotopic visual pigment tend to be longer in

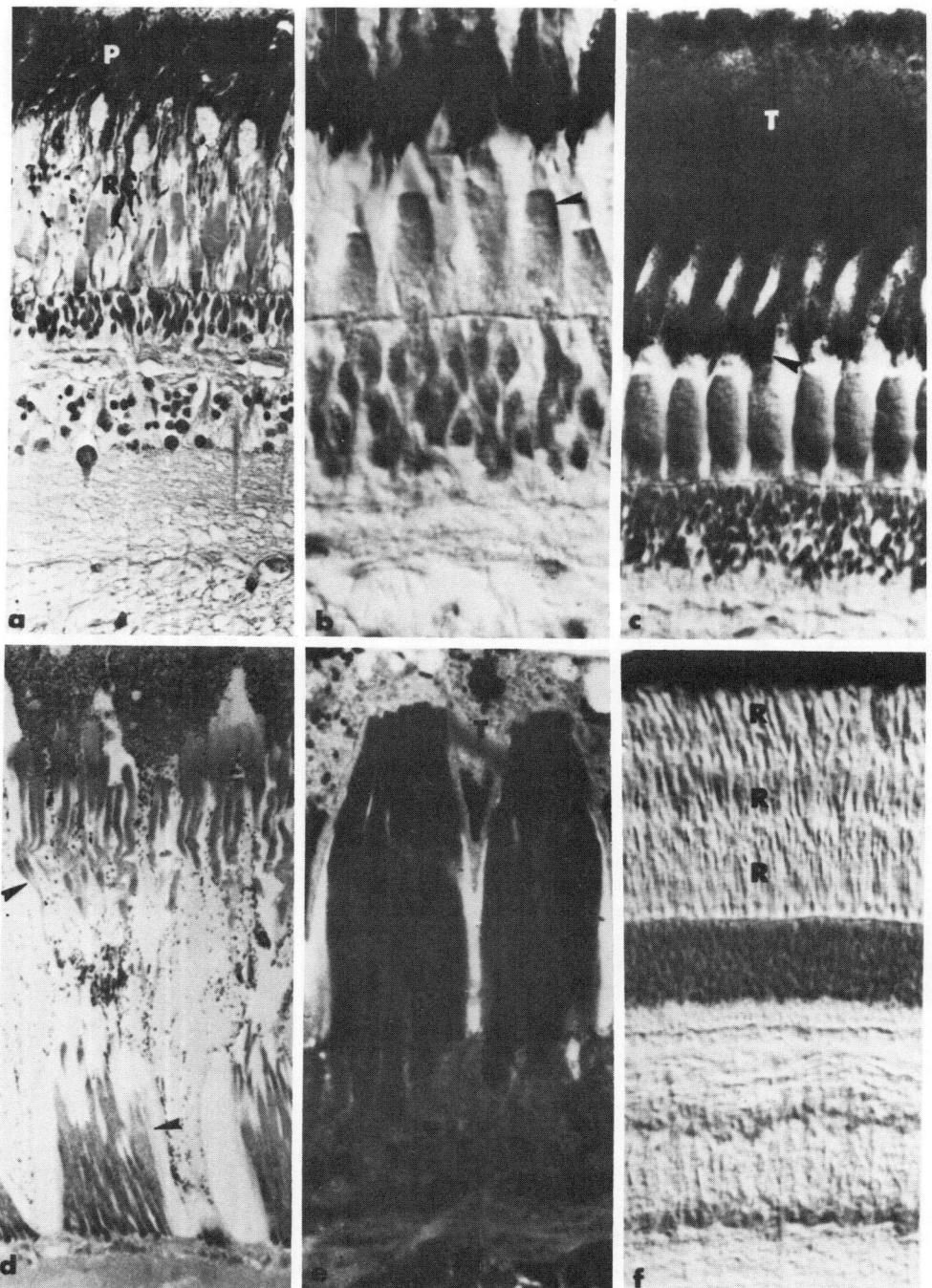

Figure 5

Fig. 5. Transverse histological sections of retinas in fishes from different environments. a, duplex retina of the brook trout, *Salvelinus fontinalis*; b, cone droplets (arrow) in retina of the surface fish, *Fundulus heteroclitus*; c, tapetum lucidum (T) and melanin pigment (arrow) in retina of the walleye, *Stizostedion vitreum*; d, rod bundles (upper arrow) and cone bundles (lower arrow) separated from adjacent bundles by tapetal processes in retina of the turbid water fish, *Hiodon tergisus*; e, two rod bundles separated by a tapetal process (T) in a deep-sea fish, *Scopelarchus quentheri*; f, superimposed layers of rods (R) in a deep-sea fish, *Argentina silus*. P, pigment epithelium; RC, layer of rods and cones.

deep-sea than in shallow-water fishes, or are short but stacked into superimposed layers (Munk, 1966). These features increase the amount of photosensitive membranes available for light capture and thus, as in nocturnal insects, enhance visual sensitivity in a dim environment. Unusually high concentrations of visual pigment have been measured in deep-sea fish pure-rod retinas (Denton and Warren, 1957). Interestingly, some cephalopods also tend to increase their photoreceptive membrane mass as they translocate to deeper oceanic waters during their life cycle (Young, 1975). These examples undoubtedly emphasise the importance taken on by this adaptive feature in the course of the evolutive process.

Converging mechanisms aimed at adapting the eye sensitivity to dimly lit environments may prove to be widespread in the animal world. Let us consider specifically the visual systems of the nocturnal fireflies (*Photuris*) and of some luminescent deep-sea fishes (Anctil, 1975). The rhabdomeres of the firefly ommatidia are arranged in two superimposed layers, the distal layer being smaller and less massive than the proximal one (Horridge, 1969). Similarly, some deep-sea fish pure-rod retinas have their photoreceptor outer segments arranged in superimposed layers (Fig. 5) (Munk, 1966; Ali and Anctil, 1976). But the comparison does not stop there. In each firefly ommatidium there are four crystalline cone cells which are situated just below the corneal lens. These elongate cells include longitudinal microfilaments at their inner periphery, which are the basis for the crystalline cone thread (Horridge, 1968). The latter is essentially a structure with presumed waveguide properties, directing light onto the photosensitive rhabdoms. In deep-sea luminescent fishes, similar optical guides direct light from one photoreceptor layer to the next, but in this case the photoreceptors themselves provide the structural basis for this ability, e.g., the rod myoid (Locket, 1970; Miller, 1976; Miller and Snyder, 1977). The rod myoids of the second layer of rod outer segments are tightly interspersed among the outer segments of the first layer, and guide the light directly on the outer segment membranes of their own cells. It is noteworthy that those myoids, as the firefly cone threads, usually contain microfilaments (Eakin, 1972; Burnside, 1976).

Superimposed layers of photoreceptors are also present in the pallial eyes of the bivalve mollusc scallop, *Pecten* (Land, 1968), but this peculiar arrangement apparently does not serve the purpose of incremental sensitivity, as discussed above. In this case, the two layers contain different photoreceptor types: (1) the distal layer (near the lens) is composed of ciliary photoreceptors which give "off" responses, are directionally sensitive and image forming, and (2) the proximal layer is composed of rhabdomeric-type photoreceptors, which give "on" responses to light and cannot resolve images. The photoreceptor axons are collected into separate optic nerves for the two layers. This remarkable organisation mediates different components of the visual behaviour of the scallop. The ciliary photoreceptors are involved in the animal's responses to shading and movement, thus protecting it from potential predators. The rhabdomeric-type receptors, on the other hand, keep the animal informed about the distribution of brightness in the immediate environment and thus may mediate the orientation behaviour displayed by scallops. It is interesting that "off" shadow responses, as discussed earlier, are mediated by a diffuse, neuronal, dermal light sense in echinoderms and molluscs (Millott, 1968), as well as by specialised photoreceptors in molluscs and, exceptionally, in a few annelid species with ciliary photoreceptors (Nicol, 1950; Krasne and Lawrence, 1966).

The shape of the eye and/or photoreceptor organisation may be modified so as to form a main retina and an accessory retina in widely divergent phyla. In all instances in which this occurs, it appears to be associated with nocturnal habits or dim environments, but it does not imply that this feature fulfills the same purpose for all those animals. It is present in the small ovoid eyes on the tentacle of the nocturnal slug *Agriolimax* (Newell and Newell, 1968), the ovoid or tubular eyes of deep-sea squids (Chun, 1903; Young, 1975), the somewhat less than hemispherical eyes of deep-sea annelid worms (Wald and Rayport, 1977) and the tubular eyes of deep-sea fishes (Brauer, 1908; Munk, 1966). The main retina coats the bottom (fundus) of the eye in all species investigated, and is usually well developed with numerous, packed and elongate photoreceptor cells. The accessory retina is an extension of the main retina and is located on the lateral walls of the tubular eyes in squids and fish, whereas it is a separate patch of tissue located near the cornea in the slug and annelid worms; this retina is poorly developed, containing fewer and shorter receptor cells than in the main part. The role of the main retina is clearly to form more or less sharp images, using often binocular vision, since light from the lens converges there. The role of the accessory retina is more debatable. From its position and behavioural patterns in *Agriolimax*, it is believed to be an infra-red receptor (Newell and Newell, 1968). From studies on spectral sensitivity, Wald and Rayport (1977) concluded that the accessory retina of deep-sea annelids, with a peak sensitivity at about 560 nm, serves to gauge depth in the water column by comparing the spectral distribution of downwelling light striking the main retina with that of the long wavelength light reflected from the fundus onto it. The apparent impossibility for the accessory retina of deep-sea squid and fish eyes to receive focused images has led to the suggestion that it detects blurred moving objects in the monocular field. The significance of the accessory retina in the sensory ecology of those animals clearly requires experimental reevaluation.

Reflecting layers (tapeta lucida) are a common occurrence in the photoreceptor organs of animals, and are usually found in crepuscular, nocturnal and deep-sea species, (Fig. 5 e) or in animals dwelling in turbid waters (Fig. 5 c,d). These structures owe their reflective properties to granules or crystals made of a variety of organic molecules depending on species, the most popular being purines, riboflavin, triglycerides, pteridines. The same kind of chemical substance may occur in the eyes of animals belonging to different phyla, such as the presence of guanine in the eyes of molluscs (Land, 1966), arthropods (Kleinholtz, 1959), and fishes (Nicol *et al.*, 1973), and that of pteridines in the eyes of crustaceans (Zyznar and Nicol, 1971) and teleostean fishes (Zyznar and Nicol, 1973; Zyznar and Ali, 1975). The mechanisms involved may be diffuse or specular reflectivity, and the shape of the reflecting organelles, as well as their position in the eye will determine whether they mediate diffuse scattering of light or rather enhance sensitivity of discrete photoreceptor units by optically isolating them from one another (specular reflexion). These reflectors are generally attributed a biological significance that readily comes to mind: that of increasing scotopic sensitivity by allowing the unabsorbed light to be reflected back on the photosensitive structures. This is indeed the most common role, but other functions cannot be excluded, as shown in the case of the scallop eye (Land, 1968), where a specular reflector intercepts the image projected by the lens and deflects it focally on the distal layer of photoreceptors. The lens, by itself, would throw the image behind the eye. Interestingly, reflecting layers lining the crystalline cones are involved in the formation of superposition images in the compound eyes of deep-sea crustaceans (Land, 1976).

Tapeta lucida are basically static features in animals exposed to a relatively stable photic environment, such as strictly nocturnal or deep-sea species. However, dynamic adaptive features appear in crepuscular or arhythmic species, especially in fishes displaying retinomotor responses to light and dark adaptation (see Ali, 1971, 1975). The tapetal organelles themselves may move inside the retinal pigment epithelium in response to light or dark (Fineran and Nicol, 1977), along with the screening melanosomes. The most common organisation, however, is well examplified in the crepuscular pikeperches (Ali and Anctil, 1977). In these species, the tapetal organelles are distributed throughout the long processes of the pigment epithelium. During light adaptation, the screening pigment moves to the tips of the epithelial processes and the scotopic rods migrate toward the epithelial cell bodies, thus shielding the rods from the incoming light and preventing light scattering. During dark-adaptation, the screening pigment moves up, thus exposing the reflector; the cones do not move but the rods contract and are distributed in bundles overlying the cone outer segments. This arrangement enhances scotopic sensitivity in a two-fold manner: (1) the cones are used as funnels converging light into the overlying rod bundles (Ali et al., 1977), and (2) the funneled light is reflected by the tapetum back on the rod bundles. Thus, a tapetal organisation used in the context of retinomotor responses provides its possessor with an unparalleled visual versatility to cope with changing environments.

Dim light environments, as shown above, have been instrumental in the development of numerous and striking adaptive ocular features. Insofar as bright light conditions are concerned, a few elements of clear adaptive value nevertheless deserve mention. For one thing, very bright light can cause glare effects, especially in fishes living in the superficial layer of the water column. Many reputed surface fishes, such as the guppy (Berger, 1966), the four-eyed fish (Borwein and Hollenberg, 1973), the killifish (Anctil and Ali, 1976) and flying-fishes (Anctil and Ali, 1970), all have colourless "oil droplets" of mitochondrial origin in their cones (Fig. 5 b). It is presumed that glare effects are attenuated by the probable light-collimating properties of these droplets, thus directing light preferentially toward the photosensitive outer segments of optically isolated cones. The coloured cone "oil droplets" of diurnal terrestrial vertebrates could have a similar function by improving contrast perception in environments where Rayleigh scattering could be a dominant factor (Walls, 1942; see Muntz, 1972 for discussion). In some fishes, reduction of infra-ocular glare is apparently not accomplished at the level of the photoreceptor cells, as shown above, but rather by iridescent multilayers in the cornea (Lythgoe, 1975).

Polarised light is an important component of the photic environment. Many animals can detect polarised light and use sky polarisation as an extension of the sun compass to orient themselves (Waterman, 1975). Arthropods, especially insects and crustaceans, use the distribution of visual pigments and photoreceptor rhabdomeric membranes in two planes perpendicular to each other in order to achieve workable polarotaxis. In vertebrates known to be sensitive to polarised light (e-vector), e.g. several fishes, an amphibian and a bird (Waterman, 1975), the geometrical arrangement of the photoreceptor membranes differs from that of anthropods in that all outer segment discs lie parallel to each other and are isotropic to light passing through their optical axis. The mechanism for polarotaxis in vertebrates must therefore lie elsewhere, and recent evidence points to optic tectum units as being responsible for polarised light sensitivity (Waterman and Hashimoto, 1974). An unusual instance of cone outer segment discs lying parallel to the optical axis has recently been described in the anchovy (Fineran and Nicol, 1976), and since the rod outer segment discs lie perpendicular to this axis, it could be argued that we have here a plausible retinal mechanism for polarotaxis in a vertebrate. However, the organisation of the retinal tapetum lucidum in these species indicates that light is nevertheless reflected in such a way as to traverse the cone outer segments in a plane perpendicular to that of the discs as in other vertebrates.

PHOTORECEPTOR SENSITIVITY AND THE SPECTRAL ENVIRONMENT

As mentioned earlier, molecules associated with photoreceptor membrane specialisations are responsible for the many forms of sensitivity to light expressed by animals. Most, if not all these pigment molecules, however, have retinal (Vitamin A) as their chromophore, which is attached to a proteinic part, the opsin. This basic arrangement is retained by both vertebrates and invertebrates, as well as by eyeless primitive organisms, and has interesting evolutive implications. It has been speculated that such rhodopsin-like molecules, as found in bacteria and fungi, were originally involved in protecting the cell's DNA against harmful radiation (see Wolken, 1975) or in mediating restoration of cellular ATP stores by light in anaerobic conditions (Danon and Stoeckenius, 1974). Vitamin A has assumed other physiological roles during animal evolution, notably in growth processes, hence its importance and ubiquity in the animal world.

The extent of absorption of light by photoreceptor visual pigments depends, among other factors, on the frequency of radiating energy, that is, on the spectral distribution of light in a given environment. Conversely, the absorption properties of visual pigments should influence the sensitivity of a photoreceptor to its photic environment. The wavelength of maximum absorption (λ_{max}) for a visual pigment may vary from species to species, and these variations have often been correlated with differences in the spectral characteristics of ambient light where the species live (see Goldsmith, 1972; Lythgoe, 1972). There are two basic mechanisms which bring about this diversity of absorption maxima. The most common one is through a modification of the proteinic component (opsin) of the rhodopsin molecule, and appears to be genetically determined. The second one involves the addition of a double bond in the chromophore or prosthetic group of rhodopsin, resulting in a porphyropsin with a shift of about 20 nm in the location of λ_{max}.

The latter mechanism appears to be environmentally and hormonally determined, although genetic influences cannot be ruled out (Bridges, 1974). In both cases, the evolutive implications are important, as shown below.

Examples of rhodopsin λ_{max} approximately matching the spectral distribution of the photic environments abound in the animal taxa. In lower invertebrates too little is known about the absorption properties of their visual pigments to warrant generalisations. In higher invertebrates and vertebrates, a wealth of data is now available and its ecological connotations can be assessed. The photic parameters of aquatic environments have been measured in more detail, and thus provide the basis for fruitful correlations. The spectral distribution of downwelling light is broad at the surface of clear water basins, but becomes narrower with depth in the water column. This narrowing is accompanied by a confinement of the spectral transmittance toward the shorter wavelengths (blue end). Correspondingly the λ_{max} of surface or epipelagic crustaceans and fishes tend to be longer than those of their deep-sea counterparts (Fig. 6) (Goldsmith, 1972; Muntz, 1975), in order to optimise sensitivity. Fresh and estuarine waters tend to transmit longer wavelengths because of high concentrations of phytoplankton and yellow substance in them, and again the measured λ_{max} of visual pigments of fresh and brackish water crustaceans and fishes tend to match this particular spectral environment (Fig. 6). These correlations are oversimplified for the sake of illustration, and several other factors are involved in the adaptive selection of specific rhodopsins, as recently shown by Lythgoe (1972) and Munz and McFarland (1973). Indeed, for the analysis of photoreceptor sensitivity to be meaningful, more precise data on the photic environment are required. In aquatic environments, this entails comparisons of spectral characteristics of downwelling, upwelling and background light in the horizontal line of sight, as well as to light reflected by a nearby bottom, whenever the case. These measurements, moreover, should be performed at different intervals of day and night, in view of significant spectral shifts during transition periods such as twilight (Munz and McFarland, 1973). The visual behaviour of the animals under study should be evaluated accordingly, especially the periods of day when vision is used and the role of vision in predator-prey relationships. Using this composite approach, Munz and McFarland (1973) came to the conclusion that tropical marine fishes have mostly rhodopins whose λ_{max} match the spectral environment available at twilight. This is a critical transition period for the diurnal and nocturnal shore faunas when there is a passage from photopic to scotopic thresholds. An enhanced spectral sensitivity at twilight would then reduce their vulnerability to crepuscular predators. It is implicit from the foregoing discussion that crepuscular predators share with their diurnal or nocturnal preys scotopic visual pigments in the same range of absorption maxima. An example

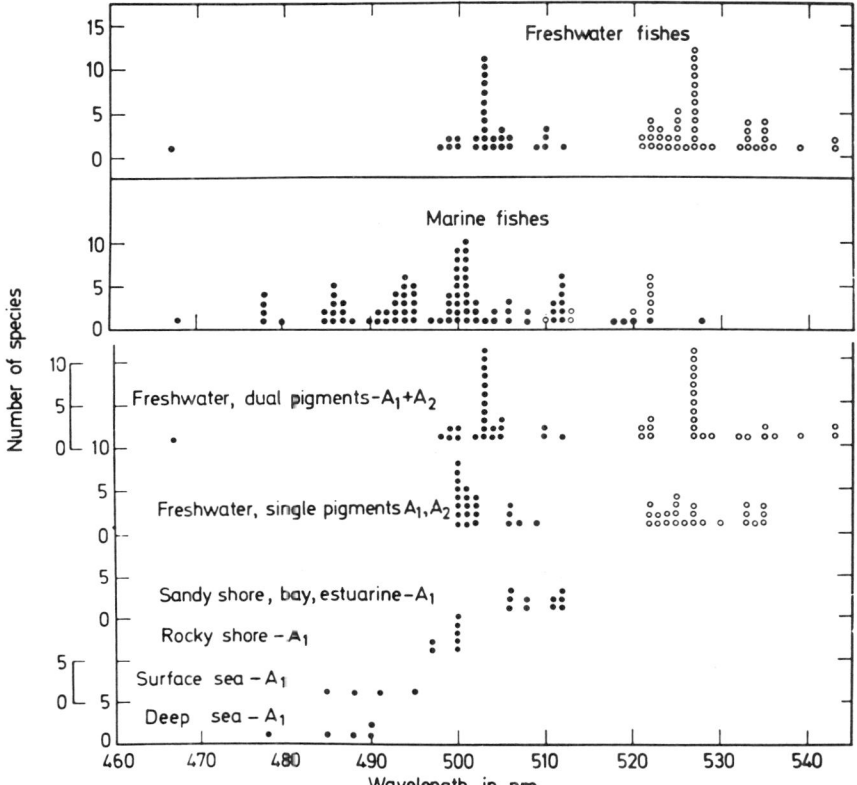

Fig. 6. Distribution of visual pigment absorption maxima among fish species from different habitats. Filled circles, rhodopsin-based pigments; plain circles, porphyropsin-based pigments. (After Crescitelli, 1972, Springer-Verlag).

involving freshwater fishes will further demonstrate this point. Yellow perch (*Perca flavescens*) are mostly diurnal fishes, whereas walleyes (*Stizostadion vitreum*) are essentially crepuscular feeders (Ryder, 1977). They both simultaneously occupy shoal areas at twilight, and have similar absorption maxima for their scotopic (λ_{max} 530 and 533 nm), as well as photopic visual pigments (Ali et al., 1977). However, the presence of a reflecting tapetum and the photoreceptor organisation, especially large cones, in the walleye's retina (Ali and Anctil, 1977) provide the basis for a superior visual performance at twilight in these species, whereas that of the yellow perch is simultaneously decreasing. Therefore, young perch tend to easily fall victims of the walleye's crepuscular feeding forays.

One major factor responsible for impairing visual capabilities of teleosts at twilight is that the retinomotor changes of the retinal pigment epithelium and photoreceptors (Ali, 1975) are incomplete during the photopic-scotopic transition period. This appears to be instrumental in triggering the seaward, downstream migration of Pacific salmon at night (Ali and Hoar, 1959), since rheotactic responses in these fishes are believed to be visually mediated. The action spectrum curve of those retinomotor responses, at least for the pigment epithelium and rods, coincides approximately with the scotopic absorption curve of the visual pigments (Ali and Crouzy, 1968). Thus, changes in the spectral properties of the environment and in the visual system at twilight carry significant ecological implications for shallow-water fishes, and deserve more comprehensive investigations.

Porphyropsin is present in freshwater or euryhaline fishes, as well as in amphibians (Schwanzara, 1967; Lythgoe, 1972), that is, in animals living in aquatic habitats where transmitted light is shifted toward the red. In amphibians switching habitat at metamorphosis, the porphyropsin system of the aquatic tadpole is completely converted to a rhodopsin system in the terrestrial adult (Crescitelli, 1972). This conversion is mediated by endocrine factors involved at large in metamorphic processes, such as thyroxine. The occurrence of mixtures of rhodopsin and porphyropsin in the visual system is fairly common among certain groups (see Bridges, 1972; Beatty, 1975). In most of those cases, changes in the proportions of the two pigments may occur in nature or be induced experimentally, and are related to the life history of the species, to environmental or endogenous factors (Bridges, 1972). In fishes undertaking spawning migrations from one aquatic habitat to another, the visual pigment mixture is gradually converted to pure rhodopsin (eels) or to a mixture largely dominated by porphyropsin (salmonids). However, those changes are occurring in anticipation of the migratory journeys and are not directly influenced by immediate environmental factors. The controlling anticipatory agents are endocrine in nature, involving both the pituitary and thyroid glands. Thus, the endogenous factors regulating sexual maturity and metamorphic events may also influence directly the conversion of one chromophore group to another. In non-migratory fishes, changes in the proportions of rhodopsin and porphyropsin may occur, and are usually related to seasonal or geographic variations in the colour and intensity of transmitted light (Bridges, 1972). The environmental agents responsible for those variations are the intensity and duration of light exposure, including the daily light cycle, and temperature. In certain species, exposure to bright or continuous light favours an increase in the proporiton of rhodopsin; dim light or darkness produces the reverse (Allen and McFarland, 1973). In others, the opposite is true (Allen et al. 1973), and yet these two fish groups respond similarly to seasonal changes in nature. Higher temperatures tend to reduce the proportion of porphyropsin in the pigment mixture, and these changes are more sharply defined when a photoperiod is introduced in the experimental protocol (McFarland

and Allen, 1977). It appears, therefore, as suggested by the latter authors, that under certain conditions, temperature effects override the influence of photic factors on the rhodopsin-porphyropsin mixtures.

The wide range of rhodopsin λ_{max} in fishes is the result of selective pressures related to the diversity of photic habitats in the aquatic environments. This evolutive process has probably been effected mostly through polymorphism in genes coding for opsin (Bridges, 1974). However, genetic factors may also be implicated in switching from a chromophore group to another, independently of environmental factors, as appears to be the case in the postglacial evolution of rhodopsin and porphyropsin in landlocked populations of the smelt (Bridges, 1974). Estuarine "free" smelts have a rhodopsin-dominated mixture, whereas those migrating upstream for spawning have mostly porphyropsin in their retina. Eyes of smelts landlocked because of the retreat of ice during the last glacial period would be expected to contain mostly porphyropsin, but instead had mixtures dominated by rhodopsin in most cases. The evolution of those different isolated populations or races of smelt would have occurred, surprisingly enough, within the short period of about 10,000 years.

Colour vision is an aspect of photoreceptor spectral sensitivity whose ecological significance has been largely ignored until recently. Retinal mechanisms of colour discrimination, especially the trichromatic processes, have been extensively reviewed by Autrum (1968) and MacNichol et al., (1973). Colour perception is widespread in the animal kingdom. Some polychaete worms seem to have the potential to discriminate colour (Yingst et al., 1972) and colour vision has been established in many insects, fishes, some amphibians, lizards, turtles, birds and higher primates using behavioural, electrophysiological and microspectrophotometric techniques. Cephalopods which display cryptic colouration, such as *Octopus*, are apparently colour-blind (Messenger et al., 1973; Messenger, 1977), but they use their chromatophores for camouflage against the background in order to avoid predation from teleost fishes that can see colours. The ability to distinguish colour is materialised only in photopic vision and thus works best in diurnal animals. Moreover, those diurnal species are required to possess a multireceptor system, in which visual pigments, with different spectral sensitivities are associated with different classes of receptor cells, according to the trichromacy theory. An attractive hypothesis, based on extensive surveys of photopic visual pigments in pelagic and shore fishes (Munz and McFarland, 1975) and of underwater photic conditions during daytime (McFarland and Munz, 1975a), concerning the origin and evolution of colour vision in vertebrates has recently been proposed (McFarland and Munz, 1975b). According to these authors, when the early fish-like benthic vertebrates became pelagic and horizontal visual fields assumed more importance for location of food, it resulted in a narrowing of the spectral distribution of background light against which potential preys are seen. Multiple visual pigments in separate cone pathways provided a mechanism to improve contrast of bright or dark targets in a variety of daytime photic conditions. Improvements in photocontrast, as provided by one matching and one or two offset cone visual pigments in relation to background light, could have evolved into behaviourally meaningful hue discrimination. In view of this and of our previous discussion on the ecological implications of scotopic visual pigments, we cannot overemphasise the importance of evolutive strategies of predator-prey relationships in shaping visual capabilities of animals.

ECOLOGICAL ASPECTS OF RETINAL ELECTROPHYSIOLOGY

Electrophysiological studies, both at the macro and microelectrode levels, gave and still could provide considerable insight into the relationship of visual adaptations to environmental constraints. Electroretinography (ERG), S-potential recordings, intracellular recordings from photoreceptor and second or third order neurons are some of the tools of the trade used to analyse functional states of photoreceptor organs (see Fuortes, 1972). A comparative approach to these electrophysiological studies would be most suited for ecologically meaningful conclusions to be made. For example, the on-off, centre-surround responses of retinal ganglion cells described by Raynauld (1975) in the goldfish could be looked at from a comparative viewpoint, since fishes occupy a multiplicity of photic environments and display a great variety of visual behaviours.

The relationship of electroretinograms (ERG's) to environmental factors has been stressed by Ali and Muntz (1975). A comprehensive survey of fish ERG's allowed Kobayashi (1962) to conclude that deepwater and nocturnal species display ERG responses that saturate at relatively low light intensities by comparison with diurnal or shallow-water fishes. Similarly, the flicker fusion frequencies measured from ERG responses were lower in nocturnal or deep-sea species. Thus ERG responses are reliable monitors of the performance or sensitivity of photoreceptor organs in animals adapted to a specific photic environment.

Spectral sensitivity as measured from ERG responses has also been used to correlate visual range with photic conditions to which animals are exposed (Kobayashi, 1962, fishes; Scott and Mote, 1974, crustaceans; Wald and Rayport, 1977, annelids). Kobayashi (1962) reported that the ERG scotopic spectral sensitivity showed a maximum shifted towards shorter wavelengths when comparing shallow-water to deep-sea species. This is in agreement with the distribution of visual pigment absorption maxima as discussed earlier. S-potentials presumably recorded from horizontal cells (luminosity response) in a variety of teleostean fishes tend to show a λ_{max} shifted towards the shorter wavelengths in the marine as opposed to the freshwater species (Niwa and Tamura, 1969), which is consonant with the ERG results reported above. Moreover, it was shown that the S-potentials recorded from the different types of fish horizontal cells reflected the trichromatic input of three cone types with different visual pigment λ_{max} (Svaetichin et al., 1965). These observations clearly indicate that the ecological implications of patterns of retinal electrophysiological activity have been little explored, although they should yield interesting results. Among other things, we may ask ourselves: how meaningful are receptive field properties of retinal functional units in terms of the light regime and visual tasks available in discrete habitats of animal communities?

SOLAR RADIATION, BIOLUMINESCENCE, AND VISUAL STRATEGIES

Bioluminescence, as well as downwelling ambient light, is an important component of the photic environment of oceanic midwater communities. (Blaxter, 1970; Lythgoe, 1972; Tett and Kelly, 1973). Ecologically meaningful discussions on the significance of bioluminescence in deep-sea animals must take into account several considerations: (1) the intensity, distribution patterns and spectral properties of solar transmitted and emitted light; (2) the life history and habits of the luminescent animals; (3) the physiological limitations underlying the luminescent behaviour of midwater animals; and (4) the visual capabilities and ecology of those animals most likely to detect bioluminescence in their habitats. Some ecological consequences of interactions between vision and bioluminescence in fish have already been reviewed (Anctil, 1975).

In situ observations of negative or positive phototropic responses in midwater animals (Backus et al., 1968; Church, 1970) indicate that they are highly sensitive to the residual light available at depths of 200-700 m. This transmitted light is highly unidirectional (downwelling) and so is the predominantly ventral bioluminescence of midwater animals (Young, 1973; Denton et al., 1972). However, the vertical or oblique orientation of lethargic, deep-sea fishes during daytime (Barham, 1970) would direct their luminescence along more horizontal lines of sight for potentially detecting animals, although this appears to occur when bioluminescence is not operating. Most luminescent forms undertake extensive daily vertical migrations in the oceanic water column, ascending at twilight periods to feed on the smaller components of the trophic web. These migrations appear to be triggered by photic stimuli, such as changes in ambient light intensity at twilight (Boden and Kampa, 1967). An important consequence of this migratory behaviour and the trophic organisation of oceanic waters is that, more often than not, the preys are located above the predators. It is not surprising, then, to find that many mesopelagic predators possess tubular eyes looking upward or lateral hemispherical eyes with a well-developed ventral retina (Munk, 1966;

Ali and Anctil, 1976). The preys may appear as dark targets contrasted against downwelling light, to the predator's eye. However, counterstrategies evolved by potential preys ensured that the predator's visual task be made difficult. In order to disrupt the ventral silhouette, preys were presumed to countershade it by silvery reflective effects (Denton and Nicol, 1966) or by ventral bioluminescence (Clarke, 1963; Nicol, 1967; Young, 1973). This presumption would require that the countershading organism possesses (1) a mechanism whereby the eyes (retina) compare the intensity of downwelling light with that of a photophore emitting light into the retina, and (2) a mechanism ensuring that the intensity of ventral bioluminescence will match that of downwelling light once the information from the monitoring retina has been evaluated by the brain. Subjective observations, suggestive morphological considerations, and experimental results all indicate that the two aforementioned mechanisms are operating in luminescent squids (Young, 1977; Young and Roper, 1976, 1977), shrimps (Young and Roper, 1977) and fishes (Lawry, 1974; Case et al., 1977). However, the exact nature of these regulating mechanisms remains obscure, and a physiological approach to the problem should now be undertaken.

Another interesting aspect of those predator-prey relationships is that the large predator fishes tend to have a less well-developed retina than that of the luminescent, countershading preys (Anctil, 1975). This might make it easier for the countershading prey to produce a disrupting effect in the eye of the predator by requiring only an approximate match of downwelling and living light. This condition might also stem from the probably higher visual capabilities required to detect predators and quickly produce an escape reaction, than what is required to spot and catch a prey. Munz and McFarland (1973) noted similar retinal differences between predators and preys among tropical shorefishes, and related these to their feeding strategies at twilight.

In terms of visual strategies, how close is the correlation between the spectral maxima of the environmental light and the visual pigments and bioluminescence of marine organisms? For one thing, the spectral maxima of downwelling light at midwater depths, at 470-490nm, provide a fairly good match for the absorption maxima of visual pigments in deep-sea crustaceans (see Goldsmith, 1972) and fishes (Lythgoe, 1972). Moreover, luminescence spectral maxima (see Lythgoe, 1972), generally match well the known visual pigment λ_{max} of midwater animals (Fernandez and Tsuji, 1976; O'Day and Fernandez, 1974, 1976). For instance, the lanternfish *Stenobrachius leucopsarus* has a scotopic visual pigment absorbing maximally at 492 nm, while the light emitted by its choice prey, the shrimp *Euphausia pacifica*, has a spectral maximum at 476 nm. The discrepancy is not significant in view of the high sensitivity of the lanternfish retina over a broad spectral range (O'Day and Fernandez, 1976). This fitness, of course, appears to be adaptive for better countershading effects of luminescence preys by matching the spectral sensitivity of potential predators to downwelling light with that of ventral bioluminescence. However, the presence of yellow lenses in several mesopelagic predators, including squids and fishes (Somiya, 1976; Muntz, 1976), provides a counterstrategy aimed at disrupting the camouflaging effect of matching background light, ventral bioluminescence and photoreceptor pigments. The yellow lens accomplishes this by contrasting the bioluminescence against a dimmer background light (Muntz, 1976).

CONCLUDING REMARKS

Summing up, the variety of habitats and photic environments to which evolving animals were exposed certainly was a decisive factor in making up the present adaptive radiation of photoreceptors, visual pigments and visual tasks. However, our analysis emphasised two relatively neglected aspects of sensory ecology of photoreception, namely evolutionary convergence of structure and function, and the dramatic role of predator-prey relationships in shaping visual strategies. Another perspective which has been merely touched on and could prove rewarding in the future is the ecological approach to the study of changing visual functions during the life cycle of a given species. Few studies have yet devoted themselves to the visual capabilities of invertebrate or vertebrate larvae and juveniles (see Blaxter, 1974 and 1975, on fish larvae). Comprehensive studies of the role of vision in ecological relationships of animals throughout their life, especially those species undergoing changes of habitat and life style, would be most welcome at this point.

REFERENCES

Adrian, E.D. (1928). The basis of sensation. Christophers, London.
Ali, M.A. (1971). Les réponses rétinomotrices: caractères et mécanismes. Vision Res. 11: 1225-1288.
Ali, M.A. (1975). Retinomotor responses. In: Vision in Fishes: New Approaches to Research. p. 313-355, ed. M.A. Ali. Plenum Press, New York.
Ali, M.A. and Anctil, M. (1976). Retinas of Fishes: An Atlas. Springer-Verlag, Heidelberg. 284p.
Ali, M.A. and Anctil, M. (1977). Retinal structure and function of the walleye (*Stizostedion vitreum*) and sauger (*S. canadense*). J. Fish. Res. Board Can. 34: 1467-1474.
Ali, M.A. and Crouzy, R. (1968). Action spectrum and quantal thresholds of retinomotor responses in the brook trout, *Salvelinus fontinalis* (Mitchill). Z. verg. Physiol. 59: 86-89.
Ali, M.A. & Hoar, W.S. (1959). Retinal responses of pink salmon associated with its downstream migration. Nature 184: 106-107.
Ali, M.A. & Muntz, W.R.A. (1975). Electroretinography as a tool for studying fish vision. In: Vision in Fishes, New Approaches to

Research. p. 159-167, ed. M.A. Ali. Plenum Press, New York.
Ali, M.A., Ryder, R.A. & Anctil, M. (1977). Photoreceptors and visual pigments as related to behavioral responses and preferred habitats of perches (*Perca spp.*) and pikeperches (*Stizostedion spp.*) J. Fish. Res. Board Can. 34: 1475-1480.
Allen, D.M. & McFarland, W.N. (1973). The effect of temperature on rhodopsin-porphyropsin ratios in a fish. Vision Res. 13: 1303-1309.
Allen, D.M., McFarland, W.N., Munz, F.W. & Poston, H.A. (1973). Changes in the visual pigments of trout. Can. J. Zool. 51: 901-914.
Anctil, M. (1975). Prospects in the study of interrelationships between vision and bioluminescence. In: Vision in Fishes: New Approaches in Research, p. 657-671, ed. M.A. Ali. Plenum Press, New York.
Anctil, M. & Ali, M.A. (1970). Retina of *Exocoetus volitans* and *Fodiator acutus* (Pisces Exocoetidae). Copeia No. 1, 43-48.
Anctil, M. & Ali, M.A. (1976). Cone droplets of mitochondrial origin in the retina of *Fundulus heteroclitus* (Pisces: Cyprinodontidae). Zoomorphol. 84: 103-111.
Ashmore, J.F. & Falk, G. (1976). Absolute sensitivity of rod bipolar cells in a dark-adapted retina. Nature 263: 248-249.
Autrum, H. (1968). Colour vision in man and animals. Naturwiss. 55: 10-18.
Backus, R.H., Craddock, J.E., Haedrich, R.L., Shores, D.L., Teal, J.M., Wing, A.S., Mead, G.M. & Clarke, W.D. (1968). *Ceratoscopelus maderensis*: peculiar sound-scattering layer identified with this myctophid fish. Science 160: 991-993.
Barham, E.G. (1970). Deep-sea fishes: lethargy and vertical orientation. In: Proc. Intern. Symp. Biol. Sound Scattering in the Ocean, p. 100-118, ed. G.B. Farquhar. Maury Center for Ocean Science, Washington.
Baylor, D.A. & Fettiplace, R. (1977). Transmission from photoreceptors to ganglion cells in turtle retina. J. Physiol. Lond. 271: 391-424.
Baylor, D.A. & Fuortes, M.G.F. (1970). Electrical responses of single cones in the retina of the turtle. J. Physiol. 207: 77-92.
Baylor, D.A. & Hodgkin, A.L. (1973). Detection and resolution of visual stimuli by turtle photoreceptors. J. Physiol. Lond. 234: 163-198.
Baylor, D.A. & Hodgkin, A.L. (1974). Changes in time scale and sensitivity in turtle photoreceptors. J. Physiol. 242: 729-758.
Baylor, D.A., Hodgkin, A.L. & Lamb, T.D. (1974a). The electrical response of turtle cones to flashes and steps of light. J. Physiol. 242: 686-727.
Baylor, D.A., Hodgkin, A.L. & Lamb, T.D. (1974 b). Reconstruction of the electrical responses of turtle cones to flashes and steps of light. J. Physiol. 242: 759-791.

Beatty, L. D. (1975). Rhodopsin - Porphysopsin changes in paired-pigment fishes. In: Vision in Fishes: New Approaches to Research. p. 635-644, ed. M.A. Ali. Plenum Press, New York.

Berger, E. R. (1966). On the mitochondrial origin of oil drops in the retinal double cone inner segments. J. Ultrastruct. Res. 14: 143-157.

Blaxter, J.H.S. (1970). Light. In: Marine Ecology, p. 213-285, ed. O. Kinne. Wiley, London.

Blaxter, J.H.S. (ed.) (1974). The Early Life History of Fish. Springer-Verlag, New York.

Blaxter, J.H.S. (1975). The eyes of larval fish. In:Vision in Fishes: New Approaches in Research, p. 427-443, ed. M.A. Ali. Plenum Press, New York.

Boden, B.P. & Kampa, E.M. (1967). The influence of natural light on the vertical migrations of animal community in the sea. Symp. Zool. Soc. Lond. 19: 15-26.

Borwein, B. & Hollenberg, M.J. (1973). The photoreceptors of the four-eyed fish, *Anableps anableps* L. J. Morphol. 140: 405-442.

Brauer, A. (1908). Die Tiefseefische. 11. Anatomische teil. B. Augen, 266p. Wissenchaftliche ergebnisse der Deutschen Tiefsee-expedition auf dem Dampler "Valdivia" 1898-1899, Bd. 15. Gustav-Fisher, Jena.

Bridges, C.D.B. (1972). The rhodopsin-porphyropsin visual system. In: Handbook of Sensory Physiology, Vol. VII/1, Photochemistry of Vision, p. 417-480, ed. H.J.A. Dartnall. Springer-Verlag, New York.

Bridges, C.D.B. & Delisle, C.E. (1974). Evolution of visual pigments. Exp. Eye Res. 18: 323-332.

Burnside, B. (1976). Microtubules and actin filaments in teleost visual cone elongation and contraction. J. Supramol. Struct. 5: 257-275.

Case, J.F., Warner, J., Barnes, A.T. & Lowenstine, M. (1977). Bioluminescence of lantern fish (*Myctophidae*) in response to changes in light intensity. Nature 265: 179-181.

Cervetto, L. (1973). Influence of sodium potassium and chloride ions on the intracellular responses of turtle photoreceptors. Nature, 241: 401-403.

Cervetto, L., Pasino, E. & Torre, V. (1977). Electrical responses of rods in the retina of *Bufo marinus*. J. Physiol. 267: 17-51.

Chun, C. (1903). Deutsche Tiefsee-Expedition 'Valdivia". Verh. Dtsch. Zool. Ges. 13: 67. Gustav Fischer, Jena, 1910-1915. English translation by Israel Program for Scientific Translations Jersalem, 1975.

Church, R. (1970). Bioluminescence: the sea's living light. Oceans Mag. 3: 20-29.

Clarke, W.D. (1963). Function of bioluminescence in mesopelagic organisms. Nature 198: 1244-1246.

Crescitelli, F. (1972). The visual cells and visual pigments of the vertebrate eye. In: Handbook of Sensory Physiology, Vol. VII/1, Photochemistry of Vision, p. 245-263, ed. H.J.A. Dartnall. Springer-Verlag, New York.

Danon, A. & Stoeckenius, W. (1974). Photophosphorylation in *Halobacterium halobium*. Proc. Nat. Acad. Sci. U.S.A. 71: 1234-1238.

Denton, E.J., Gilpin-Brown, J.B. & Wright, P.G. (1972). The angular distribution of the light produced by some mesopelagic fish in relation to their camouflage. Proc. R. Soc. Lond. B. 182: 145-158.

Denton, E.J. & Nicol, J.A.C. (1966). A survey of reflectivity in silvery teleosts. J. Mar. Biol. Ass. U.K. 46: 685-722.

Denton, E.J. & Warren, F.J. (1957). The photosensitive pigments in the retinae of the deep-sea fish. J. Mar. Biol. Ass. U.K. 36: 651-662.

Eakin, R.M. (1972). Structure of invertebrate photoreceptors. In: Handbook of Sensory Physiology, Vol. VII/1, Photochemistry of Vision, p. 625-684, ed. H.J.A. Dartnall. Springer-Verlag, New York.

Fain, G.L. (1976). Sensitivity of toad rods: dependence on wave-length and background illumination. J. Physiol. 261: 71-101.

Fernandez, H.R. & Tsuji, F.I. (1976). Photopigment and spectral sensitivity in the bioluminescent fish, *Porichthys notatus*. Mar. Biol. 34: 101-107.

Fineran, B.H. & Nicol, J.A.C. (1976). Novel cones in the retina of the anchovy (*Anchoa*). J. Ultrastruct. Res. 54: 295-303.

Fineran, B.H. & Nicol, J.A.C. (1977). Studies on the eyes of anchovies *Anchoa mitchelli* and *A. hepsetus* (Engraulidae) with particular reference to the pigment epithelium. Phil. Trans. R. Soc. Lond. B. 276: 321-350.

Fuortes, M.G.F. (1959). Initiation of impulses in visual cells of *Limulus*. J. Physiol. 148: 14-28.

Fuortes, M.F.G. (ed.) (1972). Handbook of Sensory Physiology. Vol, VII/2: Physiology of Photoreceptor organs. Springer-Verlag, New York.

Goldsmith, T.H. (1972). The natural history of invertebrate visual pigments. In: Handbook of Sensory Physiology, Vol. VII/1, Photochemistry of Vision, p. 685-719, ed. H.J.A. Dartnall. Springer-Verlag, New York.

Hagins, W.A. (1972). The visual process: excitatory mechanisms in the primary receptor cells. Ann. Rev. Biophys. Bioeng. 1: 131-158.

Hartline, H.K. (1934). Intensity and duration in the excitation of single photoreceptor units. J. Cell. Comp. Physiol. 5: 229-247.

Hartline, H.K., Wagner, H.G. & MacNichol, E.F. (1952). The perispheral origin of nervous activity in the visual system.

Cold Spring Harb. Symp. Quant. Biol. 17: 125-141.
Hodgkin, A.L. & O'Bryan, P.M. (1977). Internal recording of the early receptor potential in turtle cones. J. Physiol. 267: 737-766.
Horridge, G.A. (1968). Pigment movement and the crystalline threads of the firefly eye. Nature 218: 778-779.
Horridge, G.A. (1969). The eye of the firefly *Photuris*. Proc. R. Soc. Series B. 171: 445-463.
Kleinholz, L. (1959). Purines and pteridines from the reflecting pigment of the arthropod retina. Biol. Bull Mar. Biol. Lab. Woods Hole 116: 125-135.
Kobayashi, H. (1962). A comparative study on electroretinogram in fish, with special reference to ecological aspects. J. Shimonoseki Coll. Fish. 11: 407-538.
Korenbrot, J.I. & Cone, R.A. (1972). Dark ionic flux and the effects of light in isolated rod outer segments. J. Gen. Physiol. 60: 20-45.
Krasne, F.B. & Lawrence, P.A. (1966). Structure of the photoreceptors in the compound eyespots of *Branchiomma vesiculosum*. J. Cell. Sci. 1: 239-248.
Land, M.F. (1966). A multilayer interference reflector in the eye of the scallop (*Pecten maximus*). J. Exp. Biol. 45: 433-447.
Land, M.F. (1968). Functional aspects of the optical and retinal organisation of the mollusc eye. Symp. Zool. Soc. Lond. 23: 75-96.
Land, M.F. (1976). Superposition images are formed by reflection in the eyes of some oceanic decapod crustacea. Nature 263: 764-765.
Lawry, J.V. (1974). Lantern fish compare downwelling light and bioluminescence. Nature 247: 155-157.
Liebman, P.A. (1972). Microspectrophotometry of photoreceptors. In: Handbook of Sensory Physiology, Vol. VII/1, Photochemistry of Vision p. 481-528, ed. H.J.A. Dartnall. Springer-Verlag, New York.
Liebman, P.A. & Granda, A.M. (1971). Microspectrophotometric measurements of visual pigments in two species of turtle, *Pseudemys scripta* and *Chelonia mydas*. Vision Res. 11: 105-114.
Locket, N.A. (1970). Deep-sea fish retinas. Brit. Med. Bull. 26: 107-111.
Lythgoe, J.N. (1972). The adaptation of visual pigments to the photic environment. In: Handbook of Sensory Physiology VII/1, Photochemistry of Vision, p. 566-603, ed. H.J.A. Dartnall. Springer-Verlag, New York.
Lythgoe, J.N. (1975). The ecology function and phylogeny of iridescent multilayers in fish corneas. In: Light as an Ecological Factor: II, 16th Symposium Brit. Ecol. Soc. 26-28 Mar. 1974. Eds. G.C. Evans, R. Bainbridge and O. Rackham. Blackwell, Oxford.

MacNichol, E.F. Jr., Feinberg, R. & Harosi, F.I. (1973). Colour discrimination processes in the retina. Proc. 2nd Congress Internat. Colour Ass. Colour 73: 191-251. Adam Hilger, London.
McFarland, W.N. & Allen, D.M. (1977). The effect of extrinsic factors on two distinctive rhodopsin-porohyropsin systems. Can. J. Zool. 55: 1000-1009.
McFarland, W.N. & Munz, F.W. (1975a). The photic environment of clear tropical seas during the day. Vision Res. 15: 1063-1070.
McFarland, W.N. & Munz, F.W. (1975b). The evolution of photopic visual pigments in fishes. Vision Res. 15: 1071-1080.
Marks, W.B., Dobelle, W.H. & MacNichol, E.F. (1964). Visual pigments of single primate cones. Science, 143: 1181-1183.
Matthiews, B.H.C. (1931). The response of a single end organ. J. Physiol., Lond. 71: 64-110.
Messenger, J.B. (1977). Evidence that *Octopus* is colour blind. J. Exp. Biol. 70: 49-56.
Messenger, J.B., Wilson, A.P. & Hedge, A. (1973). Some evidence for colour-blindness in *Octopus*. J. Exp. Biol. 59: 77-94.
Miller, W.H. (1976). Optical guiding by photoreceptor cells. Fed. Proc. 35: 37-43.
Miller, W.H. & Snyder, A.W. (1977). The tiered vertebrate retina. Vision Res. 17: 239-255.
Millott, N. (1968). The dermal light sense. Symp. Zool. Soc. Lond. 23: 1-36.
Munk, O. (1966). Ocular anatomy of some deep-sea teleosts. Dana rep. 70: 1-62.
Muntz, W.R.A. (1972). Inert absorbing and reflecting pigments. In: Handbook of Sensory Physiology, Vol. VII/1, Photochemistry of Vision, p. 529-565, ed. H.J.A. Dartnall. Springer-Verlag, New York.
Muntz, W.R.A. (1975). Visual pigments and their environment. In: Vision in Fishes: New Approaches in Research. p. 565-578, ed. M.A. Ali. Plenum Press, New York.
Muntz, W.R.A. (1976). On yellow lenses in mesopelagic animals. J. Mar. Biol. Assoc. U.K. 56: 963-976.
Munz, F.W. & McFarland, W.N. (1973). The significance of spectral position in the rhodopsins of tropical marine fishes. Vision Res. 13: 1829-1874.
Munz, F.W. & McFarland, W.N. (1975). Presumptive cone pigments extracted from tropical marine fishes. Vision Res. 15: 1045-1062.
Newell, P.F. & Newell, G.E. (1968). The eye of the slug, *Agriolimax reticulatus* (Müll). Symp. Zool. Soc. Lond. 23: 97-111.
Nicol, J.A.C. (1950). Responses of *Branchiomma vesiculosum* (Montagu) to photic stimulation. J. Mar. Biol. Assoc. U.K. 29: 303-320.
Nicol, J.A.C. (1967). The luminescence of fishes. Symp. Zool. Soc. Lond. 19: 27-55.
Niwa, H. & Tamura, T. (1969). Investigation of fish vision by means of S-potential. II. Spectral sensitivity and colour vision.

Rev. Can. Biol. 28: 79-88.

O'Day, W.T. & Fernandez, H.R. (1974). *Aristostomias scintillens* (Malacosteidae): a deep-sea fish with visual pigments apparently adapted to its own bioluminescence. Vision Res. 14: 545-550.

O'Day, W.T. & Fernandez, H.R. (1976). Vision in the lanternfish *Stenobrachius leucopsarus* (Myctophidae). Mar. Biol. 37: 187-195.

Oesterhelt, D. & Stoeckenius, W. (1971). Rhodopsin-like protein from the purple membrane of *Halobacterium halobium*. Nature New Biology 233: 149-152.

Raynauld, J.-P. (1975). A model for the ganglionic receptive field organisation. In: Vision in Fishes: New Approaches in Research. p. 91-98, ed. M.A. Ali. Plenum Press, New York.

Rodieck, R.W. (1973). The Vertebrate Retina: Principles of Structure and Function. Freeman, San Francisco.

Ryder, R.A. (1977). Effects of ambient light variations on behavior of yearling subadult, and adult walleyes (*Stizostedion vitreum vitreum*). J. Fish. Res. Board Can. 34: 1481-1491.

Schwanzara, S.A. (1967). The visual pigments of freshwater fishes. Vision Res. 7: 121-148.

Scott, S. & Mote, M.I. (1974). Spectral sensitivity in some marine crustacea. Vision Res. 14: 659-663.

Somiya, H. (1976). Functional significance of the yellow lens in the eyes of *Argyropelecus affinis*. Mar. Biol. 34: 93-99.

Steven, D.M. (1963). The dermal light sense. Biol. Rev. 38: 204-240.

Svaetichin, G.K., Negishi, K. & Fatehchand, R. (1965). Cellular mechanism of a Young-Hering visual system. In: Colour Vision, Physiology and Experimental Psychology. Little, Brown, Boston, Mass. (Ciba Found. Symp. Colour Vision p. 178-207).

Tett, P.B. & Kelly, M.G. (1973). Marine bioluminescence. Oceanogr. Mar. Biol. Ann. Rev. 11: 89-173.

Tomita, T. (1965). Electrophysiological study of the mechanism subserving color coding in the fish retina. Cold Spring Harb. Symp. Quant. Biol. 30: 559-566.

Wald, G. & Rayport, S. (1977). Vision in annelid worms. Science 196: 1434-1439.

Wales, W. (1975). Extraretinal photosensitivity in fish larvae. In: Vision in Fishes: New Approaches to Research. p. 445-450, ed. M.A. Ali. Plenum Press, New York.

Walls, G.L. (1942). The vertebrate eye and its adaptive radiation. Cranbrook Institute of Science Bull. 19: 785 p.

Waterman, T.H. (1975). Natural polarized light and e-vector discrimination by vertebrates. In: Light as an Ecological Factor II, 16th Symposium of Brit. Ecol. Soc., 26-28 March 1974. Eds. G.C. Evans., R. Bainbridge and O. Rackham. Blackwell, Oxford.

Waterman, T.H. & Hashimoto, H. (1974). E-vector discrimination by the goldfish optic tectum. J. Comp. Physiol. 95: 1-12.

Wolken, J.J. (1975). Photoprocesses, Photoreceptors and Evolution. Academic Press, New York. 317 p.

Yamamoto, T., Tasaki, K., Sugaware, Y. & Tonosaki, A. (1965). Fine structure of the octopus retina. J. Cell. Biol. 25: 345-359.

Young, R.E. (1973). Information feedback from photophores and ventral countershading in midwater squid. Pacif. Sci. 27: 1-7.

Young, R.E. (1975). Transitory eye shapes and the vertical distribution of two midwater squids. Pacif. Sci. 29: 243-255.

Young, R.E. (1977). Ventral bioluminescent countershading in midwater cephalopods. Symp. Zool. Soc. Lond. 38: 161-190.

Young, R.E. & Roper, F.E. (1976). Bioluminescent countershading in midwater animals: evidence from living squid. Science 191: 1046-1048.

Young, R.E. & Roper, F.E. (1977). Intensity regulation of bioluminescence during countershading in living midwater animals. Fish. Bull. 75: 239-252.

Zyznar, E.S. & Ali, M.A. (1975). An interpretative study of the organization of the visual cells and tapetum lucidum of *Stizostedion*. Can. J. Zool. 53: 180-196.

Zyznar, E.S. & Nicol, J.A.C. (1971). Ocular reflecting pigments of some malacostraca. J. Exp. Biol. Ecol. 6: 235-248.

Zyznar, E.S. & Nicol, J.A.C. (1973). Reflecting materials in the eyes of three teleosts, *Orthopristes chrysopterus, Dorosoma cepedianum* and *Anchoa mitchilli*. Proc. R. Soc. Lond. 184: 15-27.

A SURVEY OF VERTEBRATE STRATEGIES FOR VISION IN AIR AND WATER

J. G. Sivak

Laboratory of Comparative Optometry

University of Waterloo, Waterloo, Ontario, Canada

INTRODUCTION

Possibly the most important ecological factor to affect the evolution of the vertebrate eye is the effect of air and water on the refractive power of the cornea. Whereas the cornea is usually the major refractive element of the terrestrial eye, it is of little or no refractive consequence to an aquatic vertebrate. This has been attributed to the close similarity in refractive indices of water and corneal tissue (Walls, 1942; Duke-Elder, 1958; Tansley, 1965). In fact, the refractive index of the cornea is always appreciably greater than that of water (Table 1). However, its refractive index is of little importance since the cornea is bounded by two parallel of nearly parallel surfaces which separate it from water or air in front and aqueous humor behind. The indices for water and aqueous humor *are* very similar.

A second point, noted only recently (Sivak, 1974; Sivak and Bobier, unpublished) concerns the restricted spectral range (blue, blue-green) of many aquatic habitats (Jerlov, 1968). The importance of this factor will vary with the habitat (e.g. shallow vs. deep water) and the amount of chromatic aberration exhibited by the eye in question. Since the refractive power of an eye is greatest in blue light, this factor will tend to mitigate the above effect.

Table 1. Summary of corneal refractive indices among vertebrates. In most instances indices were measured with an Abbe refractometer. Calculated indices were derived from measurements of other optical components.

Species	Mean Index (±SD)	No. Eyes	Method	Source
Fish				
blunt nose stingray (*Dasyatis sayi*)	1.373 (0.005)	6	Abbe	Sivak, 1976c
nurse shark (*Ginglymostoma cirratum*)	1.382 (0.003)	3	Abbe	" "
goldfish (*Carrassius auratus*)	1.35	–	calculated	Charman and Tucker, 1973
yellow perch (*Perca flavescens*)	1.375	2	Abbe	Sivak (unpublished)
Amphibians				
Rana esculenta	1.390		Abbe	du Pont and de Groot, 1976
Rana pipiens	1.366 (0.001)	5	Abbe	Sivak (unpublished)
Reptiles				
gecko (*Grecko gecko*)	1.36	4	Abbe	Citron and Pinto, 1973
green iguana (*Iguana iguana*)	1.35	3	Abbe	" "
yellow rat snake (*Elaphae quadravittata*)	1.370 (0.006)	3	Abbe	Sivak, 1977
black racer (*Coluber constrictor*)	1.371 (0.008)	4	Abbe	Sivak, 1977

Table 1 (continued)

Species	Mean Index (±SD)	No. Eyes	Method	Source
Birds				
double-crested cormorant (*Phalacrocorax auritus*)	1.369	2	Abbe	Sivak, *et al.*, 1977
brown pelican (*Pelicanus occidentalis*)	1.369	2	Abbe	" " "
pigeon	1.337	—	calculated	Gundlach, Chard and Skahen, 1945
Mammals				
pig	1.373	4	Abbe	Aurell and Holmgren, 1953
cow	1.382 (0.002)	14	Abbe	Aurell and Holmgren, 1953
rat	1.374 (0.006)	8	Abbe	Massof and Change, 1972
rabbit	1.376	3	Abbe	Hughes, 1972
horse	1.369 (0.003)	6	Abbe	Sivak and Allen, 1975
cat	1.376	—	calculated	Vakkur and Bishop, 1963
Weddell seal	1.376		Abbe	Wilson, 1970
fin whale	1.383		?	Matthieson, 1893 (cited by Dawson *et al.*, 1972)
man	1.376	2	Abbe	Matthieson, 1891 (cited by Gullstrand, 1909)

AQUATIC AND TERRESTRIAL EYES

Differences in vision in air and water have resulted in the evolution of a terrestrial eye which differs morphologically and physiologically from its aquatic precursor. For example, the fish eye has a spherical lens of high refractive index while the lens of the human eye is elliptical in shape, softer in consistency and of lower refractive index (Walls, 1942). Accommodation in the former is achieved by movement of the lens through contraction of an ectodermal lens muscle (Beer, 1894). The mesodermal ciliary muscle of the human eye is responsible for producing accommodative changes in lens shape. While the fish eye is abnormally myopic in air, humans are very hyperopic in water (unless a face mask is used to restore the cornea-air interface).

METHODS FOR ACHIEVING AMPHIBIOUS VISION

Researchers have long been fascinated with the study of possible mechanisms by which amphibious vertebrates of all classes may maintain adequate vision in air and water. The following list represents an attempt to classify the methods which have been reported in the literature. It is followed by more detailed descriptions which are grouped according to vertebrate class.

1. Flat corneal facets or flattened corneas - The cornea has little or no refractive power in air or water.

2. Accommodation - The refractive power of the crystalline lens is increased sufficiently to overcome the refractive loss of the cornea when the eye is in water.

3. Stenopeic slit pupil - Large depth of focus which maintains adequate image quality in air or water.

4. Nictitating membrane - A window of high refractive index in the nictitating membrane compensates for the refractive loss of the cornea when the eye is in water.

5. Multiple optic paths - Refractive media are appropriate for vision in air along one path and vision in water along another.

FISHES

Flat Corneal Facets

Baylor (1967) reported that the cornea of the eye of the Atlantic flying fish (*Cypselurus heterurus*) consists of three relatively flat triangular facets with apices that meet at its centre (Fig. 1). The flattened nature of the facets was observed

Cypselurus heterurus

Dialommus fuscus

Figure 1   The appearance of faceted corneas in two fish species. [After Baylor, 1967 and Munk, 1969].

with a keratoscope and confirmed by refractive findings (through the facets) of emmetropia in air and slight hyperopia in water. Munk (1969) suggested that the two corneal facets of *Dialommus fuscus* (a clinid) serve the same function; although some doubt existed due to the effects of preservation on corneal shape (Fig. 1). Two flat corneal facets were observed by Graham and Rosenblatt (1970) in a fresh specimen of another clinid (*Mnierpes macrocephalus*) and it appears that Munk's suggestion is correct. The natural histories of all three species indicate a need for acute vision in air as well as water.

Multiple Optic Paths

Anableps anableps (a cyrinodontid) is probably the best known example of a vertebrate with an amphibious visual capability. Anableps spends much of its time swimming at the water surface with the upper half of each eye exposed to air (Klinckowström, 1895; Schneider-v. Orelli, 1907). A horizontal stripe divides the cornea at the water line while the pupil is divided by extensions stemming from the nasal and temporal iris margins. The lens is egg shaped; the long axis corresponds to the ventral pupil axis while its short axis corresponds to the dorsal pupil axis (Fig. 2). The small difference in refractive states measured through the two pupils (with the dorsal cornea in air and the ventral one in water)

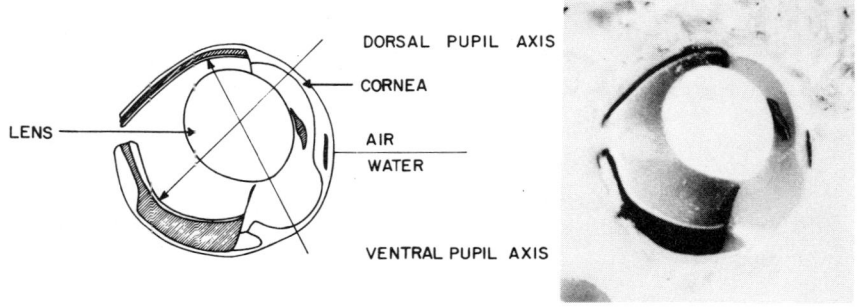

Figure 2 Drawing and frozen section of the eye of *Anableps anableps* indicating the directions of aerial and aquatic optic paths. [after Sivak, 1976].

indicates that the variation in lens shape compensates for the presence or absence of a corneal refractive contribution (Sivak, 1976a).

While not truly amphibious visually, two additional fishes have been placed in this category (Walls, 1942; Protasov, 1968). The archer fish (*Toxotes jaculatrix*) lives in calm brackish estuaries and feeds by spouting a jet of water at an overhead insect (Lüling, 1958). Since the eyes do not leave the water the fish must take into account the refractive effect of the air-water interface. It seems that the fish's accuracy develops gradually by learning. If it is considered that numerous terrestrial vertebrates hunt for aquatic prey by learning to overcome a similar refractive problem, the ability of the archer fish doesn't appear to be all that remarkable.

The mudskipper (*Periophthalmus*) clings to seashore rocks or tidal flats with its dorsally positioned eyes exposed to air (Walls, 1942). The eyes of this fish are adapted for aerial vision. For example, the lens is elliptical, not spherical, in shape. However, since no evidence exists concerning an ability to compensate for the refractive change associated with immersion of the eyes in water, acute amphibious vision cannot definitely be said to exist.

AMPHIBIANS AND REPTILES

While the amphibians represent the evolutionary transition point between aquatic and terrestrial ways of life, no evidence of adaptations for amphibious vision have been reported (Walls, 1942). Du Pont and de Groot (1976) suggest that the nictitating membrane of *Rana esculenta* may produce a pre-corneal air space when the eye is in water. However, this possibility is highly speculative (Rivamonte, 1977). It appears that amphibian eyes are either aquatic or terrestrial in form and function, depending on the predominating environment.

Accommodation

Studies of the mechanism of accommodation in reptiles carried out some time ago (Beer, 1898; cited by Walls, 1942) indicate a very large accommodative ability in certain aquatic turtles (*Emys*) and aquatic snakes (*Natrix tesselatus*). In both instances the lens deforming effect of a large ciliary muscle is supplemented by the effect of the iris sphincter. The sphincter is larger than normal and the lens is very soft in consistency. Contraction of the sphincter reduces the radius of curvature of the anterior lens surface (Fig. 3). This mechanism (it reappears in avian and mammalian forms) is interpreted by Walls to indicate that the refractive loss of the cornea when the eye is in water is made up by an accommodative increase in focal power.

There has been practically no follow up to Beer's work. However, a large accommodative ability was noted recently in freshwater turtles (*Pseudemys scripto elegans*) and it is believed that the above mechanism is responsible (Northmore, personal communication). Moreover, the findings of a psychophysical study (Dudziak, 1955) indicate equal visual acuity in air and water for a turtle (*Emys orbicularis*). Thus it appears that Walls' interpretation is correct, at least as far as turtles are concerned.

BIRDS

The extent of amphibious adaptations among aquatic birds varies from those which such adaptations are marked to those in which they are muted. Walls (1942) predicted that species which rely exclusively on the aquatic environment for such behaviour as feeding would have eyes which were adapted for underwater vision while those with minimal need for acute underwater vision would be strictly aerial visually. Adaptations for vision in air and water would be found in birds of neither extreme.

Flattened Cornea

Penguins are considered to be completely aquatic birds

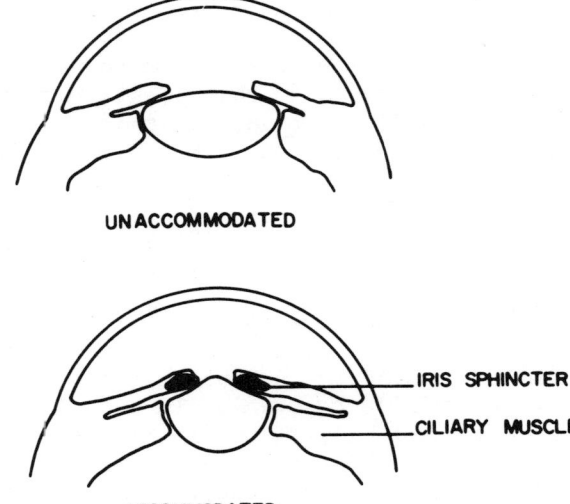

Figure 3 An accommodative mechanism involving both the ciliary muscle and the iris sphincter. [After Walls, 1942].

(Kooyman, 1975). This fact together with early popular reports (e.g. Murrary, 1910) led to the conclusion that penguins were emmetropic underwater and extremely myopic on land (Walls, 1942; Duke-Elder, 1958). However, refractive state measurements in air and simulated underwater conditions reveal emmetropia or near emmetropia in air and moderate hyperopia (8-21 dioptres) in water (Sivak, 1976b; Sivak and Millodot, 1977). The photokeratoscopic study of corneal curvature indicates that the refractive effect of movement from air to water is minimized by the existence of a relatively flat cornea (Fig. 4). Chromatic aberration, as measured by retinoscopy performed through coloured filters, will not substantially compensate for the refractive loss of the cornea even if the spectral quality of the aquatic habitat is restricted to the blue end of the visible spectrum. Possibly an accommodative ability of moderate magnitude would account for the remaining hyperopia (Sivak and Millodot, 1977).

Accommodation

Cormorants (*Phalacrocorax*, Hess, 1912; cited by Duke-Elder, 1958) and dippers (*Cinclus mexicanus*), (Goodge, 1960) can apparently compensate for the refractive loss of the cornea when the bird is in water by means of a very large accommodative ability. As noted above for aquatic turtles and snakes, focal changes of this magnitude (50 dioptres) are produced by contraction of the iris sphincter as well as the ciliary muscle.

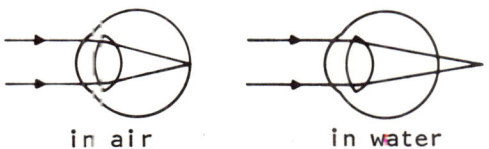
A TERRESTRIAL EYE

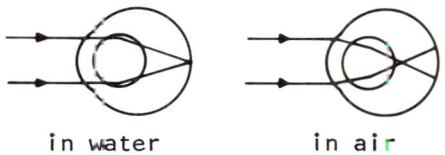
AN AQUATIC EYE

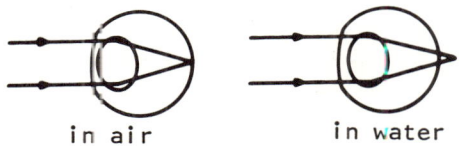
PENGUIN EYE

Figure 4   A schematic representation of the effect of a flattened cornea on aerial and aquatic refraction. [After Sivak and Millodot, 1977; with permission of Springer Verlag].

A recent study compared the refractive states of a cormorant (*Phalacrocorax auritus*) with that of a pelican (*Pelecanus occidentalis*) in air and water (Sivak et al.). While both species are emmetropic in air, the cormorant becomes only slightly hyperopic in water while the pelican is hyperopic by an amount relatively close to the refractive power of the cornea in air. As in the case of penguins, the chromatic aberration of the eye is insufficient to significantly affect refractive state, even in a blue environment. In the absence of an alternative explanation, it would appear that the accommodative mechanism attributed to Hess is confirmed. The difference found between cormorants and pelicans is consistent with descriptions of their feeding habits. While the pelican plunge dives for fish, on a hit or miss basis, cormorants actively pursue their prey while underwater (Ashmole, 1971).

Nictitating Membrane

In a series of articles, Ischreyt described the gross and microscopic appearance of the eyes of a number of diving birds, especially diving ducks (1912a, b, c, 1914). Differences in the morphology of the nictitating membrane of diving and non-diving ducks were reported (Ischreyt, 1912a). The nictitating membrane in non-diving ducks lacks transparency. It functions as a device to clean and moisten the cornea. However, a central transparent window exists in the membrane of diving ducks (e.g. *M. merganser*) and loons (*Gavia lumme*). Ischreyt believed that the membrane is pulled over the eye as a protective goggle when the birds dive into water. He suggested that the membrane might have an elevated refractive index which would help compensate for the neutralizing effect of water on the cornea.

Reviewers have accepted Ischreyt's suggestion. It is reported that loons, auks and diving ducks maintain acute vision underwater through the refractive effect of the nictitating membrane in association with moderate amounts of accommodation (Walls, 1942; Duke-Elder, 1958). In fact the existence of this mechanism is highly unlikely. Preliminary study indicates that the refractive indices of all avian nictitating membranes are approximately equal to that of the cornea (Sivak, unpublished). Furthermore, because of its thin membranous nature, both surfaces of the nictitating membrane probably assume the curvature of the cornea when the membrane is in place in front of the eye. In this situation, the refractive index of the membrane is unimportant since the positive refractive effect of the first surface (assuming its refractive index is significantly greater than that of water) is neutralized by the negative effect of the second surface.

Multiple Optic Paths

As in the case of diving ducks, the mechanism for amphibious vision described for the kingfisher (*Alcedo attis attis*) is based on circumstantial evidence. Kolmer (1924) made the following observations concerning the kingfisher eye: Two foveas exist; nasal and temporal. The lens together with its annular pad is oval in shape, the long axis being nearly parallel to the equator of the eye and in a direction coincident with the temporal fovea. Finally, the temporal ciliary body is considerably thicker than the nasal portion. Kolmer believed that the central fovea is used in air and the temporal one in water. The refractive loss of the cornea in water is made up by the increased refractive power of the lens in the direction of its long axis as well as by movement of the lens (via the ciliary muscle) away from the temporal retina. Further study is required to verify the existence of this mechanism.

MAMMALS

A multiplicity of reports on air-water vision of mammals have appeared in recent years. They are characterized by the attention paid to the coupling of optical study with behavioral evidence and by controversy arising from the variety of methods used to approach the problem. With rare exception (e.g. a preliminary study of polar bear visual optics which indicated a lack of amphibious adaptation; Sivak and Piggins, 1975) the animals studied have been otters (mustelids), seals and sea lions (pinnipeds) or dolphins (cetaceans).

Flattened Cornea and Stenopeic Pupil

Aerial refractive state measurements of the eyes of the harbour seal (*Phoca vitulina*), harp seal (*Pagophilus groenlandicus*) and the Weddell seal (*Leptonychotes weddelli*) indicate a high degree of astigmatism, the axis of which is vertical (Johnson, 1893; Piggins, 1970; Wilson, 1970; Jamieson, 1971). That this astigmatism is due to the shape of the cornea is confirmed by the fact that virtually all of it disappears when the eye is underwater (Piggins, 1970). The observation of a round pupil when the eye is underwater and a vertical slit pupil when it is in air (Johnson, 1893) has led to the following inference regarding amphibious vision (Walls, 1942): The seal eye is optically adapted for underwater vision (as indicated by the existence of a nearly spherical lens). The myopia introduced by exposure to air is minimized by way of a slit pupil which restricts the light rays entering the eye to those refracted by the flattest corneal meridian. The constricted pupil also acts as a stenopeic slit and increases depth of focus. Streamlining of the eye in an aquatic medium is the explanation offered by Jamieson (1971) as to why the cornea of pinnipeds (and cetaceans) are astigmatic in the first place.

The above inference is apparently supported by visual acuity measurements for the sea lion (*Zalophus californianus*); Schusterman and Balliet, 1970; Schusterman, 1972). While aerial and underwater visual acuities were equal in bright light, underwater was superior in dim light conditions. The latter finding was presumed to be due to the deleterious effect of corneal astigmatism and a dilated pupil in air. However, Jamieson and Fisher (1970) reported equal aerial and aquatic visual discrimination ability in the harbour seal (at 1.7M) even though light levels were relatively low and the pupil was oval in shape. Lavigne and Ronald (1972) noted that the pupil of the harp seal varied in shape and size with the level of ambient light independent of whether the seal was in air or water.

The pinniped accommodative ability has not been the subject of a specific study and the role of accommodation in the amphibious behaviour of these animals is uncertain. It is important to note that refractive studies of Johnson and Piggins were carried out while the eye was under the influence of atropine, a cycloplegic, while accommodative activity was not inhibited during behavioral studies of resolution ability.

Accommodation

According to several authors (Walls, 1942; Duke-Elder, 1958; Prince, 1956) otters overcome the effect of water on a terrestrial eye by an accommodative mechanism similar to that reported for turtles and cormorants; i.e. by the combined effects of the ciliary muscle and iris sphincter on the curvature of the lens. To place this adaptation in perspective it is necessary to note that the mammalian ciliary apparatus does not usually reach the high level of accommodative development common to avian and lacertilian eyes.

Behavioral studies carried out with *Amblonyx cineria cineria* support the above description (Balliet and Schusterman, 1971; Schusterman and Barrett, 1973). Aerial and aquatic visual acuities are approximately equal in relatively bright light while aerial acuity is superior to aquatic acuity at low luminance levels. The latter finding is attributed either to insufficient pupil dilation resulting from the accommodative function of the iris sphincter or to insufficient accommodative function of the iris sphincter resulting from pupil dilation.

Observations of dolphin behavior at marine exhibits as well as behavioural studies of visual acuity (Pepper and Simmons, 1973; Herman *et al.*, 1975) have led to the widespread belief that the animal is capable of seeing well in air and water (Dawson *et al.*, Dral, 1972, 1974, 1975; Herman *et al.*, 1975; Rivamonte, 1976). However, there is little or no agreement concerning a mechanism for air-water vision.

Dral (1972, 1974, 1975) measured the refractive state of *T. truncatus* in air and water with an ophthalmoscope. Myopia and emmetropia were found along the pupil axis in air and in simulated underwater conditions respectively. Moreover, the cornea was considerably astigmatic in the same direction. Along a ventral and rostral direction (apparently the axis of fixation for aerial viewing; Dawson *et al.*, 1972) Dral reported emmetropia in air and water and little or no corneal astigmatism. Dral attributes the persistence of emmetropia along this direction to an accommodative mechanism. Since histological study (Dral, 1975) reveals the absence of a ciliary muscle (as well as the apparent absence of an ectodermal muscle capable of moving the lens) Dral suggested that the operculum, an expandable flap of the dorsal iris, may be capable of altering the position or form of the lens. More recently Dral (personal communication) has expressed support for a mechanism involving multiple optic paths which is described below.

Stenopeic Slit

According to Dawson et al. (1972) *T. truncatus* is myopic in air by an amount similar to that found by Dral (approx. 17 dioptres), and moderately hyperopic in water. No obvious corneal astigmatism was noted. While it is believed that an accommodative ability exists it is not considered large enough to account for the refractive effect of movement from water to air. Instead, these researchers point to the possibility that pupil constriction under daylight conditions produces a stenopeic aperture with a depth of focus capable of overcoming this problem. Herman et al. (1975) note that under bright illumination the operculum overlaps the ventral iris margin leaving two vertical slits nasally and temporally. They attribute the approximately equal aerial and aquatic visual acuity of *T. truncatus* at 2.5 m as well as the finding of diminished aerial acuity and enhanced aquatic at near to the optics of double stenopeic slits.

Multiple Optic Paths

In a comprehensive review of dolphin vision research Rivamonte (1976) concluded that neither accommodation nor stenopiec slits are likely explanations for the ability of the dolphin to see well in air and water. An accommodative mechanism was ruled out on the basis of the lack of definite anatomical or physiological evidence as well as the finding that dolphin acuity varies with viewing distance. Objections to the theory of stenopeic slits include; the measurement of emmetropia in air along a rostral-ventral direction when the pupil is not completely constricted (Dral, 1972), behavioural evidence of good visual acuity at low light levels (Madsen, 1972) and the diffraction limitations of the eye.

Rivamonte has developed a schematic eye for the dolphin which includes a lens which is over-corrected for spherical aberration. The refractive index of the lens cortex is less than that of its core by an amount which cancels the refractive contribution of the cornea in air. Thus the eye is emmetropic along the axial viewing direction in water and for an oblique viewing direction in air. Although experimental confirmation is required, this mechanism is an elegant way to explain the observed equality of aerial and aquatic acuity without invoking the objections mentioned above.

CONCLUSION

The present state of knowledge in this area precludes the choice of one of the above strategies as being superior to the others. A major disadvantage is apparent in each case. Corneal facets or flattened corneas are detrimental to the field of vision. In this context it is noted that penguins are reported to be among the few vertebrates having no binocular field of vision (Duke-Elder, 1958). An accommodative mechanism (if the iris sphincter is involved) and the use of stenopeic apertures are restricted to periods of elevated light levels. The existence of a mechanism involving the nictitating membrane is very doubtful. In the case of multiple optic paths it would appear that the field of vision would be reduced in each environment for in effect part of the eye is used in air and a different portion in water. Schwassmann and Kruger (1965) demonstrated that the division of the retina of *A. microlepis* into aerial and aquatic portions is projected to the optic tectum.

It is evident from the above descriptions that despite the interest in strategies for amphibious vision, considerable additional study is required. Too often assumptions are made which are based purely on morphology. The lack of behavioural investigations (mammals excepted) is particularly glaring.

REFERENCES

Ashmole, N.P. (1971). Seabird ecology and the marine environment. In Avian Biology, Farner, D.S., King, J.R. and Parkes, K.C., pp. 223-285, Academic Press, New York.
Aurell, G. and Holmgren, H. (1953). On the metachromic staining of the corneal tissue and some observations on its transparency. Acta Ophthalmol. 31: 1-27.
Balliet, R.F. and Schusterman, R.J. (1971). Underwater and aerial visual acuity in the Asian "Clawless" Otter (*Amblonyx cineria cineria*). Nature 234: 305-306.
Baylor, E.R. (1967). Air and water vision of the Atlantic Flying Fish, *Cypselurus heterurus*. Nature 214: 307-309.
Beer, T. (1894). Die Accommodation des Fishauges. Pflügers Arch. Physiol. 58: 523-650.
Beer, T. (1898). Die Accommodation des Auges bei den Reptilien. Pflügers Arch. Physiol. 69: 507-568.
Charman, W.N. and Tucker, H. (1973). The optical system of the goldfish eye. Vision Res. 13: 1-8.
Citron, M.C. and Pinto, L.H. (1973). Retinal image: Larger and more illuminous for nocturnal than for diurnal lizard. Vision Res. 13: 873-876.
Dawson, W.W., Birndorf, L.A. and Perez, J.M. (1972). Gross anatomy

and optics of the dolphin eye (*Tursiops truncatus*). Cetology: 19, 1-12.
Dral, A.D.G. (1972). Aquatic and aerial vision in the bottle-nosed dolphin. Neth. J. Sea Res. 5: 510-513.
Dral, A.D.G. (1974). Problems in image-focusing and astigmatism in Cetacea. J. aquat. Mammals 2: 22-28.
Dral, A.D.G. (1975). Vision in Cetacea. J. Zoo Animal Med. 6: 17-21.
Dudziak, J. (1955). Ostrosc widzenia u zolwia blotnego (*Emys orbicularis* L.). (Visual acuity in the *Emys orbicularis* L. tortoise in air and water). Folia Biol. (Krakow) 3: 205-227.
Duke-Elder, S. (1958). System of Ophthalmology. Vol. I. The eye in evolution. Hemry Kimpton, London.
DuPont, Y.S. and de Groot, P.J. (1976). A schematic dioptric apparatus for the frog's eye (Rana esculenta). Vision Res. 16:803-810.
Goodge, W.R. (1960). Adaptations for amphibious vision in the Dipper (*Cinclus mexicanus*). J. Morph. 107: 79-91.
Graham, J.B. and Rosenblatt, R.H. (1970). Aerial vision: unique adaptation in an intertidal fish. Science 168: 586-597.
Gundlach, R.H., Chard, R.D. and Skahen, J.R. (1945). The mechanism of accommodation in pigeons. J. Comp. Physiol. Psychol. 38: 27-42.
Herman, L.M., Peacock, M.F., Yunker, M.P. and Madsen, C.J. (1975). Bottlenosed dolphin: double slit pupil yields equivalent aerial and underwater diurnal acuity. Science 189: 650-652.
Von Hess, C. (1912). Vergleichende Physiologie des Gesichtssinnes. In: Handbuch der Vergleichende Physiologie, Winterstein, H. Bd. 4:pp. 1-290, Gustav Fisher, Jena. Cited by Duke-Elder, 1958.
Hughes, A. (1972). A schematic eye for the rabbit. Vision Res. 12: 123-128.
Ischreyt, G. (1912a). Zur vergleichenden Morphologie des Entenauges. Arch. f. vergl. Ophthalmol. 3: 39-76.
Ischreyt, G. (1912b). Zur vergleichenden Morphologie des Entenauges. Zweiter Beitrag. Arch. f. vergl. Ophthalmol. 3: 369-379.
Ischreyt, G. (1912c). Zur Morphologie des Auges der Urinatores (Taucher). Arch. f. vergl. Ophthalmol. 3: 380-394.
Ischreyt, G. (1914). Zur vergleichenden Morphologie des Entenauges. Dritter Beitrag. Arch. f. vergl. Ophthalmol. 4: 162-181.
Jamieson, G.S. (1971). The functional significance of corneal distortion in marine mammals. Can. J. Zool. 49: 421-423.
Jamieson, G.S. and Fisher, H.D. (1970). Visual discriminations in the harbour seal *Phoca vitulina*, above and below water. Vision Res. 10: 1175-1180.
Jerlov, N.G. (1968). Optical Oceanography. Elsevier, New York.
Johnson, G.L. (1893). Observations on the refraction and vision of the seal's eye. Proc. Zool. Soc. Lond. 719-723.

Klinckoström, A. (1895). Beitrage zur Kenntniss der Augen von *Anableps tetrophthalmus*, Skand. Arch. Physiol. 5: 67-69.

Kolmer, W. (1924). Über das auge des Eisvogels (*Alcedo attis attis*). Pflügers Arch. Physiol. 204: 266-274.

Kooyman, G.L. (1975). Behavior and physiology of diving. In: The Biology of Penguins, Stonehouse, B., pp. 115-137. University Park Press, Baltimore.

Lavigne, D.M. and Ronald, K. (1972). The harp seal, *Pagophilus groenlandicus* (Erxleben 1777). XXIII. Spectral sensitivity. Can. J. Zool. 50: 1197-1206.

Lüling, K. (1958). Morpho-anatomische und histologische Untersuchungen an Auge Toxotes jaculatrix (Palls.). Z. Morph. Oekol. Tiere, 47: 529-610.

Madsen, C.J. (1972). Visual acuity in the bottle nose dolphin, *Tursiops truncatus* (Montagu, 1821). M.Sc. dissertation, McGill University.

Massof, R.W. and Chang, F.W. (1972). A revision of the rat schematic eye. Vision Res. 12: 793-796.

Matthiessen, L. (1880). Untersuchungen über dem Aplanatismus und die Periscopie der Kristallinsen in den Augen der Fische. Pflügers Arch. Physiol. 21: 287-307. Cited by Walls, G.L. (1942). The Vertebrate Eye and its Adaptive Radiation. Cranbrook Institute of Science, Bloomfield Hills, Michigan.

Matthiessen, L. (1891). Die nuren Fortschritte in unserer Kenntnis von dem optischen Baue des Auges der Wirbeltiere. Hamburg. Cited by Gullstrand, A. (1909). In: Handbuch der Physiologischen Optik. Edited by von Helmholtz, Voss, Hamburg.

Matthiessen, L. (1893). Über den physkaisch - optischen bau der augonvom knowal und finwal. Zeitsch. f. Verl. Augenheilk. 7: 94. Cited by Dawson, W.S., Birndorf, L.A. and Perez, J.M. (1972).

Murray, J. (1910). Some notes by James Murrary, biologist to the expedition. In: Shakelton, S.E. The Heart of the Antarctic. pp. 345-359. Muson, Toronto.

Munk, O. (1969). The eye of the "four-eyed" fish *Dialommus fuscus* (Pisces, Blenniodei, Clinidae). Vidensk. Medd. Dan. naturhist. Foren. 132: 7-24.

Pepper, R.L. and Simmons, J.V. (1973). In-air visual acuity of the bottlenose dolphin. Exp. Neurol. 271-276.

Piggins, D.J. (1970). Refraction of the Harp Seal, *Pagophilus groenlandicus* (Erxleben 1777). Nature 227: 78-79.

Prince, J.H. (1956). Comparative Anatomy of the Eye. Thomas, Springfield, Illinois.

Protasov, V.R. (1968). Vision and Near Orientation of Fish. Acad. Sci. USSR. Israel Program for Scientific Translations, 1970.

Rivamonte, A. (1976). Eye model to account for comparable aerial and underwater acuities of the bottlenose dolphin. Neth. J. Sea Res. 10: 491-498.

Rivamonte, A. (1977). The under-corrected lens of the frog eye (*Rana esculenta*) could yield comparable aerial and underwater vision. Vision Res. (in press).
Schneider-V. Oreilli, M. (1907). Untersuchaugan über das Auge von *Anableps tetrophtnalmus*. Mitt. Naturforsch. Ges. Bern, 1629-1664, 87-113.
Schusterman, R.J. (1972). Behavior of Marine Animals, In: Vertebrates V(II), Winn, H.E. and Olla, B.L., Plenum, New York.
Schusterman, R.J. and Balliet, R.F. (1970). Conditioned vocalizations as a technique for determining visual acuity thresholds in sea lions. Science 169: 498-500.
Schusterman, R.J. and Barret, B. (1973). Amphibious nature of visual acuity in the Asian "Clawless" Otter. Nature 244: 518-519.
Schwassman, H.O. and Kruger, L. (1965). Experimental analysis of the visual system of the four-eyed fish *Anableps microlepis*. Vision Res. 5: 269-281.
Sivak, J.G. (1974). The refractive error of the fish eye. Vision Res. 14: 209-213.
Sivak, J.G. (1976a). Optics of the eye of the "four-eyed fish" (*Anableps anableps*). Vision Res. 16: 531-534.
Sivak, J.G. (1976b). The role of a flat cornea in the amphibious behavior of the blackfoot penguin (*Spheniscus demersus*). Can. J. Zool. 54: 1341-1345.
Sivak, J.G. (1976c). Refraction and accommodation of the elasmobranch eye. In: Sensory Biology of Elasmobranchs. Hodgson, E.S. and Mathewson, R.W. U.S. Government Printing Office, (in press).
Sivak, J.G. (1977). The role of the spectacle in the visual optics of the snake eye. Vision Res. 17: 293-298.
Sivak, J.G. and Piggins, D.J. (1975). Refractive state of the eye of the Polar Bear (*Thalarctos maritimus* Phipps). Norw. J. Zool. 23: 89-91.
Sivak, J.G., Lincer, J.L. and Bobier, W.(1977). Amphibious visual optics of the eyes of the double-crested cormorant (*Phalacrocorax auritus*) and the brown pelican (*Pelecanus occidentalis*). Can. J. Zool. 55: 782-788.
Sivak, J.G. and Millodot, M. (1977). Optical performance of the penguin eye in air and water. J. Comp. Physiol. 119: 241-247.
Tansley, K. (1965). Vision in Vertebrates. Chapman and Hill, London.
Vakkur, G.J. and Bishop, P.O. (1963). The schematic eye of the cat. Vision Res. 3: 357-381.
Walls, G.L. (1942). The Vertebrate Eye and its Adaptive Radiation. Cranbrook Institute of Science, Bloomfield Hills, Michigan.
Wilson, G. (1970). Some comments on the optical system of pinnipedia as a result of observations on the Weddell seal (*Leptonychotes weddelli*). Br. Antarct. Surv. Bull. 23, 57-62.

ECOLOGICAL ASPECTS OF ELECTRORECEPTION

H. O. Schwassmann

Department of Zoology, University of Florida

Gainesville, Florida, 32611, U.S.A.

Proper placement of electroreception as an ecosensory function within the array of the different sensory modalities seems difficult; however, some close relationship with vision can be established. Among the vertebrates, visual receptors and the many accessory dioptric and related structures are most highly developed in birds and fishes, but for obviously different reasons. In day-active birds, an eye of large diametre and a high ratio of retinal ganglion cells to cones throughout the retina permit a high degree of resolving power, perhaps at the cost of sensitivity, while the trend prevailing in fishes is towards maximal utilization of the little light available in most aquatic habitats. Here, correlated adaptive features are an enormous pupillary aperture, a relatively large spherical lens, and, in some deep-sea forms, a telescopic eye reminiscent of the tubular eyes of night-active owls. In certain situations, as an existence in very turbid waters, or while changing to a night-active life style, further refinement of the visual apparatus must have proved uneconomical, and other sensory systems were relied upon. Of these, electroreception became a highly effective mechanism for near-field orientation and communication in the aquatic medium.

True electroreceptors are found in different groups of fishes. These secondary sensory cells are related to the neuromasts, lateral line mechanoreceptors; the innervation of the several sensory cells making up one electroreceptor organ is by fibres of the acoustico-lateralis, the eighth cranial nerve. In the "active" or electrogenic fishes, those having electric organs, the object-detecting function shows great similarities to the echo-locating orientating mechanism of bats which emit patterned and modulated ultra-sound bursts. Bats, like the electrosensory fish, are capable of detecting size, distance, and relative movement of objects by specific alterations of the stimulus energy characteristics. Both these large groups, the fish equipped with an active electrosensory system and the echo-locating bats, are producing the stimulus energy themselves in the form of characteristic pulses of specific frequency content.

The specialized electric organs which provide this stimulus energy are found in several groups of distinct and unrelated fishes. The electric organs are thought to have evolved independently in these groups because of their derivation from different muscles, or even nerves, in various body regions. It also appears as if natural selection has shaped the detailed arrangement of the modified electrogenic units into a variety of parallel and serial patterns which are adapted to specific resistivities of fresh water and ocean water, a process that could be considered an example of convergent parallel evolution.

That some fish could produce electric shocks was known to several ancient civilizations. The Egyptians pictured the electric catfish *Malapterurus* in connection with their symbol for lightning. Several reports, as well as archaeological data, indicate Indian tribes of Amazonia to have been familiar with the shocking powers of the puraqué (*Electrophorus*) and to have used these fish with therapeutic intent (Kellaway, 1946).

Of relatively recent recognition are the weak electric organ discharges in two large freshwater fish groups, the gymnotoidea of South America and the mormyridae and *Gymnarchus* of Africa. Demonstration of their electrogenic abilities became possible with the advent of sensitive electronic instruments, like the oscilloscope. These electrogenic properties of the "weakly" electric fishes are of special importance since they form an essential part of a highly effective "active" electrosensory system for electrolocation, orientation, and social communication.

These last two groups of fishes, one in South America and the other in Africa, are phylogenetically far apart, and any resemblance, especially between *Gymnarchus* and the gymnotoidea, must be due to parallel adaptations to a new life style utilizing the electrogenic and electrosensory abilities. Gymnotoids are considered the most highly modified characoidei of ostariophysan stock, while the African mormyridae are not related to any African characins but are placed with the osteoglossiformes (Greenwood *et al.*, 1966). It should, however, be noted that electroreception and electrogenesis may not be as recent an innovation as is suggested by its presence in the rather modern groups mentioned above. The conspicuous dorsomedial and lateral plates on the bony shield of the cephalaspids, belonging to the osteostraci, were suspected to have been electric organs (Stensiö, 1927). More recently, the lateral fields in these primitive extinct fishes were referred to as "electrosensory"

(Romer, 1966). Whatever their real significance, there is a striking resemblance between the thick presumed nerve trunks connecting the lateral plates to brain placodes and the motor nerve supply to the similarly located lateral electric organs in the torpedinid marine rays (Bennett, 1971). This similarity would make the suspected electrogenic nature of the lateral fields in the cephalaspids a more likely possibility.

Development of electroreceptors and of electric organs did not always occur side by side. Some fish are electrosensory but lack electric organs (Table I, a). In the two large groups of "active" electric fishes (b_1 and b_2 in Table I) differentiation of electrosensitive "tonic" receptors could have preceded the development of electric organs, but the appearance of "phasic" electroreceptors must have been paralleled by the modification of certain muscle groups, or spinal nerves in the Apteronotidae, into electric organs. Increasingly high water turbidity, precluding effective use of vision must have been the underlying selective pressure in this development (Lissmann, 1958). Possession of an effective electrosensory system for orientation purposes then made possible shifting of the time of activity into the night. The trend towards a nocturnal activity pattern could be considered a later adaptation at a time when many waters, with the exception of the major rivers, became clear to the advantage of day-active visually orienting predators.

The list of electric and electrosensory fishes in Table I separates those that are considered "passive" (non-electrogenic) from the "active" (electrogenic) species. The so-called "weakly electric" freshwater fishes, the gymnotoids of South America and the African mormyrids and *Gymnarchus*, are found in the subcategories b_1 and b_2.

The functions of electric organs which are mostly, but not always, encountered together with specialized electroreceptors can be summarized as follows: 1) electroparalysis - for defense or feeding purposes; 2) electrolocation - for orientation, escape, feeding purposes (this mode is found in all weakly-electric fish, with *Electrophorus* being capable of delivering strong electric shocks); and 3) electrocommunication - for purposes of social interactions within populations, especially during territorial and mating behaviour, perhaps also interspecifically.

Regarding the development of ideas and actual studies that led to our present knowledge of electrosensory function in fishes, Lissmann must be mentioned as one of the early pioneers who not only demonstrated the great sensitivity of *Gymnarchus* in detecting impedance changes (Lissmann and Machin, 1958; Machin and Lissmann, 1960) but who also postulated the existence of electroreceptor organs in these fishes as well as their derivation from lateral line mechanoreceptors (Lissmann, 1958). A few of the early extremely low threshold values may have been computed somewhat erroneously,

TABLE I. List of fishes with electrosensory ability (fishes that utilize electrical stimulus energy and possess specialized electroreceptors).

a) non-electrogenic (no electric organs - with ampullary electroreceptors)
   a_1) marine: many sharks and rays
   a_2) freshwater: several catfishes, freshwater rays, dipnoi

b) electrogenic (with electric organs, tuberous and ampullary receptors)
   b_1) wave species (hummers, or tone species, with approximately sinusoidal discharges at relatively steady, high frequency, 250 to 1,700 Hz -- but note: *Sternopygus* with 60 to 130 Hz --: *Gymnarchus*, many *spp.* of Apteronotidae, also Sternopygidae (*Sternopygus* and *Eigenmannia*)
   b_2) pulse species (buzzers, discharges as brief pulses separated by long intervals, repetition rates usually low and variable): Mormyridae (many *spp.*), Gymnotidae (*Gymnotus spp.*), Rhamphichthyidae (*Rhamphichthys, Gymnorhamphichthys, Hypopomus spp., Hypopygus, Steatogenys spp.*). Some Rajidae (marine skates) with very irreglar and weak discharges can be included in this group.
   b_3) pulses; also additional powerful shocks: *Electrophorus, Torpedo, Narcine*
   b_4) powerful shocks only; electroreceptors present: *Malapterurus*
   b_5) powerful shocks only; electroreceptors not known: *Astroscopus*

but the great majority of determinations indicating very low thresholds have been confirmed (Kalmijn, 1974).

Hagiwara and his colleagues succeeded in demonstrating part of the mechanism involved in detecting nearby objects, employing nerve fiber recordings in several species of gymnotoids (Bullock *et al.*, 1961; Hagiwara *et al.*, 1962; Hagiwara and Morita, 1963). A sensitivity of the catfish *Ameiurus* to minute electric currents had been known for a long time (Parker and Heusen, 1917) when Murray (1960, 1962) showed electrophysiological evidence that the ampullae of Lorenzini of elasmobranchs responded to weak electric fields. Kalmijn (1971) was able to measure the extremely low thresholds involved and could show in behaviour experiments the role of electroreception for prey localization in sharks and rays.

The last fifteen years have witnessed a rapid succession of discoveries concerning morphology and function of electroreceptors, their coding properties, and the central nervous pathways and mechanisms involved. A comprehensive review of the mostly electrophysiological work has appeared recently (Fessard, ed., 1974).

Electroreception is a unique sensory modality that is based on a particular sensitivity of modified "neuromast" cells to electric fields. These modified cells make up the electroreceptor organs which detect temporal changes in voltage gradients nearby. Central mechanisms evaluate the coded pattern of temporally changing intensity vectors acting on neighboring receptor organs. The noted high sensitivity seems due to a match between high skin resistance and low receptor membrane resistance; the effective stimulus is produced by a great drop in voltage across the high skin resistance. Very small alterations in the receptor cell s transmembrane potential, due to smaller than 1.0 μv/cm voltage gradients, can result in the release of transmitter substance. This last process activates the afferent sensory nerve fiber.

With respect to the afferent pathway, the possible roles of particular coding types for electrocommunication and localization are discussed in detail by Scheich and Bullock (1974); a simplified scheme of the different modes of coding is shown in Figure 1. With one possible exception, Type I, all receptor organs respond to amplitude changes of the voltage gradient. Ampullary as well as tuberous receptor organs are found in the skin of the gymnotid eels. The ampullary receptors, formerly called "tonic" because of their firing rate dependence on stimulus level (Bullock and Chichibu, 1965), respond to direct current and pulses of low frequency content at low repetition rates. They seem to be at least functionally related to the ampullae of Lorenzini and are also found in siluroids and lungfish. The tuberous receptors used to be referred to as "phasic" since they exhibit transient responses to brief changes in voltage. They are activated by high frequency components of pulses, or waves, of usually high repetition rates and depend on the fish's own organ pulses, or on pulses emitted by other electrogenic fish nearby. Details regarding the coding properties are mentioned in the legend to Figure 1.

In contrast to the many discoveries concerning sensory and central nervous system mechanisms, there exist many conspicuous gaps in our knowledge of life history, geographic distribution, and ecology of these fishes. The systematics of the superfamily gymnotoidea is in dire need of revision; many species have been neither described nor classified.

Early phases of interest in the natural history and ecology of the gymnotoid eels must be associated with the Museu Goeldi in Belém at the mouth of the Amazon River. This research station, with its former elaborate fish holding facilities and interest in fish culture, was the principal source of supplies of electric fishes for the Instituto de Biofisica in Rio de Janeiro. Here, most of the early studies on the histology and biochemistry of electrocyte function were initiated by the Fessards from France in collaboration with Brazilian colleagues, Antonio M. Couceiro, Carlos Chagas, Jr.,

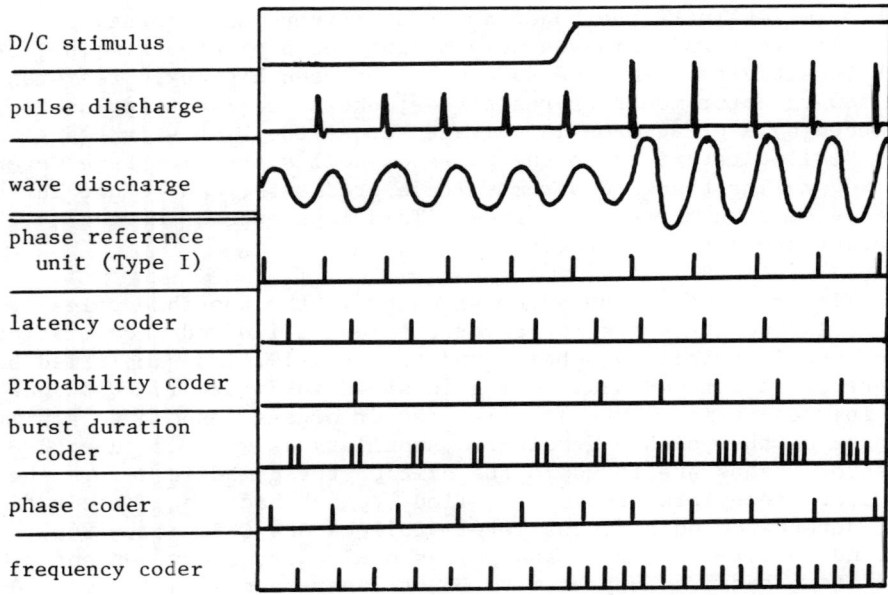

FIG. 1 Electroreceptor types and their coding characteristics in "active" electric fishes. Three possible kinds of stimulation of electroreceptor organs are shown in the upper three rows. The lower six rows illustrate the demonstrated coding modes by showing numbers and phase relations of spikes in the afferent nerves. Phase reference unit: Part of fast-conducting system, invariable and short delay, independent of amplitude variation at the receptor, could probide reference for other units (Szabo and Sakata, 1967; Fessard and Szabo, 1974). Latency coder: Latency of spike depends on stimulus pulse amplitude at receptor. Probability coder: One spike follows every one, every second, third, stimulus pulse depending on amplitude. Burst duration coder: Number of impulses in response to each stimulus pulse depends on latter's amplitude. Phase coder: Latency of spike depends on phase when threshhold is reached by sinusoidal stimulus pulse (limited to 90° of period and present in wave spp. only). Frequency coder: Firing rate in sensory fiber is function of intensity of electric field; responds to D/C and low frequency A/C.

His M. Ferreira, and others. Needless to say, the Museu Goeldi attracted many visiting scientists from foreign countries. Interest centered mostly on the electric eel, since the electrogenic nature of the other gymnotoids was demonstrated only later. In the ponds and holding tanks of the Museu at Belém, early attempts to "call" *Electrophorus* by means of electric pulses (Cox, 1938) were repeated and extended by us in 1954. These big eels were easily attracted from a distance of four metres by electric pulses stimulating the low voltage eel pulses (Bullock, 1969). Frequent stays at the Museu since 1959 provided ample opportunity for studying the gymnotoids in their natural habitat. Once, during the rainy season, a time of year when fishing is deemed very difficult, we succeeded in collecting within one day more than one hundred big eels in the flooded interior of the estuarine island of Marajó. These big fish were subsequently shipped to Rio de Janeiro.

Published accounts on the gymnotoids in their natural environment are rare. Lissmann (1961) made significant ecological observations on daily periodic movements and community composition during travels to Guyana and to the Museu Goeldi. His method of detecting presence and following movements of the continually discharging gymnotoids, by means of electrodes in the water and audio-amplifier plus ear phones, was copied and successfully employed by many of us. Findings of several field studies are presented below.

One member of the rhamphichthyid family, the "sandfish" *Gymnorhamphichthys hypostomus* produces brief pulses at highly variable rates, separated by long silent intervals. Field Marshal Rondon already mentioned the curious fact that this species spends the daytime hours buried in the sand of swift streams (Miranda Ribeiro, 1920). Studying the sandfish in the field and in the laboratory in 1964, we discovered that two distinct levels of discharge frequency were precisely correlated with the phases of activity and rest. By continuous monitoring of the electric pulses of isolated fish in controlled conditions we were able to show details of the endogenous circadian activity rhythm (Lissmann and Schwassmann, 1965; Schwassmann, 1971). Other studies added information on species specific frequencies, patterns of daily movements, behaviour in nature (Bullock, 1969; Steinbach, 1970), and on a sex difference in pulse rate in *Sternopygus* (Hopkins, 1972, 1974).

The two principal potential uses of electroreception in the electrogenic gymnotoids are considered to be electro-orientation and communication. Further significant evidence with respect to these two ecologically important aspects has come mostly from laboratory work. The "jamming avoidance response" of the wave species, or hummers, was noted first by Watanabe and Takeda (1963) in *Eigenmannia*. This adaptive frequency adjustment, which can greatly diminish electrical interference in the orientating behaviour among conspecifics, was investigated in detail later on in *Eigenmannia* and in the apteronotids and its adaptive advantage was recognized (Bullock et al., 1972; Larimer and McDonald, 1968; Gaddis, 1978). Another social signal was also found to be correlated with jamming-avoidance behaviour. This was the occurrence of brief frequency increases, called pings (Larimer and McDonald, 1968) or chirps (Bullock, 1969), which are interpreted as possible threatening signals. These two behaviours, involving EOD frequency, can be observed in nature; one such record is shown in Figure 2. Five specimens of the wave species *Apteronotus leptorhynchus*, 8.5 to 23.0 cm long, were located electronically in shelters on the bottom of a slowly flowing backwater of the Rio Guatiquia near Villavicencio, Colombia, in March 1977. It was shortly before sunset and all five fish remained stationary under leaves and hollow logs, with the exception of one medium-sized individual that had been disturbed by the electrodes and had moved to a different place nearby. The electric discharges of all five fish were recorded in succession on magnetic tape, transformed later-on into direct current levels by an integrator, and written out by a chart recorder. This record is shown in Figure 2. One fish of this group

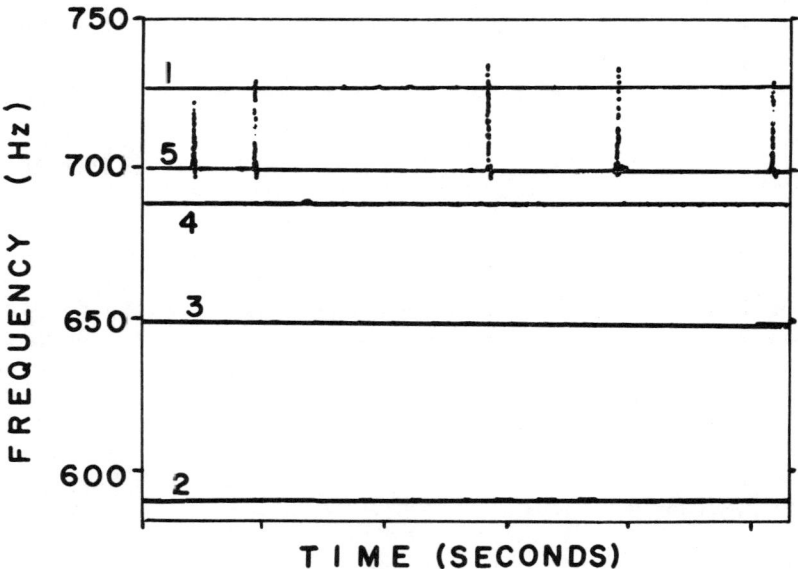

FIG. 2. Electric organ discharge frequencies of a natural group of five *Apteronotus leptorhynchus*. The recorded frequencies were integrated, written out by a chart recorder, and have been superimposed here manually. Fish No. 5 was the largest and exhibited "chirps;" the others could not be identified as to size but were all considerably smaller than 5.

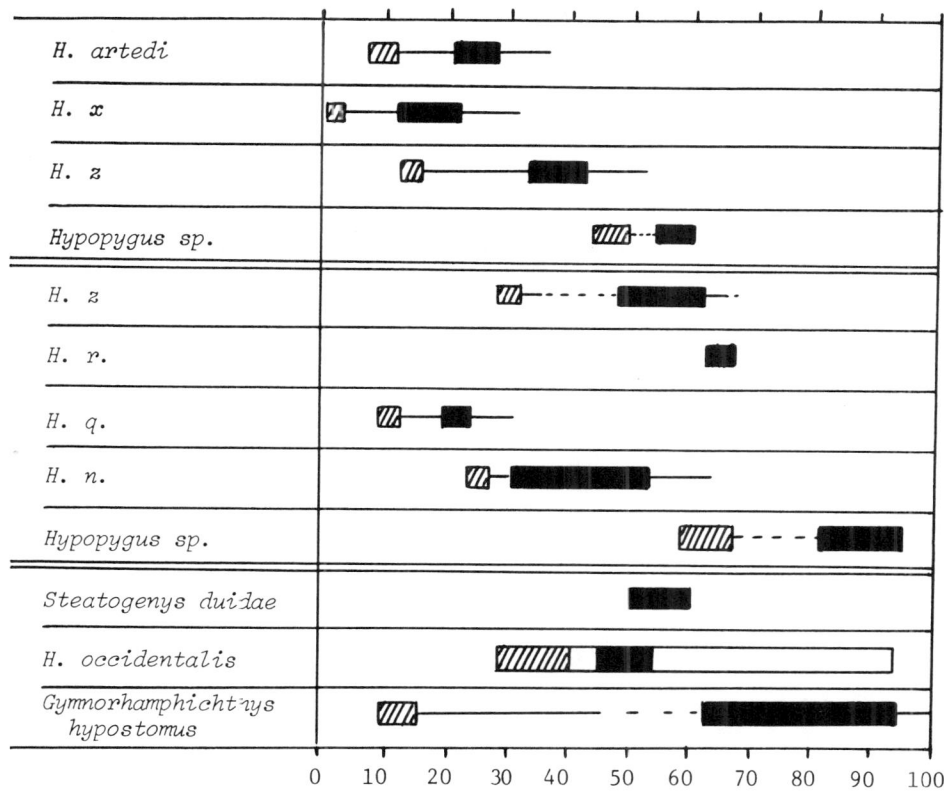

FIG. 3. Frequency partitioning in sympatric populations of *Hypopomus* and other related pulse-type species. The samples are from a floodplain lake near Manaus, Brazil (A), from the banks of the Rio Pacaas Novas and a small sidearm nearby, Rondônia, Brazil (B), and from the Caño Ema near Puerto Lopez in the Colombian llanos (C). Discharge frequencies of these fish were recorded in aquaria at 26 to 28°C during day and night time on the two days following capture. The ranges obtained from all specimens of each species are indicated as: Resting rate during the day by shaded bars, active range during the night by solid bars. Bursting (transient increases in rate in response to stimulation) is indicated by single lines; *Hypopomus occidentalis* does not burst but shows slowly occurring level changes (open bars). All *Hypopomus*, with the exception of *H. r.* exhibit different day and night levels.

exhibited frequent chirps; it was by far the largest of them all. I have made similar observations on several specimens of *Apteronotus albifrons* resting during the day in shelters and definitely in "hearing" range of each other, but I could not notice any chirping. As mentioned by Bullock (1969), chirps occur frequently during and after encounters of these fish with conspecifics, which could mean that they are to be observed predominantly at night.

Based on stomach content analysis, Knöppel (1970) reported on food organisms preferred by several gymnotoids. Only one species has been studied in greater detail in its natural habitat. The sandfish, *Gymnorhamphichthys hypostomus*, is found in many distinctly different isolated populations over a very wide area of the Amazon, Orinoco, and adjacent river systems (Schwassmann, 1976). Other studies reveal that sympatric pulse-type gymnotoids exhibit nonoverlapping ranges of discharge frequencies and of peak-power frequency, or pulse shape (Heiligenberg and Bastian, 1978; Schwassmann, 1978). Two examples of data sets showing frequency partitioning in *Hypopomus* and *Hypopomus*-related species, one group from Rondônia, one from near Manaus, are contained in Figure 3. When inspecting the pulse rates of the sympatric species, the active, or nightly rates should be of greater importance, since mutual interference with each others orienting mechanisms due to many successive coinciding pulses would then be most critical. Although this observation might appear to be an example of resource partitioning, one should then also consider the difficulty encountered by conspecifics living together in groups. Since their signals are almost identical in repetition rate and pulse frequency components, interference must be substantial. My own observations, however, seem to indicate that, at least at night, certain minimum distances between individuals of one species are faithfully observed. This is definitely the case in the sandfish which is often found in dense aggregations during the day in the very restricted suitable sand substrate, and which disperses over the entire stream bed range at night.

Summarizing, it should be noted that electroreception has evolved several times in some unrelated groups of fishes. In the active, electrogenic freshwater groups, the gymnotoids of South America and the mormyrids with *Gymnarchus* in Africa, it must be recognized as an adaptation to conditions which precluded vision as an orientation mode, and, secondarily perhaps, as an adaptation for a successful night-active life style. The areas of potential ecological usefulness can be considered as detection of objects, obstructions, and territorial limits; also, the detection of other organisms, be they conspecifics, predators, or prey. Intra- and interspecific social communication would be other adaptive features. The latter communication function could also include possible synchronization effects, mainly of a circadian nature, within populations of these fish.

REFERENCES

Bennett, M. V. L. (1971). Electric organs. In: W. S. Hoar and D. J. Randall, Eds., Fish Physiology 5: 347-491, Academic Press, New York, N.Y.

Bullock, T. H. (1969). Species differences in effect of electroreceptor input on electric organ pacemakers and other aspects of behavior in electric fish. Brain Behav. Evol. 2: 85-118.

Bullock, T. H. and Chichibu, S. (1965). Further analysis of sensory coding in electroreceptors of electric fish. Proc. Natl. Acad. Sci. 54: 422-429.

Bullock, T. H., Hagiwara, S., Kusano, K., and Negishi, K. (1961). Evidence for a category of electroreceptors in the lateral line of gymnotid fishes. Science 134: 1426-1427.

Bullock, T. H., Hamstra, R. H. Jr., and Scheich, H. (1972). The jamming avoidance response of high frequency electric fish. J. Comp. Physiol. Psychol. 77: 1-48.

Cox, R. T. (1938). The electric eel at home. Bull. New York Zool. Soc. 41: 59-65.

Fessard, A. Ed. (1974). Electroreceptors and other Specialized Receptors. Vol. III - 3 of Handbook of Sensory Physiology, Springer, Berlin.

Fessard, A. and Szabo, T. (1974). Physiology of electroreceptors. In: Handbook of Sensory Physiology. Vol. III - 3, Electroreceptors and other Specialized Receptors in Lower Vertebrates, Fessard, A., ed., pp. 59-124. Springer, Berlin.

Gaddis, P. K. (1977). Harmonization of discharge frequency of *Eigenmannia virescens*, Sternopygidae, (Pisces). Rev. Canad. Biol. 36: 317-320.

Greenwood, P. H., Rosen, D. E., Weitzmann, S. H., and Myers, G. S. (1966). Phyletic studies of teleostean fishes, with a provisional classification of living forms. Bull. Am. Mus. Nat. Hist. 131: 339-455.

Hagiwara, S., Kusano, K., and Negishi, K. (1962). Physiological properties of electroreceptors of some gymnotids. J. Neurophysiol. 25: 430-449.

Hagiwara, S. and Morita, H. (1963). Coding mechanisms of electroreceptor fibers in some electric fish. J. Neurophysiol. 26: 551-567.

Heiligenberg, W. and Bastian, J. (1978). Species specificity of electric organ discharges in sympatric gymnotoid fish of the Rio Negro and Solimões, Brazil. in press.

Hopkins, C. D. (1972). Sex differences in electric signalling in an electric fish. Science 176: 1035-1037.

Hopkins, C. D. (1974). Electric communication in the reproductive behavior of *Sternopygus macrurus* (Gymnotoidei). Z. Tierpsychol. 35: 518-535.

Kalmijn, A. J. (1971). The electric sense of sharks and rays. J. Exp. Biol. 55: 371-383.

Kalmijn, A. J. (1974). The detection of electric fields from inanimate and animate sources other than electric organs, pp. 147-200. In: Handbook of Sensory Physiology, Vol. III/3: Electroreceptors and Other Specialized Receptors in Lower Vertebrates, A. Fessard, ed., Springer, Berlin.

Kellaway, P. (1946). The part played by electric fish in the early history of bioelectricity and electrotherapy. Bull. Hist. Med. 2: 112-137.

Knöppel, H. A. (1970). Food of Central Amazonian fishes. Contributions to the nutrient ecology of Amazonian rainforest streams. Amazoniana 2: 257-352.

Larimer, J. L. and MacDonald, J. A. (1968). Sensory feedback from electroreceptors to electromotor pacemaker centers in gymnotids. Am. J. Physiol. 214: 1253-1261.

Lissmann / H. W. (1958). On the function and evolution of electric organs in fish. J. Exp. Biol. 35: 156-191.

Lissmann, H. W. (1961). Ecological studies on gymnotids. In: Bioelectrogenesis, C. Chagas, Jr. and A. Paes de Carvalho, eds., Elsevier, Amsterdam.

Lissmann, H. W. and Machin, K. E. (1958). The mechanism of object location in *Gymnarchus niloticus* and similar fish. J. Exp. Biol. 35: 451-486.

Lissmann, H. W. and Schwassmann, H. O. (1965). Activity rhythm of an electric fish, *Gymnorhamphichthys hypostomus*, Ellis. Z. Vergl. Physiol. 51: 153-171.

Machin, K. E. and Lissmann, H. W. (1960). The mode of operation of the electric receptors in *Gymnarchus niloticus*. J. Exp. Biol. 37: 801-811.

Murray, R. W. (1960). Electrical sensitivity of the ampullae of Lorenzini. Nature, (London) 187: 957.

Murray, R. W. (1962). The response of the ampullae of Lorenzini of elasmobranchs to electrical stimulation. J. Exp. Biol. 39: 119-128.

Parker, G. H. and Heusen, A. P. van (1917). The responses of the catfish, *Ameirus nebulosus*, to metallic and non-metallic rods. Am. J. Physiol. 44: 405-420.

Ribeiro, A. de Miranda (1920). Peixes (excl. Characinidae). Commissão de Linhas Tel. Estr. de Matto Grosso ao Amazônas. Zool. Hist. Nat. Ann. 58: 1-15

Romer, A. S. (1966). Vertebrate Palaeontology. Univ. Chicago Press, Chicago, Ill. 3rd. ed.

Scheich, H. and Bullock, T. H. (1974). The detection of electric fields from electric organs, pp. 201-256. In: Handbook of Sensory Physiology, Vol. III/3: Electroreceptors and Other Specialized Receptors in Lower Vertebrates, A. Fessard, ed. Springer, Berlin.

Schwassmann, H. O. (1971). Circadian activity patterns in gymnotid electric fish. In: Biochronometry, M. Menaker, ed. pp. 186-199, Natl. Acad. Sciences, Washington, D.C.

Schwassmann, H. O. (1976). Ecology and taxonomic status of different geographic populations of *Gymnorhamphichthys hypostomus* Ellis (Pisces, Cypriniformes, Gymnotoidei). Biotropica 8: 25-40.

Schwassmann, H. O. (1978). Activity rhythms in gymnotoid electric fish. In: Rhythmic Activities in Fishes, J. Thorpe, ed. Academic Press, New York, N.Y. in press.

Steinbach, A. B. (1970). Diurnal movements and discharge characteristics of gymnotid fishes in the Rio Negro, Brazil. Biol. Bull. Woods Hole 138: 200-210.

Stensiö, E. A. (1927). The Downtonian and Devonian vertebrates of Spitzbergen. I: Family Cephalaspidae. Skr. om Svalbard og Nordishavet 12, pp. 391.

Watanabe, A. and Takeda, K. (1963). The change of discharge frequency by A. C. Stimulus in a weak electric fish. J. Exp. Biol. 40: 57-66.

4. INDICES

AUTHOR INDEX

Page numbers in italics refer to the bibliography

Abelein, M. see Best, J.B. et al.
Abelsdorf, G. see Köttgen, E., Abeldorf, G.
Abeshaheen, J.P. see Burghardt, G.M., Abeshaheen, J.P.
Abraham, R. see Storch, V., Abraham, R.
Ache, B.W., Case, J. 300, 308
-- Davenport, D. 300, 308
-- Fuzessery, Z.M., Carr, W.E.S. 301, 308
Adler, J. 42, 50
-- Dahl, M.M. 50
-- see De Pamphilis, M.L., Adler, J.
-- see Larsen, S.H. et al.
Adler, K. 178, 190
-- Taylor, D.H. 178, 190
-- see Taylor, D.R., Adler, K.
Adrian, E.D. 298, 309, 474, 495
Akoev, G.N., Ilyinski, Zadan, P.M. 165, 166
Albergoni, V. see Piccinni, E. et al.
Alexandrowicz, J.S. 426, 428, 433, 434, 435
Ali, M.A. 3-8, 91-122, 485, 490, 495
-- Anctil, M. 481, 483, 486, 489, 494, 495
-- Anctil, M., Cervetto, L. 467-502
-- Croll, R.P., Jaeger, R.G. 11-29
-- Crouzy, R. 490, 495
-- Hoar, W.S. 490, 495
-- Muntz, W.R.A. 492, 495
-- Ryder, R.A., Anctil, M. 486, 489, 496
-- Wagner, H.-J. 28
-- see Anctil, M., Ali, M.A.
-- see Zyznar, E.S., Ali, M.A.

Allen, D.M., McFarland, W.N. 490, 496
-- McFarland, W.N., Munz, F.W., Poston, H.A. 490, 496
-- see McFarland, W.N. et al.
Altman, G., Warnke, U. 146, 148
Altner, H. 459, 461, 462
-- Sass, H., Altner, I. 459, 460, 461, 462
-- Tichy, H., Altner, I. 460, 462
Altner, I. see Altner, H. et al.
Anctil, M. 483, 493, 494, 496
-- Ali, M.A. 486, 496
-- see Ali, M.A., Anctil, M.
-- see Ali, M.A. et al.
Anderson, K.K., Bernstein, D.J. 307, 309
Anderson, O. see Grant, D. et al.
Anderson, P., Mackie, G.O. 19, 67, 72, 82
Anderson, R.A. see Berg, H.C., Anderson, R.A.
Ando, M. see Yokohari, F. et al.
Andres, K.H. see Hensel, H. et al.
Andrews, P. see Gorner, P., Andrews, P.
Aneshansley, D. see Eisner, T. et al.
Angaut-Petit, D., Clarac, F., Vedel, J.P. 434, 435
Anisimov, V.D. 219, 244
Arai, M.N. 68, 74, 82
Arnold, S.J. 171, 172, 181, 184, 190
Arnott, H.J. see Nicol, J.A.C. et al.
Arntz, B. 331, 337
Aronova, M. 68, 82
Arronet, N.I. 47, 50
Ash, J.F., McClure, W.O., Hirsch, J. 69, 83
Ashmole, N.P. 511, 516
Ashmore, J.F., Falk, G. 476, 496

Atema, J. 295, 302, *309*
— see Todd, J.H. et al.
Atkinson, R.J.A. see Naylor, E., Atkinson, R.J.A.
Aurell, G., Holmgren, H. 505, *516*
Austin, W.D. see Hartman, H.B., Austin, W.D.
Autrum, H. 491, *496*
— Thomas, I. 129, *148*, 257, *279*
— Zwehl, von V. 129, *148*
Aylor, D. 218, *244*, 359, 362, *370*

Backus, R.H., Craddock, J.E., Haedrich, R.L., Shores, D.L., Teal, J.M., Wing, A.S., Mead, G.M., Clarke, W.D. 493, *496*
— see Schevill, W.E. et al.
Badcock, J. 159, *166*
Bagger-Sjöböck, D. see Wersäll, J., Bagger-Sjöböck, D.
Baird, I.L. *28*
Baldaccini, N.E. see Papi, F. et al.
Balliet, R.F., Schusterman, R.J. 271, *279*, 514, *516*
— see Schusterman, R.J., Balliet, R.F.
— see Schusterman, R.J. et al.
Bamford, O.S. see Cook, A. et al.
Banbury, J.C. see Myrberg, A.A. et al.
Bang, B.G. 240, *244*
— Cobb, S. *28*, 239, *243*
Bannister, L.H. 297, *309*
Barber, S.B. *19*
Bardach, J.E. 301, *309*
— Case, J. 303, *309*
— Todd, J.H., Crickmer, R. 295, 303, *309*
— Villars, T. *28*, 302, *309*
— see Lager, K.F. et al.
— see Todd, J.H. et al.
Barham, E.G. 493, *496*
Barlow, R.B. see Kaplan E. et al.

Barnes, A.T. see Case, J.F. et al.
Barnes, R.D. *120*
Barnes, W.J.P., Spirito, C.P., Evoy, W.H. 420, *435*
Barns, S.N., Goldsmith, T.H. 137, *148*
Barrett, B. see Schusterman, R.J., Barrett, B.
Barrett, R. *28*
Barrington, E.J.W. *120*
Barth, F.G. 124, *148*
Barth, G. 427, 433, *435*
Bartholomew, C.E., Collias, N.E. 271, *279*
Bässler, U. 330, 337, 413, *435*
Bastian, J. see Heiligenberg, W., Bastian, J.
Baunacke, W. *337*
Baylor, D.A., Fettiplace, R. 476, *496*
— Fuortes, M.G.F. 475, *496*
— Hodgkin, A.L. 211, *213*, 475, *496*
— Hodgkin, A.L., Lamb, T.D. 475, *496*
— see Nicholls, J.C., Baylor, D.A.
Baylor, E.R. 506, 507, 516
Beatty, D.D. 490, *497*
Beauchamp, G.K., Doty, R.L., Moulton, D.G., Mugford, R.A. 293, *309*
— see Rogers, J.G. Jr., Beauchamp, G.K.
Beaugrand, J.P. 243, *244*
Beck, G. 359, *370*
Becker, D. 441, *463*
Becker, G. 146, *148*
Bedini, C., Ferrero, E., Lanfranchi, A. 70, 73, *83*
— Lanfranchi, A. 64, *83*
Beer, T. 506, 509, 516
Beets, M.G.J. 297, *309*
Beidler, L.M. 297, *309*
— see Moulton, D.G., Beidler, L.M.
Békésy, von G. 221, *250*
Bell, L.G.E. see Jeon, K.W., Bell, L.G.E.

AUTHOR INDEX

Bennett, C.E. 78, *83*
Bennett, M.V.L. 523, *531*
Bennet-Clark, H.C. 124, 141, *148*, 348, 366, *370*
Benvenuti, S., Fiaschi, V., Fiore, L., Papi, F. 300, *309*
— see Papi, F. et al.
Benzing, H., Hensel, H., Wurster, R. 448, *463*
Berg, H.C. *50*
— Anderson, R.A. *50*
Bergeijk, W.A. van 163, *166*, 333–335, *343*
— see Harris, G.G., Bergeijk, W.A. van
Berger, E.R. 486, *497*
Bergquist, P.R., Sinclair, M.E., Hogg, J.J. 61, 69, *83*
Bermond, N. see Corbiere-Tichané, G., Bermond, N.
Bernard, G.D. see Goldsmith, T.H., Bernard, G.D.
Bernard, J. 441, 460, *463*
— see Pinet, J.M., Bernard, J.
Bernard, R.A. see Drewes, C.D., Bernard, R.A.
Bernstein, D.J. see Anderson, K.K., Bernstein, D.J.
Best, C.G. see Nicol, J.A.C. et al.
Best, J.B., Goodman, A.B., Pigon, A. 74, *83*
— Howell, W., Riegel, V., Abelein, M. 74, *83*
— see Pigon, A. et al.
Binggeli, R.L., Paule, W.J. 229, *244*
Binyon, J. *19*
Birch, M.C. 293, 304, *310*
Birndorf, L.A. see Dawson, W.W. et al.
Bischof, H.-J. 331, *337*
Bishop, I.G. see Yingst, D.R. et al.
Bishop, P.O. see Vakkur, G., Bishop, P.O.
Blair, W.F. 174, *190*

Blanchetti, R. see Ferrara, R., Blanchetti, R.
Blaxter, J.H.S. 157, *166*, 375–409, 493, 495, *497*
— Denton, E.J. 378, 383, 385, *406*
— Tytler, P. 381, 382, 388, 389, 399, *400*
— see Denton, E.J., Blaxter, J.H.S.
— see Gibson, R.N. et al.
— see Tytler, P., Blaxter, J.H.S.
Blazevic, A. see Gogala, M. et al.
Blest, A.D., Land, M.F. *166*
Blevins, C.E., Parkins, B.J. 277, *279*
Blitz, N.M., Smyth, J.D. 79, *83*
Blough, P.M. 232, *244*
Blum, M.S. 304, *310*
Bobier, W. see Sivak, J.G. et al.
Boden, E.P., Kampa, E.M. 493, *497*
Boeckh, J., Kaissling, K.E., Schneider, D. *19*
Boettiger, E.G. see Hartman, H.B., Boettiger, E.G.
— see Wiersma, C.A.G., Boettiger, E.G.
Bohn, H., Thuber, U. 136, *148*
Bok, D. see Young, R.W., Bok, D.
Bone, Q. 162, *166*
— Chubb, A.D. 402, *406*
Bonke, D. 331, *337*
Bookman, M.A. 243, *244*
Boolootian, R.A. *120*
Borchers, I.U. see Gruschka, H.D. et al.
Borror, D.J., De Long, D.M. 123, *148*
Borwein, B., Hollenberg, M.J. 486, *497*
Bosarge, O.S. see Ferguson, D.E. et al.
Bossert, W.H., Wilson, E.O. 293, 304, *310*
— see Wilson, E.O., Bossert, W.H.
Boulton, P.S. see Horridge, G.A., Boulton, P.S.
Bovee, E.C., Jahn, T.L. 40, *50*
— see Jahn, T.L., Bovee, E.C.

Bowen, I.D., Ryder, T.A. 70, *83*
Bowmaker, J.K., Knowles, A. 235, 236, 238, *244*
Bowman, K. *see* Hensel, H., Bowman, K.
Boyle, P.R. *19*
Bradbury, J.W. 264, *279*
— Nottebohm, F. 268, *279*
Bradford, J. *see* White, D. et al.
Bradford, B. *see* Suthers, R.A. et al.
Braten, T., Hopkins, C.A. 79, *83*
Brauer, A. 484, *497*
Brauer, R.W. 377, *406*
Braveman, N.S. 303, *310*
Breumer, L. *see* Dral, A.D.G., Breumer, L.
Brewer, R.H. 69, 83
Bridgeman, C.S., Smith, K.U. 254, *279*
Bridges, C.D.B. 488, 490, 491, *497*
— Delisle, C.E. 488, *497*
Brooker, B.E. 80, *83*
Bromm, B., Hensel, H., Tagmat, A.T. 457, *463*
Brown, A.I. *see* Keeton, W.T., Brown, A.I.
Brown, A.M., Pye, J.D. *280*
Brown, C.W. 171, *190*
Brown, F.A. 66, *83*
Brown, F.K. *see* Wald, G. et al.
Brown, H.M., Ogden, T.E. 73, *83*
Brown, W.L. Jr. 292, *310*
— Eisner, T., Whittaker, R.H. 292, *310*
Brownell, P.H. 329, *337*
Budelmann, B.-U. 325, 326, 327, *337*
— Thies, G. 326, *337*
— Wolff, H.G. 327, *337*
Buetow, D.E. *50*
Bullock, T.H. 307, *310*, 527, 528, *531*
— Chichibu, S. 525, *531*
— Diecke, F.P.J. 439, 441, 444, 461, *463*

— Fox, W. 441, 442, 457, *463*
— Grinnell, A.D., Ikezono, E., Kameda, Y., Katsuki, Y., Nomoto, M., Sato, O., Suga, N., Yanagisawa, K. *280*
— Hagiwara, S., Kusano, K., Negishi, K. 528, *531*
— Hamstra, R.H. Jr., Scheich, H. 524, *531*
— Horridge, G.A. 11, *19*, 60, *84*, 116, *120*, 412, 413, *435*
— *see* Scheich, H., Bullock, T.H.
Burghardt, G.M. 173, *190*, 303, 306, *310*
— Abeshaheen, J.P. 173, *190*
— *see* Gove, D., Burghardt, G.M.
Burke, W. 420, 421, *435*
Burkhardt, D. 129, 136, *149*, 319, *338*, 455, *463*
— De la Motte, I. 131, *149*
— Gewecke, M. 413, *435*
— Schneider, G. 413, *436*
— *see* Huth, H., Burkhardt, D.
Burnside, B. 484, *497*
Bursell, E. 138, 139, 140, *149*
Bush, B.M.H. 419, 424, 426, *436*
Byrd, W.W. *see* Jordan, O.R. et al.
Byrne, B.C. *see* Byrne, B.J., Byrne, B.C.
Byrne, B.J., Byrne, B.C. 48, *51*
Byrne, J., Castellucci, V., Kandel, E.R. *19*

Cable, R.M. 80, *84*
Cadle, D.R. *see* Martin, G. et al.
Cahn, P.H. 334, *338*
Calabrese, R.L. *see* Wiese, K. et al.
Cameron, N. *see* White, D. et al.
Camorgis, G. *see* Davenport, D. et al.
Campbell, J.I. *see* Meadows, P.S., Campbell, J.I.
Campbell, R.D. 66, *84*
Canning, E.U., Wright, C.A. 78, *84*
Capranica, R.R. 28, 174, 175, *190*

AUTHOR INDEX

Capranica, R.R., Frishkoph, L.S., Nevo, E. 174, *190*
-- see Murray, M.J., Capranica, R.R.
-- see Narins, P.M., Capranica, R.R.
Caprio, J. 295, *310*
Carlile, M.J. 32, *51*
Carlson, D.E., McDaniel, O.H., Reethof, G. 359, 362, 364, *371*
Carr, A., Ogren, L. 209, *213*
Carr, W.E.S., Hall, E.R., Gurin, S. 301, *310*
-- see Ache, B.W. et al.
Carrel, D. see Eisner, T. et al.
Carricaburu, P. 124, *149*
Carter Rowley III, J. see Moran, D.T., Carter Rowley III, J.
Case, J. *19*
-- see Ache, B.W., Case, J.
-- see Bardach, J.E., Case, J.
Case, J.F., Warner, J., Barnes, A.T., Lowenstine, M. 494, *497*
-- see Ghiradella, H. et al.
Caspary, D. *280*
Castellucci, V. see Byrne, J. et al.
Cervetto, L. 475, 476, *497*
-- Pasino, E., Torre, V. 475, *497*
-- see Ali, M.A. et al.
Chamberlain, S.C. see Kaplan, E. et al.
Chambliss, O.L., Jones, C.M. 301, *310*
Chang, F.W. see Massof, R.W., Chang, F.W.
Chang, J.J., Johnson, F.H. 66, *84*
Chang, T.M.S., Macintosh, F.C., Masson, S.G. 44, *51*
Chapman, K.M. see Spinola, S.M., Chapman, S.M.
Chard, R.D. see Gundlach, R.H. et al.

Charman, W.N., Tucker, H. 504, *516*
Chase, J. 158, *166*, 266, 268, *280*
-- Suthers, R.A. 268, *280*
-- see Suthers, R.A. et al.
Cheal, M., Sprott, R.L. 304, *310*
Checcuci, A. 40, *51*
-- Colombetti, G., Ferrara, R., Lenci, F. 40, 41, *51*
Chia, F.S., Rostron, M.A. 69, *84*
Chichibu, S. see Bullock, T.H., Chichibu, S.
Childress, J.J. see Fuzessery, Z.M., Childress, J.J.
Chubb, A.D. see Bone, Q., Chubb, A.D.
Chun, C. 484, *497*
Church, R. 493, *497*
Citron, M.C., Pinto, L.H. 504, *516*
Clarac, F. 388, *406*, 413, 432-434, *436*
-- Wales, W. 432, *436*
-- Wales, W., Laverack, M.S. 431, 432, *436*
-- see Angaut-Petit, D. et al.
-- see Moulins, M., Clarac, F.
-- see Wales, W. et al.
Clark, R.B., Olive, P.J.W. 118, *120*
Clarke, G.L., Denton, E.J. 159, *166*
-- Kelly, M.G. 159, *166*
Clarke, W.D. 494, *497*
-- see Backus, R.H. et al.
Clément, P. 73, *84*
-- Pourriot, R. 73, *84*
-- see Pourriot, R., Clément, P.
Cobb, S. see Bang, B.G., Cobb, S.
Coble, J.G. see Gruschka, H.D. et al.
Cody, M.L., Diamond, J.M. 185, *190*
Coenen-Stass, D. 448, *463*
Cohen, M.J., Dykgraaf, S. *19*, 329, *338*
-- see Evoy, W.H., Cohen, M.J.
-- see Gordon, S.A., Cohen, M.J.

Cohen, N.W. *see* Stebbins, R.C., Cohen, N.W.
Cokl, A. *149*, 441, 446, *463*
— Kalmring, K., Wittig, H. 141, *149*
— *see* Gogala, M. et al.
Cole, L.P. *see* Nachman, M., Cole, L.P.
Collias, N.E. *see* Bartholmew, G.A., Collias, N.E.
Colombetti, G. *see* Checcuci, A. et al.
Cone, R.A. *see* Korenbrot, J.I., Cone, R.A.
Connor, G.I. *see* Johnson, T.B. et al.
Cook, A., Bamford, O.S., Freeman, J.D.B., Teideman, D.J. 299, *310*
Cook, S.B. 299, *310*
Cooper, G.F., Robson, J.G. 256, 257, *280*
Cooper, J.C. *see* Madison, D.M. et al.
— *see* Scholz, A.T. et al.
Cooper, T.G. *see* Meyer, D.B., Cooper, T.G.
Coppellotti, O. *see* Piccinni, E. et al.
Corbière-Tichané, G. 461, *463*
— Bermond, N. 461, *463*
Corey, D.P. *see* Hudspeth, A.J., Corey, D.P.
Corey, S. *see* Ramprashad, F. et al.
Corliss, J.D. *see* Winkler, R.N., Corliss, J.D.
Corning, W.C., Kelly, S. 60, *84*
Costa, W.R. *see* Shoemaker, V.H. et al.
Couillard, P. 31–54
— *see* Marsot, P., Couillard, P.
Coward, S.J., Johannes, K.B. 69, *84*
Cox, R.T. 527, *531*
Craddock, J.E. *see* Backus, R.H. et al.

Crawford, R.M. *see* Dodge, J.D., Crawford, R.M.
Crescitelli, F. 198, 201, 202, 203, *213*, 472, 489, 490, *498*
Crickmer, R. *see* Bardach, J.E. et al.
Crisp, D.J. 19, 69, *84*, 298, 299, *310*
— *see* Thorpe, W.H., Crisp, D.J.
Crocker, M.J., Price, A.J. 348, 354, *371*
Croll, N.A. 69, 72, 74, *84*
— Riding, J.L., Smith, J.M. 19
Croll, R.P. *see* Ali, M.A. et al.
Cronshaw, J. *see* Ghiradella, H. et al.
Crouzy, R. *see* Ali, M.A., Crouzy, R.
Crozier, W.J., Wolf, E. 203, *213*
Crump, M.L. 170, 175, *190*, *191*

Dahl, E., Emanuelsson, H., von Mecklenburg, C. 306, *310*
Dahl, M.M. *see* Adler, J., Dahl, M.M.
Dales, R.P. 112, *120*
Daley, J.M. 96, *120*
Damas, D. 78, *84*
Dando, M.R. *see* Wales, W. et al.
Danner, J. *see* Lenhoff, H.M. et al.
Danon, A., Stoeckenius, W. 487, *498*
Darian-Smith, J., Johnson, K.O., Dykes, R. 441, 451, *463*
Darian-Smith, I. *see* Johnson, K.O. et al.
Darlington, P.J. 197, *213*
Darslar, K. 329, *337*
Dartnall, H.J.A. 129, *149*, 159, *166*, 210, *213*
— *see* Knowles, A., Dartnall, H.J.A.
— *see* Lowe, E.R., Dartnall, H.J.A.
— *see* Lythgoe, J.N., Dartnall, H.J.A.
Daumer, K. 125, 133, *149*
Davenport, D. 300, *310*
— Camorgis, G., Hickok, J.F. 19

AUTHOR INDEX

Davenport, D., Norris, K.S. 300, *311*
— see Ache, B.W., Davenport, D.
— see Dimock, R.V., Davenport, D.
— see Hickok, J., Davenport, D.
Davies, J.T. 297, *311*
Davis, E.E., Sokolove, P.G. 439, 446-448, 460, 461, *463*
Dawson, W.W., Birndorf, L.A., Perez, J.M. 514, *516*
de Burlet, H.M. 164, *166*
de Groot, P.J. see DuPont, Y.S., de Groot, P.J.
de Groot, S.J. see Gibson, R.N. et al.
Delamare Deboutterville, C. 64, 67, *84*
De la Motte, I. see Burkhardt, D., De la Motte, I.
Delisle, C.E. see Bridges, C.D.B., Delisle, C.E.
Delius, J., Perchard, R., Emmerton, J. 240, *244*
De Long, D.M. see Borror, D.J., De Long, D.M.
Denison, J. see Wilson, R.A., Denison, J.
Dennis, M.J. *19*
Denton, E.J., Blaxter, J.H.S. 161, *166*
— Gilpin-Brown, J.B. 401, *406*
— Gilpin-Brown, J.B., Howarth, J.V. 401, *406*
— Gilpin-Brown, J.B., Wright, P.G. 159, *166*, 493, *498*
— Nicol, J.A.C. 494, *498*
— Warren, F.J. 483, *493*
— see Blaxter, E.J.S., Denton, E.J.
— see Clarke, G.L., Denton, E.J.
De Pamphilis, M.L., Adler, J. *51*
de Peyer, J. see Machemer, H., de Peyer, J.
Desmond, A.J. 197, *213*

Dethier, V.G. 125, 137, *149*, 298, *311*
De Valois, R.L., Jacobs, G.H. 255, *280*
Diamond, I. see Masterton, B., Diamond, I.
Diamond, J.M. see Cody, M.L., Diamond, J.M.
Dice, D.L. 244
Dickenson, S.H. 450, *463*
Diecke, F.P.J. see Bullock, T.H., Diecke, F.P.J.
Diehn, B., Feinleib, M., Haupt, W., Hildebrandt, E., Lenci, F., Nultsch, W. 32, *51*
Diercks, K.J. 263, *280*
— Trochta, R.T., Greenlaw, C.F., Evans, W.E. 276, *280*
Digby, P.S.B. 403, 404, *407*
Dijkgraaf, S. 334, *338*, 388, 407
Dimock, R.V., Davenport, D. 300, *311*
Dizon, A.E. see Madison, D.M. et al.
Dobelle, W.H. see Marks, W.B. et al.
Dobzhansky, T.G., Hecht, M.K., Steere, W.C. 467
Dodge, J.D., Crawford, R.M. 42, *51*
Dodt, E., Jessen, K.H. 203, *213*
— Scherer, E. 208, *213*
— Walther, J. 254, *278*, *280*
Doetsch, R.N. 42, *51*
Dogiel, V.A. 50, *51*
Dole, J.W. 177, *191*
Donato, R.J. 364, *371*
Donner, J. *19*
Donner, K.O. 232, *244*
Dooling, R.J. 220, *244*
— Mulligan, J.A., Miller, J.D. 220, *245*
— Saunders, J.C. 222, 223, *244*, *245*
— see Saunders, J., Dooling, R.
Doroshow, C.A. see Simmons, J.A. et al.
Doroszenski, M. 48, *51*
Doty, R.L. 304, *311*

Doty, R.L. see Beauchamp, G.K. et al.
Dougherty, E.C. 84
Døving, K.B., Nordeng, H., Oakley, B. 299, *311*
Dral, A.D.G. 274, 275, *280*, 514, 515, *517*
-- Beumer, L. *280*
Draslar, K. 455, *463*
-- see Gogala, M. et al.
Drewes, C.D., Bernard, R.A. *19*
Drury, W.H. see Nisbet, I.C.T., Drury, W.H.
Duchamp, A. see Holley, A. et al.
Dudok van Heel, W.H. 278, *280*
Dudziak, J. 209, 213, 509, *517*
Dudzinski, M.L. see Mykytowycz, R. et al.
Duke, B.O.L. 80, *84*
Duke-Elder, S. 227-229, *245*, 503, 512, 514, 516, *517*
Duncan, C.J. 48, *51*
Dunn, P.A. see Silvey, G.E. et al.
Dunn, R.F. 201, *214*
Duntley, S.Q. 158, *167*
DuPont, Y.S., de Groot, P.J. 504, 509, *517*
Durand, J.P., Gourbault, N. 67, *84*
During, M.V. see Hensel, H. et al.
Dykes, R. see Darian-Smith, J. et al.
Dykgraaf, S. see Cohen, M.J., Dykgraaf, S.

Eakin, R.M. 207, *214*, 467, 472, 477, 484, *498*
-- see Stebbins, R.C., Eakin, R.M.
Earl, P.R. 80, *85*
Eckert, R. 47, 51, 107, *120*, 323, *338*
-- Naitoh, Y. 47, *51*
-- Nautch, Y., Friedman, K. *19*
-- see Kung, C., Eckert, R.
-- see Naitoh, Y., Eckert, R.

Edwards, C.A., Lofty, J.R. *120*
Edwards, J.S., Palka, J. 329, *338*
Ehers, B., Ehers, U. 64, 85
Ehers, U. see Ehers, B., Ehers, U.
Ehrenfeld, D.W., Koch, A.L. 209, 210, *214*
Ehret, C.F. see Willee, J.J., Ehret, C.F.
Eiben, R. see Müller, W.A. et al.
Eidmann, H., Kühlhorn, F. 125, 126, *149*
Eisenberg, J.F., Kleiman, D.G. 304, 306, *311*
Eisner, T. 245, 308, *311*
-- Silberglied, R.E., Aneshansley, D., Carrel, D., Howland, H.C. 125, 133, *149*
-- see Brown, W.L. et al.
Ellins, S.R., Masterson, F.A. 268, 280
Elsner, N. 368, *371*
Emanuelsson, H. see Dahl, E. et al.
Embleton, T.F.W. 349, *371*
-- Piercy, J.E., Olson, N. 349, 364, 365, *371*
-- see Piercy, J.E. et al.
Emlen, J.M. 185, *191*
Emlen, S.T. 175, 185, 187, *191*, 240, 242, *245*
Emmerton, J. see Delius, J. et al.
Enright, J.T. 395, 396, 399, *407*
Erickson, R. 298, *311*
Esch, H. 440, *463*
Evans, F.G.C. 108, *120*
Evans, W.E. 271, *280*
-- Haugen, R.M. 271, *280*
-- see Diercks, K.J. et al.
-- see Norris, K.S. et al.
Evans, W.G. 439, 445, 463
Evoy, W.H., Cohen, M.J. 433, *436*
-- see Barnes, W.J.P. et al.
Ewer, R.F. *311*
Ewert, J.P. 176, 177, 189, *191*
-- Härter, H.-A. 177, *191*
-- Ingle, D. 176, *191*
Eyrings, C. 362, *371*

AUTHOR INDEX

Fagerlund, V.H.M. see Idler, D.R. et al.
Fain, G.L. 476, *498*
Falk, G. see Ashmore, J.F., Falk, G.
Fallis, A.M. 78, *85*
Farkas, S.R., Shorey, H.H. 293, *311*
Fatehchand, R. see Svaetichin, G.K. et al.
Fauré-Frémiet, E. 37, 38, *51*
Fauvel, P. 66, *85*
Fay, R.R. see Popper, A.N., Fay, R.R.
Feder, H.M. 307, *311*
Feeny, P.P. see Whittacker, R.H., Feeny, P.P.
Feinberg, R. see MacNichol, E.F. et al.
Feinleib, M. see Diehn, B. et al.
Feldman, M., Lenhoff, H.M. 73, *85*
Fenton, M.B. 226, *245*
— Fleming, T.H. 225, *245*
Ferguson, D.E. 177, *191*
— Landreth, H.F., McKeown, J.P. 177, *191*
— McKeown, J.P., Bosarge, O.S., Landreth, H.F. 177, *191*
— see Jordan, O.R. et al.
— see Landreth, H.F., Ferguson, D.E.
— see Taylor, D.H., Ferguson, D.E.
Ferguson, J.H. see Gorman, R.R., Ferguson, J.H.
Fernandez, H.R., Tsuji, F.I. 494, *498*
— see O'Day, W.T., Fernandez, H.R.
— see Yingst, D.R. et al.
Ferrara, R., Blanchetti, R. 40, *51*
— see Checcuci, A. et al.
Ferrero, E.A. 55-90, 72, *85*
— Wales, W. 435, *436*
— see Bedini, C. et al.
— see Galleni, L. et al.

Fessard, A. 524, *531*
— Szabo, T. 526, *531*
Fettiplace, R. see Baylor, D.A., Fettiplace, R.
Fiaschi, V., Wagner, G. 300, *311*
— see Benvenuti, S. et al.
— see Papi, F. et al.
Field, L.H., MacMillan, D.L. 96, *120*
Fields, R.B. see Ordal, G.W., Fields, R.B.
Fineran, H.R., Nicol, J.A.C. 485, 487, *498*
Fioravanti, R., Fuortes, M.C.F. 20
Fiore, L. 74, *85*
— Ioalé, P. 74, 85
— see Benvenuti, S. et al.
— see Papi, F. et al.
Fisher, F.H., Simmons, V.P. 352, *371*
Fisher, H.D. see Jamieson, G.S., Fisher, H.D.
Fite, K.V. 229, 232, *245*
— Rosenfield-Wessels, S. 229, 230, *245*
Fleming, T.H. see Fenton, M.B., Fleming, T.H.
Flock, A. 28
Flock, P. 327, 332, *338*
Flügel, H. 382, *407*
Forget, J. 32, *51*
Fox, H., Vevers, G. 124, *149*
Fox, R., Lehmkuhle, S.W., Westendorf, D.H. 232, *245*
Fox, W. see Bullock, T.H., Fox, W.
Foxton, P. *167*
Franc, A. 20
Fraenkel, G.S. 71, *85*, 292, *311*
— Gunn, D.L. 32, *51*, 71, 74, *85*
Francis, D. see Mornin, L., Francis, D.
Franz, V. 229, *245*
Fraser, D.F. 183, *191*
Fraser, F.C., Purves, P.E. 278, *281*
Fraser, P.J. 329, *338*
— Sandeman, D.C. 329, *338*

Frederiksen, R.D. see Munk, O., Frederiksen, R.D.
Fredman, S.M. see Jahan-Perwar, B., Fredman, S.M.
Freeman, J.D.B. see Cook, A. et al.
Friedman, K. see Eckert, R. et al.
Friend, J. see Griffin, D.R. et al.
Frisch, von K. 125, 131, 134, *150*, 239, 250, 307, *311*
Frishkoph, L.S. see Capranica, R.R. et al.
Fruhstorfer, H. see Kenshalo, D.R. et al.
Fuortes, M.G.F. 474, 492, *498*
— see Baylor, D.A., Fuortes, M.G.F.
— see Fioravanti, R., Fuortes, M.G.F.
Furumoto, L. see Harrison, J.B., Furumoto, L.
Fuzessery, Z.M., Childress, J.J. 295, *311*
— see Ache, B.W. et al.

Gaddis, P.K. 528, *531*
Gaffal, K.P., Tichy, H., Theiss, J., Seelinger, G. 328, *338*, 461, *464*
Gagne, H.T. see Graziadei, P.P.C., Gagne, H.T.
Galifret, Y. 229, *245*
Galleni, L., Tongiorgi, P., Ferrero, E., Salghetti, U. 69, *85*
Galley, N. see Klinke, R., Galley, N.
Galtsoff, P.S. 119, *120*, 305, *311*, *312*
Gambale, S. see Mykytowycz, R. et al.
Gamble, F.W., Keeble, F. 71, *85*
Gamow, R.I. see Harris, J.F., Gamow, R.I.

Garcia, J., Hankins, W.G. 303, *312*
Gardiner, M.S. *120*
Garrone, R. see Pavans de Ceccaty, M. et al.
Gelfand, V.I. see Zelickman, E.A. et al.
Gelperin, A. 114, *120*
Gentry, R.L., Petersen, R.S. 270, *281*
Gentry, R. see Schusterman, R.J. et al.
Gerhardt, H.C. 175, *191*
Gerschenfeld, H.M. 116, *120*
Gesteland, R.C., Lettvin, J.Y., Pitts, W.H. 297, *312*
— Lettvin, J.Y., Pitts, W.H., Rojas, A. 297, *312*
Getchell, T.V. 297, *312*
Gewecke, M. 124, *150*
— see Burkhardt, D., Gewecke, M.
Ghiradella, H., Case, J.F., Cronshaw, J. 296, *312*
Gibson, R.N. 20, 387, *407*
— Blaxter, J.H.S., de Groot, S.J. 387, *407*
Giese, A.C. 35, *52*
Gift, J.J. 440, *464*
Gilpin-Brown, J.B. see Denton, E.J., Gilpin-Brown, J.B.
— see Denton, E.J. et al.
Gilula, N.B. see Satir, P., Gilula, N.B.
Glaser, R. 207, *214*
Gleason, K.K. see Reynierse, J.H. et al.
Gleeson, R.A. 291–317
Gnatzy, W., Schmidt, K. 328, *338*
Gogala, M. 123–153, 131, *150*
— Cokl, A., Draslar, K., Blazevic, A. 142, 144, *150*
— Hamdorf, K., Schwemer, J. 136, *150*
— Razpotnik, R. 142, 143, *150*
— see Hamdorf, K., Gogala, M.
— see Schwemer, J. et al.
Goin, C.J., Goin, O.B. *28*
Goin, O.B. see Goin, C.J., Goin, O.B.

Goldsmith, T.H. 472, 487, 488, 494, *498*
-- Bernard, G.D. *150*
-- *see* Barns, S.N., Goldsmith, T.H.
Goodge, W.R. 228, *245*, 511, *517*
Goodman, A.B. *see* Best, J.B. et al.
Goodwin, G.G., Greenhall, A.M. 267, *281*
Goodwin, M. *see* Regnier, F.E., Goodwin, M.
-- *see* Thiessem, D.D. et al.
Gordon, I.E. *see* Martin, G., Gordon, I.E.
-- *see* Martin, G. et al.
Gordon, S.A., Cohen, M.J. 320, *338*
Goris, R.C., Nomoto, M. 439, 441, 443, *464*
-- Terachima, S. 441, *464*
-- *see* Terashima, S.I., Goris, R.C.
-- *see* Terashima, S.I. et al.
Gorman, A.L.F. *see* McReynolds, J.S., Gorman, A.L.F.
Gorman, R.R., Ferguson, J.H. 177, *192*
Görner, P. 329, *338*
-- Andrews, P. 329, *338*
Gourbault, N. *see* Durand, J.P., Gourbault, N.
Gourevitch, G., Hack, M. 259, *281*
Govardovskii, V.I., Zueva, L.V. 236, 239, *245*
Gove, D., Burghardt, G.M. 303, *312*
Graber, V. 414, *436*
Graf, V. 211, *214*
-- Norren, D.V. 236, *245*
Graham, J.B., Rosenblatt, R.H. 508, *517*
Granda, A.M. *214*
-- Haden, K.A. 211, *214*
-- O'Shea, P.J. 211, *214*
-- Stirling, C.E. 211, *214*

-- *see* Liebman, P.A., Granda, A.M.
Grant, D., Anderson, O., Twitty, V.C. 171, *192*
Grant, P.I. *see* Mackie, A.M., Grant, P.I.
Grant, W.C. 171, 182, *192*
Gray, J., Lissmann, H.W. 113, *120*
Graziadei, P.P.C. 296, *312*
-- Gagne, H.T. 78, *85*
Graziadei, P. *see* Kenshalo, D.R. et al.
Green, R.P. *see* Walcott, C., Green, R.P.
Greenhall, A.M. *see* Goodwin, G.G., Greenhall, A.M.
Greenlaw, C.F. *see* Diercks, K.J. et al.
Greenwood, D. 222, *245*
Greenwood, P.H. 158, *167*
-- Rosen, D.E., Weitzmann, S.H., Meyers, G.S. 522, *531*
Grell, K.F. 41, *52*
Greuet, O. 41, *52*
Griffin, D.R. 219, 221, 235, *246*, 262-264, *281*, 351, 353, *371*
-- Friend, J., Webster, F. 264, *281*
-- Hopkins, C. 219, *246*
-- Suthers, R. 219, 226, 227, *246*
Grinnell, A.D. 263, *281*
-- *see* Bullock, T.H. et al.
Grocki, K. *see* Späth, M., Grocki, K.
Grubb, J.C. 170, 177, *192*
Grubb, T.C. Jr. 240, *246*, 299, *312*
Gruber, A. 38, *52*
Gruschka, H.D., Borchers, I.U., Coble, J.G. 259, *281*
Grzimek, B. 127, *150*
Gundlach, R.H., Chard, R.D., Skahen, J.R. 505, *517*
Gundy, G.C., Ralph, C.L. 208, *214*
-- Ralph, C.L., Wurst, G.Z. 208, *214*
Gunn, D.L. *see* Fraenkel, D.S., Gunn, D.L.
Gunter, R. 254, *281*
Gurin, S. *see* Carr, W.E.S. et al.

Ha, S.A.J. see Myrberg, A.A. et al.
Hack, M. see Gourevitch, G., Hack, M.
Haden, K.A. see Granda, A.M., Haden, K.A.
Haedrich, R.L. see Backus, R.H. et al.
Hagins, W.A. 472, *498*
Hagiwara, S., Kusano, K., Negishi, K. 524, *531*
— Morita, H. 524, *531*
Hailman, J.P., Jaeger, R.G. 187, 188, *192*
— see Jaeger, R.G., Hailman, J.P.
— see Jaeger, R.G. et al.
Haldall, P. 32, *52*
Haldane, J.B.S. 290, *312*
Hall, E.R. see Carr, W.E.S. et al.
Hall, J.D. see Jacobs, D.W., Hall, J.D.
Hall, J.R. 307, *312*
Hamasaki, D.I. 208, *214*
Hamdorf, K., Gogala, M. 136, *150*
— Schwemer, J. 136, 137, *150*
— see Gogala, M. et al.
— see Höglund, G. et al.
— see Razmjoo, S., Hamdorf, K.
— see Schwemer, J. et al.
Hamner, P., Hamner, W.M. 304, *312*
Hamner, W.M. see Hamner, P., Hamner, W.M.
Hamstra, R.H. see Bullock, T.H. et al.
Hankins, W.G. see Garcia, J., Hankins, W.G.
Hara, T.J. *28*
Harosi, F.I. see MacNichol, E.F. et al.
Harris, C.M. 353, *371*
Harris, G.G. *338*
— Bergeijk, W.A. van 161, *167*, 334, *338*
Harris, J.F., Gamow, R.I. 439, 441, 445, *464*

Harrison, J.B., Furumoto, L. 220, *246*
Harrison, T. 226, *246*
Harter, H.-A. see Ewert, J.P., Harter, H.-A.
Hartline, H.K. 20, 474, *498*
— Wagner, H.G., MacNichol, E.F. 474, *498*
Hartline, P.H. 205, 206, *214*
Hartman, H.B., Austin, W.D. 418, *436*
— Boettiger, E.G. 421, 423, 424, *436*
Harvey, G.W. see Norris, K.S., Harvey, G.W.
Hashimoto, H. see Waterman, T.H., Hashimoto, H.
Haskell, P.T. 142, 143, *150*, 328, *339*
Hasler, A.D. 299, *312*
— Wisby, W.J. 299, *313*
— see Madison, D.M. et al.
— see Scholz, A.T. et al.
Haugen, R.M. see Evans, W.E., Haugen, R.M.
Haupt, W. see Diehn, B. et al.
Hawking, E. 80, *85*
Hawkins, A.D. see Schuijf, A., Hawkins, A.D.
Heagy, W. see Lenhoff, H.M. et al.
Heath, J.E. see McCrea, M.J., Heath, J.E.
Hebel, R. 254, 255, *281*
Hecht, S., Shlaer, S., Pirenne, M.H. 155, *167*
Hedge, A. see Messenger, J.B. et al.
Heffner, H., Ravizza, R., Masterson, B. 259, *281*
Heiligenberg, W., Bastian, J. 528, *531*
Heinrich, B. 440, *464*
Heise, G.E. 220, *246*
Heitkamp, U. 74, *85*
Hellon, R.F., Hensel, H., Schäfer, K. 441, 449, *464*
Henderson, S.T. 129, 131, *150*
Hensel, H. 439, 441, 444–446, 455, 457, *464*

Hensel, H., Andres, K.H., Düring, M.V. 458, 462, *464*
-- Bowman, K. 441, *464*
-- Kenshalo, D.R. 448-450, *464*
-- Zotterman, Y. 441, 445, *464*
-- see Benzing, H. et al.
-- see Brown, B. et al.
-- see Hellon, R.F. et al.
-- see Kenshalo, D.R. et al.
Herman, L.M., Peacock, M.F., Yunker, M.P., Madsen, C.J. 275, *281*, 514, *517*
Herman, P. see Wever, E.G. et al.
Hermayer, K. see Keeton, W.T. et al.
Hernandez-Nicaise, M.L. 67, *85*
Hertzler, D.R. see Wever, E.G. et al.
Hess, von C. 228, *246*, 511, *517*
Hess, W.N. 417, *436*
Hesterman, E.R. see Mykytowycz, R. et al.
Heusen, A.P. van see Parker, G.H., Heusen, A.P. van
Heusser, H. 177, *192*
Heyer, W.R. see Straughan, I.R., Heyer, W.R.
Hickok, J., Davenport, D. 300, *313*
Hickok, J.F. see Davenport, D. et al.
Highman, K.C., Hill, L. 117, *121*
Highton, R. 184, *192*
Hildebrandt, E. see Diehn, B. et al.
Hill, L. see Highman, K.C., Hill, L.
Himmelmann, J.H. 119, *121*
Hirsch, J. see Ash, J.F. et al.
Hoar, W.S. 463
-- see Ali, M.A., Hoar, W.S.
Hodgkin, A.L., O'Bryan, P.M. 472, *499*
-- see Baylor, D.A., Hodgkin, A.L.
-- see Baylor, D.A. et al.

Hodgson, E.S. 20, 137, *150*, 301, *313*
-- Roeder, K.D. *20*
Hoffmann, C. 144, 147, *150*
Hoffmann, M. see Jander, R. et al.
Hogg, J.J. see Bergquist, P.R. et al.
Höglund, G., Hamdorf, K., Rosner, G. 137, *151*
Hollenberg, M.J. see Borwein, B., Hollenberg, M.J.
Holley, A., Duchamp, A., Revial, M.F., Juge, A., MacLeod, P. 297, *313*
Hollick, F.S.J. 413, 437
Holmgren, H. see Aurell, G., Holmgren, H.
Holst, von E. 335, *343*
Holt, E.B., Lee, F.S. 34, *52*
Hopkins, C.A. see Braten, T., Hopkins, C.A.
Hopkins, C.D. 527, *531*
Hopkins, C. see Griffin, D.R., Hopkins, C.
Horch, K. 413, *437*
Horn, E. 330, *339*
-- Kessler, W. 330, *339*
-- see Jander, R. et al.
Horrall, R.M. see Madison, D.M. et al.
-- see Scholz, A.T. et al.
Horridge, G.A. 20, 57, 67, 68, 72, *85*, *86*, 114, *121*, 131, *151*, 324, 325, 327, *339*, 483, *499*
-- Boulton, P.S. 325, *339*
-- see Bullock, T.H. et al.
Howarth, J.V. see Denton, E.J. et al.
Howe, N.R., Skeikh, Y.M. 307, *313*
Howell, D.J. see Simmons, J.A. et al.
Howell, W. see Best, J.B. et al.
Howland, H.G. see Eisner, T. et al.
Howse, P.E. 415, 417, 426, *437*
-- see Wilkinson, R., Howse, P.E.
Huang, B., Pitelka, D.R. 34, *52*

Hudspeth, A.J., Corey, D.P.
332, *339*
Hughes, A. 505, *517*
Hutchinson, V.H., Kosh, R.J.
207, *214*
Huth, H., Burkhardt, D.
239, *246*
Hyman, L.H. 61, 67, *86*

Idler, D.R., Fagerlund, V.H.M.,
Mayoh, H. 307, *313*
Iggo, A. 441, 449, *464*
Ikezono, E. *see* Bullock, T.H.
et al.
Ilichev, V.D. *see*
Kartashev, N.N., Ilichev, V.D.
Ingle, D. 176, *192*
— *see* Ewart, J.P., Ingle, D.
Ioalé, P. *see* Fiore, L.,
Ioalé, P.
— *see* Papi, F. et al.
Ireland, L.C. *see* Williams,
T.C. et al.
Isaac, M.J. *see* Naylor, E.,
Isaac, M.J.
Isaacks, N. *see* Thiessen,
D.D. et al.
Ishreyt, G. 228, *246*, 512, *517*

Jacobs, D.W. 278, *281*
— Hall, J.D. 278, *281*
Jacobs, G.H. *see* De Valois,
R.L., Jacobs, G.H.
— *see* Yolton, R.L. et al.
Jacobson, M. 304, *313*
Jaeger, L.S. *see* Jaeger, R.G.
et al.
Jaeger, R.G. 169-195
— Hailman, J.P. 179, 180, 181,
188, *192, 193*
— Hailman, J.P., Jaeger, L.S.
189, *193*
— *see* Ali, M.A. et al.
— *see* Hailman, J.P., Jaeger,
R.G.
Jägers-Röhr, E. 416, *437*

Jahan-Perwar, B., Fredman, S.M.
121
Jahn, T.I., Bovee, E.C. *52*
Jahn, T.L. *see* Bovee, E.C.,
Jahn, T.L.
Jahnke, V., Lundquist, P.G.,
Wersall, J. 221, *246*
Jamieson, G.S. 513, *517*
— Fisher, H.D. 268, 269, 270,
281, 513, *517*
Jander, R., Horn, E., Hoffmann,
M. 330, *339*
Jarosch, R. *see* Mussill, M.,
Jarosch, R.
Järvilehto, T. 450, *464*
Jennings, H.S. 32, *52*
Jennings, R.A. *see* Poulter, T.C.,
Jennings, R.A.
Jeon, K.W., Bell, L.G.E. 44, *52*
Jerlov, N.G. 158, *167*, 503, *517*
Jessen, K.H. *see* Dodt, E.,
Jessen, K.H.
Johannes, K.B. *see* Coward, S.J.
Johannes, K.B.
Johnson, C.S. 269, *282*
Johnson, F.H. *see* Chang, J.J.,
Johnson, F.H.
Johnson, G.L. 278, *282*, 513, *517*
Johnson, H.M. 232, *246*
Johnson, K.O., Darian-Smith, I.,
LaMotte, C. 456, *464*
— *see* Darian-Smith, J. et al.
Johnson, L. 270, *282*
Johnson, T.B., Salisbury, F.B.,
Connor, G.I. 209, *214*
Johnston, J.W. Jr., Moulton, D.G.,
Turk, A. 304, *313*
Johnstone, B. *see* Saunders, J.,
Johnstone, B.
Jones, A.R. 20, *86*
Jones, C.M. *see* Chambliss, C.L.,
Jones, C.M.
Jordan, O.R., Byrd, W.W.,
Ferguson, D.E. 177, *193*
Judd, H.D. *see* Walls, G.L.,
Judd, H.D.
Juge, A. *see* Holley, A. et al.
Justis, C.S., Taylor, D.H.
178, *193*

AUTHOR INDEX

Kaestner, A. 125, 126, *151*
Kaissling, K.E. 137, *151*, 297, 306, *313*
— see Boeckh, J. et al.
— see Schneider, D. et al.
Kalmijn, A.J. 164, 165, *167*, 524, *531*, *532*
Kalmring, K. see Cokl, A. et al.
Kameda, Y. see Bullock, T.H. et al.
Kampa, E.M. see Boden, E.P., Kampa, E.M.
Kandel, E.R. 114, *121*
— see Byrne, J. et al.
— see Kupperman, I., Kandel, E.R.
Kaneko, H. see Naitoh, Y., Kaneko, H.
Kaplan, E., Barlow, R.B. Jr., Chamberlain, S.C., Stelzner, D.J. 20
Kare, M.R., Rogers, J.G. Jr. 28
Karlson, P., Lüscher, M. 292, *313*
Kartashev, N.N., Ilichev, V.D. 219, *246*
Kasakashian, M.W. see Kasakashian, S.J. et al.
Kasakashian, S.J., Karakashian, M.W., Rudzinska, M.A. 36, *52*
Kasang, G. see Schneider, D. et al.
Katsuki, Y., Suga, N. 329, *339*
— see Bullock, T.H. et al.
— see Terashima, S.-I. et al.
Kauer, J.S. 28, 173, *193*
— Moulton, D.G. *193*
— Shepherd, G.M. 173, *193*
Kawakatsu, M. 70, *86*
Keast, D.N. see Wiener, F.M., Keast, D.M.
Keeble, F. see Gamble, F.W., Keeble, F.
Keeton, W.T. 241, *246*, 300, *313*

— Brown, A.I. 241, *246*, 300, *313*
— Kreithen, M., Hermayer, K. 241, *246*
— see Kreithen, M.L., Keeton, W.T.
— see Larkin, T.S., Keeton, W.T.
— see Yodlowski, M. et al.
Kellaway, P. 522, *532*
Kellogg, W.N. 277, *282*
— Rice, C.E. 275, *282*
Kelly, M.G. see Clarke, G.L., Kelly, M.G.
— see Tett, P.B., Kelly, M.G.
Kelly, S. see Corning, W.C., Kelly, S.
Kennedy, C.R. 78, 80, *86*
Kennedy, D. see Mellon, D. Jr., Kennedy, D.
— see Wiese, K. et al.
Kenshalo, D.R., Hensel, H., Graziadei, P., Fruhstorfer, H. 450, 458, *465*
— see Hensel, H., Kenshalo, D.R.
Kerkut, G.A., Taylor, B.J.R. 441, *465*
Kessler, W. see Horn, E., Kessler, W.
Kiefer, S.W. see Simmons, J.A. et al.
King-Smith, P.E. 234, *246*
Kinne, O. 375, 377, *407*
Kirschfeldt, K. 156, *167*
Kittleman, L.R. see Morris, R.W., Kittleman, L.R.
Kittredge, J.S., Takahashi, F.T., Lindsey, J., Lasker, R. 308, *313*
Kleiman, D.G. see Eisenberg, J.P. Kleiman, D.G.
Kleinholz, L. 485, *499*
Klinckoström, A. 507, *518*
Klinke, R., Galley, N. 334, *339*
— Schmidt, C.L. 334, *339*
Knapp, M.F. see Lowe, D.A. et al.
Knight-Jones, E.W. 112, *121*
— Morgan, E. 68, *86*, 382, 395, *407*
— see Morgan, E. et al.
— see Qasim, S.Z. et al.

Knüppel, H.A. 528, *532*
Knowles, A., Dartnall, H.J.A. 155, *167*
-- see Bowmaker, J.K., Knowles, A.
Kobayashi, H. 492, *499*
Koch, A.L. see Ehrenfeld, D.W., Koch, A.L.
Kohn, A.J. *121*
Kolle-Kralik, U. Ruff, P.W. 324, *339*
Kolmer, W. 512, *518*
Konishi, M. 218, 220, 246, *247*, 362, *371*
Koopowitz, H. 60, *86*
-- Silver, D., Rose, G. 60, 69, *86*
Kooyman, G.L. 511, *518*
Korenbrot, J.I., Cone, R.A. 475, *499*
Korsunovakaya, O.S. see Zhantiev, R.D., Korsunovakaya, O.S.
Kort, E.N. see Larsen, S.H. et al.
Kosh, R.J. see Hutchinson, V.H., Kosh, R.J.
Koshland, D.E. *52*
Köttgen, E., Abelsdorf, G. 210, *214*
Krasne, F.B., Lawrence, P.A. 484, *499*
Kreithen, M.L., Keeton, W.T. 240, 243, 244, *247*, 381, 390, 391, 400, *407*
Kreithen, M. see Keeton, W.T. et al.
-- see Yodlowski, M. et al.
Krisch, B. 67, *86*, 325, *340*
Kruger, L. see Schusterman, R.J., Kruger, L.
Kühlhorn, F. see Eidmann, H., Kühlhorn, E.
Kühne, W. 210, *215*
Kung, C., Eckert, R. *52*
Kupperman, I., Kandel, E.R. 117, *121*
Kusano, K. see Bullock, T.H. et al.

Kusano, K. see Hagiwara, S. et al.

Lacher, V. 441, 448, 454, 460, *465*
Lager, K.F., Bardach, J.E., Miller, R.R. *28*
Lamb, T.D. see Baylor, D.A. et al.
LaMotte, C. see Johnson, K.O. et al.
Land, M.F. 484, 489, *499*
-- see Blest, A.D., Land, M.F.
Landers, W.S., Rhodes, E.W. 69, *86*
Landreth, H.F., Ferguson, D.E. 175, 177, *193*
-- see Ferguson, D.E. et al.
Lanfranchi, A. see Bedini, C., Lanfranchi, A.
-- see Bedini, C. et al.
Lang, H. 329, *340*
Larimer, J.L., MacDonald, J.A. 528, *532*
Larkin, R., Sutherland, P. 242, *247*
Larkin, T.S., Keeton, W.T. 242, *247*
Larsen, S.H., Reader, R.W., Kort, E.N., Tso, W.W., Adler, J. *52*
Lasker, R. see Kittredge, J.S. et al.
Laurent, P., Rouzeau, J.-D. 405, *407*
Lavender, B.A. see Simmons, J.A. et al.
Lavender, W.A. see Simmons, J.A. et al.
Laverack, M.S. 20, 296, *313*, 329, *340*, 388, *407*, 413, 427, 429, *437*
-- see Clarac, F. et al.
-- see Thorpe, J.P. et al.
-- see Wales, W. et al.
Lavigne, D.M., Ronald, K. 513, *518*

AUTHOR INDEX

Lawrence, B., Schevill, W.E. 277, *282*
— *see* Schevill, W.E., Lawrence, B.
Lawrence, M. *see* Wever, E.G., Lawrence, M.
Lawrence, P.A. *see* Krasne, F.B., Lawrence, P.A.
Lawry, J.V. 494, *499*
Lawson, N. *see* Thiessen, D.D. et al.
Lay, D.M. 259-261, *282*
Leask, M.J.M. 243, *247*
Lee, D.L. *20*
Leedale, G.F. *52*
Legouix, J.P., Wisner, A. 261, *282*
Lehmkuhle, S.W. *see* Fox, R. et al.
Leigh, E.G. Jr. 185, 186, 188, *193*
Lenci, F. *see* Checcuci, A. et al.
— *see* Diehn, B. et al.
Lende, R.A. *see* Poulos, D.A., Lende, R.A.
Lenhoff, H.M. 69, *86*
— Heagy, W., Danner, J. 69, *86*
Leston, D. 142, *151*
Lettvin, J.Y., Maturana, H.R., McCulloch, W.S., Pitts, W.H. 28, 176, 188, *193*
— *see* Gesterland, R.C. et al.
Levin, D.A. 303, *313*
Lewis, W.J. *see* Norlund, D.A., Lewis, W.J.
Liebman, P.A. 210, *215*, 247, 475, *499*
— Granda, A.M. 211, *215*, 472, *499*
Lincer, J.L. *see* Sivak, J.G. et al.
Lincoln, R.J. 393, *407*
Lindauer, H. *see* Martin, H., Lindauer, M.
Lindauer, M., Nedel, J.O. 330, *340*

Lindsey, J. *see* Kittredge, J.S. et al.
Lindstedt, K.J. 301, 302, *313*
Linsenmair, E.K. 125, 144, *151*
Lissman, H.W. 162, *167*, 412, 413, 437, 523, 527, *532*
— Machin, K.E. 165, *167*, 523, *532*
Lissmann, H.W., Schwassmann, H.O. 527, *532*
— *see* Gray, J., Lissmann, H.W.
— *see* Machin, K.E., Lissmann, H.W.
Little, E.E. 305, *313*
Littlejohn, M.J. 174, *193*
Livamov, M.N. *see* Vasilenko, T.R.D. Lisamov, M.N.
Livingston, R. *see* Simmons, J.A. et al.
Lock, M.A. 70, *86*
Locket, N.A. 156, *167*, 483, *499*
Lockie, J.D. 229, *247*
Loeb, J. 32, *52*, 178, *193*
Loftus, R. 439-466
Lofty, J.R. *see* Edwards, C.A., Lofty, J.R.
Lombard, R.E. 28, 174, *194*
— Straughan, I.R. 174, *194*
Lowe, D.A., Mill, P.J. 419, *437*
— Mill, P.J., Knapp, M.F. 424, 426, *437*
— *see* Mill, P.J., Lowe, D.A.
Lowe, E.E., Dartnall, H.J.A. 210, *215*
Lowenstein, O.E. 28, 320, 332-335, *340*
Lowenstine, M. *see* Case, J.F. et al.
Lüling, K. 508, *518*
Lundquist, P.G. *see* Jahnke, V. et al.
Lüscher, M. *see* Karlson, P., Lüscher, M.
Lynch, J.F. *see* Wake, D.B., Lynch, J.F.
Lyon, R.H. 347, 348, 354, *371*
Lyons, K.M. 69, 78, *87*
Lythgoe, J.N. 115-168, 211, *215* 257, *282*, 471, 486-488, 490, 493, 494, *499*

Lythgoe, J.N., Dartnall, H.J.A. 270, *282*

MacArthur, R.H. *see* May, R.M., MacArthur, R.H.
MacDonald, A.G. 68, *87*
-- *see* Sleigh, M.A., Macdonald, A.G.
MacDonald, J.A. *see* Larimer, J.L., MacDonald, J.A.
MacGinitie, G.E., MacGinitie, N. 308, *313*
MacGinitie, N. *see* MacGinitie, G.E., MacGinitie, N.
Machemer, H. 323, *340*
-- de Peyer, J. 323, *340*
Machin, K.E., Lissmann, H.W. 523, *532*
-- *see* Lissman, H.W., Machin, K.B.
MacInnis, A.J. 81, *87*
MacIntosh, F.C. *see* Chang, T.M.S. et al.
Mackie, A.M., Grant, P.T. 307, *314*
Mackie, G.O. 57, 79, *87*, 107, *121*
-- Singla, C.L. 67, *87*
-- *see* Anderson, P., Mackie, G.O.
MacLeod, P. *see* Holley, A. et al.
MacMillan, D.L. *see* Field, L.H., MacMillan, D.L.
MacNichol, E.F. Jr., Feinberg, R., Harosi, F.I. 491, *500*
-- *see* Hartline, H.K. et al.
-- *see* Marks, W.B. et al.
Madison, D.M. 170, 171, 172, 177, 178, 184, *194*
-- Scholtz, A.T., Cooper, J.C., Horrall, R.M., Hasler, A.D., Dizon, A.E. 299, *314*
-- Shoop, C.R. 171, *194*
Madsen, C.J. 515, *518*
-- *see* Herman, L.M. et al.
Manske, U., Schmidt, U. 268, *282*
Mariscal, R.N. 58, 76, *87*

Markl, H. *20*, 144, *151*, 319-344
-- Tautz, J. 329, 330, *341*
Markovich, A.M. *see* Popov, A.V. et al.
Marks, W.B., Dobelle, W.H., MacNichol, E.F. 475, *500*
Marler, P. 362, *371*
-- *see* Marten, K., Marler, P.
Marrous, M.A. *see* Mokhtar, M., Marrous, M.A.
Marshall, N.B. 159, 161, *167*, 306, *314*
Marsot, P., Couillard, P. 43, *52*
Marten, K., Marler, P. 362, 363, *371*
Martin, G. 233, *247*
-- Gordon, I.E. 232, 233, *247*
-- Gordon, I.E., Cadle, D.R. 233, *247*
Martin, H., Lindauer, M. 147, *151*
Mason, S.G. *see* Chang, T.M.S. et al.
Massof, R.W., Chang, F.W. 505, *518*
Mast, S.O. 33, 34, *52*
Masterson, B. *see* Heffner, H. et al.
Masterson, F.A. *see* Ellins, S.R., Masterson, F.A.
Masterton, B., Diamond, I. *28*
Matthiessen, L. 505, *518*
Matthews, B.H.C. 474, *500*
Maturana, H.R. *see* Lettvin, J.Y. et al.
May, R.M., MacArthur, R.H. 185, 188, *194*
Maynard, D.M. *see* Wyse, G.A., Maynard, D.M.
Mayoh, H. *see* Idler, D.R. et al.
Mayr, I. 234, *247*
Mazokhin-Porshnyakov, G.A. *20*, 125, 129, 133, 136, 137, 146, 147, *151*
McClure, W.O. *see* Ash, J.F. et al.

AUTHOR INDEX

McCormick, J.G., Wever, E.G., Palin, J., Ridgway, S.H. 281, *282*
McCormick, J.C. *see* Wever, E.G. et al.
McCoullogh, C.B. *see* Passano, L.M., McCoullogh, C.B.
McCrea, M.J., Heath, J.E. 440, *465*
McCulloch, W.S. *see* Lettvin, J.Y. et al.
McCutcheon, F.H. 379, 387, *407*
McDaniel, O.H. *see* Carlson, D.E. et al.
McDonald, D.L. *see* Van Weel, P.B., McDonald, D.L.
McFarland, W.N. 275, *282*
-- Allen, D.M. 490, 491, *500*
-- Munz, F.W. 209, *215*, 239, *247*, 257, *282*, 491, *500*
-- *see* Allen, D.M., McFarland, W.N.
-- *see* Allen, D.M. et al.
-- *see* Munz, F.W., McFarland, W.N.
McFarlane, I.D. *87*
McGavin, M. 172, 183, *194*
McIver, S.B. 460, 461, *465*
-- Siemicki, R. 461, *465*
McKeown, J.P. *see* Ferguson, D.E. et al.
McLaren, D.J. 78, *87*
McReynolds, J.S., Gorman, A.L.F. *20*
McVay, S. *see* Payne, R.S., McVay, S.
Mead, G.M. *see* Backus, R.H. et al.
Meadows, P.S., Campbell, J.I. 69, *87*
Medway, Lord 226, *247*
Meglitsch, P.A. 57, *87*
Meister, F.J. 358, *371*
-- Ruhrberg, W. 358, *371*
Mellon, D. 329, *341*

Mellon, D. Jr., Kennedy, D. *20*
Melvill Jones, G. 335, *341*
Mendelson, M. 421, 424, 426, *437*
Mendenhall, B. *see* Murphey, R.K., Mendenhall, B.
Menzel, R., Roth, F. 73, *87*
Messenger, J.B. 491, 500
-- Wilson, A.P., Hedge, A. 491, *500*
Meyer, D.B. 234, *247*
-- Cooper, T.G. 234, *248*
Meyers, G.S. *see* Greenwood, P.M. et al.
Michalski, W.J., Séguin, J.J. 455, 456, *465*
Michelsen, A. 21, 141, *151*, 345-373
-- Nocke, H. 141, *151*, 348, 366, 367, *371*
Michieli, S. 136, *151*
Mill, P.J. 412, 413, 421, 422, *437*
-- Lowe, D.A. 417, 419, 421, 422, 424-426, *437*
-- *see* Lowe, D.A., Mill, P.J.
-- *see* Lowe, D.A. et al.
Miller, B. *see* Steinbrecht, R.A., Miller, B.
Miller, J. 259, *282*
Miller, J.D. *see* Dooling, R.J. et al.
Miller, R.R. *see* Lager, K.F. et al.
Miller, R.S. 183, *194*
Miller, W.H. 483, *500*
-- Snyder, A.W. 483, *500*
-- Wolbarsht, M.L. 208, *215*
Millodot, M. *see* Sivak, J.G., Millodot, M.
Millott, N. 481, 484, *500*
Minnich, D.E. 329, *341*
Mitchell, R.K. *see* Van Haga Reinauts, H.A., Mitchell, R.K.
Mitchell, R.W. 67, *87*
Miyake, A. 45, 46, *53*
Moffett, S.B. 432, *437*
Møhl, B. 272, 273, *282*

Møhl, E., Ronald, K. 345, *372*
Mokhtar, M., Marrous, M.A. 218, *248*
Moncrieff, R.W. 296, *314*
Money, K.E. *see* Steinbach, M.J., Money, K.E.
Moore, F.R. 242, *248*
Moran, D.T., Carter Rowley III, J. 328, *341*
Morgan, E. 20, 383, 385, 396, *407*
-- Nelson-Smith, A., Knight-Jones, E.W. 379, 387, *407*
-- *see* Knight-Jones, E.W., Morgan, E.
Morita, H. *see* Hagiwara, S., Morita, H.
Morita, M. *see* Pigon, A. et al.
Mornin, L., Francis, D. 41, *53*
Morris, R.W., Kittleman, L.R. 405, *408*
Morris, V.B., Shorey, C.D. 234, *248*
Morton, E.S. 218, *248*, 360, 361, *372*
Mote, M.I. *see* Scott, S., Mote, M.I.
Moulins, M. 417, 425, 429, *437*
-- Clarac, F. 419, 430, *437*
Moulton, D.G. 298, *314*
-- Beidler, L.M. 291, 296, 297, 298, 306, *314*
-- *see* Beauchamp, G.K. et al.
-- *see* Johnston, J.W. Jr. et al.
-- *see* Kauer, J.S., Moulton, D.G.
Moushegian, G., Rupert, A.L. 262, *283*
Mozell, M.M. 298, 304, *314*
-- *see* O'Connell, R.J., Mozell, M.M.
-- *see* Müller-Schwarze, D., Mozell, M.M.

Mrosovsky, N. 209, *215*
Mugford, R.A. *see* Beauchamp, G.K. et al.
Muller, W.A., Wicker, F., Eiben, R. 69, *87*
Müller-Schwarze, D. 293, 304, *314*
-- Mozell, M.M. 304, *314*
Mulligan, J.A. *see* Dooling, R.J. et al.
Munk, O. 159, *167*, 483, 493, 500, 507, *518*
-- Frederiksen, R.D. 156, *167*
Muntz, W.R.A. 178, 181, 186, *194*, 197-216, 232, 234, 236, *248*, 254, 256, *283*, 486, 488, 495, *500*
-- Sokol, S. 211, *215*
-- *see* Ali, M.A., Muntz, W.R.A.
-- *see* Sokol, S., Muntz, W.R.A.
Munz, F.W. *see* Allen, D.M. et al.
-- McFarland, W.N. 488, 491, *500*
-- *see* McFarland, W.N., Munz, F.W.
Murphey, R.K., Mendenhall, B. 329, *341*
-- Zaretsky, M.D. 141, *151*
Murray, J. 513, *518*
Murray, M.J., Capranica, R.R. 181, *194*
Murray, R.W. 28, *524, 532*
Mussill, M., Jarosch, R. 42, *53*
Mykytowycz, R., Hesterman, E.R., Gambale, S., Dudzinski, M.L. 306, *314*
Myrberg, A.A., Ha, S.A.J., Walewski, S., Banbury, J.C. *168*

Nachman, M., Cole, L.P. 302, *314*
Nagy, K.A. *see* Shoemaker, V.H. et al.
Naitoh, Y. 47, *53*
-- Eckert, R. 323, *341*
-- Kaneko, H. *53*
-- *see* Eckert, R., Naitoh, Y.
Nakamura, S. *see* Yamada, S. et al.
Narins, P.M., Capranica, R.R. 175, *194*

AUTHOR INDEX

Nautch, Y. see Eckert, R. et al.
Naylor, E., Atkinson, R.J.A. 393, 396, 397, 408
— Isaac, M.J. 383, 384, 408
Necker, R. 441, 465
Nedel, J.O. see Lindauer, M., Nedel, J.O.
Negishi, K. see Bullock, T.H. et al.
— see Hagiwara, S. et al.
— see Svaetichin, G.K. et al.
Nekrutenko, Y.P. 133, 152
Nelson, J.S. 29
Nelson-Smith, A. see Morgan, E. et al.
Neville, A.C. 124, 139, 152
Nevo, E. see Capranica, R.R. et al.
Newell, G.E. see Newell, P.F. Newell, G.E.
Newell, P.F., Newell, G.E. 484, 500
Newell, R.C. 68, 87
Nicholls, J.C., Baylor, D.A. 21
Nicol, J.A.C. 484, 494, 500
— Arnott, H.J., Best, C.G. 156, 168
— see Denton, E.J., Nicol, J.A.C.
— see Fineran, H.R., Nicol, J.A.C.
— see Zyznar, E.S., Nicol, J.A.C.
Nisbet, I.C.T., Drury, W.H. 398, 408
Nishihira, M. 69, 87
Niwa, H., Tamura, T. 493, 500
Nixon, J. see Schusterman, R.J. et al.
Noble, G.K. 29
— Schmidt, A. 206, 215
Nocke, H. 363, 368, 372
— see Michelsen, A., Nocke, H.
Nomoto, M. see Bullock, T.H. et al.
— see Goris, R.C., Nomoto, M.
Norberg, A. 224, 248
Nordeng, H. 299, 314
— see Døving, K.B. et al.
Norlund, D.A., Lewis, W.J. 292, 314
Norren, D.V. 236, 248
— see Graf, V., Norren, D.V.
Norris, K.S. 276, 277, 283
— Evans, W.E., Turner, R.N. 275, 283
— Harvey, G.W. 277, 279, 283
— see Davenport, D., Norris, K.S.
Nottebohm, F. see Bradbury, J.W., Nottebohm, F.
Novick, A. 226, 248
Nultsch, W. see Diehn, B. et al.
Nye, P.W. 229, 248
Nyholm, K.-G. 61, 88

Oakley, B. see Døving, K.B. et al.
Obara, Y. 133, 152
O'Bryan, P.M. see Hodgkin, A.L., O'Bryan, P.M.
O'Connell, R.J. 297, 314
— Mozell, M.M. 297, 314
O'Day, W.T., Fernandez, H.R. 494, 501
Oehme, H. 229, 248
Oesterhelt, D., Stoeckenius, W. 467, 501
Offutt, G.C. 329, 341
Ogden, T.E. see Brown, H.M., Ogden, H.M.
Ogren, L. see Carr, A., Ogren, L.
Ogura, A., Takahashi, K. 48, 53
Okajima, A. see Sandeman, D.C., Okajima, A.
Olive, P.J.W. see Clark, R.B., Olive, P.J.W.
Olivo, R.F. 21
Olson, N. see Embleton, T.F.W. et al.
Omodeo, P. see Piccinni, E., Omodeo, P.
Oosawa, F. see Seji, M., Oosawa, F.
Ordal, G.W., Fields, R.B. 42, 53

Ordy, J.M., Samorajski, T. 255, 256, *283*
Oreilli, M. *see* Schneider, V., Oreilli, M.
Organ, J. 171, 182, *194, 195*
O'Shea, P.J. *see* Granda, A.M., O'Shea, P.J.
Ostarello, G.L. 69, *88*
Otte, D. 304, *314*
Ottemann, R. *see* Reynierse, J.H. et al.
Otto, I. 206, *215*
Ottoson, D. *29*

Palin, J. *see* McCormick, J.G. et al.
— *see* Wever, E.G. et al.
Palka, J. *see* Edwards, J.S., Palka, J.
Palm, T. 446, *465*
Papi, F. 241, *248*
— Fiore, L., Fiaschi, V., Benvenuti, S. 300, *314*
— Ioalé, P., Fiaschi, V., Benvenuti, S., Baldaccini, N.E. 300, *315*
— *see* Benvenuti, S. et al.
Pardy, R.L. 76, *88*
Parker, G.H., Heusen, A.P. van 524, *532*
Parkins, B.J. *see* Blevins, C.E., Parkins, B.J.
Parsons, T.S. 29, 296, *315*
Partridge, B.L. *see* Pitcher, T.J. et al.
Pasino, E. *see* Cervetto, L. et al.
Passano, L.M. 67, *88*
— McCoullogh, C.B. 72, *88*
Paul, D.H. *21*
Paule, W.J. *see* Binggeli, R.L., Paule, W.J.
Pavans de Ceccaty, M. 107, *121*
— Thiney, Y., Garrone, R. 57, *88*
Payne, R.S. 224, 225, *248*

— McVay, S. 277, *283*
— Webb, B. 277, *283*, 345, 348, 352, 353, 355, *372*
Peacock, M.F. *see* Herman, L.M. et al.
Pearse, V.B. 76, *88*
Pedersen, S.B. *see* Skovmand, O., Pedersen, S.B.
Pedler, C., Tilley, R. 254, *283*
Peiponen, V.A. 234, *248*
Penard, E. 50, *53*
Pepper, R.L., Simmons, J.V. Jr. 275, *283*, 514, *518*
Perchard, R. *see* Delius, J. et al.
Perez, J.M. *see* Dawson, W.W. et al.
Perez-Miravetz, A. 33, *53*
Petersen, R.S. *see* Gentry, R.L., Petersen, R.S.
Peteya, D.J. 66, *88*
Petrovskaya, E.D. 141, *152*
— Rojkova, G.J., Tokareva, V.S. 329, *341*
Pfaffmann, C. 298, *315*
Phillips, D.W. 307, *315*
Piccinni, E., Albergoni, V., Coppellotti, O. *53*
— Omodeo, P. 40, *53*
Piddington, R.W. 334, *341*
Piercy, J.E., Embleton, T.F.W., Sutherland, L.C. 351, 352, 363, 364, *372*
Piercy, J.E. *see* Embleton, T.F.W. et al.
Piggins, D.J. 267, *281*, 513, *518*
— *see* Sivak, J.G., Piggins, D.J.
Pigon, A., Morita, M., Best, J.B. 74, *88*
— *see* Best, J.B. et al.
Pinet, J.M., Bernard, J. 138, *152*
Pirenne, M.H. 203, 204, *215*
— *see* Hecht, S. et al.
Pitcher, T.J., Partridge, B.L., Wardle, C.S. 163, *168*
Pitts, W.H. *see* Gesterland, R.C. et al.
— *see* Lettvin, J.Y. et al.
Platt, J.R. 49, *53*

AUTHOR INDEX

Polak, E.H. 302, *315*
Polyak, S.L. 230, *248*
Popov, A.V. 367, *376*
— Shuvalov, I.D., Svetlogorskaya, I.D., Markovich, A.M. 364, *372*
Popper, A.N. 164, *168*
— Fay, R.R. 164, *168*
Porter, K.R. *29*, 169, *195*
Poston, H.A. see Allen, D.M. et al.
Poulos, D.A., Lende, R.A. *455*, *465*
Poulter, T.C. 271, *283*
— Jennings, R.A. 271, *283*
— see Shaver, H.N., Poulter, T.C.
Pourriot, R., Clément, P. 73, *88*
— see Clément, P., Pourriot, R.
Powers, J.B., Winans, S.S. 306, *315*
Prescott, D.M. *53*
Price, A.J. see Crocker, M.J., Orice, A.J.
Price, L. 259, *283*
Priesner, E. 137, *152*
Prince, J.H. 514, *518*
Pringle, J.W.S. 412, *425*, *438*
Prosser, C.L. 11, *21*
Protasov, V.R. 508, *518*
Pumphrey, R.J. 161, 163, *168*, 219, 228, 231, 232, *248*, *249*, 367, *372*
— Rawdon-Smith, A.F. 329, *341*
Purves, P.E. 277, 278, *284*
— see Fraser, F.C., Purves, P.E.
Putter, A. 268, *284*
Pye, D. see Sales, G., Pye, D.
Pye, J.D. 353, *372*
— see Brown, A.W., Pye, J.D.

Qasim, S.Z., Rice, A.L., Knight-Jones, E.W. 394, *408*
Qutob, Z. 388, *391*, 402, *408*

Rabb, G.B., Rabb, M.S. 170, *195*
Rabb, M.S. see Rabb, G.B., Rabb, M.S.
Raikova, E.V. 77, *88*
Ralph, C.A. see Gundy, G.C., Ralph, C.A.
Ralph, C.L. see Gundy, G.C. et al.
Rampitsch, J. 61, *88*
Ramprashad, F., Corey, S., Ronald, K. 271, *284*
Rasmont, R. 73, *88*
Ravizza, R. see Heffner, H. et al.
Rawdon-Smith, A.F. see Pumphrey, R.J., Rawdon-Smith, A.F.
Ray, C. see Schevill, W.E. et al.
Raynauld, J.-P. 492, *501*
Rayport, S. see Wald, G., Rayport, S.
Razmjoo, S., Hamdorf, K. 137, *152*
Razpotnik, R. see Gogala, M., Razpotnik, R.
Reader, R.W. see Larsen, S.H. et al.
Reethof, G. see Carlson, D.E. et al.
Regbier, F.E. see Thiessen, D.D. et al.
Regen, J. 141, *152*
Regnier, F.E., Goodwin, M. 294, 295, *315*
Reille, A. 243, *249*
Renz, J. see Yolton, R.L. et al.
Repenning, C.A. 271, *284*
Revial, M.F. see Holley, A. et al.
Reynierse, J.H., Gleason, K.K., Ottemann, R. 74, *88*
Reynoldson, T.B. 70, *89*
Reysenbach de Haan, F.W. 278, *284*
Rheinlaender, J. 368, *372*
Rhodes, E.W. see Landers, W.S., Rhodes, E.W.
Ribeiro, A. de Miranda 527, *532*
Rice, A.L. 378, 382, 393, *408*
— see Qasim, S.Z. et al.

Rice, C.E. *see* Kellogg, W.N., Rice, C.E.
Rice, M. *see* Thiessen, D.D. et al.
Richard, G. 417, *438*
Richards, D.G. *see* Wiley, R.H., Richards, D.G.
Ridgway, S.H. *see* McCormick, J.G. et al.
— *see* Wever, E.G. et al.
Riding, J.L. *see* Croll, N.A. et al.
Riegel, V. *see* Best, J.B. et al.
Ripps, H., Weale, R.A. 155, 157, *168*
Rivamonte, A. 509, *518*, *519*
Rivamonte, L.A. 276, *284*
Roberts, B.L., Russell, I.J. 162, *168*, 334, *341*
— *see* Russell, I.J., Roberts, B.L.
Roberts, L.S. 79, *89*
Robson, J.G. *see* Cooper, G.F., Robson, J.G.
Rockstein, M. 125, 129, *152*
Rodieck, R.W. 156, 157, *168*, 254, *284*, 472, *501*
Roeder, K.D. *21*
— *see* Hodgson, E.S., Roeder, K.D.
Rogers, J.G. Jr., Beauchamp, G.K. 293, 305, *315*
Rogers, J.G. *see* Kare, M.R., Rogers, J.G. Jr.
Rogers, W.P. 72, *89*
Rojas, A. *see* Gesterland, R.C. et al.
Rojkova, G.J. *see* Petrovskaya, E.D. et al.
Romer, A.S. 523, *532*
Römer, H. 368, *372*
Romeskie, M., Yager, D. 236, *249*
Romoser, W.S. 123, 125, 126, *152*
Ronald, K. *see* Lavigne, D.M., Ronald, K.
— *see* Møhl, E., Ronald, K.

— *see* Ramprashad, F. et al.
— *see* Terhune, J.M., Ronald, K.
Ropartz, P. 305, *315*
Roper, F.E. *see* Young, R.E., Roper, F.E.
Rose, G. *see* Koopowitz, H. et al.
Rosen, D.E. *see* Greenwood, P.H. et al.
Rosenblatt, R.H. *see* Graham, J.B., Rosenblatt, R.H.
Rosenfield-Wessels, S. *see* Fite, K.V., Rosenfield-Wessels, S.
Rosner, G. *see* Höglund, K. et al.
Ross, D.M. 75, *89*
— Sutton, L. 75, *89*
Rostron, M.A. *see* Chia, F.S., Rostron, M.A.
Roth, F. *see* Menzel, R., Roth, F.
Rottman, S.J., Snowdon, C.T. 306, *315*
Rouzeau, J.-D. *see* Laurent, P., Rouzeau, J.-D.
Rudzinska, M.A. *see* Kasakashian, S.J. et al.
Ruff, P.W. *see* Kolle-Kralik, U., Ruff, P.W.
Ruhrberg, W. *see* Meister, F.J., Ruhrberg, W.
Rupert, A.L. *see* Moushegian, G., Rupert, A.L.
Rupprecht, R. 142, *152*
Russell, F.S. 67, 72, *89*
Russell, I.J. 160, 161, *168*, 334, *341*
— Roberts, B.L. 334, *341*
— Sellick, P.M. 160, *168*
— *see* Roberts, B.L., Russell, I.J.
Ryder, R.A. 489, *501*
— *see* Ali, M.A. et al.
Ryder, T.A. *see* Bowen, I.D., Ryder, T.A.
Ryland, J.S. *21*

St. John, S. *see* Schusterman, R.F. et al.
Saji, M., Oosawa, F. 36, *53*

AUTHOR INDEX

Sales, G., Pye, D. 144, 152, 225, 249, 263, 276, 284, 366, 372
Salghetti, U. see Gelleni, L. et al.
Salisbury, F.B. see Johnson, T.B. et al.
Samorajski, T. see Ordy, J.M., Samorajski, T.
Sandeman, D.C. 329, 335, 342
-- Okajima, A. 329, 342
-- see Fraser, P.J., Sandeman, D.C.
-- see Silvey, G.E. et al.
Sass, H. see Altner, H. et al.
Satir, P., Gilula, N.B. 416, 438
Sato, M. 298, 315
Sato, O. see Bullock, T.H. et al.
Saunders, J., Dooling, R. 222, 249
-- Johnstone, B. 219, 249
Saunders, J.C. see Dooling, R.J., Saunders, J.C.
Savage, J.M. 170, 195
Scallet, A.C. see Simmons, J.A. et al.
Schäfer, K. see Hellon, R.F. et al.
Schaller, D. 460, 465
Scheich, H., Bullock, T.H. 165, 168, 525, 532
-- see Bullock, T.H. et al.
Scherer, E. see Dodt, E., Scherer, E.
Schevill, W.E., Lawrence, B. 278, 284
-- Watkins, W.A., Backus, R.H. 271, 278, 284
-- Watkins, W.A., Ray, C. 271, 284
-- see Lawrence, B., Schevill, W.E.
-- see Watkins, W.A., Schevill, W.E.
Schiøtz, A. 174, 195

Schlichter, D. 76, 89
Schmidt, A. see Noble, G.K., Schmidt, A.
Schmidt, C.L. see Klinke, R., Schmidt, C.L.
Schmidt, K. 328, 342
-- see Gnatzy, W., Schmidt, K.
Schmidt, U. see Manske, U., Schmidt, U.
Schmook, J. see Schusterman, R.J. et al.
Schneider, D. 304, 315
-- Kasang, G., Kaissling, K.-E. 304, 315
-- Steinbrecht, R.A. 296, 304, 316
-- see Boeckh, J. et al.
Schneider, F. 148, 152
Schneider, G. see Buckhardt, D., Schneider, G.
Schneider-V. Oreilli, M. 507, 519
Schoener, T.W. 183, 185, 188, 189, 195
Scholz, A.T., Horrall, R.M., Cooper, J.C., Hasler, A.D. 299, 316
Scholtz, A.T. see Madison, D.M. et al.
Schöne, H. 320, 327, 342
Schoonhoven, L.M. 441, 465
Schuijf, A., Hawkins, A.D. 333, 334, 342, 345, 372
Schusterman, R.J. 269, 274, 284, 285, 513, 519
-- Balliet, R.F. 270, 271, 285, 513, 519
-- Balliet, R.F., Nixon, J. 274, 285
-- Balliet, R.F., St. John, S. 271, 285
-- Barrett, B. 271, 285, 514, 519
-- Gentry, R., Schmook, J. 271, 285
-- see Balliet, R.F., Schusterman, R.J.
Schwanzara, S.A. 490, 501
Schwartz, E. 162, 168, 334, 342

Schwartzkopff, J. *29*, 141, 153, 219–221, *249*, 328, *342*
Schwassman, H.O. 521–533
— Kruger, L. 516, *519*
Schwassmann, H.O. *see* Lissmann, H.W., Schwassmann, H.O.
Schwemer, J., Gogala, M., Hamdorf, K. 136, *153*
— *see* Gogala, M. et al.
— *see* Hamdorf, K., Schwemer, J.
Scott, S., Mote, M.I. 492, *501*
Seelinger, G. *see* Gaffal, K.P. et al.
Séguin, J.J. *see* Michalski, W.J., Séguin, J.J.
Seitz, G. 124, *153*
Sekhon, S.S. *see* Silifer, E.H., Sekhon, S.S.
Sellick, P.M. *see* Russell, I.J., Sellick, P.M.
Sever, D.M. 171, *195*
Shafer, R. 297, *316*
Shaver, H.N., Poulter, T.C. 271, *285*
Shaw, E.A.G. 357, *372*
Shaw, S.R. *21*
Sheikh, Y.M. *see* Howe, N.R., Sheikh, Y.M.
Sheldon, G.A.P. *see* Thorpe, J.P. et al.
Shepherd, G.M. *see* Kauer, J.S., Shepherd, G.M.
Sherrington, C.S. 411, 413, *438*
Shifrin, M.A. *see* Zelickman, E.A. et al.
Shlaer, R. 233, *249*
Shlaer, S. *see* Hecht, S. et al.
Shoemaker, V.H., Nagy, K.A., Costa, W.R. 440, *466*
Shoop, C.R. *see* Madison, D.M., Shoop, C.R.
Shores, D.L. *see* Backus, R.H. et al.

Shorey, C.D. *see* Morris, V.B., Shorey, C.D.
Shorey, H.H. 291, 302, 304, *316*
— *see* Farkas, S.R., Shorey, H.H.
Shuvalov, I.D. *see* Popov, A.V. et al.
Siemicki, R. *see* McIver, S.E., Siemicki, R.
Silberglied, R.E. *see* Eisner, T. et al.
Sillman, A.J. *29*, 227, 229, 234, *249*
Silver, D. *see* Koopowitz, H. et al.
Silverstein, R.M., Young, J.C. 292, *316*
Silvey, G.E., Dunn, P.A., Sandeman, D.C. *342*
Simmons, J.A. 263, *285*
— Howell, D.J., Suga, N. 263, 264, *285*
— Lavender, W.A., Lavender, B.A., Doroshow, C.A., Kiefer, S.W., Livingston, R., Scallet, A.C. 264, *285*
— *see* Wever, E.G. et al.
Simmons, J.V. Jr. *see* Pepper, R.L., Simmons, J.V. Jr.
Simmons, V.P. *see* Fisher, F.H., Simmons, V.P.
Sinclair, M.E. *see* Bergquist, P.R. et al.
Singla, C.L. 67, 68, *89*
Sivak, J.G. 228, *249*, 503–519
— Lincer, J.L., Bobier, W. 228, *249*, 505, 511, *519*
— Millodot, M. 228, *249*, 510, *519*
— Piggins, D.J. 513, *519*
Skahen, J.R. *see* Gundlach, R.H. et al.
Skovmand, O., Pedersen, S.B. 369, 370, *372*
Sleigh, M.A., Macdonald, A.G. 377, 408
Slifer, E.H., Sekhon, S.S. 460, *466*

Slijper, E.J. 274, *285*
Smith, C.A. *see* Takasaka, T., Smith, C.A.
Smith, D.C. 36, *53*
Smith, F.H. *see* Wald, G. et al.
Smith, J.E. 110, *121*
Smith, J.C. *see* Tucker, D., Smith, J.C.
Smith, J.M. *see* Croll, N.A. et al.
Smith, K.U. *see* Bridgeman, C.S., Smith, K.U.
Smyth, J.D. *21*
-- *see* Blitz, N.M., Smyth, J.D.
Snow, D.W. 225, 226, *249*
Snowdon, C.T. *see* Rottman, S.J., Snowdon, C.T.
Snyder, A.W. *see* Miller, W.H., Snyder, A.W.
Sokol, S., Muntz, W.R.A. 211, *215*
-- *see* Muntz, W.R.A., Sokol, S.
Sokolove, P.G. *see* Davis, E.E., Sokolove, P.G.
Solomon, D.J. 299, *316*
Somiya, H. 495, *501*
Southern, W.E. 241, *249, 250*
Späth, M. 456, *466*
-- Grocki, K. 456, *466*
Spaulding, J.G. 77, *89*
Spinola, S.M., Chapman, K.M. 328, *342*
Spirito, C.P. *see* Barnes, W.J.P. et al.
Spong, P. *see* White, D. et al.
Spray, D.C. 441, *466*
Sprott, R.L. *see* Cheal, W., Sprott, R.L.
Stager, K.E. 240, 241, *250*
Stavenga, D.G. 137, *153*
Stebbins, R.C., Cohen, N.W. 207, *215*
-- Eakin, R.M. 207, *215*
Steinbach, A.B. 527, *533*
Steinbach, M.J., Money, K.E. 229, *250*

Steinbrecht, R.A. 296, 297, *316*
-- Miller, B. 460, *466*
-- *see* Schneider, D., Steinbrecht, R.A.
Stelzner, D.J. *see* Kaplan, E. et al.
Stensiö, E.A. 522, *533*
Steven, D.M. 61, *89*, 481, *501*
Stirling, C.E. *see* Granda, A.M., Stirling, C.E.
Stoeckenius, W. *see* Danon, A., Stoeckenius, W.
-- *see* Oesterhelt, D., Stoeckenius, W.
Stone, J. 253, *285*
Storch, V., Abraham, R. 70, *89*
Straughan, I.R. 174, 175, *195*
-- Heyer, W.R. 175, *195*
-- *see* Lombard, R.E., Straughan, I.R.
Strother, G.K. 234, *250*
Suga, N. *see* Bullock, T.H. et al.
-- *see* Katsuki, Y., Suga, N.
-- *see* Simmons, J.A. et al.
Sugaware, Y. *see* Yamamoto, T. et al.
Sulkin, S.D. 395, 396, *408*
Sutherland, L.C. *see* Piercy, J.E. et al.
Sutherland, P. *see* Larkin, R., Sutherland, P.
Suthers, R.A. 217-287
-- Chase, J., Braford, B. 268, *286*
-- Wallis, N. 265, *286*
-- *see* Chase, J., Suthers, R.A.
Suthers, R. *see* Griffin, D.R., Suthers, R.
Sutton, L. *see* Ross, D.M., Sutton, L.
Svaetichin, G.K., Negishi, K., Fatehchand, R. 493, *501*
Svetlogorskaya, I.D. *see* Popov, A.V. et al.
Swedmark, B. *89*
Szabo, T. *see* Fessard, A., Szabo, T.

Tagmat, A.T. see Brown, H. et al.
Takahashi, F.T. see Kittredge, J.S. et al.
Takahashi, K. see Ogure, A., Takahashi, K.
Takasaka, T., Smith, C.A. 219, 222, *250*
Takeda, K. see Watanabe, A., Takeda, K.
Takeoka, S. see Yamada, S. et al.
Tamura, T. see Niwa, H., Tamura, T.
Tansley, K. 201, 203, *215, 216*, 503, *519*
Tardent, P. 72, *89*
Tartar, V. 34, *53*
Tasaki, K. see Yamamoto, T. et al.
Tateda, H. see Yokohari, F., Tateda, H.
Täuber, U. see Bohn, H., Täuber, U.
Tautz, J. 329, 330, *342*
-- see Markl, H., Tautz, J.
Tavolga, W.N. *29*
Taylor, B.J.R. see Kerkut, G.A., Taylor, B.J.R.
Taylor, D.H., Adler, K. *195*
-- Ferguson, D.E. 177, *195*
-- see Adler, K., Taylor, D.H.
-- see Justis, C.S., Taylor, D.H.
Taylor, R.C. 426, *438*
Teal, J.M. see Backus, R.H. et al.
Teideman, D.J. see Cook, A. et al.
Terashima, S.I., Goris, R.C. 206, *216*
-- Goris, R.C., Katsuki, Y. 442, 444, 457, *466*
Terachima, S. see Goris, R.C., Terachima, S.
Terborgh, J., Weske, J.S. 185, *195*
Terhune, J.M., Ronald, K. 273, 274, *286*

Tett, P.B., Kelly, M.G. 493, *501*
Theiss, J. see Gaffal, K.P. et al.
Thies, G. see Budelmann, B.-U., Thies, G.
Thiessen, D.D., Regnier, F.E., Rice, M., Goodwin, M., Isaacks, N., Lawson, N. 293, *316*
Thiney, Y. see Pavans de Ceccaty, M. et al.
Thomas, I. see Autrum, H., Thomas, I.
Thorpe, J.P., Sheldon, G.A.B., Laverack, M.A. *21*
Thorpe, W.H. 109, *121*
-- Crisp, D.J. 402, 404, *408*
Thurm, U. 323, 328, *343*, 455, 461, *466*
Thurow, G. 182, 185, *195*
Thurston, M.H. 159, *168*
Tichy, H. 441, 450, 454, *466*
-- see Altner, H. et al.
-- see Gaffal, K.P. et al.
Tilley, R. see Pedler, C., Tilley, R.
Tobach, E. 291, *316*
Todd, J.H., Atema, J., Bardach, J.E. 306, *316*
-- see Bardach, J.E. et al.
Tokareva, V.A. see Petrovskaya, E.D. et al.
Tominaga, Y. see Yokohari, F. et al.
Tomita, T. 475, *501*
Tongiorgi, P. see Galleni, L. et al.
Tonosaki, A. see Yamamoto, T. et al.
Torelle, E. 178, *196*
Torre, V. see Cervetto, L. et al.
Trainer, J.E. 220, *250*
Tristram, D.A. 172, 182, *196*
Trochta, R.T. see Diercks, K.J. et al.
Tso, W.W. see Larsen, S.H. et al.
Tsuji, F.I. see Fernandez, H.R., Tsuji, F.I.

Tsvetkov, V.I. 392, 408
Tucker, D., Smith, J.C. 29
Turk, A. see Johnston, J.W. et al.
Turner, R.N. see Norris, K.S. et al.
Twitty, V.C. 171, 184, 196
-- see Grant, D. et al.
Tyler, S. 64, 89
Tytler, P., Blaxter, J.H.S. 390, 395, 402, 408
-- see Blaxter, J.H.S., Tytler, P.

Ulmer, M.J. 81, 90
Underwood, G. 29, 201, 202, 216

Vakkur, G., Bishop, P.O. 254, 286, 505, 519
Vandel, A. 62, 67, 90
Van Bergeijk, W.A. see Bergeijk, van, W.A.
-- see Harris, G.C., Van Bergeijk, W.A.
Van Haga Reinauts, H.A., Mitchell, R.K. 21
Van Tyne, J., Berger, A.J. 29
Van Wagtendonk, W.J. 36, 53
Van Weel, P.B., McDonald, D.L. 446, 466
Vasilenko, T.R.D., Livanov, M.N. 388, 408
Vedel, J.P. see Angaut-Petit, D. et al.
Vernberg, F.J. see Vernberg, W.B., Vernberg, F.J.
Vernberg, W.B., Vernberg, F.J. 68, 90
Vevers, G. see Fox, H., Vevers, G.
Viaud, G. 72, 73, 90
Vidaver, W. 390, 408

Villars, T. see Bardach, J.E., Villars, T.
Vinnikov, Y.A. 296, 297, 316, 323-326, 332, 343
Vinogradov, M.E. 68, 72, 90
Von Békésy, G. see Békésy, von G.
Von Frisch, K. see Frisch von K.
Von Hess, C. see Hess von, C.
Von Holst, E. see Holst, von E.
Von Mecklenburg, C. see Dahl, E. et al.
Voyelle, J. 298, 316

Waelchi, G. 234, 250
Wagner, G. see Fiaschi, V., Wagner, G.
Wagner, H.G. see Hartline, H.K. et al.
Wagner, H.-J. see Ali, M.A., Wagner, H.-J.
Wake, D.B. 169, 196
-- Lynch, J.F. 169, 196
Walcott, C., Green, R.P. 242, 250
Wald, G. 236, 250
-- Brown, F.K., Smith, P.H. 211, 216
-- Rayport, S. 21, 96, 121, 484, 485, 492, 501
Waldow, U. 441, 448, 454, 460, 466
Wales, W. 411-438, 481, 501
-- Clarac, F., Dando, M.R. Laverack, M.S. 417, 418, 427, 429, 434, 438
-- Clarac, F., Laverack, M.S. 430, 438
-- see Clarac, F., Wales, W.
-- see Clarac, F. et al.
-- see Ferrero, E., Wales, W.
Walewski, S. see Myrberg, A.A. et al.
Walker, E.P. 29
Walker, R.J. 116, 121
Wallace, H.R. 66, 74, 90
Wallis, N. see Suthers, R.A., Wallis, N.

Walls, G.L. 156, 157, 160, 168, 198-201, 209, 216, 227-231, 250, 253, 254, 256-268, 271, 286, 486, 501, 503, 506, 508-512, 514, 519
— Judd, H. 234, 235, 250, 256, 286
Walsby, A.E. 391, 392, 408
Walther, J. see Dodt, E., Walther, J.
Ward, S.N. 69, 90
Wardle, T.L. see Pitcher, T.J. et al.
Warner, J. see Case, J.F. et al.
Warnke, U. 146, 147, 153
— see Altman, G., Warnke, U.
Warren, F.J. see Denton, E.J., Warren, F.J.
Watanabe, A., Takeda, K. 527, 533
Watanabe, T. see Yamada, S. et al.
Waterman, T.H. 21, 486, 501
— Hashimoto, H. 487, 501
Watkins, W.A., Schevill, W.E. 277, 286
— see Schevill, W.E. et al.
Weale, R.A. 254, 255, 256, 286
— see Ripps, H., Weale, R.A.
Webb, B. see Payne, R.S., Webb, B.
Webster, D.B. 29, 258-261, 286
— Webster, M. 258-262, 286
Webster, F. see Griffin, D.R. et al.
Webster, M. see Webster, D.B., Webster, M.
Wehner, R. 131, 136, 153
Weitzmann, S.H. see Greenwood, P.H. et al.
Wells, M.J. 21, 106, 115, 118, 121, 327, 343
Wendler, G. 330, 343

Wenzel, B.M. 240, 250, 303, 316
Wersäll, J., Bagger-Sjöbäck, D. 336, 343
— see Jahnke, V. et al.
Weske, J.S. see Terborgh, J., Weske, J.S.
Westendorf, D.H. see Fox, R. et al.
Westfall, J.A. 60, 90
Wever, E.G., Herman, P., Simmons, J.A., Hertzler, D.R. 220, 250
— Lawrence, M. 287
— McCormick, J.C., Palin, J., Ridgway, S.H. 278, 287
— see McCormick, J.G. et al.
Wheeler, J.W. 294, 295, 316
White, D., Cameron, N., Spong, P., Bradford, J. 275, 287
Whitear, M. 417, 419, 424, 426, 438
Whitehead, J.A. 49, 53
Whittacker, R.H., Feeny, P.P. 292, 317
Whittaker, R.H. see Brown, W.L. Jr. et al.
Whitten, W.K. 305, 317
Whitteridge, D. 254, 287
Whittow, G.C. 29
Wicker, F. see Müller, W.A. et al.
Wiener, F.M., Keast, D.N. 354, 356, 362, 372
Wiersma, C.A.G. 424, 438
— Boettiger, E.G. 420, 421, 423, 424, 438
Wiese, K. 329, 343
— Calabrese, R.L., Kennedy, D. 329, 344
Wieser, W. 66, 81, 90
Wiewandt, T.A. 175, 187, 196
Wigglesworth, V.B. 125, 153
Wilbur, K.M., Yonge, C.M. 121
Wiley, R.H., Richards, D.G. 346, 372
Wilkinson, R., Howse, P.E. 223, 251
Willee, J.J., Ehret, C.F. 50, 53

Williams, A.O. 355, *373*
Williams, G.B. 61, 73, *90*
Williams, J.M. *see*
 Williams, T.C.,
 Williams, J.M.
— *see* Williams, T.C. et al.
Williams, T.C., Ireland,
 L.C., Williams, J.M.
 265, *287*
— Williams, J.M. 265, *287*
Wilson, A.P. *see* Messenger,
 J.B. et al.
Wilson, D.P. 298, *317*
Wilson, E.O. 291, 307, *317*
— Bossert, W.H. 292, *317*
— *see* Bossert, W.H.,
 Wilson, E.O.
Wilson, G. 505, 513, *519*
Wilson, R.A., Denison, J.
 81, *90*
Wiltschko, R. *see*
 Wiltschko, W., Wiltschko,
 R.
Wiltschko, W., Wiltschko, R.
 242, 243, *251*
Winans, S.S. *see* Powers, J.B.,
 Winans, S.S.
Winet, H. 49, *53*
Wing, A.S. *see* Backus, R.H.
 et al.
Winkler, R.N., Corliss, J.D.
 37, *54*
Wisby, W.J. *see* Hasler, A.D.,
 Wisby, W.J.
Wisner, A. *see* Legouix, J.P.,
 Wisner, A.
Wolbarsht, M.L. *see* Miller,
 W.H., Wolbarsht, M.L.
Wolf, E. *see* Crozier, W.J.,
 Wolf, E.
Wolff, H.G. 325, *344*
— *see* Budelmann, B.-U.,
 Wolff, H.G.
Wolff, T. 66, *90*
Wolken, J.J. 21, 39, 40,
 54, 467, 477, 481, 487, *502*
— Shin, E. 39, *54*
Wood, C.A. 229, *251*
Wood, D.C. 35, *54*

Wood, L. 96, *121*
Woods, J.D. *168*
Wottog, H. *see* Cokl, A. et al.
Wright, C.A. 80, 81, *90*
— *see* Canning, E.U., Wright, C.A.
Wright, P.G. *see* Denton, E.J.
 et al.
Wright, R.H. 138, *153*
Wurst, G.Z. *see* Gundy, G.C.
 et al.
Wurster, R. *see* Benzing, H.
 et al.
Wyse, G.A., Maynard, D.M.
 418, *438*

Yager, D. *see* Romeskie, M.,
 Yager, D.
Yamada, S., Watanabe, T.,
 Nakamura, S., Yokoyama, H.,
 Takeoka, S. 351, *373*
Yamamoto, T., Tasaki, K.,
 Sugaware, Y., Tonosaki, A.
 481, *502*
Yanagisawa, K. *see* Bullock, T.H.
 et al.
Yingst, D.R., Fernandez, H.R.,
 Bishop, I.G. 96, *122*, *491*
Yodlowski, M., Kreithen, M.,
 Keeton, W. 221, *251*
Yokohari, F., Tateda, H.
 459, *466*
— Tominaga, Y., Ando, M.,
 Tateda, H. 454, 459, *466*
Yokoyama, H. *see* Yamada, S. et al.
Yolton, D.P. *see* Yolton, R.L.
 et al.
Yolton, R.L., Yolton, D.P.,
 Renz, J., Jacobs, G.H. 256,
 287
Yonge, C.M. *see* Wilbur, K.M.,
 Yonge, C.M.
Young, D. 415-417, 419, 426,
 438
Young, J.C. *see* Silverstein,
 R.M., Young, J.C.
Young, J.Z. 29, 115, *122*
Young, R.E. 483, 484, 493, 494,
 502

Young, R.E., Roper, F.E.
  494, *502*
Young, R.W., Bok, D.
  235, *251*
Yunker, M.P. *see* Herman,
  L.M. et al.

Zablotskaya, M.M. 221, *251*
Zadan, P.M. *see* Akoev, G.N.
  et al.
Zaretsky, M.D. *see*
  Murphey, R.K.,
  Zaretsky, M.D.
Zelickman, E.A., Gelfand,
  V.I., Shifrin, M.A.
  67, *90*

Zhantiev, R.D., Korsunovakaya,
  O.S. 368, *373*
Zotterman, Y. *see* Hensel, H.,
  Zotterman, Y.
Zueva, L.V. *see* Govardovskii,
  V.I., Zueva, L.V.
Zug, G.R. 170, *196*
Zwehl, von V. *see* Autrum, H.,
  Zwehl, von V.
Zweifel, R.G. 174, *196*
Zyznar, E.S., Ali, M.A.
  485, *502*
— Nicol, J.A.C. 485, *502*

SUBJECT INDEX

Absorption 238, 350, 353, 358
-- maxima 274
-- peaks 136, 270
-- spectrum 234, 236
Abyssal species 66
Acanopus carbo 162
Acanthocephala 59, 76
Acceleration, Angular 163
Accommodation 228, 268, 270, 275, 506, 509, 510, 514
Acetes siboqae australis
Acetylcholine 116
Acmaea 307
Acmaea limatula 307
Acmaea scutum 307
Acoelomate 57
Acoustic prey detection *See* Prey detection Acoustic
Acoustical orientation 223, 226, 267
-- signals 141, 346
Acoustico-lateris system 144, 160
-- Mechanical 160
Acris 174
Actinia equina 69
Actinians 72
Acuity 230, 231
-- *See also* Visual acuity
Adaptation 6
Adaptive radiation *See* Radiation, Adaptive
Adaptive radiation, Proprioceptors *See* Proprioceptors Adaptive radiation
Adhesive glands 64
-- organ 78
-- structures 139
Advanced systems 108
Aedes 460, 467
Aedes aeqypti 446, 447
Aeqolius funereus 234
Aequorea 67
African honeyguides 241
-- serpent eagle 233

Aggregation 69
Agnatha 164
Agriolimax 484
Air 352, 503
-- borne odors 173
-- flow receptor 320
-- movement receptor 141
-- pressure 244
-- turbulence 367
Alarm pheromones 306
-- reaction 6
Albatrosses 240
Alcedo attis attis 512
Algae 392
Allelochemics 292
Allomones 292
Allopora 69
Amazon dolphin 278
Ambient light 470, 493
-- Sunrise 209
-- sea noise 277
-- temperature *See* Temperature, Ambient
Ambiguity of response 445, 448, 452
Amblonyx cineria 270, 514
American kestrel 232
Amino acid 69
Amoeba 5, 324
-- *Light* 33,
-- *proteus* 33, 43-45
Amphibia 13, 24, 57, 169, 170, 233, 490, 491, 508
Amphibious vision 228, 513
Amphioxus 164
Amphipod 383, 395
Amphiprion 300
Amplitude modulation 367, 368
Anabaena flos-aqua 392
Anableps anableps 507, 508
-- *microlepis* 516
Anchovy 157

569

Anemoreceptor 125
Anemotaxis 303, 304
Angular acceleration 335
Animal behaviour 108
Animal calls, Evolution 362
-- Luminescent 493
-- Midwater 493
-- photoreceptor, Photopigments 467
-- phyla 5
Annelida 16, 96, 112, 116
-- See Also Polychaete annelid
Annelid, Deep-sea 485
-- Learning capacity 114
-- Nervous system 112
-- worms 477, 484
Anolis carolinensis 207, 208
Anoura 268
-- *qeoffroyi* 266
Anser caerulescens 231
Antagonistic hygroreceptors 459
-- responses 446, 448, 449, 450, 459
-- thermoreceptors 446, 448, 559, 560
Antarctic nototheneid 405
Antennae 57
Anthopleura 76
-- *elegantissima* 307
Anuran 169-189
Aotes trivirqatus 255
Aphakic space 156
Aphelocheirus aestivalis 402
Aphotic habitats 66
Apical organ 67
Aplysia 114, 117
Apteronotus 165
-- *albifrons* 530
-- *leptorhynchus* 528
Apteryx 240, 303
Aquatic bug 331
-- habitats 209, 490
Aqueous humor 503
Aquila chrysaetos 230

Arachnida 123, 128, 136, 147
Archer fish 508
Ardea cinerea 235
Area 228, 229
-- centralis 254, 255, 256
Argentina silus 483
Arhythmic form 6
-- species 485
Arthropod eye 137
Arthropoda 17, 96, 116, 327, 485, 486
Artificial exposure 118
Ascalaphus 131, 136
-- *macaronius* 131, 132
Asian clawless otter 270
Asio flammeus 229
Association, Formation 56
Asterias rubens 97
Asteroidea 106, 110
Astigmatism 513
Astyanax jordani 164
Athene brahma 259
Atlantic flying fish 506
Atmospheric attenuation 264, 265
-- pressure, Detection of changes 243, 244
Attenuation See Excess attenuation, Sound attenuation
Auditory bullae 161, 259
-- meatus 271
-- nervous system 174
-- sense 275
-- sensitivity 220, 221, 244, 274
-- Seasonal changes 221
-- Specializations 258
Auditory stimuli control 265
--system 174, 362
--temporal discrimination 278
-- threshold 220, 221, 225, 258, 259, 272, 273
-- tuning 220

SUBJECT INDEX

Auricular muscles 271
Autotomy 430
-- plane 430
Aversion 303
Aves 25, 217-250
Avoiding reaction 323
Axial diameter 228
Axon 107
Axon terminals 458

Bacteriorhodopsin 467
Balaenoptera physalus 277
Balanus balanoides 299
Baleen whale 274, 277
Bar magnets 241
Barathra brassicae 142
Barbel taste buds 295
Barn owl 224, 225
Barnacle 299
Barometric pressure 243
-- receptors 244
Baroreception 375-409
Basilar membrane 221, 222
Basilar papilla 219
Bat 346, 353
Bathypelagic fishes 306
-- zone 159
Bats 264, 265, 267
-- *See also* Echolating bats, Vampire bat
Bearded seals 269
Bees 146, 147
Beeswax 241
Beetle 138, 308
-- *See also* Cucumber beetle, Cave beetle
Behaviour 109
-- *See also* Scent marking behaviour, Feeding behaviour
-- Adaptive 109
-- Animal *See* Animal behaviour
-- Avoidance 177
-- Courtship 184
-- Homing 183, 184, 187, 299
-- Instinctive 108

Behaviour, Microchiropteran 265
--Photokinetic 67
-- Photonegative 179
--Photopositive 179
-- Phototactic 178, 181, 187, 189
-- Protozoa 32
-- Tapping 183
-- Visual 488
-- Indicator 172
-- Interactions 176
Benthic forms 96
Betaeus 300
-- *hartfordi* 300
-- *macginnitae* 300
Bibio 137
Bilateral symmetry 60
Bimodal receptors 454, 456
Binocular field 516
Bioconvection patterns 49
-- *See also by subject Tetrahymena*, Bioconvection patterns
Bioelectric response 147
Biological clock 178
Bioluminescence 66, 156, 158, 493-495
Bioluminescent squids 270
Birds 12, 14, 217-250, 346, 360, 361, 362, 367, 491
-- *See also* Diving birds, Migrating birds, Song birds, Oilbirds, Humming birds, Diurnal birds of prey, Echolocating birds
-- Aquatic 509-511
-- Communicating 360
-- Eye 227
-- Navigation 239
-- of prey 232
-- Sensory ecology 217-250
-- Song in different habitats 360-363
-- Vision 226-239
Bittern 229
Bivalvia 106, 114, 117
Blepharisma 35, 36, 45, 46

Blepharisma, gamones 45
Blepharisma intermedium 45
Blepharismin 34, 35
Blind cave dwelling species 138
-- fish 164
Blood sucking insects 138
Blowfly 131, 298, 301
Blowhole 277
Blue haze 239
Blue response 178, 179, 181
Boidae 205
Bombyx mori 304
Bone conduction 219, 272
Bottle-nosed dolphin 275
Boundaries of discontinuity 68
Brachiopoda 18, 96
Brachyurans 427
Broadcast range 217
Brucke's muscle 228
Brotulids 160
Bryozoa 18
Bubo virginianus 220
Bufo bufo 176
-- *marinus* 175, 189, 476
-- *typhonius* 189
Bug 140, 142, 143, 144, 402, 446, 455
Bug perceivers 176
Bulla 260, 279
Bullfinch 220, 223
Buprestide beetle 139
Burrowing 112
-- life 66
Bush baby 255
Bushcricket 367, 368
Bushes 358
Buteo borealis 231
Butterfly 133-135, 239
Buzzer 165, 166, 524

Caecilian 13, 169
Caffeine 35
California se lion 269, 274
Call notes 217
Calliactis 69
-- *parasitica* 75

Callinectes sapidus 305, 384
Calliphora erythrocephala 131
Callithrix jacchus 255
Callorhinus ursinus 277
Camouflage 491, 495
Canal, Semi-circular 164
Canary 220
Cancer pagurus 424, 425, 426
Canthophorus 142
Cap sensillae 427
Caprimulgus vociferus 240
Carausius 450, 452, 460, 467
Carcinus maenas 430, 431, 434
Cardiac conditioning 381
Cardiac responses, Conditioned 221
-- responses of cod 389
-- responses of pigeons 391
Carduelis caniceps 221
-- *carduelis* 221
-- *chloris* 223
Carnivores 254, 273
Carollia 268
-- *perspicillata* 268
Carotenoid pigment 233, 467
Carp 388
Cat 257, 260
Categorization, Simplified 12
Caterpillar 142
Catfishes 160, 295, 306
Cathartes aura 240
Caudate 169, 170
Cave animals 138
-- beetle 461
-- life 56
-- swiftlets 226
Celestial cues 187
Cell *See also* Muscle cell, Nerve cell, Sensory cell, Ganglion cell, Neuromast cell
Cells, Horizontal 493
-- membranes 477

Cells of Hensen *See* Hensen's cells
Central control 111
Centropristus striatus 388
Cephalaspidomorphi 22
Cephalization 60
Cephalopod 106, 116, 483, 491
-- Evolution 115
-- photoreceptor 481
Cercaria 81
Cerebral eyes 73
Cerebral ganglion *See* Ganglion, cerebral
-- photoreceptors *See* Photoreceptors, Cerebral
Ceriantheopsis 66
Ceriatoid angler fishes 161
Cestoda 116
Cetacea 274
Cetaceans 272, 275, 513
Chaetognath 477
Changing conditions 106
Channel, Sound in *See* Sound channel
Chelicerata 125
Chelonia mydas 210, 211, 212
Chemical communication 66, 295
Chemical cues, 171, 181, 182
Chemical factor 75
Chemical recognition 75
Chemical sense, Common 295 296
-- signals 182, 291, 304
-- stimulation 12, 302
-- trail 303
-- transmission 291
Chemoreception 12, 13, 96, 115, 119, 169-173, 183 295, 297
Chemoreceptor 13, 293, 295 297, 298
-- Metazoan 296
-- Outer segments 297
Chemosensory information 303

Chemosensory interactions 137
-- structures and mechanisms 295-298
-- systems 291-317
-- Adaptive functions 298-308
Chemotaxis 303, 304
Chicken 221, 236
Chinchilla 259
Chiroderma villosum 266
Chiroptera 265
Chirps 369 528
Chitinous exoskeleton 124, 142
-- structures 124, 138
Chlorella 38
Chondrichthyes 22
Chordate, primative 164
Choroid 228, 254, 270, 274
Chordotonal organs 414
Chromatic aberration 203, 235, 256, 257, 503
Chromaticity 477
Chromophore 472, 491
Chrysemys picta 211
Ciliary body 228, 209
Ciliary locomotion 61
-- membrane 467
-- muscle 264, 275, 506
-- reversal 34
Ciliated cells 323, 325
Ciliates 323
Cinclus mexicanus 228, 510
Circadian activity rhythm 527
-- cycles 72
-- rhythm 181, 188
Clava 61
Clemmys insculpta 209
Clown fish 300
Cnidaria 15, 58, 60, 477
Cnidocytes 76, 325
Cochlear fluid 260, 279
-- microphonic 220, 225 258, 261

Cochleas 221, 261, 271, 272, 278, 279
Cockroach 447, 448, 451
Cod 390
Coder, Burst duration 526
-- Frequency 526
-- Latency 526
-- Phase 526
-- Probability 526
Coding types 525
Coelenterata 66, 108, 110, 324
Coelomate invertebrates 91-122
Coleoptera 123, 128
Cold spots 458
Colibri serrirostris 239
Collocalia 225, 226
-- *esculenta* 226
-- *vanikorensis* 226
Colostethus nubicola 189
Colour-blind 491
Colour perception 491
-- vision 136, 157, 233, 257, 491
-- system 134
Columba livia 219, 390
Columbidae 227
Columella 219, 221
Comb-plates 61
Communication 362
-- *See also* Sound communication
-- Acoustical 144
-- Range 366, 367
Community ecology, Salamandar 186
Compound eyes 131, 132, 133 481, 485
-- Insect 481
Concrement vacuoles 50
Conditioning abilities 60
Condor 240
Cone 6, 229, 230, 231, 233 234-237, 253, 254, 256, 492
-- *See also* Turtle cones
-- Accessory 234

Cone, Crystalline 485
-- Double 234, 237
-- Green 475
-- outer segments 486
Conjugation 45
Connective chordotonal organs 413, 415, 419, 420, 426, 427, 430, 433
-- tissue support 419
Contact chemoreceptor 64
-- receptor 79
Contractile vacuole 33
Convection currents 356
Convoluta psammophila 70, 72
-- *roscoffensis* 71
Coragyps atratus 240
Cormorant 228, 231, 235, 511
Cornea 228, 239, 257, 265, 268, 269, 274, 275, 503, 510, 511
-- Curvature 512
-- Flat facets 506, 508
-- iridescent multilayers 486
Corophium 386
-- *volutator* 383
Corvus monedula 236
Corymorpha 66
Cosmic rays 468
Coturnix 233
Courtship cues 175
Crab 397, 427, 431, 434
-- *See also* Horseshoe crab
-- Blue 305
Crampton's muscle 228
Crangon 428
Craspedacusta 61
Crassostrea angulata 69
-- *virginica* 305
Crayfish 305
Creeping habits 61
-- movement 61
Crenobia alpina 70
Crepscular condition
-- species 485
Cricetidae 258

SUBJECT INDEX 575

Cricket 363, 364, 366, 367, 368
-- See also Molecricket, Bushcricket
Crinoid 96
Crista 326
Cristallography 472
Critical ratio 222
Crocodile 209
Crotalidae 205
Crustacean 5, 75, 328, 424, 485, 494
-- Amphipod 306
Ctenophora 15, 58, 61, 325, 477
Cucumber beetle 301
Cupular 160, 161
Curlew 231
Currents 64
Cut off filters 234
Cuticle 57
-- Soft 430
Cuticular stress detectors 430
Cyclothone spp. 306
Cylindrical spreading 348
Cynidae 142
Cynomys ludovicianus 257
Cyprinus carpio 388
Cypselurus heterurus 506
Cytochromes 33

Dalyellia 73
Damping 366, 367
Dark adaptation 486
Darkness 490
Daw 236
Decticus 141
Deep-sea fish 270
-- forms 96, 485
-- teleosts 481
Deep-water species 492
Deer, Black-tailed 293
-- mouse 224
Deinocerites cancer 461
Demospongiae 61
Dendritic field 107
Dendrobates auratus 189

Dendrobates granuliferus 175
Depolarization 107, 474
Depth 355
-- of focus 265
-- regulation 383
Dermal light sense 67
Desert rodents 258, 261, 262
Deserts 218
Desmodontidae 266, 267
Desmodus rotundus 268
Deuterostomes 92
Diabrotica 301
-- *undecimpunctata* 301
Didelphis virginiana 254
Diel rhythms 378
Diel vertical migration 396
Difference limens (DL) 222
Differential manometer 331
Diffraction 356, 515
Diffuse reception 5
Digenean 74
Dileptus 48
Dinoflagellates 5
Dinosaur 197
Diopsidae 136
Dioptric system 275
Dipodomys 261, 262, 259, 260
-- *merriami* 260, 262
Dipole source 347
Dipper 228, 510
Diptera 123, 128
Dipteran species 140
Direction finding 239, 241
Directional cues 184
-- resolution 225
Directionality of hearing See Hearing, Directionality
Dispersal 64
Dissipation 350, 352
Dissipation in air 351
-- Sound energy, Water 353
Distance for communication 352
Diurnal passerines 232
-- raptors 233

Diurnal rhythms 208
-- species 481
-- vision 253
Diving birds 219, 234, 512
Dog 254
Dogfish 165, 476
Dolichovespula media 142
Dolphins 274, 513, 514, 515
-- Eye 276
-- See also Amazon dolphin, Bottle-nosed dolphin, Ganges river dolphin, Pacific white-sided dolphin
-- Pupil 276
Domestic fowl 229
Dopamine 117
Doppler shift 264
Drosophila 141, 147
Dryotriorchis spectabilis 233
Duck 235, 512
Ducks, Diving 228, 512
Dugong 274

Eagle 229, 231
-- See also African serpent eagle, Martial eagle
-- Golden 230, 231
Ear 258, 259, 260, 262, 271, 273, 346, 357, 362, 368
-- Avian 219
-- Cetacean 278
-- External 224
-- drum 224
-- Inner 258, 260, 278, 279
-- Middle 258, 260, 272, 278, 279
-- Middle, Ossicles 272
-- Middle, Volume 262
-- Outer 272
Earth's electromagnetic field 160
Earthworm 113
Echinodermata 18, 108, 110
Echinoderm 111, 477, 481, 484
Echinoidea 106
Echiura 17
Echolating bats 261, 262, 264, 265
-- birds 225, 226
Echolocation 158, 225, 226, 261, 264, 265, 271, 276, 277, 278, 350, 353
-- Microchiropetran 263
Echolocative function 275
Ecological acoustics 217
Ecological functions 124
Ecosensory functions in insects 123-153
-- Lower invertebrates 55-90
Ecosensory relations 125, 126
Ecosystems 308
-- Terrestrial 345
Ectoneural nerve net 111
Ectoneural plexus 111
Ectoparasite 78
Eels 490
Eggs, Release 79
Eigenmannia 165, 524
Electric fields 146, 147, 160
-- Interference 147
Electric fishes 164
-- "Weakly" 522
Electric ray 165
-- sense 164, 165
-- shocks 522
-- stimuli 146
Electrical recordings 476
Electrocommunication 523
Electrolocation 523
Electromagnetic energy 468
-- field 242
-- Earth 165
-- spectrum 197, 469
Electromechanical 160
Electrophorus 165, 522, 527
Electrophysiological investigations 141
-- measurements 475

Electrophysiological
 recordings 206, 208,
 420
-- techniques 129, 491
Electrophysiology 472
-- Retinal, Ecological
 aspects 492-493
Electroreception 126, 169
 521-533
Electroparalysis 523
Electroreceptors 521,
 523, 524
-- "Phasic" 523
Electroretinogram 236,
 492
-- studies 211, 492
Electrosensory system,
 "Active" 522
Electro-thermal units
 456
Eleodes 308
Elephants 254
Eleutherodactylus 181
-- *coqui* 175
Elongate bodies 162
Elongation, Sensitive 422,
 432
Emballonuridae 266, 267
Emmetropia 507
Emu 240
Emys 509
-- *orbicularis* 209, 510
Endocrine activity 117
-- system 221
Endoparasite 77
Energy See also Electromagnetic energy, Quantal energy, Radiant energy, Scattering of energy
-- Absorption 468
-- attenuation 470
-- Emission 468
Engraulis mordax 157
Enhydra lutris 270
Enterobius vermicularis 79
Enterozoa 107
Entobdella soleae 78
Entoprocta 18, 59

Environment 106
-- Aquatic 478
-- Dim light 486
-- Light 494
-- Natural 477
-- Photic 487, 488
-- Physical 186
-- Spectral 488, 490
-- Turbid
Environmental conditions
 4, 7
-- constraints 5
-- factors 7
-- filtering 175
-- parameters 291
-- physiology 4
-- stimuli 62, 107, 114,
 137
Ephemerella 131, 132
Epipelagic zone 159
Epithelial pigment 6, 235,
 254
-- processes 486
Epitoky 118
Eptesicus fuscus 268
Equilibrium sense 13
Erithacus rubecula 242
Erythropsis 5
-- *Ocellum* 41
-- *pavillardi* 41
Escape to water hypothesis
 178
Escherichia coli 42
-- *Temporal sensing*
-- *Tumbling* 42
Esox lucius 162
Eudendrium 69
Euglena 5, 38-40, 477
Eumetopias jubatus 269
Eunice fucata 118
Euphausia pacifica 494
European robin 242, 243
-- water ousel 235
Excess attenuation 218
Excitation 107
-- level 136
Exposure 109
External auditory canal
 278, 279

Eye 64, 265, 274, 275, 506
-- See also Arthropod eye, Cerebral eye, Compound eye, Nocturnal eye, Parietal eye, Pineal eye, Pinniped eye, Refracting eye, Third eye
-- Aperture 156
-- Geometry 156
-- movement 229
-- Ovoid 484
-- spot 5
-- Streamlining 513
-- Structural peculiarity 115
-- Terrestrial 506
-- Tubular 493
-- Vertebrate 503
Exposure 109

Facial pit organs 442, 443, 444
Fagopyrum 35
Falciform birds 230
Falco sparvorius 232
Far field 161, 321, 347
Faunal break 158, 159
Feeders, Crepuscular 489
Feeding 96, 100-102, 301, 494
-- behaviour 176, 177, 301, 302
Fence lizard 207
Fertilization, External 305
Fibrils, intra-cytoplasmic 107
File 367, 368
Filter, Low-pass 365
Filtering 4, 346
Fin whales 277
Firefly 483
Fish eye, Myopic 506
Fish larvae 481
Fishes 12, 75, 233, 277, 345, 485, 491
-- See also Bathypelagic fish, Clown fish,

Fishes (continued) Crayfish, Electric fish, Flat fish, Flying fish, Sand fish, Fish by name (Common and scientific)
-- Deepsea lethargic 493
-- Deep-sea, Luminescent 483
-- Euryhaline 490
-- Diurnal 489
-- Four-eyed 486
-- Fresh-water 490
-- Non-migratory 490
-- Predator 494
-- Shallow-water 490, 492
-- Shore 491, 494
-- Surface 486
-- Teleostean 492
-- Tropical marine 488
Fissioning 74
Fixation 231
Flagella, Bacterial 42
Flagellate 5
Flagellin 42
Flagellum 473
Flamingo 231
Flatfish 388, 405
Flavins 33, 40
Flicker fusion, Critical 203
Flicker fusion frequency 157, 368
-- response 203
Flowers 133, 239
Fluctuation of sound pressure See Sound pressure, Fluctuation
Fluorscence 239
Fly, Stalked-eye 136
Flying fish 486
Flying squirrel 256, 257
FM-sweeps 370
"f" number 156, 233
Focal distance 253
Focal length 254, 265
Fog 353
Foliage 359
Food cup 43

Foraging behaviour 226
-- habitats 188
Forest 360, 361
Forest floor 218
forests, Tropical *See* Tropical forests
Forms of resistance 70
Fossorial habitat 208
Fovea 228-231, 255, 256
-- Central 229-231
-- Convexiclivate 230-233, 255
-- Temporal 229
Foveal location 230
Frequency, ΔF 222, 223
-- cutoffs, High 220
Frequency discrimination 221, 222, 273, 278, 368, 369, 370
-- filtering 349, 350, 367, 370
-- information 370
-- modulation 142
-- modulation, Band 142
-- partitioning 524
-- ranges 6
-- resolution 222
-- sensitivity 174
-- spectrum 369
Freshwater 472
-- habitats 70
-- turtle 209, 211
Frog 13, 169-176, 186, 187, 362, 367
-- *See also* bullfrog, migrating frog, poison-arrow frog, treefrog
-- Chorus 219
-- Ecological environment 186
Fruit bats 254
-- flies 348
Fulmarus glacialis 229
Functional adaptation 420
-- diversity 291
Fundulus heteroclitus 483
Fur seals 271

Gadus morhua 390
Galago crassicaudatus 255
Galliformes 227
Gallus domesticus 227
Galvanotaxis 71
Gamma radiation 71
-- rays 468
Gammarus duebeni 306
Gamones 45, 119
Ganges river dolphin 274
Ganglion cell 229, 230, 255, 256, 267, 476
-- Cerebral 113, 114
-- Visceral 115
Gannet 235
Gastropod 12, 106, 114
-- Marsh 307
Gastrotricha 16, 59, 64
Gavia lumme 572
Geckoe 201, 202, 203
Geographic variations 490
Geomagnetism 242
Geometric spreading 347
Geotaxis 49, 67
Gerbillinae 262
Gerbelline rodents 258
Gerbil 259, 260
Glacial period 491
Gland, Thyroid 490
Gland, Pituitary 490
Glare effects 486
Glenodinium 42
Gnathophausia ingens 295
Gnathostomulida 16
Goat sucker 234
Goggle 512
Goldfinch, Black-headed 221
-- Grey-headed 221
Goldfish 492
Gonadal cycle 221
Gonepteryx rhamni 133-135
Grass 349, 365
Grasshopper 347, 367, 368, 369
Grasslands 218, 360, 361
Gravireception 66
Gravitational waves 148
Gravity 67, 160, 163

Gravity receptor 320, 324, 330
-- sensors 335
-- waves 162
Grazing angle 348 364
Green response 178
Greenfinch 223
Grey squirrel 256, 257
Ground 349
-- absorption 358
-- impedance 363-365
-- squirrel 256
-- wave 365
Gryllus campestris L. *363*
Guanine 485
Guinea hen 229
Gull chicks, Ring-billed 242
Gulls 234
Guppy 486
Gustation 295
Gymnarchus 522, 524
Gymnorhamphichthys hypostomus 527, 530
Gymnotidae 165, 522
Gymnotus 165
Gymnotus spp. 524

Habitat 7, 92-95
-- See also Aphotic habitats Aquatic habitats, Foraging habitats Freshwater habitats, Fossorial habitats
-- Photic 491
-- selection 298
-- of larval, Aquatic 127
-- of larval, Terrestrial 127
Haddock
Haemonchus contortus 80
Hair cell 163, 219, 222, 279, 331, 332, 335
-- plate organs 428
-- plates 330, 412
-- sensilla 141
Halcampa 61
Halobates 128

Harbour seal 269, 272, 273, 274, 513
Harmonics 370
Harp 366
-- seal 269, 273, 274, 513
Hawk 229, 231
Hearing 164, 278
-- Directionality 357
-- hairs 329
-- organs 346
Heart rate response, Conditioned 243
Helicotrema 261
Hemichordata 108, 110, 116
Hemidactylus turcicus 203
Hemiptera 123
Hen 232
Hensen's cells 261
Herbivores 301
Hermit crab 75
Herodactylus 197
Heron 235
Herring gull 231
Heterodera schachtii 60
Heteroptera 128
Hieracyum silvaticum 133
High intensities 186
Hiodon tergisus 483
Hirudinea 96, 106
Homarus 432
-- *gammarus 428*
Homeothermic animals 138, 197, 204
Homing 265
Homing See also Behaviour, Homing
Homoptera 128
Honey bees 134
Hormone secretion 109
Horns 366
Horse 254
Horseshoe crab 474
Host, Final 72
-- Intermediate 72
Hue discrimination 237, 239
Humans 164, 257, 273
-- Hyperopic in water 506
Humidity 70, 264, 352, 353 356

Humidity gradient 133
-- receptors 138
Hummer 165, 524
Hummingbird 229, 239
Humpback whale 277
Hyaline mass 261
Hybridication 184
Hydra 73
-- littoralis 73
Hydra viridis 76
Hydractinia echinata 75
Hydromedusae 67
Hydropolyps 72
Hydrostatic pressures 66, 355
Hygroreception 126, 138
Hygroreceptive function 138
Hygroreceptor 138, 454, 459
Hyla chrysoscelis 174
-- *cinerea* 175
-- *versicolor* 174
Hymenolepis diminuta 79
Hymenoptera 123
Hypericin 34, 35
Hypercum 35
Hyperpolarization 47
Hyperpolarizing wave 474
Hypopomus 165, 524, 530, 524
Hypopygus 524

Ictalurus natalis 306
-- *punctatus* 295
Illumination 131, 188, 474, 476
-- Optimum ambient 179
Image formation, Aquatic reptiles 209
Immunology 472
Impedance 364
-- Acoustical 348
-- transducers 146
Impulse frequency 369
-- rate 368
-- response 366
Incus 279
Independent effectors 57

Independent units, Mosaic 476
Indicatoridae 241
Indigo bunting 242
Inea geoffrensis 278
Infauna 64
Information, Nature of 346
Infrared detector 203, 205
-- Sensitivity 206
-- to optic tectum, Projection 206
-- energy 205
-- radiation, Detection 205
-- rays 468
-- receptor 206, 484
Ingestive conditioning 69
Inner ear, Clupeid 161
-- segment 233
Insects 5, 12, 123-153 346, 366, 367, 486, 491
-- Blood sucking See Blood sucking insects
Intensity differences limen 223
-- Interaural 224
-- discrimination 223
Interaural time difference 272
Interference 350, 356
-- competition 182
Interspecific associations 75
-- competition 184-186
Interstitial animals 64
Intertidal zone 68
Intracellular recordings 492
Intramembranous plaques 48
Intraspecific aggregations
-- communication 304
-- competition 182, 187
-- territories 185
Invertebrate larvae 495
-- photoreceptor 474
Invertebrates, Higher 488

Invertebrates, Multicellular *See* Multicellular invertebrates
Iquana iquana 208
Iranian blind fish 158
Iris sphincter 509
-- muscle 228
Irradiancy 129

Jamming avoidance response 527
Japanese quail 233
Johnston's organ 413
Jungle fowls

Kairomones 292
Kangaroo rat 253, 259, 260, 261
Keratoscope 507
Kestrel 333
Killer whale 275, 278
Killifish 486
Kinetodesmata 107
Kingfisher 229, 234, 512
Kinocilia 325, 326, 332, 333
Kinorhyncha 16, 59, 64
Kiwi 240, 303
Klinophototaxis 66
Klinotactic orientation 71

Laboratory white rat 259
Labyrinth sense 163, 164
Labyrinthine proprioceptors 413
Lacerta sicula campestris 208
Lachrymal glands 274
Lagena 335
Lagenorhyncus obliquidens
Lagodon rhomboides 388
Lambert-Bouguer law 470
Lamellibranchs 114
Lanternfish 494
Larus argentatus 237
-- *delawarensis* 242
Larval stages 61
Larynx 263, 277

Lateral eyes 13, 14, 177, 201, 208
-- line 160, 161
-- canal systems 334
-- Dogfish 162
-- line organ 333
Learning abilities 60
Leech 78, 113, 118
Lemming 225
Lemur catta 256
Lemur, Ring-tailed 256
Lens 228, 239, 253, 257, 265, 268, 275, 276, 495
-- Muscle 506
Leopard 254
Lepidochiton 109
-- *cinereus* 108
Lepidoptera 123, 128
Leptodactylus pentadactylus 189
Leptodirus hochenwarti 138, 139
Leptodius floridanus 397
Leptonychotes weddelli 270, 513
Leuciscus rutilus 388
Leuckartiara 72
Life, Burrowing *See* Burrowing life
Life cycles 70
-- modes 106
Light 66, 467, 468
-- *See also* Ambient light, Sunlight, Ultraviolet light Reflecting light, White light
-- Absorption 487
-- Adaptation 486
-- Attenuation 478
-- Background 492, 495
-- Bright 109, 486, 513
-- dark adaptation 73
-- Dermal 481
-- Detection 136
-- Diffuse 158
-- Dim 155, 490
-- Downwelling 485, 488, 493, 494, 495

Light, Energy 136, 205
-- Funneled 486
-- Intensity 180, 186
-- Monochromatic 136, 178, 475
-- Patterns 131
-- Paucity
-- Penetration 96
-- Propagation 468
-- Receptors 5
-- Scattered 158, 486
-- sense, Dermal See Dermal light sense
-- sensitivity 472, 481, 487
-- sources, Extended 156
-- stimuli 136
-- transmission 169, 471, 490, 493
-- Upwelling 488
-- Visible 468
Limb muscles 433
Limpets 299, 307
Limulus 96, 474, 476
Line spectrum 370
Liomys 260, 262
Lipid 233, 277
Lithostyles 325
Littoral 64
Littorina irrorata 307
Lizard 199, 201, 207, 491
-- See also Fence lizard, Night lizard
Loa loa 50
Lobster 428
Local reflexes 60
Locomotion 106
Locomotor motions 113
-- reflexes 113
Locomotory modes 96
-- reaction 109
-- strategies 64
Locust 141, 368, 460
Locust See also Migrating locust
Loons 228, 512
Lorenzini, Ampullae 456, 524
-- dorsal ampullae 165

Low intensities 186
Lower animal 7
Loxodes 50
Lumbricus 118
Lutein 256

Macropipus holsatus 384
Macula 326
-- lutea 256
Magnetic cues 242
-- disturbances 242
-- field 12, 71, 147, 242, 243
-- See also Electro-magnetic field
-- field, Detecting 243
-- Earth 147, 241, 242
Magnetic fluctuations, Natural 242
-- orientation 148
-- stimuli 146
magnetoreception 126
Malapterurus 522
Malleus 279
Mammals 14, 26
-- See also Terrestrial mammals, Nocturnal mammals, Marine mammals
-- Sensory ecology 253-287
Man 223, 232, 233, 239, 259, 278
Manatee 274
Manx shearwater 229
Marine mammals 274-279
Marine toad 175
Marmoset 255
Marmota monax 257
Marsupial 233
Martial eagle 233
Masking effect 351
Mass attraction 320
Mating call 174, 175
Mazocraes alosae 80
Mechanical properties 59
-- stimulation 107, 319, 324
Mechano-cold units 455
Mechanoreception 96, 126, 319-344

Mechanoreceptive hairs 327
-- proprioreceptor 141, 320
Mechanoreceptor 319, 323, 412
-- Ciliated 323
-- Cold-sensitive 455, 456
-- diversification 323-336
-- functions 13, 96, 138, 140, 147
Mechanotaxis 46
Mechano-thermal units 455
Medusae 72
Megachiroptera 254
Megaptera novaeangliae 277
Melanogrammus aeglefinus 390
Melanophila 439
-- *acuminata* 445, 446
Melanosomes, Screening 485
Melon 277
Melopsittacus undalatus 220
Membrane See also Tympanic membrane, Cilia membrane, Rhabdomeric membrane
Membrane 106
-- Basilar 261, 278
-- disturbance 107
-- Nictitating 228, 506, 512
-- permeability 475
-- Photosensitive 481
-- potential 474
-- Pro-otic 161
-- resistance 474
-- Surface 106
Memory time 157
Meriones hurrianae 262
Meriones libycus 259, 260
-- *unquiculatus* 294
Mesopelagic fishes 159
Mesozoa 16, 58, 76, 291
Metabolic changes 12
Metazoans 291
-- See also Chemoreceptor, Metazoans
Metridium 69
Microchiroptera 158, 262, 266, 268
Microclimate 358
Microdipodops 260, 261, 262

Microfilariae 79
Micrometeorological parameters 356
Micronycteris megalotis 267
Microspectrophotometry 234, 236, 472, 491
Migrating birds 219
-- frog 175
-- locust 368
-- passerines 242
Migration 79, 265
-- See also Spawning migrations, Upstream migration, Vertical migration
Migratory orientation 242
-- restlessness 242, 243
Minimum audible angle 272
-- separable angles 232, 268, 270
Minnow schools 307
Miracidium 80
Mirounga leonina 270
Mnemiopsis 66
Mnierpes macrocephalus 507
Modes of life 7, 117-119
Molecricket 366, 368
Molecules 136, 301, 471, 472
Mollusca 16, 96, 114, 116, 296, 481, 484, 485
Molluscan nervous system 114-119
Molossidae 266
Mongolian gerbil 294
Monkey 254, 255
-- See also Nocturnal monkey, owl monkey Squirrel monkey
Monocular fixation 229
Monotreme 233
Moray eel 308
Mormoopidae 266, 267
Mormyridae 522
Morphological correlate of function 457
Mosquito 446

SUBJECT INDEX

Moth, Silk 304
Motor system 111
Mouse 224
Movement receptor, Air particle 144
Mucus trails 299
Mudskipper 508
Muller's vesicles 50
Multicellular animals 107
-- invertebrates 13
Multiple scattering 358
Muscle *See also* Auricular muscles, Limb muscles
-- cell 106
-- receptor organ 433
-- Stapedius 262
Mustelids 270
Mutatis mutandis 142
Myochordotonal organ 413, 433, 434
Myotis lucifugus 263, 268
Mysticeti 274, 277, 278
Mytilus 105
-- *gallo-provincialis* 69
Myxine 164, 334

Nasal anatomy 240
Natalidae 266
Natrix 303
Natrix tesselatus 509
Nautilus 115
Navigation 241
Near field 161, 321, 329, 334, 347
-- medium vibration 329
Nematocyst 60
Nematode 57, 59, 64
-- Marine 90
Nematomorpha 16, 59
Nemertinea 58, 67, 116
Neotropics 179
Nepa 331, 402
Nephtys 399, 401
Nereid worms 112
Nereis 118
-- *diversicolor* 118
-- *succinea* 119
Nerve cell 106

Nervecords, Ventral 112
-- fibres, Giant 116
-- Vertebrate 116
Nervous system 107, 108
-- *See also* Annelid nervous system, Molluscan nervous system, Auditory nervous system
-- Central 108, 346, 468
-- Hemichordate 111
-- Hyponeural 110
-- Metameric 112
-- Primitive 108, 110-112
Neural control 110
-- pathway 108
Neurohaemal organs 107
Neurohormones 117
Neurohumours 107, 116, 117
Neuromast 160, 332, 521
-- cells 525
Neurosecretion 117, 119
Neurosecretory "by cells" 117
-- cells 117
Neuroid conduction 57
Neuron 107
-- photosensitive *See* Photosensitive neuron
Neuroptera 128
Newt 171
Niche dimensionality 169-196
Night lizard 207
Nocturnal animal 157
-- eye 265
-- mammals 268
-- monkey 254
-- species 485, 492
-- vision 253
Noemacheilus smithi 158
Noise 4, 163, 347
-- band 222
Non-resonant singers 367
Northern fulmar 229
Nuclear layer, outer 270
Nycera fuliquia 235
Nymphon gracile 385

Obstacle avoidance 226
Oceanodroma leucorrhoa 299
Oceans 355
Ocellar pigment 64
-- potential 73
Ocelli 64
Octopus 78, 115, 326, 327, 491
Ocypode 427
Odocoileus hemionus columbianus 293
Odontocete whales 275, 278
Odontocetes 277, 278
Odontoceti 274
Odor generalists 297
-- specialists 297
Odorant response patterns 298
Oesophagus 102
Oil droplet distribution, Ecological correlates 235
Oil-droplets 211, 233, 234, 236, 475, 486
Oilbird 223, 225, 226, 241
Olfaction 181, 240, 265, 295, 300
-- mediated behaviour 295
Olfactory communication 17, 183, 184, 293
-- cues 177, 184, 186, 241, 303
-- function 308
-- lobes 240
-- ratio 240
-- receptors 296, 297, 298
-- sense 241, 302
-- systems 295, 298, 303
Oligochaeta 96
Omocestus viridulis L. 368
Oncopeltus 446
-- *fasciatus* 139, 140
Oncorhynchus 6, 307
Onychophora 17
Operculum 275, 276
Ophrydium versatile 37
Ophthalmoscope 514
Opossum 254

Opsin 474, 491
Optic disc 255, 256, 257
-- paths, Multiple 506, 507
Optical aberation 231
Orcinus orca 275
Organ of Corti 261
Organic stimuli 302
Orientation clicks 227
Orthoptera 123
Oscillation velocity 348
Ossicular chain 278
Osteichthyes 22
Otolith organs 335
Otolithes, Crystalline 164
Otter 270, 513
-- *See also* Sea otter, Asian "clawless" otter
Outer segment 235, 270, 475
-- *See also* Chemoreceptor, Outer segments
-- Membranes 483
Owl 223-225, 229, 234, 258, 259
-- Great grey 225
-- Great horned 220, 230
-- monkey 255
-- Tawny 232, 233
Owlfly 131
Ox 254
Oxygen tension 64
Oyster 69, 305

Pacemaker system 60
Pacific salmon 6, 490
-- white-sided dolphin 279
Pagophilus groenlandicus 269, 513
Pagurus bernhardus 75
Pandalus borealis 429
Panulirus argus 429
Papillae 78
Paraflagellar body 39, 473
-- rod 40
Parakeet 220-223
Paramecium 46-48, 107, 323, 324

Paramecium aurelia 45
-- *bursaria* 36, 37, 38, 48
-- *caudatum* 46, 48
Parameters, Available for information 367
Parasite 78
-- Intestinal 78
Parasitic interaction 56
Parasitical associations 76
Parasitism 91
Parazoa 57, 107
Parielectomy, Effects of
Parietal eye 207-209
Parophrys vetulus 405
Pasiphae 428
Passerina cyanea 242
Pathways, Synaptic 476
Pattern vision 239
Peachia 72
Pecten 102, 484
-- *oculi* 228
Pedicellariae, Movement 110
Pelican 511
-- Brown 228
Pelicaniformes 240
Pelicanus occidentalis 228, 511
Penguin 219, 228, 509-511
-- Black-footed 220
Pentastomida 17, 92, 96
Pentatomidae 142
Perca 6
-- *flavescens* 489
Perception 4, 170-173, 181
Perch 6
-- Yellow 489
Percidae 6
Periophthalmus 508
Peripheral control 111
-- excitation 111
Periplaneta 447, 448, 450, 451, 452, 453, 459, 460, 461
-- *americana* 451
Permeability, Ionic 474
Perognathus 260, 262

Perognathus penicillatus 262
Peromyscus leucopus 259
Petrels 240
Petromastoid bone 278
Petromyzon 334
PH 70
Phagocytosis 43
Phagosome 43
Phalacrocorax 511
-- *aristotelis* 235
-- *auritus* 228, 511
-- *carbo* 228
Phase reference unit 525
Phasmidia 57
Pheromone 74, 292, 293, 304
-- See also Alarm pheromones
-- concept 293
-- Sex 305
-- Sex, Detection 306
Phoca barbata 269
Phoca vitulina 269, 512
Phormia 110
-- *regina* 298, 307
Phoronida 17
Photic cues 182
Photic environment 124, 239, 468, 477, 492
-- factors 491
-- pelagic district 62
-- stimuli 468
Photobiological processes 469
Photocapture 36
Photochemical processes 471-474
Photoisomerisation 476
Photokinesis 39
Photokinetic behaviour See Behaviour, Photokinetic
Photolabile pigments 467
Photons 155, 471, 475
-- Absorption 474
-- Detection 476
Photoperiod 73, 490
Photophores 159, 494
Photopic system 233

Photopigment 96, 136, 236
Photoreception 124, 126, 129, 169, 176-181, 187 467-502
Photoreceptive organ 125
Photoreceptor 5, 129, 136 468, 472, 476, 491, 492, 495
-- See also Animal photoreceptor, Invertebrate photoreceptor
-- axon 484
-- Cerebral 64
-- Ciliary 477, 484
-- Extraoptic 178
-- Iamellae 235
-- organelles 473
-- organization 489
-- rhabdomeric membranes 486
-- Sensitivity 488
-- --, Spectral environment 487-492
-- visual pigments 487
Photoregeneration 137
Photosensitive neuron 72
-- structures 485
Photosensitivity 5
Photosynthesis 159
Photosynthetic pigments 471
Phototaxis 178, 179, 180, 181, 188
Phototropic fungi 467
-- response 72, 467, 493
Photuris 483
Phoxinus laevis 307

Phyletic groups 58, 59
Phyllostomidae 267
Phyllostomus hastatus 268
Phylogenetic distribution 11
-- survey 11
Physeter catodon 277
Physiological study 148
Phytoplankton 488
-- bloom, Spring 119
Piezodorus lituratus 143

Pig 254
Pigeon 219, 223, 229, 232, 233, 234, 236, 239, 242, 390
Pigeon, Homing 221, 239, 241, 243
-- Navigation 300
Pigment See also Epithelial pigment, Ocellar pigment, Photosynthetic pigments, Visual pigment
-- granules 473
-- mixture 490
--oil droplet combination 238
Pike 162
Pikeperche 6, 486
Pineal eye 207
Pinfish 388
Pinna 271
Pinnipeds 268, 269, 271, 272, 513
-- Eyes
Pipa pipa 170
Pisaster ochraceus 307
Pisces 116
Pitch discrimination 278
Pit-organ map 206
Pit-viper 206
Plaice 165
Planarian 60
Plants 358
Platanista gangetica 274
Platyctenea 61
Platyhelminthe 16, 58, 69, 112
Platynereis dumerilii 118
Plecotus 263
-- *townsendii* 263
Plethodon 184, 185
-- *cinereus* 172, 173, 182, 185
-- *jordani* 171, 172, 184
-- *nettingi shenandoah* 172, 173, 185
Pleurobrachia 68
Pleuronectes platessa 165, 388

SUBJECT INDEX

Plexus, Fibres 112
Pogonophora 17
Poikilotherm 139, 197
Poikilothermic existence,
    Adaptation 203-209
Point source 156
Polarization 106, 107, 136
-- See also Depolarization
-- parameters 136
-- pattern of the sky 240
-- sky 486
Polarized light 239, 240, 486
Pollachius virens 163, 390
Polycelis felina 70
-- *niora* 70
Polychaete 66, 106, 298, 300, 395
-- annelid 5
-- annelid See also Annelid
Polychaete worms 491
Polypodium 72
Pond-living triclads See Triclads, Pond-living
Population See also Wild rodent population
-- density 353, 362, 367
-- ecology 4
-- growth 74
-- Landlocked 491
Porifera 15, 55, 58
Porous medium 364
Porphyropsin 472, 490, 491
Porpoises 277
Potassium 106, 474
Prairie dog 256, 257
Praunus flexuosus 404
Pre-adaptation 160
Predator avoidance 307
Predators, Crepuscular 488
Preretinal filters 235
Presentation time 157
Pressure, Altitude perception 397
-- Behavioural responses 382
-- changes, Different depths and altitudes 376

Pressure differences 161
-- Ecological implications 396
--, Effect of depth and altitude 376
--, Fechner principle 395
-- gradient 348
-- High 376
-- Liminal time 395
-- Methods 378
-- Physiological responses 388
-- Quantitative aspects 392
-- receiver 79, 141
-- --, Gradient 141
-- receptor 322, 383
-- --, Charged membrane 403
--, Gas-filled structures 400
-- Piezo-electric effects 405
-- Swimbladder 401
-- Tissue compressibility 405
Pressure, Response of plants 391
-- Response of Terrestrial animals 390
-- Spontaneous behaviour 387
-- units 375
-- Thresholds 393, 394, 398
Prey capture 225
-- catching behaviour 176
-- detection, Acoustic 224
Priapulida 17
Primates 231, 254, 255, 256, 269
-- Higher 491
Procellariiform seabirds 240
Project Seafarer antenna 242
Propagation See also Sound propagation
-- losses 218
-- of Impulses 112

Proprioception 411
Proprioceptive input 420
-- organ 419
-- sense 13
Proprioceptors 66, 13, 330, 419
-- See also Labyrinthine proprioceptors
--Adaptive radiation 411-438
Prosimians 254
Protamine sulfate 44
Proterythropsis 5
Protochordares 477
Protoplasm 472
Protostome 96
Protozoa 15, 31-50, 107, 477
Psammommys obesus 262
Pseudemys 210, 211
Pseudemys scripta 203, 211, 212
-- *scripta elegans* 474, 509
Pseudocoelomata 55, 57
Pseudosystoles 33
Psychophysical experiments 476
Pteraspidomorphi 22
Pteridines 485
Pteronotus 265
-- *parnelli* 266
Pterygote insects 124
Puffinus puffinus 229
Pulsating spheres 347
Pulse species 524
Pupil 233, 253
-- See also Slit pupil, Stenopeic pupil
-- Stenopeic slit 506
Pure-rod retinas 481
Pusa hispida 273
Pyrrhocoris 455
-- *apterus* 144
Pyrrhula pyrrhula 220

Quail 229
-- See also Japanese quail
Quanta 468
Quantal energy 468

Quantum 471
-- efficiency 476
-- yield 471
Quit point 158
Q-value 366

Rabbit 254
Radial diameter 228
-- symmetry 61
Radiant energy 468, 469
Radiata 57
Radiation 468
Radiation, Adaptive 5, 6, 106, 114, 413, 424, 435, 441, 461, 495
Raja clavata 165
Rana esculenta 509
-- *catesbeiana* 175
-- species 187
-- *temporaria* 178
Range of cummunication 364
Raptor 230
Rattlesnake 259, 260
Rays 165
-- See also Electric ray
Reception 181
-- See also Tactile reception, Diffuse reception
Receptor 93, 231
-- See also Air-flow receptor, Bimodal receptor, Dynamic receptor, Gravity receptor, Pressure receptor, Static receptor, Touch receptor Vibration receptor, Water-flow receptor, Thermal receptor
-- Ampullary 525
-- cell densities 267
-- Coincidence ratio 230
-- Cold 442, 447-456, 458-460
-- density 229, 230
-- Infrared 439, 442
-- level 476
-- Mechanism 147

SUBJECT INDEX

Receptor membrane 297
--, Movement See Movement receptor
-- organs 126
-- -- See also Muscle receptor organ
-- Light 134
Receptor potentials, Stentor 35
-- Reptile 200
-- Specialized 141
-- "tonic" 523
-- Tuberous 525
-- Warm 442, 447, 451, 454, 459, 460
Red field 234
Redundancy 370
-- Signal 346
Reflecting light 133
Reflection 348, 349
-- coefficient 364
Reflex actions, Inborn 108
-- activities 114
-- response 108
Refracting eyes 481
Refraction 230, 231, 268, 353, 354
Refractive index 230, 503
-- power 228, 265, 268, 275
Relationships, Ecological 495
Relaxation, Sensitive 432
-- -- movement cells 422
Reproduction 304, 305
Reproductive capacity 305
-- cycle 73
Reptile 14, 24, 197-213, 233
-- Primitive 198
-- Terrestial 198
-- Visual pigment 202
Reservoir 39
Resolving power 233
Resonance 366
Response 4
-- Klinokinetic 69, 74

Response mechanisms 468
--, Off shadow 484
-- Ortho 69, 74
-- phototropic See Phototropic response
-- Thigmokinetic 69, 74
Reticular lamina 260
Retina 253, 254, 255, 256, 265, 270, 494
Retina, Bird 229, 232, 233
-- Chicken 234, 237, 238, 239
-- Frog 235
-- Walleye 489
Retinal illumination 233
-- image 228, 232
-- receptor, Geckoe 202
Retinol 235
Retinomotor changes 489
-- responses 73, 485
Retinoscopy 511
Rhabdocoel 73
Rhabdomeres 481, 483
Rhabdomeric membrane 467
Rhabdoms 136
Rhabdoms, Photosensitive 483
Rhamphichthys 524
Rheoreception 68
Rheotaxis 70, 303, 304
Rhodopsin 270, 472, 488, 490
-- Postglacial evolution 491
Rhombomys opimus 262
Rhynchocoela 16
Rhythms 367
Ringed seal 273
Roach 388
Rod 6, 230, 234, 236, 253, 265, 270
-- Behaviour 475
-- bundles 486
-- outer segments 481, 487
Rodent 262
Rotifera 16, 59, 73
RSMC See Relaxation sensitive movement cells 422
Rudd 388

Sabellaria 298
Sacculus 335
Saimiri sciureus 255, 455
Saithe 163, 390
Salamander 13, 169-186
-- communities 185
-- Plethodontid 182
-- Terrestrial 183
Salamandra salamandra 178
Salinity 64, 355
Salmon 307, 490
Salvelinus fontinalis 483
Sampling time 157
Sand fishes 527
Sarcoramphus papa 240
Sargassum 69
Sarsia 68
Saturation frequency 423
Sauger 6
Scabbard fish 162
Scallop 484, 485
Scanning electron microscopy 138, 144
Scardinius erythrophthalmus 388
Scattering 129, 356, 357, 470
-- See also Multiple scattering
-- centre 470
-- of energy 468, 470
-- particles 470
Sceloporus occidentalis 207
Scent marking behaviour 294
Schistosoma 78
-- *haematobium* 79
-- *japonicum* 79
Schooling distance 163
Sciurus carolinensis 257
Sclera 274
Scleral ossicles 228
Scolopale 417
Scolopidium 414, 415
Scopelarchus quentheri 483
Scotoma 256
Scotopic rods 486
-- system 233
Scraper 367, 368
Sculling 162
Scyliorhinus 165
Sea 350

Sea anemones 75, 300, 307
-- bass 388
-- bird See also Procellariiform seabirds
-- bird 229, 241
-- breams 157
-- cows 274
-- lamprey 164
-- lilies 96
-- lions 268, 271, 513
-- -- See also California sea lion, Stellar sea lion
-- otter 270
-- snake 209
-- turtle 209, 210
-- urchin 119
-- water 352
Seals 268-274, 345, 513
-- Sea also Harbour seal, Fur seal, Bearded seal, Harp seal, Elephant seal, Ringed seal, Weddell seal
Secretion 107
-- See also Hormone secretion
Sehkolben 64
Selection 362
Semicircular canals 334
Sense organ 12, 78, 97-99, 126, 345
-- -- infrared 445
Sensilla 125, 428
-- perforated 460
Sensitive elongation See Elongation, Sensitive
-- movement 421
-- movement cells See also Relaxation sensitive movement cells
-- position 421
-- predators 141
-- relaxation See Relaxation, Sensitive
-- structures 106
-- vibration 414
Sensitivity 319, 467
-- High 476
-- Higher absolute 137
-- peaks 134
-- spectral See Spectral sensitivity

SUBJECT INDEX

Sensory adaptation 128
-- apparatus 306
-- capabilities 11, 14
-- capacities 346
-- cell 136, 147, 161, 416, 417, 423, 424, 427
-- centres 468
-- ecology 4, 5
-- functions 7, 11, 12, 139, 169
-- -- Distribution 15-18, 22-27
-- hairs 322, 327
-- interactions 137
-- mechanisms 5, 243
-- modalities 11, 12, 13
-- organ 125, 140
-- physiology, comparative 11
-- pits 73
-- relationship 124
-- response, Transduced 426
-- specialisation 128
-- structure 6
-- system 7
Sepia 326
-- *officinalis* 401
Serinus canarius 220
Sertularella miurensis 59
Setae 57
Settling 61
Sexual maturation 118
Shad 80
Shadow zone 353, 364
Shallow-water teleosts 481
Shearwater 231, 240
Sheep 254, 255
Shoal areas 489
Shrimp 300, 494
-- Planktonic 304
Sipuncula 17
Skunk 308
Slit pupil 269
Slug 114, 484
Smelt 491
Snail 12, 80
Snake 199, 201, 205, 258, 303
-- See also Sea-snake
-- Aquatic 509
Snow goose 231

Social interactions 306
Sodium permeability 474
Sofar-channel 355
Solar radiation 209, 467, 493-495
Sonar 265, 271
-- application 6
-- system 263, 264
Sond cruise 159
Song 217, 277
-- Antiphonally 223
Songbird 222, 223
Sound 182, 187, 277
-- *See also* Underwater sound, Velocity of sound
-- attenuation 218
-- channel 218, 279, 348
-- communication, Strategies 365-368
-- Environments 345-373
-- localization 224, 225, 272
-- pressure, Fluctuation 355
-- propagation 219, 347, 365
-- reception 5, 169, 173-176, 322
-- window 362
Spatial resolution 477
Spawning 6, 106, 472
-- migrations 299, 490
Species diversity 185
-- packing 185, 188
Spectral absorbance curves 131
-- analysis 198
-- bandwidth 237
-- filters 235
-- irradiance 131
-- range 205, 503
-- response 136, 179
-- -- U-shaped 179
-- sensitivity 129, 132, 135, 179, 208, 211, 233, 236, 485, 492, 495
-- -- curves 131
-- -- Photoreceptor 475
-- Univariance 475
Speophyes lucidulus 461
Sperm whales 277
Sphaerodactylus 203
-- *inaquae* 203
Sphalloplana zeschi 67

Spheniscus demersus 220
Spherical aberration 515
— spreading 348
Spincter muscle 269
Spine, Movement 110
Spirula spirula 402
Spizaetus bellicosus 233
Sponge 57, 63
Spot receptor 5
S-potential recordings 492
Squid 115, 484, 485, 494
— See also Bioluminescent squid
Squirrel 256, 257
— See also Ground squirrel, Grey squirrel, Flying squirrel
— monkey 255, 455
Stapes 279
Star orientation 243
Starfish 110, 307
Starling 220
Stars 156
Static reception 322
Statocysts 64, 325
— Cephalopod 326
— Gastropod 325
— Sac 328
Statolith organs 335
Steatogenys spp. 524
Steatornis caripensis 225, 226
Stellar sea lion 269
Stenobrachius leucopsarus 494
Stenopeic pupil 269
— slits, Double 515
— theory 270
Stenostomum 61
Stentor 34, 36
— *coeruleus* 34
— *polymorphus* 34
Stentorin 34
Stereocillia 161, 325, 332, 333
Sternopygus 165, 524
Stigma 39
Stimulus conducting mechanism 459, 460

Stimulus Physical nature 468–471
Stizostedion 6
— *canadense*
— *vitreum* 6, 483, 489
Stoichactis 300
Stomotoca 67
Storm-petrel 299
Stridulation 368
Stridulatory organ 367, 368
Strix aluco 232, 234
— *nebulosa* 225
Strombidium mirabile 38
— *oculatum* 37, 38
Structural colours 124
Structures, Sensitive *See* Sensitive structures
Sturnus vulgaris 220
Stylochus mediterraneus 69
Substrate marking 173
Sula bassana 235
Summation area 156
— time 157
Summative interactions 476
Sunlight 470, 486
Sun radiation 140
Sunlight spectrum 129
Superficial nerve net 60
— receptors 57
Suppressant 301
Surface waves 329
Surrogate senses 155
Swallow 229, 234
Swiftlets 223, 225
— Cave 226
Swifts 234, 241
Swimbladder 161, 401, 402
Symbiosis 75
Symbiotic relationships 300
— zoochlorellae 71
— zooxanthellae
Synaptulid 66
Synchelidium 395, 396, 399
Synomone 292
Syphonophore 72

Tactile cues 184
— hair 328
— reception 181

SUBJECT INDEX

Tactile receptor 13, 14
— stimulation 113
Tadpole 181, 490
Tangoreceptors 64
Tapeta 270
Tapetal organelles 485, 486
Tapetum 156, 489
— cellulosum 270
— lucidum 254, 274, 483, 485, 487
Tapeworm 79
Tapping 172
Tardigrada 17
Tarentola mauritanica 203
Taricha rivularis 171
Taste stimuli 303
Telotrochs 37
Temperature 64, 139, 353, 354, 355, 356, 490, 491
— Ambient 14
— change, Rate of 452
— —, Response to 447, 450, 452
— gradients 139, 353, 354
— Stationary 451
— Steady 447
Temporal patterns 362
— resolution 223
Tengmalm's owl 224
Tern 229, 234
Terrestrial animals 366
— mammals 268
— triclads See Triclads, terrestrial
— vertebrate communities 185
Territorial defense 183
— enforcement 184
— partitioning of habitats 184
Territorialism 183
Test odorants 297
Tetrahymena Bioconvection patterns 49
Thermal preferences 207
— radiation, Inside eye 204
— receptors 439-466
Thermocline 162

Thermodetection 14
Thermopreferendum 67
Thermoreception 126, 139
Thermoreceptive axon terminals 458
— dendrites 459, 461
Thermoreceptor See also Antagonistic thermoreceptors
— Antennal 139
— Sensitive 140
Thermoregulation 13, 140, 208, 439, 440
— Ecological 140
Theromyzon rude 118
Thigmotaxis 64
Third eye 208
Thread hairs 328, 329
Threshold 351
— Absolute 232, 254
— High 295
— Low 295
— Scotopic 488
— shift 221, 222
Thyroid activity 207
Tiaropsis 62
Tidal activity 382
— cycles 388
— rhythm 71, 397
Time of arrival differences 224
Toad 169-175, 367
— rods 476
Toothed whale 274
Torpedo marmorata 165
Touch chemoreceptors 64
— receptor 320
Toxocara canis 80
Toxotes jaculatrix 508
Transformer ratio, Effective 260
— — Impedance 260
Transmission 360
— spectra 234
Transmitter Internal 475
Transmutation hypothesis 198
— theory 201
Travelling wave 222
Treefrog 175
Tree shrew 256, 259
Trees 358

Trematomus bernacchii 405
Trichobothria 328, 329
Triclads Pond-living 61
-- terrestrial 61
Trimerosurus flavoviridis 443
Tritomegas bicolor 143
Triturus cristatus 178
Tropical forests 353, 360
-- rain forests 264
Trout brook 483
Trygon pastinaca 165
Tuning curves 262
-- -- Single unit 220
Tupaia glis 256
Turbellaria 60
Turbid water 6, 164
Turgor pressure and buoyancy 392
Turkey 229
-- vulture 240
Tursiops 276, 277
-- *gilli* 278
-- *truncatus* 275, 514, 515
Turtle 203, 209, 474, 475, 491
-- *See also* Freshwater turtle, Sea turtle
-- aquatic 509
-- cones 472, 475, 476
-- rods 475
Twilight 488, 489, 490, 494
Tympanic membrane 174, 258, 272, 279
Typhlonarke 164
Tyto alba 224, 234

Ultraviolet light 471
-- radiation 471
-- rays 239, 468, 470
Underwater sound 273
-- vision 158, 228, 268, 513
Undulations 162
Ungulates 254
Unicells 31
Upstream migration 6
Utricular macula 161
Utriculus 335
u.v. Receptors 131, 136

Vampire bat 268
Vector 72
Vegetation 358-363
Velocity of sound 353
-- of surface waves 162
-- sensitive 423
Venus mercenaria 117
Vertebrates 5, 201, 477, 488
-- benthic 491
-- diurnal terrestrial 486
-- larvae
-- strategies 503-519
Vertical migration 66, 382, 481
-- -- *See also* Diel vertical migration
Vespertilionidae 263, 265, 267
Vibration 66, 141, 160, 412, 420
-- receptor 67, 321, 324, 335
Vibrational communicative signals 124
-- signals 141
-- stimuli 141
Vibrations acoustical 142
Vibroreceptor 141
Visible spectrum 469
Vision 265, 268, 274
-- *See also* Amphibious vision, Underwater vision
Visual acuity 228, 232, 253, 255, 256, 269, 270, 275, 514
-- adaptation 198
-- angle 255
-- -- Minimum separable 255
-- axis 268
-- cues 184
-- fields Horizontal 491
-- integration 232
-- noise 158
-- perception 176, 177
-- performance 232-239
-- pigment 136, 137, 210, 236, 237, 274, 491, 492, 495
-- releasers 176
-- response 467, 472
-- resolution 232
-- strategies 493-495
-- streak 254, 255, 256
-- system Avian 227
-- -- Chelonian 210
-- Threshold 233

Visual threshold Absolute 233
Vitamin A 233
Vitreous body 239
Vocal communication 217
Vocalization 221, 271-274
Vomeronasal system 306
Vulture *See also* Turkey vulture
-- black 240
-- king 240

Walleye 6, 489
Walruses 268, 269
Warbler 243
Warnowiidae 41
Wasp 142, 147
Water 355, 488, 503
-- balance 138
-- currents 161
-- displacement 160
-- -- sense 160-163
-- flow receptor 320
-- striders 136
-- turbid *see* Turbid water 164
Wave guide 278, 483
-- height perception 400
-- species 524, 528
Wavelenght 136, 203, 357, 365, 468, 470, 471, 492
-- discrimination 237, 257
-- optimal 475
-- short 131, 211

Weather change Perception 399
Weber fraction 273
-- ratios 278
Weddell seal 270, 513
Whales 254, 277, 345, 350
-- *See also* Baleen whale, Toothed whale, Killer whale, Odontocete whales, Humpback whale, Sperm whale, Fin whale
Whip-poor-will 240
White light 178
Wild rodent population 305
Wind 351, 354
-- speed Gradients 356
Woodchuck 257
Woodpecker 228

Xanthophyll 256
Xantusia vigilis 207
X-ray diffraction 472

Yellow field 234

Zalophus californianus 269, 513
Zonation patterns 70
Zona pectinata 261
Zoochlorellae 36